Applied Mathematics

Applied Mathematics

Fourth Edition

J. David Logan
Willa Cather Professor of Mathematics
University of Nebraska, Lincoln
Department of Mathematics
Lincoln, NE

For general information on our other products and services please contact our Customer Care Department within the United States at (800) 762-2974, outside the United States at (317) 572-3993 or fax (317) 572-4002.

Wiley also publishes its books in a variety of electronic formats. Some content that appears in print, however, may not be available in electronic formats. For more information about Wiley products, visit our web site at www.wiley.com.

Library of Congress Cataloging-in-Publication Data:

Logan, J. David (John David)
  Applied mathematics / J. David Logan. — 4th ed.
    pages cm
  Includes bibliographical references and index.
  ISBN 978-1-118-47580-5 (hardback) — ISBN 978-1-118-51492-4 — ISBN 978-1-118-51493-1 — ISBN 978-1-118-51490-0 1. Mathematics—Textbooks. I. Title.
  QA37.3.L64 2013
  510—dc23   2013001305

Printed in the United States of America.

10 9 8 7 6 5 4 3 2 1

To my parents
George Edd Logan (1919-1996)
Dorothy Elizabeth Wyatt Logan (1923-2009)

Contents

Preface

The fourth edition of *Applied Mathematics* shares the same goals, philosophy, and style as its predecessors—to introduce key ideas about mathematical methods and modeling, along with the important tools, to mature seniors and graduate students in mathematics, science, and engineering. The emphasis is on how mathematics interrelates with the applied and natural sciences. Prerequisites include a good command of concepts and techniques of calculus, and sophomore-level courses in differential equations and matrices; a genuine interest in applications in some area of science or engineering is a must.

Readers should understand the limited scope of this text. Being a broad introduction to the methods of applied mathematics, it cannot cover every topic in depth. Indeed, each chapter could be expanded into a one, or even several, full-length books. In fact, readers can find elementary books on all of the topics; some of these are cited in the references at the end of the chapters. Secondly, readers should understand the mathematical level of the text. Some books on applied mathematics take a highly practical approach and ignore technical mathematical issues completely, while others take a purely theoretical approach; both of these approaches are valuable and part of the overall body of applied mathematics. Here, we seek a middle ground by providing the physical basis and motivation for the ideas and methods, and we also give a glimpse of deeper mathematical ideas.

There are major changes in the fourth edition. The material has been rearranged and basically divided into two parts. Chapters 1 through 5 involve models leading to ordinary differential equations and integral equations, while Chapters 6 through 8 focus on partial differential equations and their applications. Motivated by problems in the biological sciences where quantitative methods are becoming central, Chapter 9 deals with discrete-time models, which include some material on random processes. Sections reviewing elementary methods for solving systems of ordinary differential equations have been added in Chapters 1 and 2. Many additional examples and figures are included in this edition, and several new exercises appear throughout. Some exercises from the last edition have been revised for better clarity, and many new exercises are included. The length of the text has expanded over 160 pages. The Table of Contents details the specific topics covered.

Note that equations are numbered within sections. Thus, equation label (3.2) refers to the second numbered equation in Section 3 of the current chapter.

My colleagues in Lincoln, who have often used the text in our core sequence in applied mathematics, deserve special thanks. Glenn Ledder, Richard Rebarber, and Tom Shores have provided me with an extensive *errata*, and they supplied several exercises from graduate qualifying examinations, homework, and course exams. Former students Bill Wolesensky and Kevin TeBeest read parts of the earlier manuscripts and both were often a sounding board for suggestions. I am extremely humbled and grateful to those who used earlier editions of the book and helped establish it as one of the basic textbooks in the area; many have generously given me corrections and suggestions, and many of the typographical errors from the third edition have been resolved. Because of the extensive revision, some new ones, but hopefully not many, have no doubt appeared. I welcome suggestions, comments, and corrections, and contact information is on the book's website: `http://www.unl.edu/~jlogan1/applied-math.htm`. Solutions to some of the exercises and an errata will appear when they become available.

My editor at Wiley, Susanne Steitz-Filler, along with Jackie Palmieri, deserves praise for her continued enthusiasm about this new revision and her skill in making it an efficient, painless process. Finally, my wife, Tess, has been a constant source of support for my research, teaching, and writing, and I again take this opportunity to publicly express my appreciation for her encouragement and affection.

Suggestions for use of the text. The full text cannot be covered in a two-semester, 3-credit course, but there is a lot of flexibility built into the text. There is significant independence among chapters, enabling instructors to design special one- or two-semester courses in applied mathematics that meet their specific needs.

Portions of Chapters 1 through 5 can form the basis of a one-semester course involving differential and integral equations and the basic core of applied mathematics. Chapter 4 on the calculus of variations is essentially independent from the others, so it need not be covered. If students have a strong background in differential equations, then only small portions of Chapters 1 and 2 need to be covered.

A second semester, focused around partial differential equations, could cover Chapters 6, 7, and 8. Students have the flexibility to take the second semester, as is often done at the University of Nebraska, without having the first, provided small portions of Chapter 5 on Fourier-type expansions is covered.

Chapter 9, like Chapter 3, is independent from the rest of the book and can be covered at any time.

The text, and its translations, have been used in several types of courses: applied mathematics, mathematical modeling, differential equations, mathematical biology, mathematical physics, and mathematical methods in chemical or mechanical engineering.

J. David Logan, Lincoln, Nebraska
April 2013

1

Dimensional Analysis and One-Dimensional Dynamics

The techniques of dimensional analysis and scaling are basic in the theory and practice of mathematical modeling. In every physical setting a good grasp of the possible relationships and comparative magnitudes among the various dimensioned parameters nearly always leads to a better understanding of the problem and sometimes points the way toward approximations and solutions. In this chapter we introduce some of the basic concepts from these two topics. A statement and proof of the fundamental result in dimensional analysis, the Pi theorem, is presented, and scaling is discussed in the context of reducing problems to dimensionless form. The notion of scaling points the way toward a proper treatment of perturbation methods, especially boundary layer phenomena in singular perturbation theory as well as algebraic equations with small parameters.

The first part of Section 1.3 is a review of ordinary differential equations. This material may be perused or used as a reference by readers familiar with the basic concepts and elementary solution methods. The last part includes a discussion of stability and bifurcation; it may be less familiar.

Applied Mathematics 4th ed.,
By J. David Logan
Copyright © 2013 John Wiley & Sons, Inc.

1.1 Dimensional Analysis

1.1.1 The Program of Applied Mathematics

Applied mathematics is a broad subject area in the mathematical sciences dealing with those topics, problems, and techniques that have been useful in analyzing real-world phenomena. In a very limited sense it is a set of methods that are used to solve the equations that come out of science, engineering, and other areas. Traditionally, these methods were techniques used to examine and solve ordinary and partial differential equations, and integral equations. At the other end of the spectrum, applied mathematics is applied analysis, or the theory that underlies the methods. But, in a broader sense, applied mathematics is about mathematical modeling and an entire process that intertwines with the physical reality that underpins its origins.

By a **mathematical model** we mean an equation, or set of equations, that describes some physical problem or phenomenon having its origin in science, engineering, economics, or some other area. By **mathematical modeling** we mean the process by which we formulate and analyze the model. This process includes introducing the important and relevant quantities or variables involved in the model, making model-specific assumptions about those quantities, solving the model equations by some analytic or numerical method, and then comparing the solutions to real data and interpreting the results. This latter process, confronting the model with data, is often the most difficult part of the modeling process. It involves determining parameter values from the experimental data. This book does not address these important issues, and we refer to texts on data-fitting techniques. This confrontation may lead to revision and refinement until we are satisfied that the model accurately describes the physical situation and is predictive of other similar observations. This process is depicted schematically in Fig. 1.1. Thus, the subject of mathematical modeling involves physical intuition, formulation of equations, solution methods, analysis, and data fitting. A good mathematical model is simple, applies to many situations, and is predictive.

In summary, in mathematical modeling the overarching objective is to make quantitative sense of the world as we observe it, often by inventing caricatures of reality. Scientific exactness is sometimes sacrificed for mathematical tractability. Model predictions depend strongly on the assumptions, and changing the assumptions changes the model. If some assumptions are less critical than others, we say the model is robust to those assumptions. They help us clarify verbal descriptions of nature and the mechanisms that make up natural law, and they help us determine which parameters and processes are important, and which are unimportant.

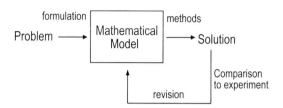

Figure 1.1 Schematic of the modeling process.

Another issue is the level of complexity of a model. With modern computer technology it is tempting to build complicated models that include every possible effect we can think of, with large numbers of parameters and variables. Simulation models like these have their place, but computer runs do not always allow us to discern which are the important processes and which are not. Of course, the complexity of the model depends upon the data and the purpose, but it is usually a good idea to err on the side of simplicity and then build in complexity as it is needed or desired.

Finally, authors have tried to classify models in several ways—stochastic vs. deterministic, continuous vs. discrete, static vs. dynamic, quantitative vs. qualitative, descriptive vs. explanatory, and so on. In this book we are interested in modeling the underlying reasons for the phenomena we observe (explanatory) rather than fitting the data with formulas (descriptive) as is often done in statistics. For example, fitting measurements of the size of an animal over its lifetime by a regression curve is descriptive, and it gives some information. But describing the dynamics of growth by a differential equation relating growth rates, food assimilation rates, and energy maintenance requirements tells more about the underlying processes involved.

Models are a blend of physical laws, such as conservation of mass, energy, etc., experimental results that lead to constitutive relations, or equations based on experiment, and even ad hoc assumptions when more specific evidence is lacking. The reader is already familiar with many models. In an elementary science or calculus course we learn that Newton's second law, force equals mass times acceleration, governs mechanical systems like falling bodies; Newton's inverse-square law of gravitation describes the motion of the planets; Ohm's law in circuit theory dictates the voltage drop across a resistor in terms of the current; the law of mass action in chemistry describes how fast chemical reactions occur; or the logistic equation, an intuitive ad hoc model of growth and competition in a population.

The first step in modeling is to select the relevant variables (independent and dependent) and parameters that we need to describe the problem. Usually these are based on available experimental data and natural laws. Physical quantities have *dimensions* like time, distance, degrees, and so on, or corresponding *units* like seconds, meters, and degrees Celsius. The equations we write down as models must be dimensionally correct. Apples cannot equal oranges. Verifying that each term in our model has the same dimensions is the first task in obtaining a correct equation. Also, checking dimensions can often give us insight into what a term in the model might be. We always should be aware of the dimensions of the quantities, both variables and parameters, in a model, and we should always try to identify the physical meaning of the terms in the equations we obtain. A general rule is to always let the physical problem and the data available drive the mathematics, and not vice-versa.

It would be a limited view to believe that applied mathematics consists only of developing techniques and algorithms to solve problems that arise in a physical or applied context. Applied mathematics deals with all the stages of the modeling process, not merely the formal solution. It is true that an important aspect of applied mathematics is studying, investigating, and developing procedures that are useful in solving mathematical problems: these include analytic and approximation techniques, numerical analysis, and methods for solving differential and integral equations. It is more the case, however, that applied mathematics deals with all phases of the problem. Formulating the model and understanding its origin in empirics are crucial steps. Because there is a constant interplay between the various stages, the scientist, engineer, or mathematician must understand each phase. For example, the solution stage sometimes involves making approximations that lead to a simplification; the approximations often come from a careful examination of the physical reality, which in turn suggests what terms may be neglected, what quantities (if any) are small, and so on. The origins and analysis are equally important. Indeed, physical insight forces us toward the right questions and at times leads us to the theorems and their proofs. In fact, mathematical modeling has been, and remains, one of the main driving forces for mathematics itself.

In the first part of this chapter our aim is to focus on the first phase of the modeling process. Our strategy is to formulate models for various physical systems while emphasizing the interdependence of mathematics and the physical world. Through study of the modeling process we gain insight into the equations themselves. In addition to presenting some concrete examples of modeling, we also discuss two techniques that are useful in developing and interpreting the model equations. One technique is *dimensional analysis*, and the other is *scaling*. The former permits us to understand the dimensional (meaning length, time, mass, etc.) relationships of the quantities in the equations and the resulting implications of dimensional homogeneity. Scaling is a technique that helps us understand the magnitude of the terms that appear in the model equations by comparing the quantities to intrinsic reference quantities that appear naturally in the physical situation. A side benefit in scaling differential equations, for example, is in the great economy it affords; more often than not, the number of independent parameters can be significantly reduced.

1.1.2 Dimensional Methods

One of the basic techniques that is useful in the initial, modeling stage of a problem is the analysis of the relevant quantities and how they must relate to each other in a dimensional way. Simply put, apples cannot equal oranges plus bananas; equations must have a consistency to them that precludes every possible relationship among the variables. Stated differently, equations must be dimensionally homogeneous. These simple observations form the basis of the subject known as **dimensional analysis**. The methods of dimensional analysis have led to important results in determining the nature of physical phenomena, even when the governing equations were not known. This has been especially true in continuum mechanics, out of which the general methods of dimensional analysis evolved.

The cornerstone result in dimensional analysis is known as the **Pi theorem**. The Pi theorem states that if there is a physical law that gives a relation among a certain number of dimensioned physical quantities, then there is an equivalent law that can be expressed as a relation among certain dimensionless quantities (often noted by $\pi_1, \pi_2, \ldots$, and hence the name). In the early 1900s, E. Buckingham formalized the original method used by Lord Rayleigh and gave a proof of the Pi theorem for special cases; now the theorem often carries his name. Birkhoff (1950) can be consulted for a bibliography and history.

Example 1.1

(**Atomic explosion**) To communicate the flavor and power of this classic result, we consider a calculation made by the British applied mathematician G. I. Taylor in the late 1940s to compute the yield of the first atomic explosion after viewing photographs of the spread of the fireball. In such an explosion a large amount of energy E is released in a short time (essentially instantaneously) in a region small enough to be considered a point. From the center of the explosion a strong shock wave spreads outward; the pressure behind it is on the order of hundreds of thousands of atmospheres, far greater than the ambient air pressure whose magnitude can be accordingly neglected in the early stages of the explosion. It is plausible that there is a relation between the radius of the blast wave front r, time t, the initial air density ρ, and the energy released E. Hence, we *assume* there is a physical law

$$g(t, r, \rho, E) = 0, \tag{1.1}$$

which postulates a functional relationship among these four dimensioned quantities. The Pi theorem states that there is an equivalent physical law between the independent dimensionless quantities that can be formed from t, r, E, and ρ. We note that t has dimensions of time, r has dimensions of length, E has dimensions of mass $\cdot$ length^2 $\cdot$ time^{-2}, and ρ has dimensions of mass $\cdot$ length^{-3}. Hence, the quantity $r^5 \rho / t^2 E$ is dimensionless, because all of the dimensions cancel out of this quantity (this is easy to check). It is not difficult to observe, and we shall show it later, that no other independent dimensionless quantities can be formed from t, r, E, and ρ. The Pi theorem then guarantees that the physical law (1.1) is equivalent to a physical law involving only the dimensionless quantities; in this case

$$f\left(\frac{r^5 \rho}{t^2 E}\right) = 0, \tag{1.2}$$

because there is only one such quantity, where f is some function of a single variable. From (1.2) it follows that the physical law must take the form (a root of (1.2))

$$\frac{r^5 \rho}{t^2 E} = C,$$

or

$$r = C \left(\frac{E t^2}{\rho}\right)^{1/5}, \tag{1.3}$$

where C is a constant. Therefore, just from dimensional reasoning it has been shown that the radius of the wave front depends on the two-fifths power of time. Experiments and photographs of explosions confirm this dependence. The constant C depends on the dimensionless ratio of the specific heat at constant pressure to the specific heat at constant volume. By fitting the curve (1.3) to experimental data of r versus t, the initial energy yield E can be computed, since C and ρ are known quantities. (See Exercise 3.) Although this calculation is only a simple version of the original argument given by Taylor, we infer that dimensional reasoning can give crucial insights into the nature of a physical process and is an invaluable tool for the applied mathematician, scientist, or engineer.   □

Example 1.2

(**Height of a projectile**) Let us imagine going outside and asking how high we can throw a baseball vertically upward. What would the maximum height h depend on? We may conjecture that it depends on the mass m of the ball, the acceleration of gravity g, and the velocity v with which we propel it. For now let's ignore the air resistance, which could be another issue. We can learn a lot by dimensional methods. Assume a physical law of the form

$$f(m, g, v, h) = 0$$

relating the four quantities we selected. All of them can be written in terms of fundamental dimensions M (mass), L (length), and T (time). If Π is a dimensionless quantity that can be formed from m, g, v, and h, then

$$\Pi = m^a g^b v^c h^d,$$

for some powers a, b, c, and d. This means

$$\Pi = M^a (LT^{-2})^b (LT^{-1})^c L^d = M^a L^{b+c+d} T^{-2b-c}.$$

Because Π is dimensionless, the dimensions must cancel and all the exponents must be zero; we obtain the homogeneous system of equations

$$a = 0, \ b + c + d = 0, \ -2b - c = 0.$$

Clearly, $a = 0$, and the last two equations have one degree of freedom (three unknowns and only two equations). So, pick $c = -2$, which gives $b = d = 1$. (Note that any multiple of this solution is also a solution.) We have shown that a dimensionless quantity Π has the form

$$\Pi = m^0 g^1 v^{-2} h^1.$$

There is only one such independent dimensionless variable (modulo the exponents being multiplied by some constant). We conclude that the physical law can be written in terms of this single dimensionless variable, or

$$g^1 v^{-2} h^1 = C,$$

where C is some constant. This means

$$h = C \frac{v^2}{g},$$

so the maximum height depends on the square of the velocity and the inverse of gravity. Based on minimal assumptions and dimensional reasoning we have learned a lot. For example, if we double the velocity, the height is quadrupled. How would we determine the constant C? A single experiment that measures h and v suffices (g is known). $\square$

As an aside, there is another aspect of dimensional analysis that is important in engineering, namely the design of small laboratory-scale models (say of an airplane or ship) that behave like their real counterparts. A discussion of this important topic is not treated here, but can be found in many engineering texts, especially books on fluid mechanics.

Example 1.3

(**Air resistance**) Those who bike have certainly noticed that the force F (mass times length, per time-squared) of air resistance appears to be positively related to their speed v (length per time) and cross-sectional area A (length-squared). But force involves mass, so it cannot depend only upon v and A. Therefore, let us add fluid, or air, density ρ (mass per length-cubed), and assume a relation of the form

$$F = f(\rho, A, v),$$

for some function f to be determined. What are the possible forms of f? To be dimensionally correct, the right side must be a force. What powers of ρ, A, and v would combine to give a force, that is,

$$F = k\rho^x A^y v^z,$$

where k is some constant without any units (i.e., dimensionless). If we denote mass by M, length by L, and time by T, then the last equation requires

$$MLT^{-2} = (ML^{-3})^x (L^2)^y (LT^{-1})^z.$$

Equating exponents of M, L, and T gives

$$x = 1, \quad -3x + 2y + z = 1, \quad -z = -2.$$

Therefore $x = 1$, $y = 1$, and $z = 2$. So the only relation must be

$$F = k\rho A v^2.$$

Consequently, the force must depend upon the square of the velocity, a fact often used in problems involving fluid resistance, and it must be proportional to density of air and area of the object. Again, substantial information can be obtained from dimensional arguments alone. □

EXERCISES

1. A pendulum executing small vibrations has period P and length l, and m is the mass of the bob. Can P depend only on l and m? If we assume P depends on l and the acceleration g due to gravity, then show that $P = $ constant $\cdot \sqrt{l/g}$.

2. A detonation wave, or shock front, moves through a high explosive with velocity D, thereby initiating an explosion that releases a specific energy (energy per unit mass) E. If there is a relationship between e and D, what can be concluded?

3. In the blast wave problem take $C = 1$ (a value used by Taylor) in (1.3), and use $\rho = 1.25$ kg/m^3. Some of the radius (m) vs. time (milliseconds) data for the Trinity explosion is given in the following table:

t	0.10	0.52	1.08	1.5	1.93	4.07	15.0	34.0
r	11.1	28.2	38.9	44.4	48.7	64.3	106.5	145

   Using these data, estimate the yield of the Trinity explosion in kilotons (1 kiloton equals $4.186(10)^{12}$ joules). Compare your answer to the actual yield of approximately 21 kilotons.

   Least Squares: Least squares is a popular method of choosing parameters in a function, or curve, in order to give the best fit to a set of data points. The philosophy is that the curve of best fit is the one that minimizes the sums of squares of the vertical distances between the curve and the data points. Let (x_i, y_i), $i = 1, 2, \ldots, n$ be n data points, and, for example, let $y = F(x, a, b)$ be a fitting function with two unknown parameters a and b. Then the least squares method is to determine a and b such that

   $$D(a, b) = \sum_{i=0}^{n} (F(x_i, a, b) - y_i)^2 \to \text{ min.}$$

   From calculus, a necessary condition for D to be minimum is that

   $$\frac{\partial D}{\partial a} = 0, \quad \frac{\partial D}{\partial b} = 0,$$

   which can be solved for a and b. With certain caveats, the least squares method can be extended to any number of parameters. Texts on data fitting can be consulted for additional information.

4. In the blast wave problem in Example 1.1, assume, instead of (1.1), a physical law of the form

   $$g(t, r, \rho, e, P) = 0,$$

   where P is the ambient air pressure. By inspection, find two independent dimensionless parameters formed from t, r, ρ, e, and P. Naming the two dimensionless parameters π_1 and π_2 and assuming the law is equivalent to

   $$f(\pi_1, \pi_2) = 0,$$

   does it still follow that r varies like the two-fifths power of t?

5. The law governing the distance x an object falls in time t in a field of constant gravitational acceleration g with no air resistance is $x = \frac{1}{2}gt^2$. How many independent dimensionless quantities can be formed from t, x, and g? Rewrite the physical law in terms of dimensionless quantities. Can the distance a body falls depend on the mass m as well? That is, can there be a physical law of the form $f(t, x, g, m) = 0$?

6. A ball tossed upward with initial velocity v reaches a height at time t given by $x = -\frac{1}{2}gt^2 + vt$, where g is the acceleration due to gravity. Show this law is dimensionally consistent. Find two independent dimensionless variables (call them y and s) and rewrite the law only in terms of the dimensionless quantities. Plot the dimensionless law in the sy plane, and make an argument that the single, dimensionless plot contains all the information in the graphs of all the plots of $x = -\frac{1}{2}gt^2 + vt$ (x vs. t), for all possible parameter values g and v.

1.1.3 The Pi Theorem

The Pi theorem states that it is generally true that a physical law

$$f(q_1, q_2, q_3, \ldots, q_m) = 0 \qquad (1.4)$$

relating m dimensional quantities $q_1, q_2, \ldots, q_m$ is equivalent to a physical law

$$F(\pi_1, \pi_2, \ldots, \pi_k) = 0$$

that relates the k dimensionless quantities $\pi_1, \pi_2, \ldots, \pi_k$ that can be formed from $q_1, q_2, \ldots, q_m$. There are tremendous values in a dimensionless formulation of a problem. One, the formula is independent of the set of units used; two, there are fewer dimensionless quantities than the dimensioned ones, and thus there is economy in the formulation; finally, important relations can be discovered among the quantities. Before making a formal statement, we introduce some basic terminology.

First, the m dimensional quantities $q_1, q_2, \ldots, q_m$, which are like the quantities t, r, ρ, and e in the blast wave example, Example 1.1, are dimensioned quantities. This means that they can be expressed in a natural way in terms of a minimal set of fundamental dimensions $L_1, L_2, \ldots, L_n$ ($n < m$), appropriate to the problem being studied. In the blast wave problem, time T, length L, and mass M can be taken to be the fundamental dimensions, since each quantity t, r, ρ, and e can be expressed in terms of T, L, and M. For example, the dimensions of the energy e are ML^2T^{-2}. In general, the dimensions of q_i, denoted by the square brackets notation $[q_i]$, can be written in terms of the fundamental dimensions as

$$[q_i] = L_1^{a_{1i}} L_2^{a_{2i}} \cdots L_n^{a_{ni}} \tag{1.5}$$

for some choice of exponents $a_{1i}, \ldots, a_{ni}$. If $[q_i] = 1$, then the quantity q_i is said to be **dimensionless**.

We proceed as follows. If π is a quantity of the form

$$\pi = q_1^{p_1} q_2^{p_1} \cdots q_m^{p_m},$$

a monomial in the dimensioned quantities, we want to find all exponents $p_1, \ldots, p_m$ for which π is dimensionless, or $[\pi] = 1$. Then,

$$
\begin{aligned}
[\pi] &= [q_1]^{p_1} [q_2]^{p_1} \cdots [q_m]^{p_m} \\
&= (L_1^{a_{11}} L_2^{a_{21}} \cdots L_n^{a_{n1}})^{p_1} \cdots (L_1^{a_{1m}} L_2^{a_{2m}} \cdots L_n^{a_{nm}})^{p_m} \\
&= 1.
\end{aligned}
$$

The powers of the L_i must sum to zero, and thus we obtain a homogeneous system of n equations in m unknowns $p_1, \ldots, p_m$, given in matrix form by

$$A\mathbf{p} = \mathbf{0},$$

where $\mathbf{p} = [p_1, \ldots, p_m]^{\mathrm{T}}$ is a column vector of the unknown exponents, and A is the $n \times m$ matrix

$$
A = \begin{pmatrix}
a_{11} & \cdots & a_{1m} \\
a_{21} & \cdots & a_{2m} \\
\vdots & & \vdots \\
a_{n1} & \cdots & a_{nm}
\end{pmatrix}
$$

called the **dimension matrix**. The elements in the ith column give the exponents for q_i in terms of the powers of $L_1, \ldots, L_n$. It follows from a key result in linear algebra that the number of independent solutions is $m - r$, where r is the rank of A. We recall that the rank of a matrix is the number of linearly independent rows, which, when the matrix is reduced to row echelon form, is the number of nonzero rows. So, the number of independent dimensionless variables that can be formed from $q_1, \ldots, q_m$ is $m - r$.

What assumptions are needed to show equivalence of the dimensionless form of the physical law? The fundamental assumption regarding the physical law (1.4) goes back to the simple statement that apples cannot equal oranges. We assume that (1.4) is *unit free* in the sense that it is independent of the particular set of units chosen to express the quantities $q_1, q_2, \ldots, q_m$. We are distinguishing the word **unit** from the word **dimension**. By units we mean specific physical units like seconds, hours, days, and years; all of these units have dimensions of time. Similarly, grams, kilograms, slugs, and so on are units of the dimension mass. Any fundamental dimension L_i has the property that its units can be changed upon multiplication by the appropriate conversion factor $\lambda_i > 0$ to obtain $\overline{L}_i$ in a new system of units. We write

$$\overline{L}_i = \lambda_i L_i, \quad i = 1, \ldots, n.$$

The units of derived quantities q can be changed in a similar fashion. If

$$[q] = L_1^{b_1} L_2^{b_2} \cdots L_n^{b_n},$$

then

$$\overline{q} = \lambda_1^{b_1} \lambda_2^{b_2} \cdots \lambda_n^{b_n} q$$

gives q in the new system of units. The physical law (1.4) is said to be independent of the units chosen to express the dimensional quantities $q_1, q_2, \ldots, q_m$, or *unit-free*, if $f(q_1, \ldots, q_m) = 0$ and $f(\overline{q}_1, \ldots, \overline{q}_m) = 0$ are equivalent physical laws. More formally:

Definition 1.4

The physical law (1.4) is **unit-free** if for all choices of real positive numbers $\lambda_1, \ldots, \lambda_n$ we have $f(\overline{q}_1, \ldots, \overline{q}_m) = 0$, if, and only if, $f(q_1, \ldots, q_m) = 0$, $\quad\Box$.

Example 1.5

The physical law

$$f(x, t, g) \equiv x - \tfrac{1}{2}gt^2 = 0 \tag{1.6}$$

relates the distance x a body falls in a constant gravitational field g to the time t. In the cgs system of units, x is given in centimeters (cm), t in seconds, and g in cm/sec^2. If we change units for the fundamental quantities x and t to inches and minutes, then in the new system of units

$$\overline{x} = \lambda_1 x, \quad \overline{t} = \lambda_2 t,$$

where $\lambda_1 = \frac{1}{2.54}$ (in/cm) and $\lambda_2 = \frac{1}{60}$ (min/sec). Because $[g] = LT^{-2}$, we have

$$\overline{g} = \lambda_1 \lambda_2^{-2} g.$$

Then

$$f(\overline{x}, \overline{t}, \overline{g}) = \overline{x} - \tfrac{1}{2}\overline{g}\,\overline{t}^2 = \lambda_1 x - \tfrac{1}{2}(\lambda_1 \lambda_2^{-2} g)(\lambda_2 t)^2 = \lambda_1 \left(x - \tfrac{1}{2}gt^2 \right).$$

Therefore (1.6) is unit-free. $\Box$

We can now give a formal statement of the Pi theorem.

Theorem 1.6

(**Pi theorem**) Let

$$f(q_1, q_2, \ldots, q_m) = 0 \qquad (1.7)$$

be a unit-free physical law that relates the dimensional quantities $q_1, q_2, \ldots, q_m$. Let $L_1, \ldots, L_n (n < m)$ be fundamental dimensions with

$$[q_i] = L_1^{a_{1i}} L_2^{a_{2i}} \cdots L_n^{a_{ni}}, \quad i = 1, \ldots, m,$$

and let $r = \operatorname{rank} A$, where A is the dimension matrix (1.5). Then there exists $m-r$ independent dimensionless quantities $\pi_1, \pi_2, \ldots, \pi_{m-r}$ that can be formed from $q_1, \ldots, q_m$, and the physical law (1.7) is equivalent to an equation

$$F(\pi_1, \pi_2, \ldots, \pi_{m-r}) = 0, \qquad (1.8)$$

expressed only in terms of the dimensionless quantities. $\qquad \square$

The existence of a physical law (1.7) is an assumption. In practice one must conjecture which are the relevant variables in a problem and then apply the machinery of the theorem. The resulting dimensionless physical law (1.8) must be checked by experiment, or whatever, in an effort to establish the validity of the original assumptions.

We prove the Pi theorem after more examples.

Example 1.7

(**Heat transfer**) At time $t = 0$ an amount of heat energy e, concentrated at a point in space, is allowed to diffuse outward into a region with temperature zero. If r denotes the radial distance from the source and t is time, the problem is to determine the temperature u as a function of r and t . This problem can be formulated as a boundary value problem for a partial differential equation (the heat equation, as in Chapter 6), but here we see what can be learned from a careful dimensional analysis of the problem. The first step is to make an assumption regarding which quantities affect the temperature u. Clearly t, r, and e are relevant quantities. It also is reasonable that the heat capacity c, with dimensions of energy per degree per volume, of the region plays a role, as will the rate at which heat diffuses outward. The latter is characterized by the thermal diffusivity k of dimensions length-squared per time. Therefore, we conjecture a physical law of the form

$$f(t, r, u, e, k, c) = 0,$$

which relates the six quantities, t, r, u, e, k, c. The next step is to determine a minimal set of fundamental dimensions $L_1, \ldots, L_n$ by which the six-dimensional quantities can be expressed. A suitable selection would be the four quantities T (time), L (length), Θ (temperature), and E (energy[1]). Then

$$[t] = T, \quad [e] = E, \quad [r] = L,$$
$$[k] = L^2 T^{-1}, \quad [u] = \Theta, \quad [c] = E\Theta^{-1} L^{-3},$$

and the dimension matrix A is given by

$$
\begin{array}{c}
\\ T \\ L \\ \Theta \\ E
\end{array}
\begin{array}{cccccc}
t & r & u & e & k & c \\
\left(\begin{matrix} 1 & 0 & 0 & 0 & -1 & 0 \\ 0 & 1 & 0 & 0 & 2 & -3 \\ 0 & 0 & 1 & 0 & 0 & -1 \\ 0 & 0 & 0 & 1 & 0 & 1 \end{matrix}\right)
\end{array}
$$

Here $m = 6$, $n = 4$, and the rank of the dimension matrix is $r = 4$. Consequently, there are $m - r = 2$ dimensionless quantities that can be formed from t, r, u, e, k, and c. To find them we proceed as follows: If π is dimensionless, then for some choice of $\alpha_1, \ldots, \alpha_6$

$$
\begin{aligned}
1 = [\pi] &= [t^{\alpha_1} r^{\alpha_2} u^{\alpha_3} e^{\alpha_4} k^{\alpha_5} c^{\alpha_6}] \\
&= T^{\alpha_1} L^{\alpha_2} \Theta^{\alpha_3} E^{\alpha_4} (L^2 T^{-1})^{\alpha_5} (E\Theta^{-1} L^{-3})^{\alpha_6} \\
&= T^{\alpha_1 - \alpha_5} L^{\alpha_2 + 2\alpha_5 - 3\alpha_6} \Theta^{\alpha_3 - \alpha_6} E^{\alpha_4 + \alpha_6}.
\end{aligned}
$$

Therefore, the exponents must vanish and we obtain four homogeneous linear equations for $\alpha_1, \ldots, \alpha_6$, namely

$$\alpha_1 - \alpha_5 = 0,$$
$$\alpha_2 + 2\alpha_5 - 3\alpha_6 = 0,$$
$$\alpha_3 - \alpha_6 = 0,$$
$$\alpha_4 + \alpha_6 = 0.$$

The coefficient matrix of this homogeneous linear system is just the dimension matrix A. From elementary matrix theory the number of independent solutions equals the number of unknowns minus the rank of A. Each independent solution will give rise to a dimensionless variable. Now the method unfolds and we can see the origin of the rank condition in the statement of the Pi theorem. By standard methods for solving linear systems we find two linearly independent solutions

$$\alpha_1 = -\tfrac{1}{2}, \quad \alpha_2 = 1, \quad \alpha_3 = \alpha_4 = 0, \quad \alpha_5 = -\tfrac{1}{2}, \quad \alpha_6 = 0,$$

---

[1] Alternately, because energy E involves mass M, the fundamental dimensions T, L, Θ, and E can be replaced by T, L, Θ, and M.

and

$$\alpha_1 = \frac{3}{2}, \quad \alpha_2 = 0, \quad \alpha_3 = 1, \quad \alpha_4 = -1, \quad \alpha_5 = \frac{3}{2}, \quad \alpha_6 = 1.$$

These give two dimensionless quantities

$$\pi_1 = \frac{r}{\sqrt{kt}}$$

and

$$\pi_2 = \frac{uc}{e}(kt)^{3/2}.$$

Therefore, the Pi theorem guarantees that the original physical law $f(t, r, u, e, k, c) = 0$ is equivalent to a physical law of the form

$$F(\pi_1, \pi_2) = 0.$$

Solving for π_2 gives

$$\pi_2 = g(\pi_1)$$

for some function g, or

$$u = \frac{e}{c}(kt)^{-3/2} g\left(\frac{r}{\sqrt{kt}}\right). \tag{1.9}$$

Again, without solving an equation governing the diffusion process, we are able via dimensional analysis to argue that the temperature of the region varies according to (1.9). For example, we can conclude that the temperature near the source $r = 0$ falls off like $t^{-3/2}$. □

Example 1.8

(**Blood flow in arteries**) Used in conjunction with experiments, dimensional analysis can be a powerful tool in obtaining theoretical results in all areas of science and engineering. In this example we study how blood flows in veins and arteries, depending on the radius of the artery r, the length of the artery l, the density and viscosity of blood, ρ and μ, and the pressure change ΔP over that length segment. Our interest is the volumetric flow rate Q (volume of blood per second) and how those quantities affect it. Of course, what we say applies to any context of viscous flow in a pipe.

There are six quantities. It is fairly clear that the dimensions of five of them are

$$[r] = [l] = L, \ [Q] = L^3 T^{-1}, \ [\Delta P] = ML^{-1}T^{-2}, \ [\rho] = ML^{-3},$$

but the viscosity may be less uncertain for many. Therefore we give a brief, heuristic explanation of viscosity that reveals its dimensions. We all have an intuitive idea about viscosity. Honey is very viscous, oil perhaps a little less, and alcohol less yet. Let us imagine a viscous fluid between two infinite plates

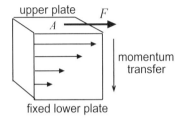

Figure 1.2 A viscous fluid between a fixed lower plate and an upper plate. A shear force on the upper plate transfers momentum downward into the fluid.

made of metal, for example. See Fig. 1.2. The lower plane is fixed, and we pull on the upper plate in a direction tangent to it, giving a shear force per unit area. The basic property of a viscous fluid is its 'stickiness' at a boundary; it adheres to a boundary, stationary or moving, with the same velocity as the boundary. So, the movement of the upper plate drags along the fluid at some velocity; at the lower boundary the velocity of the fluid is zero. Therefore, as we drag along the plate, the shear force imparts momentum to the fluid. As shown in the figure, the velocity profile decreases from the top plate to zero at the lower plate. Intuitively, it appears that the shear force should therefore be related to the vertical change of the velocity. As a constitutive assumption, we take a linear relationship with proportionality constant μ. That is,

$$\text{force per area} = \mu \cdot \text{change in velocity per unit length.}$$

Dimensionally,

$$\frac{MLT^{-2}}{L^2} = [\mu]\frac{LT^{-1}}{L},$$

or

$$[\mu] = ML^{-1}T^{-1}.$$

Now we can perform a dimensional analysis with L, T, and M as fundamental dimensions.

We make the assumption there is a relationship among these six quantities, i.e.,

$$f(r, l, Q, \Delta P, \rho, \mu) = 0.$$

To find the dimensionless variables we set

$$\begin{aligned}
\pi &= r^a l^b Q^c \Delta P^d \rho^e \mu^f \\
&= L^a L^b (L^3 T^{-1})^c (ML^{-1}T^{-2})^d (ML^{-3})^e (ML^{-1}T^{-1})^f.
\end{aligned}$$

Therefore,

$$[\pi] = L^{a+b+3c-d-3e-f} T^{-c-2d-f} M^{d+e+f} = 1,$$

and so the exponents for the six quantities must satisfy the homogeneous system

$$
\begin{aligned}
a + b + 3c - d - 3e - f &= 0, \\
-c - 2d - f &= 0, \\
d + e + f &= 0.
\end{aligned}
$$

The dimension matrix is

$$
\begin{pmatrix}
1 & 1 & 3 & -1 & -3 & -1 \\
0 & 0 & -1 & -2 & 0 & -1 \\
0 & 0 & 0 & 1 & 1 & 1
\end{pmatrix},
$$

and it is already in reduced echelon form. Clearly, all three rows are linearly independent and therefore the matrix has rank 3. Thus, $6 - 3 = 3$ dimensionless variables can be found. Also, from the properties of the matrix, a, c, and d are basic variables (the pivot positions) and b, e, and f are free; so we can solve for the three basic variables in terms of the free ones. Using backward substitution, the solution of the system can be written

$$
d = -e - f, \quad c = 2e + f, \quad a = -b - 4e - 3f.
$$

The solution can be written in vector form as

$$
\begin{pmatrix} a \\ b \\ c \\ d \\ e \\ f \end{pmatrix} = b \begin{pmatrix} -1 \\ 1 \\ 0 \\ 0 \\ 0 \\ 0 \end{pmatrix} + e \begin{pmatrix} -4 \\ 0 \\ 2 \\ -1 \\ 1 \\ 0 \end{pmatrix} + f \begin{pmatrix} -3 \\ 0 \\ 1 \\ -1 \\ 0 \\ 1 \end{pmatrix}, \quad d, e, f \in \mathbb{R}.
$$

These three linearly independent vectors define the three sets of six exponents of the independent pi variables, and we have

$$
\pi_1 = \frac{r}{l}, \quad \pi_2 = \frac{Q^2 \rho}{r^4 \Delta P}, \quad \pi_3 = \frac{Q \mu}{r^3 \Delta P}.
$$

Note that various combinations (products and powers) of pi variables are also dimensionless, and we need to make the appropriate choice to obtain what we want, namely a formula for Q. There are two quantities that depend on Q, and two that depend on ΔP. We can eliminate Q by forming the new variable

$$
\pi_4 = \frac{\pi_2}{\pi_3^2} = \frac{r^2 \rho \Delta P}{\mu^2}.
$$

Then, the Pi theorem guarantees that the physical law is equivalent to

$$
g(\pi_1, \pi_3, \pi_4) = 0,
$$

or

$$g\left(\frac{r}{l}, \frac{Q\mu}{r^3 \Delta P}, \frac{r^2 \rho \Delta P}{\mu^2}\right) = 0,$$

where g is an arbitrary function. We can solve for the first second argument in terms of the other two to get

$$\frac{Q\mu}{r^3 \Delta P} = G\left(\frac{r}{l}, \frac{r^2 \rho \Delta P}{\mu^2}\right),$$

where G is again arbitrary. This gives

$$Q = \frac{r^3 \Delta P}{\mu} G\left(\frac{r}{l}, \frac{r^2 \rho \Delta P}{\mu^2}\right).$$

At this point it appears we are at an impasse. However, we can perform experiments to get additional information; this is not uncommon. If we hold all of the quantities fixed except Q and ΔP, we find that the volumetric flow rate Q depends *linearly* on the pressure gradient ΔP. Then the arbitrary function G cannot depend on its last argument and we obtain

$$Q = \frac{r^3 \Delta P}{\mu} H\left(\frac{r}{l}\right),$$

where H is arbitrary. Next, by performing another experiment, we can find how H depends on the ratio r/l. Holding all the quantities fixed but r and l, we can plot $Q\mu/r^3 \Delta P$ vs. R/l, and we find that H is linear. Hence,

$$Q = C\frac{r^3 \Delta P}{\mu} \frac{r}{l} = C\frac{r^4 \Delta P}{\mu l},$$

where C is a constant. In conclusion, we have deduced from dimensional reasoning and experiment that the volumetric flow rate of blood depends linearly on the pressure change and the fourth power of the radius of the artery.

It is interesting to note that the theoretical equations of viscous fluid dynamics, which we examine in the last chapter, imply that the volumetric flow rate is

$$Q = \frac{\pi}{8} \frac{r^4 \Delta P}{\mu l}.$$

Of course, we used a lot of foresight in making our dimensional calculations, but it is fairly impressive that we reached almost the same result!   □

One of the fundamental techniques for finding solutions to differential equations has it basis in dimensional analysis. Among the different variables and parameters in the problem it is sometimes possible to identify new variables that reduce the equation to a much simpler one. For example, it may be possible to reduce a partial differential equation to an ordinary differential equation, or transform a second-order differential equation into a first-order equation. This method works because of natural symmetries in the problem itself, for example, invariance under a stretching transformation. A detailed treatment of symmetries and the discovery of variables that simplify a problem can be found in Logan (2008), as well as in a chapter in the first edition of the present text. Here we give a simple example of how dimensionless variables can be used in this type of reduction.

Example 1.9

(**Diffusion of a pollutant**) Imagine a long canal occupying the region $0 \leq x < \infty$ and filled with pollutant-free water. At its boundary, $x = 0$, suppose there is a constant, continual concentration u_0 of a toxicant that diffuses into the canal over time. The problem is to determine the concentration $u = u(x,t)$ of the toxicant at any $x > 0$ for all time $t > 0$. It can be shown (see Chapter 6) that $u(x,t)$ satisfies the partial differential equation

$$u_t = D u_{xx}, \quad 0 < x < \infty, \ t > 0,$$

and the boundary conditions

$$u(0,t) = u_0, \quad u(\infty,t) = 0, \ t > 0,$$

along with an initial condition

$$u(x,0) = 0, \quad 0 < x < \infty.$$

This partial differential equation is called the **diffusion equation**. As we observe later, it also appears in the context of heat conduction and random walks. The constant D is called the diffusion constant and has dimensions length-squared per time; it is a measure of how fast the toxicant diffuses through the water medium.

    If the preceding problem has a solution, then it is plausible that there is a physical law, or relationship, among the five quantities involved, or $f(x,t,D,u,u_0) = 0$. Let us perform the dimensional analysis. If π is dimensionless, then

$$
\begin{aligned}
[\pi] &= \left[x^a t^b D^c u^d u_0^e \right] = L^a T^b (L^2 T^{-1})^c (ML^{-3})^d (ML^{-3})^e \\
&= L^{a+2c-3d-3e} T^{b-c} M^{d+e} = 1.
\end{aligned}
$$

Therefore,

$$
\begin{aligned}
a + 2c - 3d - 3e &= 0, \\
b - c &= 0, \\
d + e &= 0.
\end{aligned}
$$

The dimension matrix is

$$
\begin{pmatrix}
1 & 0 & 2 & -3 & -3 \\
0 & 1 & -1 & 0 & 0 \\
0 & 0 & 0 & 1 & 1
\end{pmatrix}.
$$

The matrix is already in row echelon form and has three linearly independent rows; so the rank is 3. Thus, there are $5 - 3 = 2$ independent dimensionless variables. Clearly, basic variables are a, b and e, and the free variables are c and e. Thus, a, b and e can be determined in terms of c and e. Easily, by back substitution,

$$
d = -e, \quad b = c, \quad a = -2c.
$$

In vector form, the solution to the homogeneous system is

$$
\begin{pmatrix} a \\ b \\ c \\ d \\ e \end{pmatrix} = c \begin{pmatrix} -2 \\ 1 \\ 1 \\ 0 \\ 0 \end{pmatrix} + e \begin{pmatrix} 0 \\ 0 \\ 0 \\ -1 \\ 1 \end{pmatrix},
$$

where c and e are arbitrary real numbers. Therefore, two dimensionless quantities are

$$
\pi_1 = x^{-2} Dt, \quad \pi_2 = u^{-1} u_0.
$$

Of course, dimensionless quantities may be chosen in many ways; we can take powers, products, and quotients and still maintain independent quantities. Here, for convenience in subsequent calculations, we take

$$
\pi_1 = \frac{x}{\sqrt{Dt}}, \quad \pi_2 = \frac{u}{u_0}.
$$

From the Pi theorem we then have the equivalent law $g(\pi_1, \pi_2) = 0$, or solving for π_2,

$$
\frac{u}{u_0} = F\left(\frac{x}{\sqrt{Dt}}\right),
$$

where F is an unknown function. This is the dependence given by the Pi theorem.

However, we do not stop here. We have a partial differential equation, the diffusion equation, that can be used to determine F, and thus $u = u(x, t)$. To this end, we write

$$
u = u_0 F(s), \quad s = \frac{x}{\sqrt{Dt}}.
$$

We are thinking now that the dimensionless quantity s is an independent variable. We can substitute this expression into the diffusion equation as follows. By the chain rule,

$$u_t = u_0 F'(s) s_t = u_0 F'(s) \left(-\frac{x}{2\sqrt{D}} t^{-3/2} \right), \quad u_x = u_0 F'(s) s_x = u_0 F'(s) \frac{1}{\sqrt{Dt}},$$

and

$$u_{xx} = u_0 F''(s) \frac{1}{Dt}.$$

When these expressions are substituted into the diffusion equation, we obtain

$$F'(s) \left(-\frac{x}{2\sqrt{D}} t^{-3/2} \right) = D F''(s) \frac{1}{Dt},$$

or

$$-\frac{1}{2} s F'(s) = F''(s).$$

Amazingly enough, the diffusion equation reduced to a ordinary differential equation for $F(s)$! This cancellation of the original variables x and t signals the success of the method. The variable s is called a **similarity variable** and the solutions we obtain subsequently are called **similarity solutions**.

This differential equation can be solved easily by letting $G(s) = F'(s)$. Then $G'(s) = -(s/2)G(s)$, immediately giving $G(s) = C_1 \exp\left[-s^2/4 \right]$. Then

$$F(s) = C_1 \int_0^s e^{-\tau^2/4} d\tau + C_2.$$

To determine the constants of integration we apply the auxiliary conditions. We have $u(0,t) = u_0 F(0) = u_0$ and so $F(0) = 1$. Also $u(x,0) = u_0 F(\infty) = 0$, or $F(\infty) = 0$. Therefore $C_2 = 1$ and

$$C_1 \int_0^\infty e^{-\tau^2/4} d\tau + 1 = 0.$$

Using the well known integral formula

$$\int_0^\infty e^{-\beta \tau^2} d\tau = \frac{1}{2} \sqrt{\frac{\pi}{\beta}},$$

we get

$$C_1 = -\frac{1}{\sqrt{\pi}}.$$

Finally we can write the solution as

$$F(s) = -\frac{1}{\sqrt{\pi}} \int_0^s e^{-\tau^2/4} d\tau + 1 = \left[1 - \frac{2}{\sqrt{\pi}} \int_0^{s/2} e^{-r^2} dr \right].$$

In terms of the original variables,

$$u(x,t) = u_0 \left[1 - \frac{2}{\sqrt{\pi}} \int_0^{x/\sqrt{4Dt}} e^{-r^2} dr \right]. \quad \square$$

1.1.4 Proof of the Pi Theorem

To prove the Pi theorem we demonstrate two propositions.

(i) Among the quantities $q_1, \ldots, q_m$ there are $m - r$ independent dimensionless variables that can be formed, where r is the rank of the dimension matrix A.

(ii) If $\pi_1, \ldots, \pi_{m-r}$ are the $m - r$ dimensionless variables, then (1.7) is equivalent to a physical law of the form $F(\pi_1, \ldots, \pi_{m-r}) = 0$.

The proof of (i) has been outlined earlier—the general argument proceeds exactly like the construction of the dimensionless variables in the last several examples. It makes use of the familiar result in linear algebra that the number of linearly independent solutions of a set of n homogeneous equations in m unknowns is $m - r$, where r is the rank of the coefficient matrix. For, let π be a dimensionless quantity. Then

$$\pi = q_1^{p_1} q_2^{p_2} \cdots q_m^{p_m} \tag{1.10}$$

for some $p_1, p_2, \ldots, p_m$. In terms of the fundamental dimensions $L_1, \ldots, L_n$,

$$\pi = (L_1^{a_{11}} L_2^{a_{21}} \cdots L_n^{a_{n1}})^{p_1} (L_1^{a_{12}} L_2^{a_{22}} \cdots L_n^{a_{n2}})^{p_2} \cdots (L_1^{a_{1m}} L_2^{a_{2m}} \cdots L_n^{a_{nm}})^{p_m}$$
$$= L_1^{a_{11}p_1 + a_{12}p_2 + \cdots + a_{1m}p_m} \cdots L_n^{a_{n1}p1 + a_{n2}p2 + \cdots + a_{nm}p_m}.$$

Because $[\pi] = 1$, the exponents vanish, or

$$a_{11}p_1 + a_{12}p_2 + \cdots + a_{1m}p_m = 0$$

$$\vdots$$

$$a_{n1}p_1 + a_{n2}p_2 + \cdots + a_{nm}p_m = 0 \tag{1.11}$$

By the aforementioned theorem in linear algebra, the homogeneous system (1.11) has exactly $m - r$ independent (vector) solutions $[p_1, \ldots, p_m]$. (These solutions form a basis for the nullspace, or kernel, of A.) Each solution gives rise to a dimensionless variable via (1.10), and this completes the proof of (i). The independence of the dimensionless variables is in the sense of linear algebraic independence. Note that we can always, for example, multiply a vector by a constant, say $1/2$, and get an equivalent dimensionless variable, the square root of the dimensionless quantity.

The proof of (ii) makes strong use of the hypothesis that the law is unit-free. The argument is subtle, but it can be made almost transparent if we examine a particular example.

Example 1.10

Consider the unit-free law

$$f(x, t, g) = x - \tfrac{1}{2}gt^2 = 0 \qquad (1.12)$$

for the distance a particle falls in a gravitational field. If length and time are chosen as fundamental dimensions, a straightforward calculation shows there is a single dimensionless variable given by

$$\pi_1 = t^2 \frac{g}{x}.$$

The remaining variable g can be expressed as $g = x\pi_1/t^2$, and we can define a function G by

$$G(x, t, \pi_1) \equiv f(x, t, x\pi_1/t^2).$$

Clearly the law $G(x, t, \pi_1) = 0$, or

$$x - \tfrac{1}{2}\left(\frac{x\pi_1}{t^2}\right)t^2 = 0, \qquad (1.13)$$

is equivalent to (1.12) and is unit-free because f is. Then F is defined by

$$F(\pi_1) \equiv G(1, 1, \pi_1) = 1 - \tfrac{1}{2}\pi_1 = 0,$$

which is equivalent to (1.13) and (1.12). $\square$

We present the proof of (ii) in the special case when $m = 4$, $n = 2$, and $r = 2$. The notation will be easier, and the general argument for arbitrary m, n, and r proceeds in exactly the same manner. Therefore we consider a unit-free physical law

$$f(q_1, q_2, q_3, q_4) = 0, \qquad (1.14)$$

with

$$[q_j] = L_1^{a_{1j}} L_2^{a_{2j}}, \quad j = 1, \dots, 4,$$

where L_1 and L_2 are fundamental dimensions. The dimension matrix

$$A = \begin{pmatrix} a_{11} & a_{12} & a_{13} & a_{14} \\ a_{21} & a_{22} & a_{23} & a_{24} \end{pmatrix}$$

is assumed to have rank $r = 2$. If π is a dimensionless quantity, then

$$\pi = q_1^{p_1} q_2^{p_2} q_3^{p_3} q_4^{p_4},$$

where the exponents p_1, p_2, p_3, p_4 satisfy the homogeneous system (1.11), which in this case is

$$\begin{pmatrix} a_{11} \\ a_{21} \end{pmatrix} p_1 + \begin{pmatrix} a_{12} \\ a_{22} \end{pmatrix} p_2 + \begin{pmatrix} a_{13} \\ a_{23} \end{pmatrix} p_3 + \begin{pmatrix} a_{14} \\ a_{24} \end{pmatrix} p_4 = \begin{pmatrix} 0 \\ 0 \end{pmatrix}. \qquad (1.15)$$

We wish to determine p_1, p_2, p_3, p_4, and the form of the two dimensionless variables. Without loss of any generality, we can assume the first two columns of A are linearly independent. This is because we can rearrange the indices on the q_j so that the two independent columns appear as the first two. Then columns three and four can be written as linear combinations of the first two, or

$$\begin{pmatrix} a_{13} \\ a_{23} \end{pmatrix} = c_{31} \begin{pmatrix} a_{11} \\ a_{21} \end{pmatrix} + c_{32} \begin{pmatrix} a_{12} \\ a_{22} \end{pmatrix}, \qquad (1.16)$$

$$\begin{pmatrix} a_{14} \\ a_{24} \end{pmatrix} = c_{41} \begin{pmatrix} a_{11} \\ a_{21} \end{pmatrix} + c_{42} \begin{pmatrix} a_{12} \\ a_{22} \end{pmatrix} \qquad (1.17)$$

for some constants c_{31}, c_{32}, c_{41}, and c_{42}. Substituting into (1.15) gives

$$(p_1 + c_{31}p_3 + c_{41}p_4) \begin{pmatrix} a_{11} \\ a_{21} \end{pmatrix} + (p_2 + c_{32}p_3 + c_{42}p_4) \begin{pmatrix} a_{12} \\ a_{22} \end{pmatrix} = \begin{pmatrix} 0 \\ 0 \end{pmatrix}.$$

The left side is a combination of linearly independent vectors, and therefore the coefficients must vanish,

$$p_1 + c_{31}p_3 + c_{41}p_4 = 0,$$
$$p_2 + c_{32}p_3 + c_{42}p_4 = 0.$$

Therefore, we can solve for p_1 and p_2 in terms of p_3 and p_4 and write

$$\begin{pmatrix} p_1 \\ p_2 \\ p_3 \\ p_4 \end{pmatrix} = p_3 \begin{pmatrix} -c_{31} \\ -c_{32} \\ 1 \\ 0 \end{pmatrix} + p_4 \begin{pmatrix} -c_{41} \\ -c_{42} \\ 0 \\ 1 \end{pmatrix}.$$

The two vectors on the right represent two linearly independent solutions of (1.15); hence, the two dimensionless quantities are

$$\pi_1 = q_1^{-c_{31}} q_2^{-c_{32}} q_3,$$
$$\pi_2 = q_1^{-c_{41}} q_2^{-c_{42}} q_4.$$

Next define a function G by

$$G(q_1, q_2, \pi_1, \pi_2) \equiv f(q_1, q_2, \pi_1 q_1^{c_{31}} q_2^{c_{32}}, \pi_2 q_1^{c_{41}} q_2^{c_{42}}).$$

The physical law

$$G(q_1, q_2, \pi_1, \pi_2) = 0 \qquad (1.18)$$

holds if, and only if, (1.14) holds, and therefore (1.18) is an equivalent physical law. Since $f = 0$ is unit-free, it easily follows that (1.18) is unit-free (note that $\bar{\pi}_1 = \pi_1$, $\bar{\pi}_2 = \pi_2$ under any change of units; that is, dimensionless variables have the same value in all systems of units). For the final stage of the argument we show that (1.18) is equivalent to the physical law

$$G(1, 1, \pi_1, \pi_2) = 0, \tag{1.19}$$

which will give the result, for (1.19) implies $F(\pi_1, \pi_2) = 0$, where $F(\pi_1, \pi_2) \equiv G(1, 1, \pi_1, \pi_2)$. Because (1.18) is a unit-free, we must have

$$G(\bar{q}_1, \bar{q}_2, \pi_1, \pi_2) = 0, \tag{1.20}$$

where

$$\bar{q}_1 = \lambda_1^{a_{11}} \lambda_2^{a_{21}} q_1, \quad \bar{q}_2 = \lambda_1^{a_{12}} \lambda_2^{a_{22}} q_2$$

for *every* choice of the conversion factors $\lambda_1, \lambda_2 > 0$. We select λ_1 and λ_2 so that $\bar{q}_1 = \bar{q}_2 = 1$. We are able to make this choice because

$$\lambda_1^{a_{11}} \lambda_2^{a_{21}} q_1 = 1, \quad \lambda_1^{a_{12}} \lambda_2^{a_{22}} q_2 = 1 \tag{1.21}$$

implies

$$a_{11} \ln \lambda_1 + a_{21} \ln \lambda_2 = -\ln q_1,$$
$$a_{12} \ln \lambda_1 + a_{22} \ln \lambda_2 = -\ln q_2. \tag{1.22}$$

And, because the coefficient matrix

$$\begin{pmatrix} a_{11} & a_{21} \\ a_{12} & a_{22} \end{pmatrix}$$

in (1.22) is nonsingular (recall the assumption that the first two columns of the dimension matrix A are linearly independent), the system (1.22) has a unique solution $(\ln \lambda_1, \ln \lambda_2)$, from which λ_1 and λ_2 can be determined to satisfy (1.21). Thus (1.19) is an equivalent physical law and the argument is complete. $\square$

The general argument for arbitrary m, n, and r can be found in Birkhoff (1950), from which the preceding proof was adapted. This classic book also provides additional examples and historical comments.[2]

In performing a dimensional analysis on a problem, two judgments are required at the beginning.

1. The selection of the pertinent variables.

2. The choice of the fundamental dimensions.

---

[2] A simple proof of the Pi theorem in a linear algebraic setting is contained in J. D. Logan, et al., 1982. Dimensional analysis and the Pi theorem, *Linear Algebra and Its Applications* **47**, 117–126.

The first is a matter of experience and may be based on intuition or experiments. There is, of course, no guarantee that the selection will lead to a useful formula after the procedure is applied. Second, the choice of fundamental dimensions may involve tacit assumptions that may not be valid in a given problem. For example, including mass, length, and time but not force in a given problem assumes there is some relation (Newton's second law) that plays an important role and causes force not to be an independent dimension. As a specific example, a small sphere falling under gravity in a viscous fluid is observed to fall, after a short time, at constant velocity. Since the motion is un-accelerated we need not make use of the proportionality of force to acceleration, and so force can be treated as a separate, independent fundamental dimension. In summary, intuition and experience are important ingredients in applying the dimensional analysis formalism to a specific physical problem.

Table 1.1 gives dimensions for common quantities, including key relations that remind us of their origins. The reader should review the mks units for each of these quantities (e.g., energy is given in Joules), some of which are listed.

Quantity (symbol)	Dimensions	Relation
velocity (v)	LT^{-1}	length per time (m/s)
acceleration (a)	LT^{-2}	velocity per time-squared (m/s^2)
momentum (p)	MLT^{-1}	mass $\cdot$ velocity
mass density (ρ)	ML^{-3}	mass per volume
force (F)	MLT^{-2}	mass $\cdot$ acceleration (N)
energy (E), work (W)	ML^2T^{-2}	force $\cdot$ distance (Joules)
power (P)	ML^2T^{-3}	energy per time (Watts)
pressure (P), stress (σ)	$ML^{-1}T^{-2}$	force per area (Pascals)
frequency (ω)	T^{-1}	per time (Hertz)
internal energy (U)	ML^2T^{-2}	energy
specific heat (c)	$L^2T^{-2}\Theta^{-1}$	energy per mass per degree
heat flux (ϕ)	MT^{-3}	energy per time per area
thermal conductivity (K)	$MLT^{-3}\Theta^{-1}$	flux per length per degree
diffusivity (k)	L^2T^{-1}	$K/\rho c$
viscosity (μ)	$ML^{-1}T^{-1}$	mass per length per time
heat loss coefficient (h)	$MT^3\Theta^{-1}$	energy per volume $\cdot$ deg $\cdot$ time

Table 1.1 Dimensions of common quantities in mechanics and thermodynamics. Symbols M, L, T, and Θ (temperature) are fundamental dimensions.

EXERCISES

1. (Waves) The speed v of a wave in deep water is determined by its wavelength λ and the acceleration g due to gravity. What does dimensional analysis imply regarding the relationship between v, λ, and g?

2. (Allometry)An ecologist postulated that there is a relationship among the mass m, density ρ, volume V, and surface area S of certain animals. Discuss this conjecture in terms of dimensionless analysis.

3. (Mechanics) A small sphere of radius r and density ρ is falling at constant velocity v under the influence of gravity g in a liquid of density ρ_l and viscosity μ (given in mass per length per time). It is observed experimentally that

$$ v = \frac{2}{9} r^2 \rho g \mu^{-1} \left(1 - \frac{\rho_l}{\rho} \right). $$

Show that this law is unit-free.

4. A physical phenomenon is described by the quantities P, l, m, t , and ρ, representing pressure, length, mass, time, and density, respectively. If there is a physical law

$$ f(P, l, m, t, \rho) = 0 $$

relating these quantities, show that there is an equivalent physical law of the form $G(l^3 \rho / m, t^6 P^3 / m^2 \rho) = 0$. Find P in terms of an arbitrary function.

5. A physical system is described by a law $f(E, P, A) = 0$, where E, P, and A are energy, pressure, and area, respectively. Show that $P A^{3/2}/E = \text{const.}$

6. (Allometry) The length L of an organism depends upon time t, its density ρ, its resource assimilation rate a (mass per area per time), and its resource use rate b (mass per volume per time). Show that there is a physical law involving two dimensionless quantities.

7. (Mechanics) A piece of shrapnel of density ρ is driven off an explosive device at velocity v. The density of the explosive is ρ_e, and E is its Gurney energy (joules/kilogram), or the specific energy available in the explosive to do work. Determine how the velocity of the shrapnel depends upon E.

8. (Mechanics) In an indentation experiment, a slab of metal of thickness h is subjected to a constant pressure P on its upper surface by a cylinder of radius a. The technician then measures the vertical displacement of U of the indentation. The displacement also depends upon two material properties, Poisson's ratio v, which is dimensionless, and the Lamé constant μ, which has dimensions M/L^3T^2, where M is mass, L is length, and T is time.

Determine a set of dimensionless variables and show that the functional form of U is

$$U = aG\left(\frac{P}{\mu a^2}, \frac{h}{a}, v\right).$$

9. (Digestion) In modeling the digestion process in insects, it is believed that digestion yield rate Y, in mass per time, is related to the concentration C of the limiting nutrient, the residence time T in the gut, the gut volume V, and the rate of nutrient nutrient breakdown r, given in mass per time per volume. Show that for fixed T, r, C, the yield is positively related to the gut volume.

10. (Reactors) A chemical $\mathbf{C}$ flows continuously into a reactor with concentration C_i and volumetric flow rate q (volume/time). While in the reactor, which has volume V, the substances are continuously stirred and a chemical reaction $\mathbf{C} \rightarrow$ products, with rate constant k (1/time), consumes the chemical. The mixture exits the reactor at the same flow rate q. The concentration of $\mathbf{C}$ in the reactor at any time t is $C = C(t)$, and $C(0) = C_0$. Use dimensional analysis to deduce that

$$C = C_i F(tk, a, b),$$

where $a = C_0/C_i$ and $b = Vk/q$ are dimensionless constants and F is some function.

11. (Mechanics) The problem is to determine the power P that must be applied to keep a ship of length l moving at a constant speed V. If it is the case, as seems reasonable, that P depends on the density of water ρ, the acceleration due to gravity g, and the viscosity of water ν (in length-squared per time), as well as l and V, then show that

$$\frac{P}{\rho l^2 V^3} = f(\mathrm{Fr}, \mathrm{Re}),$$

where Fr is the Froude number and Re is the Reynolds number defined by

$$\mathrm{Fr} \equiv \frac{V}{\sqrt{lg}}, \quad \mathrm{Re} \equiv \frac{Vl}{\nu}$$

12. A spherical gas bubble with ratio of specific heats γ is surrounded by an infinite sea of liquid of density ρ_l. The bubble oscillates with growth and contraction periodically with small amplitude at a well-defined frequency ω. Assuming a physical law

$$f(P, R, \rho_l, \omega, \gamma) = 0,$$

where P is the mean pressure inside the bubble and R is the mean radius, show that the frequency must vary inversely with the mean radius R. The constant γ is dimensionless.

13. (Mechanics) Did you ever wonder how fast a long line of dominos topple over? Let us imagine an experiment where we set up a line of dominos with spacing d between them. Further, we assume a typical domino has height h and thickness τ. We seek a formula that relates these quantities, the gravitational constant g, and the velocity v.

   a) Use dimensional analysis to show that

$$v = \sqrt{gh}\, F\left(\frac{d}{h}, \frac{\tau}{h}\right).$$

   Assume τ/h is very small and can be neglected. What does the law become?

   b) Experiments have been performed that show the graph of $v/\sqrt{gh}$ vs. d/h is approximately constant, 1.5, for d/h varying over the range 0 to 0.8. Using $h = 0.05$ meters, what is the velocity of the toppling dominos?

14. A perfect gas in equilibrium has specific energy E (energy per mass), temperature T, and Boltzmann constant k (specific energy per degree). Derive a functional relationship of the form $E = f(k, T)$.

15. (Environment) In tests for fuel economy, cars are driven at constant speed V on a level highway. With no acceleration, the force of propulsion F must be in equilibrium with other forces, such as the air resistance, and so on. Assume that the variables affecting F are the velocity V, the rate C that fuel is burned, in volume per time, and the amount of energy K in a gallon of fuel, in mass per length per time-squared. Determine F as a function of V, C, and K.

16. (Biology) Across a cell membrane of thickness w the mass concentration of a biochemical molecule is C_i on the inside of the cell, and C_o on the outside. The diffusion constant in the membrane is d, measured in length-squared per time. Use dimensional analysis (not guessing) to obtain the most general relationship for the flux φ of the molecules (mass/(area·time)) through the membrane in terms of the other quantities.

1.2 Scaling

1.2.1 Characteristic Scales

Another procedure useful in formulating a mathematical model is **scaling**. Roughly, scaling means selecting new, usually dimensionless variables and reformulating the problem in terms of those variables. Not only is the procedure useful, but it often is a necessity, especially when comparisons of the magnitudes of various terms in an equation must be made in order to neglect small terms. This idea is particularly crucial in the application of perturbation methods to identify small and large parameters. Further, scaling usually reduces the number of parameters in a problem, thereby leading to great simplification, and it identifies what combinations of parameters are important.

For motivation let us suppose that time t is a variable in a given problem, measured in units of seconds. If the problem involved the motion of a glacier, clearly the unit of seconds is too fast because significant changes in the glacier could not be observed on the order of seconds. On the other hand, if the problem involved a nuclear reaction, then the unit of seconds is too slow; all of the important action would be over before the first second ticked. Evidently, every problem has an intrinsic **time scale**, or **characteristic time** t_c, which is appropriate to the given problem. This is the shortest time for discernible changes to be observed in the physical quantities. For example, the characteristic time for glacier motion would be of the order of years, whereas the characteristic time for a nuclear reaction would be of the order of microseconds. Some problems have multiple time scales. A chemical reaction, for example, may begin slowly and the concentration changes little over a long time; then, the reaction may suddenly go to completion with a large change in concentration over a short time. There are two time scales involved in such a process. Other examples are in the life sciences, where multi-scale processes are the norm. Spatial scales vary over as much as 10^{15} orders of magnitude as we progress from processes involving genes, proteins, cells, organs, organisms, communities, and ecosystems; time scales vary from times that it takes for protein to fold to times for evolution to occur. Several scales can occur in the same problem. Yet another example occurs in fluid flow, where the processes of heat diffusion, advection, and possible chemical reaction all have different scales.

Once a characteristic time has been identified, at least for a part of a process, then a new dimensionless variable $\bar{t}$ can be defined by

$$\bar{t} = \frac{t}{t_c}$$

If t_c is chosen correctly, then the *dimensionless time* $\bar{t}$ is neither too large nor too small, but rather of order unity. The question remains to determine the time scale t_c for a particular problem, and the same question applies to other variables in the problem (e.g., length, concentration, and so on). The general rule is that the characteristic quantities are formed by taking combinations of the various dimensional constants in the problem and should be roughly the same order of magnitude of the quantity itself.

After characteristic scales, which are built up from the parameters in the model, are chosen for the independent and dependent variables, the model can then be reformulated in terms of the new dimensionless variables. The result will be a model in dimensionless form, where all the variables and parameters in the problem are dimensionless. This process is called **non-dimensionalization**, or **scaling** a problem. By the Pi theorem, it is guaranteed that we can always non-dimensionalize a consistent, unit-free problem. The payoff is a simpler model that is independent of units and dimensions and that has fewer parameters, which is often a worthwhile economy of complication.

Example 1.11

(**Population growth**) Let $p = p(t)$ denote the population of an animal species located in a fixed region at time t. The simplest model of population growth is the classic **Malthus model**,[3] which states that the per capita growth rate $\frac{1}{p}\frac{dp}{dt}$ is constant. This means that the growth rate $\frac{dp}{dt}$ is proportional to the population p, or $\frac{dp}{dt} = rp$, where r is the *per capita growth rate*, given in dimensions of inverse-time. Easily, the Malthus model predicts that the population will grow exponentially for all time, that is, $p = p_0 e^{rt}$, where p_0 is the initial population. Many books on an impending world population explosion have been written with the Malthusian model as a premise. Clearly, however, as a population grows, intraspecific competition for food, living space, and natural resources limits the growth. We can modify the Malthus model to include a competition term. The simplest approach is to notice that if there are p individuals in the system, then the number of encounters, which is a measure of competition, is approximately p^2; therefore, we subtract a term proportional to p^2 from the growth rate to obtain the model

$$\frac{dp}{dt} = rp\left(1 - \frac{p}{K}\right), \quad p(0) = p_0. \tag{2.1}$$

This is saying that the per capita growth rate is not constant, but decreases linearly with population. The parameter K is the *carrying capacity*, which is the number of individuals that the ecosystem can sustain. When $p = K$ the

---

[3] Thomas Malthus (1766–1834) was an English essayist who was one of the first individuals to address demographics and food supply.

growth rate is zero. The model (2.1) is called the **logistic model**[4]; as the population grows, the negative p^2 term will kick in and limit the growth. To reduce (2.1) to dimensionless form we select new dimensionless independent and dependent variables. The time scale and population scale are formed from the constants in the problem, r, K, and p_0. Of these, only r contains the dimensions of time, and therefore we scale time by $1/r$ giving a new, dimensionless time τ defined by

$$\tau = rt.$$

There are two choices for the population scale, K or p_0. Either will work, and so we select K to obtain a dimensionless population P given by

$$P = \frac{p}{K}$$

Thus, we measure population in the problem relative to the carrying capacity. Using these dimensionless variables, it is straightforward to obtain

$$\frac{dP}{d\tau} = P(1 - P), \quad P(0) = \alpha, \tag{2.2}$$

where $\alpha \equiv p_0/K$ is a dimensionless constant. The scaled model (2.2) has only one constant (α), a significant simplification over (2.1) where there are three constants (r, K, p_0). The constant α represents a scaled, initial population. There is a single combination of the parameters in the original that is relevant to the dynamics. The initial value problem (2.2) can be solved by separating variables to obtain

$$P(\tau) = \frac{\alpha}{\alpha + (1 - \alpha)e^{-\tau}}. \tag{2.3}$$

Clearly

$$\lim_{\tau \to \infty} P(\tau) = 1.$$

It follows that, confirming our earlier statement, the limiting population p is equal to the carrying capacity K. We observe that there are two **equilibrium populations**, or constant solutions of (2.1), $p = K$ and $p = 0$. The population $p = K$ is an **attractor**; that is, regardless of the initial population, the population $p(t)$ tends to the value K as time gets large.   □

---

[4] The logistic model was introduced by P. Verhulst (1804–1849) as a model for the population of France.

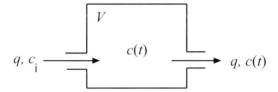

Figure 1.3 Continuously stirred tank reactor.

1.2.2 A Chemical Reactor Problem

A basic problem in chemical engineering is to understand how concentrations of chemical species vary when undergoing a reaction in a chemical reactor. To illustrate the concept of non-dimensionalization and scaling we analyze a simple model of an isothermal, continuously stirred, tank reactor (see Fig. 1.3). The reactor has a fixed volume V, and a chemical $\mathbf{C}$ of fixed concentration c_i, given in mass per volume, enters the reactor through the feed at a constant flow rate q, given in volume per time; initially the concentration of the chemical in the reactor is c_0. When the chemical enters the reactor, the mixture is perfectly stirred while undergoing a chemical reaction, and then the mixture exits the reactor at the same flow rate q. At any time t, we denote the concentration of the chemical $\mathbf{C}$ in the reactor by $c = c(t)$. The reactant chemical is assumed to disappear, that is, it is consumed by reaction, with a rate R, given in mass per unit volume, per unit time. We are thinking of a simple reaction of the form $\mathbf{C} \rightarrow$ products. Usually reaction rates depend on temperature, but here we assume that r depends linearly only on the concentration c; this is what makes our problem isothermal. The *perfectly stirred* assumption is an idealization and implies that there are no concentration gradients in the reactor; otherwise, c would also depend on a spatial variable.

To obtain a mathematical model of this problem we look for a physical law. A common principle that is fundamental to all flow problems is mass balance. That is, the time rate of change of the mass of the chemical inside the reactor must equal the rate mass flows in (qc_i), minus the rate that mass flows out (qc), plus the rate that mass is consumed by the reaction (VR). At any given time the mass of the chemical in the reactor is Vc. In symbols, the mass balance equation is

$$\frac{d}{dt}(Vc(t)) = qc_i - qc(t) - VR.$$

Observe that the factor V is required on the reaction term to obtain consistent dimensions. To fix the idea, let us take the reaction rate R to be proportional to c, that is $R = kc$, where k is the *rate constant* having dimensions of inverse time. Recall that this is the law of mass action in elementary chemistry, which states that the rate of a reaction is proportional to the product of the concentrations of the reactants. Then, the mathematical model is given by the initial value problem

$$\frac{dc}{dt} = \frac{q}{V}(c_i - c) - kc, \quad t > 0, \tag{2.4}$$

$$c(0) = c_0. \tag{2.5}$$

To non-dimensionalize the problem we choose dimensionless independent and dependent variables. This means we must select a characteristic time and a characteristic concentration by which to measure the real time and concentration. These characteristic values are formed from the constants in the problem: c_i, c_0, V, q, k. Generally, we measure the dependent variable relative to some maximum value in the problem, or any other value that represents the order of magnitude of that quantity. There are two constant concentrations, c_i and c_0, and either one of them is a suitable concentration scale. Therefore, we define a dimensionless concentration C by

$$C = \frac{c}{c_i}.$$

Therefore, all concentrations in the problem are measured relative to c_i, the concentration of the feed. To select a time scale we observe that there are two quantities with dimensions of time that can be formed from the constants in the problem, V/q and k^{-1}. The former is based on the flow rate, and the latter is based on the reaction rate. So the choice of a time scale is not unique. Either choice leads to a correct dimensionless problem. Let us hold up on making a selection and define a dimensionless time τ by

$$\tau = \frac{t}{T},$$

where T is either V/q or k^{-1}. We recast the model in dimensionless form. By the chain rule

$$\frac{dc}{dt} = \frac{c_i}{T}\frac{dC}{d\tau},$$

and therefore the model becomes

$$\frac{dC}{d\tau} = \frac{qT}{V}(1 - C) - kTC, \quad \tau > 0,$$

$$C(0) = \gamma,$$

where γ is a dimensionless constant given by the ratio

$$\gamma = \frac{c_0}{c_i}.$$

If we choose $T = V/q$, then we obtain the dimensionless model

$$\frac{dC}{d\tau} = 1 - C - \beta C, \quad \tau > 0, \tag{2.6}$$

$$C(0) = \gamma, \tag{2.7}$$

where

$$\beta = \frac{kV}{q}$$

is a dimensionless constant representing a ratio of the two time scales. Therefore, the problem has been reduced to dimensionless form, and any results that
are obtained are free of any specific set of units that we select. Moreover, the
number of parameters has been decreased from five to two, and the problem is
simpler. If we choose the other time scale, $T = k^{-1}$, then the problem reduces
to the dimensionless form

$$\frac{dC}{d\tau} = \frac{1}{\beta}(1 - C) - C, \quad \tau > 0, \tag{2.8}$$

$$C(0) = \gamma. \tag{2.9}$$

Is there any advantage of one over the other? Yes. In some problems, where
the terms differ in order of magnitude, it is important to choose the *correct*
scale, so that each term reflects the correct magnitude. For example, suppose
reaction occurs on a slow time scale compared to the flow through the reactor.
Then k is small compared to the flow rate q/V. We write $k \ll q/V$. Therefore
the dimensionless parameter β is small, or $\beta \ll 1$. Of the two dimensionless
models (2.6)–(2.7) or (2.8)–(2.9), which best reflects this assumption? We expect the reaction term to be small compared to the flow rate term, so the best
choice is (2.6)–(2.7), which means we should scale time on the more dominant
flow rate. If reaction is fast compared to the flow rate, then β is large ($\beta \gg 1$),
and we should choose a reaction time scale, giving (2.8)–(2.9); this choice puts
the small term $1/\beta$ correctly on the flow rate term.

Making the correct choice when terms have different orders of magnitude is
especially essential when we want to make an approximation by deleting small
terms in the equation. This is a very common strategy in equations that we
cannot solve, for example, most nonlinear equations. For example, if $\beta \ll 1$
and we scale by the flow rate, the model is (2.6)–(2.7). Ignoring the small term
gives the approximation

$$\frac{dC}{d\tau} = 1 - C, \quad C(0) = \gamma,$$

which gives approximation $C(\tau) = 1 + (\gamma - 1)e^{-\tau}$. This approximation is
believable; the reaction is slow and the concentration approaches the concentration of the feed. On the other hand, the model (2.8)–(2.9) gives $\beta\frac{dC}{d\tau} =$

$(1 - C) - \beta C$, which, upon neglecting the small terms, provides the approximation $C = 1$. This approximation does not even satisfy the initial condition.

Remark 1.12

In summary, *if approximations are to be made by deleting small terms, it is important how we non-dimensionalize the problem.* The common scaling strategy is to non-dimensionalize the problem using a generic time scale T that is chosen later to make the coefficients of the terms in the equation reflect their size or reflect what terms balance in the process. Ultimately, proper scaling is learned through experience and careful analysis.   □

1.2.3 The Projectile Problem

The projectile problem, as first pointed out by Lin and Segel (1974), is a good illustration of the importance of choosing correct scales, particularly when it is desired to make a simplification by neglecting small quantities. Terms in an equation that appear small are not always as they seem, and proper scaling is essential in determining the orders of magnitude of the terms.

We analyze the motion of a projectile thrust vertically upward from the surface of the earth. At time $t = 0$ on the surface of the earth, with radius R and mass M, an object of mass m is given a vertical upward velocity of magnitude V. To be determined is the height h above the earth's surface that the mass reaches at time t (see Fig. 1.4). Generally, forces on the object are the gravitational force and the force due to air resistance. But we assume the force due to air resistance is negligible in the particular problem we are considering. In general, as a first approximation it is common to neglect what are believed to be small effects, since in that case the equations are more tractable for analysis. Should the analytic results compare unfavorably with experiment or should a more detailed description be required, then additional effects can be included.

The governing equation, or mathematical model, comes from a physical law. Newton's universal gravitational law, which states that the force between the two objects is proportional to the product of the masses and inversely proportional to the square of the distance between them, where the mass of each object can be regarded as concentrated at its center. By Newton's second law the force equals the mass times acceleration, or

$$m\frac{d^2h}{dt^2} = -G\frac{Mm}{(h+R)^2},$$

where G is the proportionality constant in the universal gravitational law.

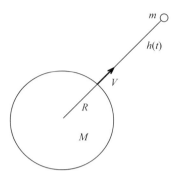

Figure 1.4 The projectile problem.

When $h = 0$, that is, at the earth's surface, the gravitational force equals $-mg$, and therefore

$$\frac{GM}{R^2} = g.$$

Thus

$$\frac{d^2h}{dt^2} = -\frac{R^2g}{(h+R)^2}, \tag{2.10}$$

with initial conditions

$$h(0) = 0, \quad \frac{dh}{dt}(0) = V. \tag{2.11}$$

The initial value problem (2.10) and (2.11) is the mathematical model for the problem.

At this point we undertake a dimensional analysis of the problem and gain considerable insight without actually attempting a solution. From our model the relevant dimensional quantities are t, h, R, V, and g having dimensions

$$[t] = \text{time } (T), \qquad [V] = \text{velocity } (LT^{-1}), \tag{2.12}$$
$$[h] = \text{length } (L), \qquad [g] = \text{acceleration } (LT^{-2}), \tag{2.13}$$
$$[R] = \text{length } (L). \tag{2.14}$$

We can use T (time) and L (length) as fundamental dimensions. Following the procedure described earlier, if π is a dimensionless combination of t, h, R, V, and g, then

$$[\pi] = [t^{\alpha_1} h^{\alpha_2} R^{\alpha_3} V^{\alpha_4} g^{\alpha_5}] = T^{\alpha_1 - \alpha_4 - 2\alpha_5} L^{\alpha_2 + \alpha_3 + \alpha_4 + \alpha_5} = 1.$$

Therefore,

$$\alpha_1 - \alpha_4 - 2\alpha_5 = 0,$$
$$\alpha_2 + \alpha_3 + \alpha_4 + \alpha_5 = 0. \tag{2.15}$$

This system has rank two and so there are three independent dimensionless variables. Either by inspection or solving (2.15), we find dimensionless quantities

$$\pi_1 = \frac{h}{R}, \quad \pi_2 = \frac{t}{R/V}, \quad \pi_3 = V/\sqrt{gR}. \tag{2.16}$$

By the Pi theorem, if there is a physical law relating t, h, R, V, and g (and we assume there must be, since in theory we could solve (2.10) and (2.11) to obtain that law), then there is an equivalent law that can be expressed as

$$\frac{h}{R} = f\left(\frac{t}{R/V}, \frac{V}{\sqrt{gR}}\right) \tag{2.17}$$

for some function $f(\pi_2, \pi_3)$.

   Actually there is considerable information in (2.17). For example, suppose we are interested in finding the time $t_{\max}$ that is required for the object to reach its maximum height for a given velocity V. Then differentiating (2.17) with respect to t and setting $h'(t)$ equal to zero gives

$$\frac{\partial f}{\partial \pi_2}\left(\frac{t_{\max}}{R/V}, \frac{V}{\sqrt{gR}}\right) = 0,$$

or

$$\frac{t_{\max}}{R/V} = F\left(\frac{V}{\sqrt{gR}}\right) \tag{2.18}$$

for some function F. Remarkably, with little analysis beyond dimensional reasoning, we have found that the time to maximum height depends on the dimensionless combination $V/\sqrt{gR}$. The value in knowing this kind of information lies in the efficiency of (2.18); a *single* graph of $t_{\max}/(R/V)$ vs. $V/\sqrt{Rg}$ contains all of the data of the graphs of $t_{\max}$ versus V for *all* choices of g and R. For example, an experimenter making measurements on different planets of $t_{\max}$ versus V would not need a separate plot of data for each planet. An entire atlas can be replaced by a single map when we plot dimensionless quantities.

   Now we select characteristic time and length scales and recast the problem represented by (2.10) and (2.11) into dimensionless form. The problem is more subtle than it originally appears. The general method requires us to choose a new dimensionless dependent variable $\bar{h}$ and independent variable $\bar{t}$ by

$$\bar{t} = \frac{t}{t_c}, \quad \bar{h} = \frac{h}{h_c}, \tag{2.19}$$

where t_c is an intrinsic time scale and h_c is an intrinsic length scale; the values of t_c and h_c should be chosen by taking combinations of the constants in the problem, which in this case are R, V, and g. This problem presents several choices. For a length scale h_c we could take either R or V^2/g. Possible time scales are R/V, $\sqrt{R/g}$, and V/g. Which choice is the most appropriate? Actually, equations (2.19) represent a legitimate transformation of variables for any choice of t_c and h_c; after the change of variables an equivalent problem results. From a scaling viewpoint, however, one particular choice is advantageous. The three choices

$$\bar{t} = \frac{t}{R/V}, \quad \bar{h} = \frac{h}{R}, \tag{2.20}$$

$$\bar{t} = \frac{t}{\sqrt{R/g}}, \quad \bar{h} = \frac{h}{R}, \tag{2.21}$$

and

$$\bar{t} = \frac{t}{Vg^{-1}}, \quad \bar{h} = \frac{h}{V^2 g^{-1}} \tag{2.22}$$

lead to the following three dimensionless problems, which are equivalent to (2.10) and (2.11):

$$\varepsilon \frac{d^2\bar{h}}{d\bar{t}^2} = -\frac{1}{(1+\bar{h})^2}, \quad \bar{h}(0) = 0, \quad \frac{d\bar{h}}{d\bar{t}}(0) = 1, \tag{2.23}$$

$$\frac{d^2\bar{h}}{d\bar{t}^2} = -\frac{1}{(1+\bar{h})^2}, \quad \bar{h}(0) = 0, \quad \frac{d\bar{h}}{d\bar{t}}(0) = \sqrt{\varepsilon}, \tag{2.24}$$

and

$$\frac{d^2\bar{h}}{d\bar{t}^2} = -\frac{1}{(1+\varepsilon\bar{h})^2}, \quad \bar{h}(0) = 0, \quad \frac{d\bar{h}}{d\bar{t}}(0) = 1, \tag{2.25}$$

respectively, where ε is a dimensionless parameter defined by

$$\varepsilon = \frac{V^2}{gR}.$$

To illustrate how difficulties arise in selecting an incorrect scaling, let us modify our original problem by examining the situation when ε is known to be a small quantity; that is, V^2 is much smaller than gR. Then one may be tempted, in order to make an approximation, to delete the terms involving ε in the scaled problem. Problem (2.23) then becomes

$$(1+\bar{h})^{-2} = 0, \quad \bar{h}(0) = 0, \quad \frac{d\bar{h}}{d\bar{t}}(0) = 1,$$

which has no solution, and problem (2.24) becomes

$$\frac{d^2\bar{h}}{d\bar{t}^2} = -\frac{1}{(1+\bar{h})^2}, \quad \bar{h}(0) = 0, \quad \frac{d\bar{h}}{d\bar{t}}(0) = 0,$$

which has no physically valid solution (the graph of $\bar{h}(\bar{t})$ passes though the origin with zero slope and is concave downward, thereby making $\bar{h}$ negative).

Therefore it appears that terms involving small parameters cannot be neglected. This is indeed unfortunate because this kind of technique is common practice in making approximations in applied problems. What went wrong was that (2.20) and (2.21) represent *incorrect* scalings; in these cases, terms that appear small may not in fact be small. For example, in the term $\varepsilon d^2 \overline{h}/d\overline{t}^2$, the parameter ε is small but $d^2 \overline{h}/d\overline{t}^2$ may be large, and hence the term may not be negligible compared to other terms in the equation.

If, on the other hand, the term $\varepsilon \overline{h}$ is neglected in (2.25), then $d^2 \overline{h}/d\overline{t}^2 = -1$, or $\overline{h} = \overline{t} - \overline{t}^2/2$, after applying the initial conditions. Therefore

$$h = -\tfrac{1}{2}gt^2 + Vt,$$

and we have obtained an approximate solution that is consistent with our experience with falling bodies close to the earth. In this case we are able to neglect the small term and obtain a valid approximation because the scaling is correct. That (2.22) gives the correct time and length scales can be argued physically. If V is small, then the body will be acted on by a constant gravitational field; hence, launched with speed V, it will uniformly decelerate and reach its maximum height in V/g units of time, which is the characteristic time. It will travel a distance of about (V/g) times its average velocity $\tfrac{1}{2}(V+0)$, or $V^2/2g$. Hence V^2/g is a good selection for the length scale. Measuring the height relative to the radius of the earth is not a good choice for small initial velocities.

In general, if a correct scaling is chosen, then terms in the equations that appear small are indeed small and may be safely neglected. Formally:

Principle of Scaling. The goal of scaling is to select intrinsic, characteristic reference quantities so that each term in the dimensional equation transforms into a term in such a manner that the dimensionless coefficient in the transformed term represents the order of magnitude or approximate size of that term. Schematically,

$$\left[\begin{pmatrix} \text{Dimensional} \\ \text{term} \end{pmatrix} \right] \rightarrow \begin{bmatrix} \text{coefficient} \\ \text{representing} \\ \text{the order of} \\ \text{magnitude} \\ \text{of the term} \end{bmatrix} \cdot \left[\begin{pmatrix} \text{dimensionless} \\ \text{factor} \\ \text{of order unity} \end{pmatrix} \right].$$

By *order of unity* we mean a term that is neither extremely large nor small. □

EXERCISES

1. Let $u = u(t)$, $0 \leq t \leq b$, be a given smooth function. If $M = \max |u(t)|$, then u can be scaled by M to obtain the dimensionless dependent variable

$U = u/M$. A time scale can be taken as $t_c = M/\max|u'(t)|$, the ratio of the maximum value of the function to the maximum slope. Find M and t_c for the following functions:

a) $u(t) = A \sin \omega t, \quad t > 0$.

b) $u(t) = Ae^{-\lambda t}, \quad t > 0$.

c) $u(t) = Ate^{-\lambda t}, \quad 0 \le t \le 2/\lambda$.

2. Consider a process described by the function $u(t) = 1 + e^{-t/\varepsilon}$ on the interval $0 \le t \le 1$, where ε is a small number. Use Exercise 1 to determine a time scale. Is this time scale appropriate for the entire interval $[0, 1]$? (Sketch a graph of $u(t)$ when $\varepsilon = 0.05$.) Explain why two time scales might be required for a process described by $u(t)$.

3. The growth rate of an organism is often measured using carbon biomass as the "currency." The **von Bertalanffy growth model** is

$$m' = ax^2 - bx^3,$$

where m is its biomass, x is some characteristic length of the organism, a is its biomass assimilation rate, and b is its biomass use rate. Thus, it assimilates nutrients proportional to its area, and it uses nutrients proportional to its volume. Assume $m = \rho x^3$ and rewrite the model in terms of the length x. Determine the dimensions of the constants a, b, and ρ. Select time and length scales ρ/b and a/b, respectively, and reduce the problem to dimensionless form. If $x(0) = 0$, find the length x at time t. Does this seem like a reasonable model?

4. A mass hanging on a spring is given a positive initial velocity V from equilibrium. The ensuing displacement $x = x(t)$ from equilibrium $(x = 0)$ is governed by

$$mx'' = -ax|x'| - kx,$$
$$x(0) = 0, \quad x'(0) = V,$$

where $-ax|x'|$ is a nonlinear damping force and $-kx$ is a linear restoring force of the spring. First make a table of the quantities and their dimensions. What are possible time scales, and on what physical processes are they based. What are possible length scales? If the restoring force is small compared to the damping force, choose appropriate time and length scales and non-dimensionalize the model so that the small term appears in the damping force.

5. The dynamics of a nonlinear spring–mass system are described by

$$mx'' = -ax' - kx^3,$$
$$x(0) = 0, \quad mx'(0) = I,$$

where x is the displacement, $-ax'$ is a linear damping term, and $-kx^3$ is a nonlinear restoring force. Initially, the displacement is zero and the mass m is given an impulse I that starts the motion.

a) Determine the dimensions of the constants I, a, and k.

b) Recast the problem into dimensionless form by selecting dimensionless variables $\tau = t/T$, $u = ax/I$, where the time scale T is yet to be determined.

c) In the special case that the mass is very small, choose an appropriate time scale T and find the correct dimensionless model. (A small dimensionless parameter should occur on the terms involving the mass in the original model.)

6. In a classic work modeling the outbreak of the spruce bud worm in Canada's balsam fir forests, researchers proposed that the bud worm population $n = n(t)$ was governed by the law

$$\frac{dn}{dt} = rn\left(1 - \frac{n}{K}\right) - p(n),$$

where r and K are the growth rate and carrying capacity, respectively, and $p(n)$ is a bird *predation* term given by

$$P(n) = \frac{bn^2}{a^2 + n^2},$$

where a and b are positive constants.

a) Determine the dimensions of the parameters in the problem.

b) Select time and population scales T and N_0 so that the model equation becomes, in dimensionless form,

$$\frac{dN}{d\tau} = sN\left(1 - \frac{N}{q}\right) - \frac{N^2}{1 + N^2} \equiv G(N) - P(N),$$

where $q \equiv K/a$ and $s \equiv ar/b$.

c) Researchers are often interested in equilibrium populations, or constant solutions of the differential equation. Working with the dimensionless model, show graphically that there is always at least one nonzero equilibrium population, and that there may be two or three, depending on the values of the parameters. (Hint: Plot the per capita growth rate $G(N)/N$ and the per capita predation rate $P(N)/N$ on the same set of N, $1/N$ axes.)

d) Find the equilibrium populations when $q = 12$ and $s = 0.25$ and when $q = 35$ and $s = 0.4$.

e) In the case $q = 35$ and $s = 0.4$, use a numerical differential equations solver (e.g., in MATLAB) to plot population curves $N = N(\tau)$ when initial populations are given by $N(0) = 40, 25, 2,$ and 0.05, respectively.

7. A rocket blasts off from the earth's surface. During the initial phase of flight, fuel is burned at the maximum possible rate α, and the exhaust gas is expelled downward with velocity β relative to the velocity of the rocket. The motion is governed by the following set of equations:

$$m'(t) = -\alpha, \quad m(0) = M,$$

$$v'(t) = \frac{\alpha\beta}{m(t)} - \frac{g}{(1 + x(t)/R)^2}, \quad v(0) = 0,$$

$$x'(t) = v(t), \quad x(0) = 0,$$

where $m(t)$ is the mass of the rocket, $v(t)$ is the upward velocity, $x(t)$ is the height above the earth's surface, M is the initial mass, g is the gravitational constant, and R is the radius of the earth. Reformulate the problem in terms of dimensionless variables using appropriate scales for m, x, v, t. (Hint: Scale m and x by obvious choices; then choose the time scale and velocity scale to ensure that the terms in the v equation and x equation are of the same order. Assume that the acceleration is due primarily to fuel burning and that the gravitational force is relatively small.)

8. In the chemical reactor problem assume that the reaction is $\mathbf{C} + \mathbf{C} \rightarrow$ products, and the chemical reaction rate is $r = kc^2$, where $c = c(t)$ is the concentration and k is the rate constant. What is the dimension of k? Define dimensionless variables and reformulate the problem in dimensionless form. Solve the dimensionless problem to determine the concentration.

9. The temperature $T = T(t)$ of a chemical sample in a furnace at time t is governed by the initial value problem

$$\frac{dT}{dt} = qe^{-A/T} - k(T - T_f), \quad T(0) = T_0,$$

where T_0 is the initial temperature of the sample, T_f is the temperature in the furnace, and q, k, and A are positive constants. The first term on the right side is the heat generation term, and the second is the heat loss term given by Newton's law of cooling.

a) What are the dimensions of the constants q, k, A?

b) Reduce the problem to dimensionless form using T_f as the temperature scale and choosing a time scale to be one appropriate to the case when the heat loss term is large compared to the heat generated by the reaction.

10. A ball of mass m is tossed upward with initial velocity V. Assuming the force caused by air resistance is proportional to the square of the velocity of the ball and the gravitational field is constant, formulate an initial value problem for the height of the ball at any time t. Choose characteristic length and time scales and recast the problem in dimensionless form.

11. A particle of mass m moves in one dimension on the x-axis under the influence of a force

$$F(x,t) = -\frac{k}{x^2}e^{-t/a},$$

where k and a are positive constants. Initially the particle is located at $x = L$ and has zero velocity.

a) Set up an initial value problem for the location $x = x(t)$ of the particle at time t.

b) Identify the length and time scales in the problem, and introduce dimensionless variables in two different ways, reformulating the problem in dimensionless form in both cases.

12. An aging spring is sometimes modeled by a force $F(x,t) = -kxe^{-t/a}$, where k is the stiffness of the spring and a is constant. Set up and non-dimensionalize a model for oscillation of a mass m on an aging spring if $x(0) = L$ and $x'(0) = V$.

13. (Fishery management) A fishing industry on the East Coast has fished an area to near depletion and has stopped its activity (assume this occurs at time zero with fish population x_0). The management assumes that the fish will recover and grow logistically with growth rate r and carrying capacity K. The question is: when should fishing resume? Let $x(t)$ be the fish population, $b(t)$ the number of fishing boats, and q be the catchability, given in units of per boat per time, so that the harvesting rate is $h(t) = qb(t)x(t)$.

a) If t_f is the time fishing is resumed, find differential equations that govern the fish population for $t > t_f$ and $t < t_f$.

b) Assume that the fleet is constant, $b(t) = B$, and non-dimensionalize the model in Part (a), choosing the growth rate as the time scale and the carrying capacity as the population scale.

c) Determine the (dimensionless) time τ that fishing begins as a function of the dimensionless fleet size $\beta = qB/r$, subject to the condition that the fish population should remain at constant for times greater than τ. Translate this relation into dimensioned form to find $t_f = t_f(B)$.

d) Under the conditions of Part (c), what is the optimum number of boats and the maximum sustainable catch, i.e., the maximum harvesting rate?

e) Suppose p is the selling price per fish, c_b is the cost of maintenance per boat per time, w is a fisherman's wage per time, and n is the number of fishermen per boat. Argue that the rate money is earned from the fishing is $ph(t) - b(t)(c_b + nw)$. If the value of money is discounted at rate R over time, conclude that the net profit can be written as a function of the fleet size as

$$J(b(t)) = \int_0^\infty e^{-Rt}[ph(t) - b(t)(c_b + nw)]\, dt.$$

f) Under the condition of a constant fleet size, $b(t) = B$, at what time t_f should the fleet resume fishing to maximize the profit $J(B)$? What is the optimal fleet size?

g) Use parameter values $R = 0.7$, $p = 2$, $q = 0.5$, $K = 50$, $r = 0.55$, $x_0 = 2$, $c_b + nw = 2$, to plot $J(B)$ in Part (f), and find the maximum starting time, the optimal fleet size, and the maximum profit.

14. A pendulum of length l with a bob of mass m executes a (dimensionless) angular displacement $\theta = \theta(t)$ from its attachment point, with $\theta = 0$ when the pendulum is vertically downward. See Fig. 1.5. Use Newton's law to derive the equation of motion for the pendulum by noting that the acceleration is d^2s/dt^2, where $s = l\theta$ is the length of the circular arc traveled by the bob; the force is $-mg$ vertically downward, and only the component tangential to the arc of oscillation affects the motion. The model is

$$\frac{d^2\theta}{dt^2} + \frac{g}{l}\sin\theta = 0.$$

If the bob is released from a small angle θ_0 at time $t = 0$, formulate a dimensionless initial value problem describing the motion. (Observe that the radian measurement θ is dimensionless, but one can still scale θ by θ_0.)

15. The initial value problem for the damped pendulum equation is

$$\frac{d^2\theta}{dt^2} + k\frac{d\theta}{dt} + \frac{g}{l}\sin\theta = 0,$$
$$\theta(0) = \theta_0, \quad \theta'(0) = w_0.$$

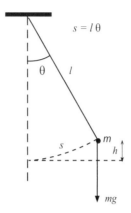

Figure 1.5 Pendulum.

a) Find three time scales and comment upon what process each involves.

b) Non-dimensionalize the model with a time scale appropriate to expecting damping to have a small contribution.

c) Non-dimensionalize the model with a time scale based on the fact that damping has an effect.

16. In a simple model of predation a fraction of the prey take refuge and are not subject to predation. If $H = H(t)$ is the number of prey, and $P = P(t)$ is the number of predators, the model takes the form

$$\frac{dH}{dt} = rH - a(H - H_r)P, \qquad \frac{dP}{dt} = -kp + b(H - H_r)P,$$

where r is the prey growth rate, k is the predator mortality rate, H_r is the number of prey in refuge (constant), and a and b are predation rates. Find dimensionless variables so that the model reduces to the dimensionless form

$$h' = h - (h - 1)p, \quad p' = -\alpha p + \beta(h - 1)p,$$

for appropriate choices of α and β.

1.3 Differential Equations

The first subsection below contains a very brief review of elementary solution techniques for first- and second-order differential equations. This review may be safely omitted for those who have fresh knowledge of elementary methods.

The Exercises review standard applications with which the reader should be familiar; working through some of these may introduce the reader to new applications and scaling methods. [Differential equations can be reviewed in Logan (2010), for example, or numerous other elementary textbooks.]

1.3.1 Review of Elementary Methods

In calculus, one focus is on finding derivatives of given functions. In differential equations the focus is on the reverse problem, namely, given a relationship between an unknown function $u = u(t)$ and some of its derivatives, find the function itself. The given relationship is called a differential equation. The simplest differential equation is the **pure time** equation

$$u' = g(t), \tag{3.1}$$

where g is a given continuous function. We recognize, by definition, that a solution $u = u(t)$ must be an antiderivative of g. By the fundamental theorem of calculus, all antiderivatives are given by

$$u(t) = \int_a^t g(s)\, ds + C, \tag{3.2}$$

where C is an arbitrary constant of integration and a is any lower limit of integration. Thus (3.2) gives all solutions of the pure time equation (3.1).

Antiderivatives are sometimes denoted by the usual indefinite integral sign

$$u(t) = \int g(t)dt + C,$$

but this representation is only symbolic and useless if we cannot find an expression for $\int g(t)\, dt$.

Example 1.13

If $g(t) = \exp(-t^2)$, then there is no simple formula for the antiderivative, and thus we cannot find, for example, the antiderivative evaluated at a specific value of t. Therefore we write the antiderivative as

$$\int_0^t \exp(-s^2)\, ds + C.$$

A multiple of this function is the common special function

$$\operatorname{erf}(t) = \frac{2}{\sqrt{\pi}} \int_0^t \exp(-s^2)\, ds,$$

which is the error or "erf" function.   □

In general, a first-order differential equation has the form

$$u' = f(t, u), \tag{3.3}$$

where t ranges over some specified interval I of time. A solution is a differentiable function $u = u(t)$ that, when substituted into (3.3), satisfies it identically for all t in the interval I. Often the differential equation is accompanied by an initial condition of the form

$$u(t_0) = u_0, \tag{3.4}$$

which specifies that the solution curve pass through the point (t_0, u_0), or that the system is in state u_0 at time t_0. The problem of solving (3.3) subject to (3.4) is called the **initial value problem** associated with the differential equation. If $f(t, u)$ is a continuous function in some open region of the tu-plane containing the point (t_0, u_0), then one can show that there is a solution to the initial value problem in some interval containing t_0. If, in addition, the partial derivative $f_u t, u$ is continuous in the domain, then the solution is unique. See, for example, Kelley & Peterson (2010).

Separable Equations. Some differential equations of first order can be solved, at least in principle. A differential equation is **separable** if it has the form

$$u' = h(u)g(t).$$

In this case we can divide both sides by $h(u)$ and integrate over t to obtain

$$\int \frac{u'(t)dt}{h(u(t))} = \int g(t)\, dt + C.$$

Here we have written explicit dependence on t. Also, as noted above, sometimes the antiderivatives must be written as definite integrals with a variable upper limit of integration. Next we can change variables in the integral on the left: $u = u(t)$, $du = u'(t)\, dt$. Then

$$\int \frac{du}{h(u)} = \int g(t)\, dt + C,$$

which, when the integrals are resolved, gives the solution u implicitly as a function of t. One may, or may not, be able to solve for u and find an explicit form for the solutions. An autonomous equation, where the right side of the differential equation does not depend on time t,

$$u' = f(u),$$

has implicit solution

$$\int \frac{1}{f(u)} du = t + C.$$

An example of a simple, separable equation is the **growth-decay** equation, $u' = \lambda u$, from which one easily obtains the general solution $u = Ce^{\lambda t}$. The solution models exponential growth if $\lambda > 0$ and exponential decay if $\lambda < 0$.

Linear Equations. A **first-order linear equation** has the form

$$u' + p(t)u = q(t). \tag{3.5}$$

Multiplying this equation by an integrating factor $e^{\int p(t)dt}$ turns the left side into a total derivative and (3.5) becomes

$$\frac{d}{dt}\left(u e^{\int p(t)dt}\right) = q(t)e^{\int p(t)dt}.$$

Now, both sides can be integrated to determine u. We illustrate this procedure with an example.

Example 1.14

Find an expression for the solution to the initial value problem

$$u' + 2tu = 2\sqrt{t}, \quad u(0) = 3.$$

The integrating factor is $\exp(\int 2t\, dt) = \exp(t^2)$. Multiplying both sides of the equation by the integrating factor makes the left side a total derivative, or

$$(u e^{t^2})' = 2\sqrt{t}e^{t^2}.$$

Integrating from 0 to t (while changing the dummy variable of integration to s) gives

$$u(t)e^{t^2} - u(0) = 2\int_0^t \sqrt{s}e^{s^2}\, ds.$$

Solving for u gives

$$u(t) = e^{-t^2}\left(3 + 2\int_0^t \sqrt{s}e^{s^2}\, ds\right) = 3e^{-t^2} + 2\int_0^t \sqrt{s}e^{s^2-t^2}\, ds.$$

As is often the case, the integrals in this example cannot be performed easily, if at all, and we must write the solution in terms of integrals with variable limits. □

Example 1.15

Some first-order equations can be solved by substitution. For example, **Bernoulli equations** are differential equations having the form

$$u' + p(t)u = q(t)u^n.$$

The transformation of dependent variables $w = u^{1-n}$ turns a Bernoulli equation into a linear equation for $w = w(t)$.   □

Second-Order Linear Equations. The **linear equation with constant coefficients**,

$$au'' + bu' + cu = 0, \qquad (3.6)$$

where a, b, and c are constants, occurs frequently in modeling RCL circuits and spring–mass–damping systems. We recall that if $u_1(t)$ and $u_2(t)$ are two independent solutions to a second-order, linear, homogeneous equation, then the general solution is $u = c_1 u_1(t) + c_2 u_2(t)$, where c_1 and c_2 are arbitrary constants. Equation (3.6) has solutions of the form $u = e^{mt}$, where m is to be determined. Substitution of this exponential form into the equation leads to the **characteristic equation**

$$am^2 + bm + c = 0.$$

This quadratic will have two roots, m_1 and m_2. Three possibilities can occur: unequal real roots, equal real roots, and complex roots (which must be complex conjugates).

(a) m_1, m_2 real and unequal. Two independent solutions of (3.6) are $e^{m_1 t}$ and $e^{m_2 t}$.

(b) m_1, m_2 real and equal, i.e., $m_1 = m_2 \equiv m$. Two independent solutions of (3.6) are e^{mt} and te^{mt}.

(c) $m = \alpha \pm i\beta$ are complex conjugate roots. Two real, independent solutions of (3.6) are $e^{\alpha t} \sin \beta t$ and $e^{\alpha t} \cos \beta t$. This follows from an application of Euler's formula,

$$e^{(\alpha \pm i\beta)t} = e^{\alpha t}(\cos \beta t) \pm i \sin(\beta t)),$$

and the fact that the real and imaginary parts of a complex solution are both real solutions

Example 1.16

Of particular importance are the two equations, $u'' + a^2 u = 0$, which has oscillatory solutions $u = c_1 \cos at + c_2 \sin at$, and $u'' - a^2 u = 0$, which has exponential solutions $u = c_1 e^{-at} + c_2 e^{at}$; we can also write this latter solution in terms of hyperbolic functions as $u = c_1 \cosh at + c_2 \sinh at$. These two equations occur so frequently that it is best to memorize them.   □

It is difficult to find analytic solution formulas for second order linear equations with variable coefficients,

$$a(t)u'' + b(t)u' + c(t)u = 0.$$

Often power series methods are applied, and they lead to solution representations as power series that define special functions like Bessel functions, Airy functions, Legendre and Hermite polynomials, and so on.

Example 1.17

There is a special variable coefficient equation that can be solved with simple formulas, namely the **Cauchy–Euler equation**:

$$at^2 u'' + btu' + cu = 0. \tag{3.7}$$

This equation admits power functions as solutions of the form $u = t^m$, where m is to be determined. Upon substitution, we obtain the characteristic equation

$$am(m-1) + bm + c = 0.$$

This quadratic equation has two roots, m_1 and m_2, and there are three cases:

(a) m_1, m_2 are real and unequal. Two independent solutions of (3.7) are t^{m_1} and t^{m_2}.

(b) m_1, m_2 are real and equal, i.e., $m_1 = m_2 \equiv m$. Two independent solutions of (3.7) are t^m and $t^m \ln t$.

(c) $m = \alpha \pm i\beta$ are complex conjugate roots. Two real, independent solutions, determined from the real and imaginary parts of the complex solution $t^{\alpha + i\beta}$ (3.7) are $t^\alpha \sin(\beta \ln t)$ and $t^\alpha \cos(\beta \ln t)$.

Nonhomogeneous Equations. The general solution of the **linear nonhomogeneous equation**

$$u'' + p(t)u' + q(t)u = f(t) \tag{3.8}$$

is

$$u(t) = c_1 u_1(t) + c_2 u_2(t) + u_P(t),$$

where u_1 and u_2 are independent solutions of the homogeneous equation (i.e., when $f(t) \equiv 0$), and u_P is any particular solution to (3.8). This structure result is valid for all linear equations of any order: The general solution is the sum of the general solution to the homogeneous equation and a particular solution to the nonhomogeneous equation. For constant coefficient equations a particular solution can sometimes be inferred, or judiciously guessed, from the form of the forcing term $f(t)$ when it is a polynomial, sine or cosine function, exponential function, or sums and products of these functions. This is the *method of undetermined coefficients*. In any case, however, there is a general formula, called the **variation of parameters** formula, that gives a particular solution in terms of the linear independent solution of the homogeneous equation. In the case of the second-order equation (3.8) with linearly independent solutions u_1 and u_2, the formula is

$$u_P(t) = u_2(t) \int_a^t \frac{u_1(s)f(s)}{W(s)} ds - u_1(t) \int_a^t \frac{u_1(s)f(s)}{W(s)} ds,$$

where $W(s) = u_1(s)u_2'(s) - u_2(s)u_1'(s)$ is the Wronskian. Any elementary differential equation book has a detailed discussion of these methods (e.g., Logan, 2010).

Nonlinear Equations. Some second-order, nonlinear equations can be immediately reduced to a first-order equation. For example, we consider Newton's second law of motion, a dynamical equation of the form

$$mx'' = F(t, x, x'),$$

where the dependent variable is $x = x(t)$, representing position; m is the mass and F is the force. If the force does not depend on position x, then the substitution $y = x'$ immediately reduces the equation to the first-order equation in the velocity y:

$$my' = F(t, y').$$

Once y is obtained, integration gives the solution x. If the force does not depend explicitly on the time t, that is, it has the form

$$mx'' = F(x, x'),$$

then we again define $y = x'$. Then, using the chain rule, $x'' = \frac{d}{dt}x' = \frac{dy}{dt} = \frac{dy}{dx}\frac{dx}{dt} = \frac{dy}{dx}y$, and the equation becomes

$$my\frac{dy}{dx} = F(x, y),$$

which is a first-order equation for the velocity in terms of position, $y = y(x)$. Then we can obtain $x = x(t)$ by solving

$$\frac{dx}{dt} = y(x).$$

There is one additional important case when the force depends only on position and is **conservative**, that is, when there is a potential function $V = V(x)$ for which

$$F(x) = -\frac{dV}{dx}.$$

Then the equation of motion becomes

$$my\frac{dy}{dx} = F(x),$$

or, upon separating variables,

$$mydy = F(x)dx.$$

Integrating both sides gives

$$\frac{1}{2}my^2 + V(x) = E, \qquad \text{(Conservation of Energy)}$$

where the constant of integration in denoted by E. This last equation expresses the fact that the total energy in the system (kinetic energy plus potential energy) is constant. Therefore, if the force $F(x)$ is conservative, then total energy is conserved. The energy constant E can be computed from the initial position and velocity, $x(0)$ and $y(0)$. Later, these models are discussed in detail.

EXERCISES

1. Find the general solution of the following differential equations:

   a. $u' + 2u - e^{-t}$.
   b. $u'' + 4u = t\sin 2t$.
   c. $u' - tu = t^2u^2$.
   d. $t^2u'' - 3tu' + 4u = 0$.
   e. $u'' + 9u = 3\sec 3t$.
   f. $x'' = -tu'^2$
   g. $u'' - 3u' - 4u = 2\sin t$.

   h. $xx'' - x'^3 = 0$.
   i. $2t^2u'' + 3tu' - u = 0$.
   j. $u' - 2tu = 1$.
   k. $u'' + 5u' + 6u = 0$.
   l. $u'' + u' + u = te^t$.
   m. $u'' + \omega^2u = \sin\beta t, \quad \omega \neq \beta$.
   n. $u'' + \omega^2u = \cos\omega t$.

2. A differential equation of the form $u' = f\left(\frac{u}{t}\right)$ is called a homogeneous equation. (The word homogeneous in this case is used differently from that of an equation whose right side is zero.) Show that the change of variables $y = u/t$ transforms the equation into a separable equation for y.

3. (Mechanics) From the conservation of energy law for a conservative force, show that the position $x = x(t)$ satisfies the equation

$$\pm\sqrt{\frac{m}{2}}\int\frac{dx}{\sqrt{E - V(x)}} = t + C,$$

where C is a constant of integration.

4. (Reactor dynamics) Consider a chemical reactor of constant volume V where a chemical $\mathbf{C}$ is pumped into the reactor at constant concentration and constant flow rate q. While in the reactor it reacts according to $\mathbf{C} + \mathbf{C} \rightarrow$ products. The law of mass action dictates that the rate of the reaction is $r = kC^2$, where k is the rate constant. If the concentration of $\mathbf{C}$ in the reactor is given by $C(t)$, then mass balance leads to the governing equation

$$(VC)' = qc_{\text{in}} - qC - kVC^2, \quad C(0) = c_0.$$

Non-dimensionalize the model by selecting the concentration scale to be c_{in}, the input concentration, and a time scale based on the flow-through rate. Determine the equilibria, or constant solutions, and find a formula for the concentration as a function of time.

5. (Heat transfer) The goal of this exercise is to derive a differential equation for the changing temperature $T = T(t)$ of a small, uniform object of initial temperature T_0 placed in an environment of temperature T_e. (Think of an object placed in an oven.) The result is **Newton's law of cooling**, and the reader is asked to follow through the steps. To this end, let U be the internal energy of the object. Then

$$U = \rho c_v T,$$

where ρ is the density (mass per volume) and c_v is the specific heat at constant volume (energy per mass, per degree). We assume, based on empirical data, that the rate of change of energy in the object is proportional to the difference between the temperature of the object and its environmental temperature, or

$$\rho c_v \frac{dT}{dt} = -h(T - T_e),$$

where h is the heat loss coefficient. Show that $[h] =$ energy per volume, per degree per time. Why is there a minus sign on the right side of the equation? Finally,

$$\frac{dT}{dt} = -k(T - T_e), \quad T(0) = T_0, \qquad \text{(Newton's law of cooling)}$$

where $k = h/\rho c_v$. Note that this argument depends on uniform conditions in the object, no temperature gradients, and so on.

   a) Determine the temperature $T = T(t)$ of the object and sketch a generic plot of temperature vs. time if $T_0 < T_e$.

   b) Find an analytic formula for the temperature if the environmental temperature varies according to $T_e = \theta(t)$.

c) Next, think of a small exothermic animal (a lizard, or snake) emerging from a cool nest at midday. In addition to exchanging heat with its environment, it also heated by solar radiation S (power per area) falling on its effective body area. Then, its temperature is governed by

$$\rho c_v \frac{dT}{dt} = -h(T - T_e) + aS.$$

What are the dimensions of the constant a?

6. (Advertising) This exercise develops a simple model to assess the effectiveness of an advertising campaign. Let $S = S(t)$ be the monthly sales of an item. In the absence of advertising it is observed from sales history data that sales decrease at a constant per capita rate a. If there is advertising, then sales rebound and tend to saturate at some maximum value $S = M$, because there are only finitely many consumers. The rate of increase in sales due to advertising is jointly proportional to the advertising rate $A(t)$ and to the degree the market is not saturated, that is,

$$rA(t)\left(\frac{M - S}{M}\right).$$

The constant r measures the effectiveness of the advertising campaign, and $\frac{M-S}{M}$ is a measure of the market share that has still not purchased the product. Combining natural sales decay and advertising gives the model

$$S' = -aS + rA(t)\left(\frac{M - S}{M}\right).$$

If $S(0) = S_0$ and the advertising is constant over a fixed time period T and is then removed, that is,

$$A(t) = \begin{cases} a, & 0 \le t \le T, \\ 0, & t > T \end{cases}$$

find a formula for the sales $S(t)$.

7. (Biogeography) The **MacArthur–Wilson model** of the dynamics of species (e.g., bird species) that inhabit an island located near a mainland was developed in the 1960s. Let N be the number of species in the source pool on the mainland, and let $S = S(t)$ be the number of species on the island. Assume that the rate of change of the number of species is

$$S' = \chi - \mu,$$

where χ is colonization rate and μ is the extinction rate. In the MacArthur–Wilson model,

$$\chi = I(1 - \frac{S}{N}) \quad \text{and} \quad \mu = \frac{E}{N}S,$$

where I and E are the maximum colonization and extinction rates, respectively.

a) Over a long time, what is the expected equilibrium for the number of species inhabiting the island?

b) Given $S(0) = S_0$, find an analytic formula for $S(t)$.

c) Suppose there are two islands, one large and one small, with the larger island having the smaller maximum extinction rate. Both have the same colonization rate. Show that the smaller island will eventually have fewer species.

8. (Chemical reactors) A large industrial retention pond of volume V, initially free of pollutants, was subject to the inflow of a contaminant produced in the factory's processing plant. Over a period of b days the EPA found that the inflow concentration of the contaminant decreased linearly (in time) to zero from its initial initial value of a (grams per volume), its flow rate q (volume per day) being constant. During the b days the spillage rate to the local stream was also q. What is the concentration in the pond after b days? Take $V = 6000$ cubic meters, $b = 20$ days, $a = 0.03$ grams per cubic meter, and $q = 50$ cubic meters per day. With these data, how long would it take for the concentration in the pond to get below a required EPA level of 0.00001 grams per cubic meter if fresh water is pumped into the pond at the same flow rate, with the same spill over? [Use software to perform the calculations.]

9. (Circuits) An electrical (RCL) circuit (Fig. 1.6) has an inductor, resistor, and capacitor connected in series, along with an electromotive force that supplies energy. The governing equation, obtained by Kirchhoff's law, is

$$Lq'' + Rq' + \frac{1}{C}q = E(t),$$
$$q(0) = q_0, \quad q'(0) = I_0,$$

where $q = q(t)$ is the charge on the capacitor, L is the inductance, R is the resistance, C is the capacitance, I_0 is the initial current, and $E(t)$ is the electromotive force (voltage) supplied by a battery or generator. Each term in Kirchhoff's is the work required to move a unit charge across the corresponding circuit element Table 1.2 shows the dimensional relations for the various parameters.

a) (Units) In rhe mks system of units time is measured in seconds, length in meters, and mass in kilograms. Then force is measured in newtons, energy in joules, and charge in coulombs. Using the dimensional relations in Table 1.2, find the mks units of current, voltage, resistance,

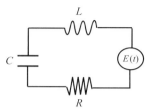

Figure 1.6 An RCL circuit containing an inductor, capacitor, and resistor in series with electromotive force $E(t)$.

Quantity (symbol)	Dimensions	Relation
charge (q)	Q	
current (I)	QT^{-1}	charge per time
voltage (V)	$ML^2T^{-2}Q^{-1}$	energy (work) per charge
capacitance (C)	$Q^2T^2M^{-1}L^{-2}$	charge per voltage
resistance (R)	$ML^2T^{-2}Q^{-1}$	voltage per current
inductance (L)	ML^2Q^{-1}	voltage per current per time

Table 1.2 Fundamental dimensions of various quantities occurring in electrical circuits. Charge Q is a fundamental dimension. Note that the quantities can be written in terms current I instead of Q; also, mass may be replaced by appropriate expressions involving force F and/or energy E.

    capacitance, and inductance (these are called amperes, volts, ohms, farads, and henrys, respectively). Verify that each term in Kirchhoff's law has dimensions of work per charge.

b) If $E(t) = V_0$ (volts) is constant, what is the behavior, or dynamics, of the circuit in the three cases: $R^2 > 4L/C$, $R^2 = 4L/C$, and $R^2 < 4L/C$. Answer this question using terms like decay, oscillatory, frequency, and amplitude.

10. In an RC circuit the charge $q = q(t)$ on a capacitor is governed by the first-order equation

$$Rq' + \frac{1}{C}q = E(t),$$

If $q(0) = q_0$, find the charge as a function of time. If the electromotive force is constant, $E(t) = V$, what is the long-time charge on the capacitor?

11. (Electricity) The basic law of electricity is **Coulomb's law**, an experimental result that states, in a vacuum, the force F on two charged particles q_1

and q_2 separated by a distance r is

$$F = \frac{q_1 q_2}{4\pi\varepsilon_0 r^2},$$

where ε_0 is an experimental constant, depending on the units, called the (electric) *permittivity of free space*. In mks units, $\varepsilon_0 = 8.85 \cdot 10^{-12}$ farads per meter. (a) Find the dimensions of ε_0 in terms of mass, length, time, and current. (b) Calculate the force required to maintain a separation distance of 1 meter between two positive, 1 coulomb charges. Is this a large force? What does this result say about the magnitude of 1 coulomb?

12. (Thermistors) A thermistor is a resistor that depends on temperature; as it gets hotter, resistance increases. They are used in fuses and and in electrical appliances to prevent devices from overheating. A typical thermistor may be a dime-shaped ceramic having metal contacts on the flat surfaces. Let us assume that a constant voltage V is applied across the thermistor, and the temperature in the thermistor is uniform and depends only on time, or $T = T(t)$. On one hand, ohmic heating causes the thermistor to heat up, and on the other hand heat is lost to the constant temperature environment T_e (Newton's law of cooling). Show that energy balance in the thermistor leads to the model

$$mcT' = \frac{V^2}{R(T)} - h(T - T_e),$$

where m and c are the mass and specific heat of the thermistor, h is the heat loss coefficient, and $R(T)$ is the resistance. Assuming that $R(T)$ is a linearly increasing function of temperature, show that there is equilibrium, or constant, solution to the model.

13. (Ecology) A population $N = N(t)$ is governed by the logistic law

$$\frac{dN}{dt} = rN\left(1 - \frac{N}{K(t)}\right),$$

where the carrying capacity varies according to $K = K(t)$. Find an integral form of the solution $N = N(t)$.

1.3.2 Stability and Bifurcation

First-order **autonomous** equations arise frequently in applications; they have the form

$$u' = f(u), \tag{3.9}$$

where there is no explicit dependence on time. Because (3.9) is separable, we can immediately obtain an implicit, general solution of the form

$$\int \frac{1}{f(u)}\, du = t + C.$$

In only the simplest equations can the integral be resolved, and so we almost always resort to a geometric, or qualitative, analysis of (3.9), which gives the essential behavior of the solutions. If the qualitative behavior is important, then we can perform numerical methods (e.g., a Runge–Kutta method).

Example 1.18

This example (see Strogatz, 1994) well illustrates that the analytic solution to a differential equation is not often transparent, and actually sometimes useless in understanding its qualitative features. Consider the simple initial value problem

$$u' = \sin u, \quad u(0) = a.$$

Separating variables and integrating immediately leads to the implicit form

$$t = \ln \left| \frac{\csc a + \cot a}{\csc u + \cot u} \right|.$$

What are the qualitative features of the solution for different values of the initial condition a? This question would draw a collective groan from a class. There is a better way to address this question.   □

Key to the qualitative analysis is to determine the equilibria of an equation and the dynamics near the equilibria. An **equilibrium solution** of (3.9) is a constant solution, which is clearly a root of the algebraic equation $f(u) = 0$. Thus, $u(t) = u^*$ is an equilibrium solution if, and only if, $f(u^*) = 0$. Geometrically, the values u^* are the points where the graph of $f(u)$ versus u intersects the u axis. We always assume the equilibria are **isolated**, i.e., if u^* is an equilibrium, then there is an open interval containing u^* that contains no other equilibria. Fig. 1.7 shows a generic plot of $f(u)$, where the equilibria are, say, $u^* = a, b, c$. In between the equilibria we observe the values of u for which u is increasing ($f(u) > 0$) or decreasing ($f(u) < 0$). We may place arrows on the u axis, or **phase line**, in between the equilibria showing direction of the *flow* of $u = u(t)$ (increasing or decreasing) as time increases. If desired, the information from the phase line can be translated into time series plots of $u(t)$ versus t (Fig. 1.8).

On the phase line, if arrows on both sides of an equilibrium point toward that equilibrium point, then we say the equilibrium point is an **attractor**, or

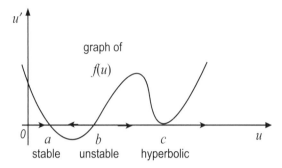

Figure 1.7 Generic phase line plot. The u axis is interpreted as a one-dimensional state, or phase, space on which the solution $u = u(t)$ moves in time. The arrows indicate the direction of movement.

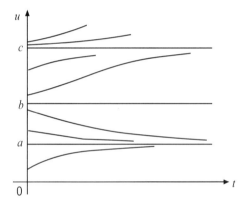

Figure 1.8 Time series plots for various initial conditions of the equation whose phase line plot is shown in the phase line diagram.

a **locally asymptotically stable equilibrium**. If both of the arrows point away, the equilibrium is a **repeller** and is **unstable**. The terms *sink* and *source* are also used. If one arrow points toward the equilibrium and one points away, the equilibrium is **hyperbolic**; some authors call these equilibria semi-stable. Repellers and hyperbolic equilibria are **unstable equilibria**. Suppose an equilibrium is an attractor and the system is in that constant equilibrium state; if it is given a small *perturbation* (i.e., a change or "bump") to a nearby state, then it returns to that state as $t \to +\infty$. In one-dimensional systems we often us the word **stable** in place of the wordy expression 'locally asymptotic stable'; in higher-dimensional systems we must distinguish between these two terms. It seems plausible that meaningful physical and biological systems will seek out stable states. Repellers and hyperbolic points are unstable because a small perturbation can cause the system to go to a different equilibrium, or even blow up.

One point of clarification: if the domain for the dependent variable is, for example, $u \geq 0$, as in a population model, and $u = 0$ is an equilibrium representing extinction, then we use the words stable and unstable in a one-sided sense, ignoring the phase line for $u < 0$.

When we state that an equilibrium u^* is locally asymptotically stable, the understanding is that it is with respect to *small* perturbations. To fix the idea, consider a population of fish in a lake that is in a locally asymptotically stable equilibrium u^*. A small death event, say caused by some toxic chemical that is dumped into the lake, will cause the population to drop. Local asymptotic stability means that the system will return the original state u^* over time. If many fish are killed by the pollution event, then the perturbation is not small and there is no guarantee that the fish population will return to the original state u^*; for example, it may go to some other equilibrium state or become extinct. If the population returns to the state u^* for all perturbations, no matter how large, then the state u^* is called **globally asymptotically stable**. A precise definition of local asymptotic stability can be given as follows.

Definition 1.19

An isolated equilibrium state u^* of (3.9) is **locally asymptotically stable** if there is an open interval I containing u^* with $\lim_{t \to +\infty} u(t) = u^*$ for any solution $u = u(t)$ of (3.9) with $u(0)$ in I. That is, each solution starting in I converges to u^*. $\square$

Graphically, stability can be determined from the slope of the tangent line $f'(u^*)$ to the graph of $f(u)$ vs. u at the equilibrium state. If $f'(u^*) < 0$, the graph of $f(u)$ through the equilibrium falls from positive to negative values, giving an arrow pattern of an attractor. Similarly, if $f'(u^*) > 0$, we obtain the arrow pattern of a repeller. If $f'(u^*) = 0$, then u^* may be an attractor, repeller, or hyperbolic.

Using linearization, we can confirm these graphical criteria analytically. If (3.9) is in an equilibrium state u^* and $U = U(t)$ represents a small perturbation from that state, then $u(t) = u^* + U(t)$ and, using Taylor's theorem,

$$u' = U' = f(u^* + U) = f(u^*) + f'(u^*)U + \frac{1}{2}f''(u^*)U^2 + \cdots,$$

where the dots denote higher-order terms in U. Ignoring all but the linear terms, we get

$$U' = f'(u^*)U,$$

which is called the **linearization**, or the perturbation equation, about the equilibrium u^*. This is a linear growth-decay equation and its solution is simply $U(t) = Ce^{\lambda t}$, where $\lambda = f'(u^*)$, a constant. If $\lambda < 0$, the perturbation decays, and if $\lambda > 0$, the perturbation grows, implying local asymptotic stability and instability, respectively.

If $\lambda = 0$, then we must analyze higher-order terms in the Taylor expansion to determine stability. For example, the term $f''(u^*)$ relates to concavity. Or, we can just plot the function $f(u)$ and observe its behavior.

Remark 1.20

As an exercise, the reader should now consider answering the question posed in Example 1.18.   □

Example 1.21

(**Logistic model**) The logistic model of population growth is

$$\frac{dp}{dt} = rp\left(1 - \frac{p}{K}\right),$$

where r is the growth rate and K is the carrying capacity. The equilibria are $p^* = 0$, K, found by setting the right side $f(p) = rp\left(1 - \frac{p}{K}\right) = 0$. The equilibrium $p^* = K$ is asymptotically stable because $f'(K) = -r < 0$, and the zero population is unstable because $f'(0) = r > 0$.   □

Example 1.22

(**Uniqueness**) The simple initial value problem

$$u' = u^{1/3}, \quad u(0) = 0$$

reveals another important issue, namely uniqueness. There are clearly two so-lutions to this problem, $u = 0$ and $u = \left(\frac{2}{3}t\right)^{3/2}$, both valid for all $t \geq 0$. The phase line diagram gives incomplete and even confusing information. The plot of $f(u) = u^{1/3}$ has an infinite derivative at $u = 0$ and is not defined. What would a particle do if it started at $u = 0$? Remain at u=0 or follow the nonzero solution? This issue is avoided when there exist unique solutions to an initial value problem. It is proved in more advanced texts that unique solutions exist for the initial value problem $u' = f(u)$, $u(0) = u_0$ when f and f' are continuous in some open interval containing u_0; the unique solution $u = u(t)$ is guaranteed to exist in some interval $-\tau < t < \tau$.   □

Remark 1.23

(**Oscillatory solutions**) Notice that uniqueness implies that a solution $u = u(t)$ plotted on the phase line cannot enter an equilibrium at finite time. It also ensures that oscillatory solutions cannot exist; they cannot ever turn around because of the monotonicity of the of direction arrows.   □

The qualitative behavior of the one-dimensional dynamical equation $u' = f(u)$ seems straightforward. But what is interesting, and not always straightforward, is how the behavior of solutions can change as one or more parameters in the model equation changes. A simple example illustrates the point.

Example 1.24

(**Harvesting**) More often than not, differential equations contain parameters. Let us modify the logistic model by adding a *harvesting* term. That is, consider a population $p(t)$ (e.g., fish) that is governed by logistic growth, while at the same time it is harvested (fishing) at a constant rate $H > 0$. Then the dynamics is

$$\frac{dp}{dt} = rp\left(1 - \frac{p}{K}\right) - H. \tag{3.10}$$

There are three parameters in this problem, r, K, and H, and we want to examine how the solutions change as the harvesting rate H varies, with r and K held fixed. To simplify the analysis we non-dimensionalize the problem by introducing a scaled population u and a scaled time τ defined by

$$u = \frac{p}{K}, \quad \tau = rt.$$

Therefore, we measure population relative to the carrying capacity and time relative to the reciprocal of the growth rate. Then equation (3.10) becomes

$$\frac{du}{d\tau} = u(1 - u) - h \tag{3.11}$$

where $h \equiv H/rK$ is the dimensionless harvesting parameter. Treating h as a control parameter, we inquire if equilibrium populations exist. These are easily found by setting $du/dt = 0$, or

$$u(1 - u) - h = 0.$$

Solving for u gives equilibria

$$u^* = \frac{1 \pm \sqrt{1 - 4h}}{2} \tag{3.12}$$

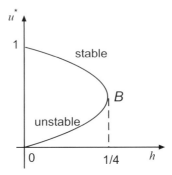

Figure 1.9 Bifurcation diagram showing equilibrium solutions u^* vs. the parameter h. There are two branches, the upper branch, which is stable, and the lower branch, which is unstable.

Therefore, if the harvesting parameter exceeds $\frac{1}{4}$, then no equilibrium populations exist. If $0 < h < \frac{1}{4}$, then there are two positive equilibrium states. We can sketch the locus (3.12) on a hu^*-coordinate system (with h the independent variable) to graphically indicate the dependence of the equilibrium solutions on the parameter h. Such a graph is shown in Fig. 1.9 and is called a **bifurcation diagram**; in this context, the harvesting parameter h is called a **bifurcation parameter**. We observe that if h is too large (i.e., there is too much harvesting of the population), then no equilibrium populations exist and (3.11) shows that $du/d\tau < 0$, so that the population dies out. As harvesting is decreased, there is a critical value of h ($h = 1/4$) below which two equilibrium populations can exist; in these two populations, there is a balance between growth and harvesting. The point B on the graph where the two solutions appear is called a bifurcation point. Next we ask if nature prefers one of these equilibria over the other. The resolution of this question lies in the stability properties of those states. We would expect nature to select out a stable state over an unstable one; for, if a small change is made in an unstable state (and small deviations are always present), then the system would leave that state permanently; if the state were stable, then small perturbations will decay away. We could perform a linearized stability analysis as in the preceding subsection to determine which, if any, of the two equilibrium populations given by (3.12) is unstable. But let us proceed a different way and plot the phase line of the differential equation (3.11) when $0 < h < \frac{1}{4}$. Let u^+ and u^- denote the two equilibrium populations in (3.12) with the plus and minus signs, respectively. A plot of the right side of (3.11) for a fixed $0 < h < \frac{1}{4}$ is shown in Fig. 1.10. The phase line clearly indicates that the larger equilibrium u^+ is stable, and the smaller u^- is unstable. Conse-

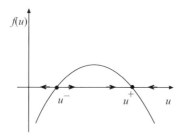

Figure 1.10 Plot of $f(u)$ vs. u and the phase line corresponding to Fig. 1.9 for a fixed $h < 1/4$. As h increases, the parabola moves down until eventually no equilibria exist.

quently, nature would select out the larger equilibrium population for a given harvesting rate $0 < h < \frac{1}{4}$. In terms of the bifurcation diagram in Fig. 1.8, the upper branch is stable and the lower branch is unstable. A bifurcation is said to occur at the point B. Note that we can also use the analytic derivative criterion, the sign of $f'(u^*)$, to determine stability. □

Example 1.25

(**Ecology**) In this example we set up a simple model of a plant–herbivore system and study its bifurcation properties. Bifurcations are highly important in all physical and biological phenomena, and theoretical ecology offers easily accessible examples without the details of complicated physical and chemical concepts. Let $P = P(t)$ be the plant biomass and assume the plant grows logistically with growth rate r and carrying capacity K. Thus, without herbivores,

$$\frac{dP}{dt} = rP\left(1 - \frac{P}{K}\right).$$

In this ecological system we assume there are H herbivores and they consume plant biomass at the rate

$$\frac{aP}{1 + bP}$$

per herbivore. Notice that this consumption rate saturates at the value a/b, reflecting satiation on the part of herbivores. The constant a is measured in dimensions of 'per time per herbivore,' and b is 'per plant biomass.' The governing model for the plant biomass is therefore

$$\frac{dP}{dt} = rP\left(1 - \frac{P}{K}\right) - \frac{aP}{1 + bP}H.$$

Because of the number of parameters (5), it is wise to scale the equation. We introduce dimensionless quantities

$$\tau = rt, \quad N = \frac{P}{K}.$$

It is easily checked that the dimensionless model becomes

$$\frac{dN}{d\tau} = N(1 - N) - \frac{hN}{1 + cN}, \tag{3.13}$$

with introduction of the dimensionless parameters

$$h = \frac{aH}{r}, \quad c = Kb.$$

Generally, the carrying capacity is large and so we assume for the analysis that $c > 1$, and c is fixed. Our interest is in examining the model when the herbivore population varies; so, h is the bifurcation parameter.

The equilibria N^* are determined by setting $dN/d\tau = 0$, and we obtain

$$N^* \left[(1 - N^*) - \frac{h}{1 + cN^*} \right] = 0.$$

Thus $N^* = 0$ (for all values of h) and

$$(1 - N^*)(1 + cN^*) - h = 0.$$

Here, it is difficult to plot equilibria as a function of h, but it is easy to plot h vs. N^*. Solving for h,

$$h = (1 - N^*)(1 + cN^*),$$

which is a concave down parabola with positive root $N^* = 1$. By simple calculus we find the maximum value occurs at

$$h = \frac{c - 1}{2c}, \quad N^* = \frac{(c + 1)^2}{4c}.$$

We can turn this parabola on its side to obtain a plot of the bifurcation diagram with N^* vs. h; this is shown in Fig. 1.11 along with the zero equilibrium.

We can determine the stability of the various branches in Fig. 1.11 using a phase line. Figure 1.12 shows the three plots for the three possible cases. We graph the fixed growth rate $N(1 - N)$ and the consumption rate $hN/(1 + cN)$ in each. They show that the upper branch of the parabola is stable, and the lower branch is unstable. □

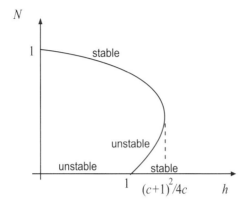

Figure 1.11 Bifurcation diagram for the model equation (3.13). As h increases from a small value, the two equilibria bifurcate at $h = 1$ into three; then, at $h = (c+1)^2/2c$ the two nonzero equilibria coalesce and only the zero equilibrium remains.

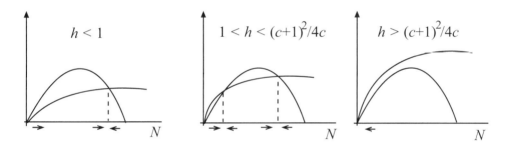

Figure 1.12 Phase line diagram for the equilibria in model equation (3.13).

Remark 1.26

(**Functional responses**) Ecology provides a rich source of problems in non-linear dynamics involving predation, cooperation, and disease transmission. Many models relating to resource dynamics (plants–herbivores, predator–prey) involve a consumption, or predation, rate. To fix the context, let x be the number of prey and y be the number of predators.[5] The number of total possible encounters is xy, and a fraction of those, axy, are effective; that is, they result in predation. This effective contact rate is also called *mass action*, and later we observe it is the reaction rate of two reacting chemical species. So, the rate of

---

[5] We are thinking of x and y as population numbers, but we can also regard them as *population densities*, or animals per area. There is always an underlying fixed area where the dynamics are occurring.

predation (prey per time, per predator) is proportional to the number of prey, i.e., ax. Thinking carefully about this leads to concerns. Increasing the prey density indefinitely leads to an extremely high per predator consumption rate, which is clearly impossible for any consumer. It seems more reasonable that the rate of predation would have a limiting value as prey density gets large. In the late 1950s, C. Holling developed a functional form that has this limiting property by partitioning the time budget of the predator. He reasoned that the number N of prey captured by a single predator is proportional to the number x of prey and the time T_s allotted for searching. Thus $N = aT_s x$, where the proportionality constant a is the effective encounter rate. But the total time T available to the predator must be partitioned into search time and total handling time T_h, or $T = T_s + T_h$. The total handling time is proportional to the number captured, $T_h = hN$, where h is the time for a predator to handle a single prey. Hence $N = a(T - hN)x$. Solving for N/T, which is the predation rate (prey per time, per predator), gives

$$\frac{N}{T} = \frac{ax}{1 + ahx}.$$

This function for the predation rate is called a Holling type II response, or the Holling disk equation. Note that $\lim_{x \to \infty} ax/(1 + ahx) = 1/h$, so the rate of predation approaches a constant value. This quantity, N/T, is measured in prey per time, per predator, so multiplying by the number of predators y gives the predation rate for y predators.

If the encounter rate a is a function of the prey density (e.g., a linear function $a = bx$), the the predation, or feeding, rate is

$$\frac{N}{T} = \frac{bx^2}{1 + bhx^2},$$

which is called a Holling type III response. Figure 1.13 compares different types of predation rates used by ecologists. For a type III response the predation is turned on once the prey density is high enough; this models, for example, predators that must form a "prey image" before they become aware of the prey, or predators that eat different types of prey. At low densities prey go nearly unnoticed; but once the density reaches an upper threshold the predation rises quickly to its maximum rate. In ecology literature one often encounters these types of consumption rates. □

Example 1.27

(**Non-isothermal tank reactor**) In this section we set up a model for determining the temperature $\bar{\theta}$ and concentration $\bar{c}$ of a heat-releasing, chemically

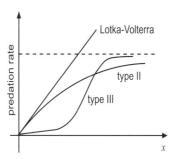

Figure 1.13 Three types of functional responses, or consumption rates, studied in ecology. The consumption rate is measured in prey per time, per predator.

reacting substance in a continuously stirred tank reactor. (We are using overbars in the variables in anticipation of using unbarred dimensionless variables later.) The reaction takes place in a tank of volume V (see Fig. 1.2) that is stirred continuously in order to maintain uniform temperature and concentration. It is fed by a stream of constant flow velocity q, constant reactant concentration c_i, and constant temperature θ_i. After mixing and reacting, the products are removed at the same volume rate q. We assume that the exothermic reaction $\mathbf{C} \rightarrow$ products is first-order and irreversible, and the reactant disappears at the rate

$$-k\bar{c}e^{-A/\bar{\theta}},$$

where A and k are positive constants. The temperature-dependent factor $e^{-A/\bar{\theta}}$ in the rate is the Arrhenius factor, where A is proportional to the activation energy (A is the activation energy divided by the gas constant R), or the amount of kinetic energy needed for the reaction to occur. The amount of heat released is assumed to be given by

$$hk\bar{c}e^{-A/\bar{\theta}},$$

where h is a positive constant, the specific heat of reaction, measured in energy per mass.

We are able to write a system of two differential equations that govern the concentration $\bar{c}(\bar{t})$ and the temperature $\bar{\theta}(\bar{t})$ of the reactant in the tank. One equation comes from balancing the mass of the reactant, and the other arises from conservation of (heat) energy. First, mass balances gives (see Section 1.2.2)

$$V\frac{d\bar{c}}{d\bar{t}} = qc_i - q\bar{c} - Vk\bar{c}e^{-A/\bar{\theta}}. \tag{3.14}$$

This equation states that the rate of change of mass of reactant in the tank equals the mass in, less the mass out, plus the rate at which the mass of the reactant disappears in the chemical reaction. Next, heat balance gives

$$VC \frac{d\bar{\theta}}{d\bar{t}} = qC\theta_i - qC\bar{\theta} + hVk\bar{c}e^{-A/\bar{\theta}}, \tag{3.15}$$

where C is the heat capacity of the mixture, measured in energy per volume per degree. These equations can be reduced to dimensionless form by introducing dimensionless variables

$$t = \frac{\bar{t}}{V/q}, \quad \theta = \frac{\bar{\theta}}{\theta_i}, \quad c = \frac{\bar{c}}{c_i}$$

and dimensionless constants

$$\mu = \frac{q}{kV}, \quad b = \frac{hc_i}{C\theta_i}, \quad \gamma = \frac{A}{\theta_i}.$$

In this case the mass and energy balance equations (3.14) and (3.15) become

$$\frac{dc}{dt} = 1 - c - \frac{c}{\mu} e^{-\gamma/\theta}, \tag{3.16}$$

$$\frac{d\theta}{dt} = 1 - \theta + \frac{bc}{\mu} e^{-\gamma/\theta}. \tag{3.17}$$

This pair of equations can be reduced to a single equation using the following argument. If we multiply (3.16) by b and add it to (3.17), we get

$$\frac{d}{dt} (\theta + bc) = 1 + b - (\theta + bc),$$

which integrates to

$$\theta + bc = 1 + b + De^{-t},$$

where D is a constant. Assuming $\theta + bc$ at $t = 0$ is $1 + b$ gives $D = 0$; then $\theta + bc = 1 + b$. Consequently the heat balance equation becomes

$$\frac{d\theta}{dt} = 1 - \theta + \frac{1 + b - \theta}{\mu} e^{-\gamma/\theta},$$

which is a single differential equation for θ. Introducing $u = \theta - 1$, this equation becomes

$$\frac{du}{dt} = -u + \frac{b - u}{\mu} e^{-\gamma/(u+1)} = \frac{1}{\mu} \left(-\mu u + (b - u) e^{-\gamma/(u+1)} \right), \tag{3.18}$$

which is the final form that we shall study.

By definition, the parameter μ in (3.18) can be identified as the ratio of the flow rate to the reaction rate, and it acts as the bifurcation parameter in our analysis because we wish to study equilibrium solutions as a function of flow rate. The equilibrium solutions are given by solutions to the equation

$$\mu u^* = (b - u^*)e^{-\gamma/(u^*+1)}. \tag{3.19}$$

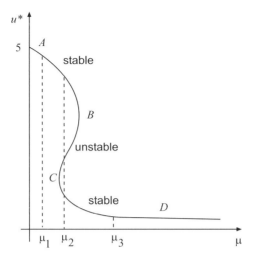

Figure 1.14 Bifurcation diagram. The top branch from A to B and the bottom branch from C to D are stable; the middle branch B to C is unstable.

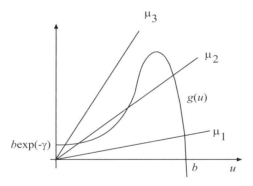

Figure 1.15 Plot of the left and right sides of equation (3.19). Intersections show the equilibria and how they change as μ changes.

Because (3.19) cannot be solved in analytic form, we resort to graphical techniques. We can plot μ vs. u^* easily and then reflect the plot through $u^* = \mu$ to obtain a bifurcation diagram with abscissa μ. Figure 1.14 shows one case ($b = 5$, $\gamma = 8$, not drawn to scale). For a small value $\mu = \mu_1$ there is one large temperature equilibrium. As μ increases slowly, to maintain equilibrium, there is a bifurcation at point C to two, then three, equilibria. As μ increases still further, say up to $\mu = \mu_3$, there is a bifurcation at point B back to a single, lower, equilibrium temperature. To determine the stability of the branches AB, CB, and CD let us write $u' = \frac{1}{\mu}\left(-\mu u + g(u)\right)$, where $g(u) = (b - u)\, e^{-\gamma/(u+1)}$. The equilibria occur when the graphs of $g(u)$ and μu intersect. Figure 1.15 shows three cases: $\mu = \mu_1, \mu = \mu_2, \mu = \mu_3$. For μ values where $g(u) > \mu u$ we have $u' > 0$, and μ values where $g(u) < \mu u$ we have $u' > 0$. This means the upper temperature branch AB must be asymptotically stable, the middle branch BC must be unstable, and the lower temperature branch must be asymptotically stable.

Physically, if the flow rate μ is fast, say $\mu = \mu_3$, the reactor is operating at a low temperature. If it is slowly decreased, the temperature changes along the lower, stable branch CD until it reaches the point C. At that time a bifurcation must occur, and further decrease in μ will cause the temperature to jump suddenly (and dynamically) to the upper, hotter branch. We say *ignition* has occurred. Conversely, if the flow rate is increased from μ_1, on reaching state B there is a sudden jump to the lower temperature state along the stable branch CD. This bifurcation is called *quenching* the reaction. In summary, the flow rate through the reactor controls its performance; when the flow rate is large it operates at a higher temperature, and at a low rate it operates at a low temperature. It is not difficult to understand why bifurcation phenomena are important in the operation of nuclear reactors.   □

EXERCISES

1. (Ecology) In the usual Malthus growth law $p' = rp$ for a population of size p, assume the growth rate is a linear function of food availability F; that is, $r = bF$, where b is the conversion factor of food into newborns. Assume that F_T is the total, constant food in the system with $F_T = F + cp$, where cp is amount of food already consumed. Write down a differential equation for the population p. What is the population as t gets large? Find the population as a function of time if $p(0) = p_0$.

2. (Biology) A model of tumor growth is the Gompertz equation

$$R' = -aR\ln\left(\frac{R}{k}\right),$$

where $R = R(t)$ is the tumor radius, and a and k are positive constants. Find the equilibria and analyze their stability. Find the solution $R = R(t)$ and plot several solution curves.

3. (Ecology) When a population gets small, the members may find it difficult to locate mates, causing the growth rate to be negative. This phenomenon is called the Allee effect. Explain how the model

$$P' = rP\left(1 - \frac{P}{K}\right)(P - a), \quad 0 < a < K,$$

explains this effect.

4. (Harvesting) A fish population in a lake is harvested at a constant rate H, and it grows logistically. The growth rate is 0.2 per month, the carrying capacity is 40 (thousand), and the harvesting rate is 1.5 (thousand per month). Write down the model equation, find the equilibria, and classify them. Will the fish population ever become extinct? What is the most likely long-term fish population?

5. (Harvesting) A deer population grows logistically, but it is harvested at a rate proportional to its population size. Conclude that the dynamics of population growth is given by

$$P' = rP\left(1 - \frac{P}{K}\right) - HP,$$

where H is the *per capita* harvesting rate. Non-dimensionalize the model and use a bifurcation diagram to explain the effects on the equilibrium deer population when H is slowly increased from a small value.

6. Draw a bifurcation diagram for the model $u' = u^3 - u + h$, where h is the bifurcation parameter, using h as the abscissa. Label branches of the curves as stable or unstable. (Hint: first plot h vs. u.)

7. For the following equations, find the equilibria and sketch the phase line. Determine the type and stability of all the equilibria.

   a) $u' = u^2(3 - u)$.

   b) $u' = 2u(1 - u) - \frac{1}{2}u$.

   c) $u' = (4 - u)(2 - u)^3$.

   d) $u' = u - \cos u$. (Hint: plot u and $\cos u$ separately.)

   e) $u' = e^u - \cos u$.

8. The following models contain a parameter h. Find the equilibria in terms of h and determine their stability. Construct a bifurcation diagram showing how equilibria depend upon h, and label the branches of the curves as unstable or stable.

   a) $u' = hu - u^2$.

   b) $u' = hu - u^3$.

   c) $u' = (1 - u)(u^2 - h)$.

   d) $u' = h - u - e^{-u}$.

e) $u' = -u + h \tanh u$.

9. Consider the model $u' = (\lambda - b)u - au^3$, where a and b are fixed positive constants and λ is a parameter that varies.

    a) If $\lambda < b$ show that there is a single equilibrium and that it is asymptotically stable.

    b) If $\lambda > b$ find all equilibria and determine their stability.

    c) Sketch the bifurcation diagram showing how equilibria vary with λ. Label each branch of the curves shown in the bifurcation diagram as stable or unstable.

10. (Ecology) An animal species of population u grows exponentially with growth rate r. At the same time it is subjected to predation at a rate $\frac{au}{1+bu}$, which depends upon its population. The constants a and b are positive, and the dynamics are

$$u' = ru - \frac{au}{1 + bu}.$$

    a) Non-dimensionalize the model so that there is a single dimensionless parameter h.

    b) Sketch a bifurcation diagram, and determine the stability of the equilibria as a function of h; indicate the results on the diagram.

11. Repeat the last problem for the model

$$u' = ru - \frac{au^2}{1 + bu^2}.$$

12. (Ecology) The biomass P of a plant grows logistically with intrinsic growth rate r and carrying capacity K. At the same time it is consumed at a rate

$$\frac{a^2 P^2}{1 + b^2 P^2},$$

per herbivore, where a and b are positive constants. The model is therefore

$$P' = rP(1 - \frac{P}{K}) - \frac{aP^2}{1 + b^2 P^2},$$

with r, K, a, and b fixed parameters.

    a) Non-dimensionalize the model to obtain

$$\frac{dN}{d\tau} = RN\left(1 - \frac{N}{k}\right) - \frac{N^2}{1 + N^2},$$

    for appropriate dimensionless parameters R and k.

b) On the same set of axes, plot the per capita growth and consumption rate as functions of N, indicating the possible equilibria that can occur depending on the values of R and k.

c) For fixed $R < 0.5$, sketch a bifurcation diagram of the equilibria vs. the bifurcation parameter k. Label each branch as stable or unstable.

d) Explain in words what happens in the model when k is slowly increased from a small value to a large value.

13. A one-dimensional system is governed by the dynamical equation

$$u' = 4u(a - u) - he^{-u},$$

where a and h are positive constants. Holding h constant, draw a bifurcation diagram with respect to the parameter a. Indicate the stable and unstable branches. Hint: Plot both terms on the same set of axes.

14. Consider the two-parameter model

$$x' = h + rx - x^3,$$

where h and r are real numbers. Carry out each step to understand completely the dynamics of this model.

a) For $r < 0$, sketch the bifurcation diagram of x^* vs. h. (Hint: plot $y = h$ and $y = x^3 - rx$ on the same set of axes to find equilibria.)

b) For $r > 0$, sketch the bifurcation diagram of x^* vs. h. (Hint: plot $y = h$ and $y = x^3 - rx$ on the same set of axes, and let h vary to find equilibria.)

c) From part (b), calculate the critical values of $h = h_c(r)$ (depending on r) where bifurcations occur. Answer:

$$h = \pm \frac{2r}{3}\sqrt{\frac{r}{3}}.$$

d) In the rh plane, sketch the regions where 1, 2, or 3 equilibria exist, and find the equation(s) for the boundaries of these regions.

REFERENCES AND NOTES

Birkhoff, G. 1950. *Hydrodynamics: A Study in Logic, Fact, and Similitude,* Princeton University Press, Princeton, NJ.

Fowler, A. C. 1997. *Mathematical Models in the Applied Sciences,* Cambridge University Press, Cambridge.

Hirsch, M. W., Smale, S. & Devaney. R. L. 2004. *Differential Equations, Dynamical Systems, and an Introduction to Chaos,* 2nd ed., Academic Press, San Diego.

Holmes, M. H. 2009. *An Introduction to the Foundations of Applied Mathematics*, Springer, NY.

Illner, R., Bohun, C. S., McCollum, S. & van Roode, T. 2005. *Mathematical Modelling: A Case Studies Approach*, American Mathematical Society, Providence.

Kelley, W. G. & Peterson, A. C. 2010. *The Theory of Differential Equations*, 2nd ed., Springer, NY.

Lin, C. C. & Segel, L. A. 1974. *Mathematics Applied to Deterministic Problems in the Natural Sciences*, Macmillan, New York [reprinted by SIAM (Soc. Industr. & Appl. Math.), Philadelphia, 1988].

Logan, J. D. 2006. *Applied Mathematics*, 3rd ed., Wiley Interscience, New York.

Logan, J. D. 2010. *A First Course in Differential Equations*, 2nd ed., Springer, NY.

Strogatz, S. H. 1994. *Nonlinear Dynamics and Chaos*, Addison-Wesley, Reading, MA.

2

Two-Dimensional Dynamical Systems

It is easily imagined that most models in science and engineering require more than one state variable to describe them. In this chapter we study two-dimensional, or planar, systems of linear and nonlinear differential equations. The last two sections introduce dynamics of chemical reactions and the propagation of pathogens, or viruses and other infections, in disease dynamics.

2.1 Phase Plane Phenomena

We extend the analysis of differential equations to a system of two coupled, simultaneous equations

$$x' = P(x, y), \quad y' = Q(x, y), \tag{1.1}$$

in two unknowns $x = x(t)$ and $y = y(t)$. The functions P and Q are assumed to have continuous partial derivatives of all orders in a domain D of the xy plane. A system of the type (1.1) in which the independent variable t does not appear in P and Q is said to be **autonomous**. Under these assumptions on P and Q, it can be shown that there is a unique solution $x = x(t)$, $y = y(t)$ to the **initial value problem**

$$x' = P(x, y), \quad y' = Q(x, y), \tag{1.2}$$
$$x(t_0) = \xi_0, \quad y(t_0) = \eta_0,$$

Applied Mathematics 4th ed.,
By J. David Logan
Copyright © 2013 John Wiley & Sons, Inc.

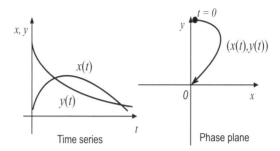

Figure 2.1 Time series representation of a solution curve and and its corresponding orbit in the phase plane.

where t_0 is an instant of time and $(\xi_0, \eta_0) \in D$. The solution is defined in some interval $\alpha < t < \beta$ containing t_0. (See Kelley and Peterson, 2010.)

There are two ways to graphically represent solutions. They can be plotted as **time series** in state space (x and y vs. t). Or, if $x(t)$ and $y(t)$ are not both constant functions, then $x = x(t)$ and $y = y(t)$ define **parametric equations** of a curve that can be plotted in the xy plane, called the **phase plane**. Such a curve is called an **orbit**, path, or trajectory, of the system (1.1). Examples of both the time series and a corresponding orbit are shown in Fig. 2.1. The orbit is directed in the sense that it is traced out in a certain direction as t increases; this direction is indicated by the direction of the tangent vector (x', y') along the curve. Because the solution to the initial value problem is unique, it follows that at most one orbit passes through each point of the phase plane, and all of the orbits cover the entire phase plane without intersecting each other.

A constant solution $x(t) = x_0$, $y(t) = y_0$ of (1.1) is called an **equilibrium**, a **steady-state** solution, or a **critical point**. Such a solution does not define a curve in the phase plane but plots as a single point. Clearly, critical points occur where both P and Q vanish, that is,

$$P(x_0, y_0) = Q(x_0, y_0) = 0.$$

It is evident that no orbit can pass through a critical point; otherwise, uniqueness would be violated. The totality of all the orbits, their directions, and critical points graphed in the phase plane is called the **phase plane diagram** of the system (1.1). Indeed, the qualitative behavior of all the orbits in the phase plane is determined to a large extent by the location of the critical points and the local behavior of orbits near those points. One can prove the following results, which essentially form the basis of the Poincaré–Bendixson theorem:[1]

(i) An orbit cannot approach a critical point in finite time; that is, if an orbit approaches a critical point, then necessarily $t \to \pm\infty$.

(ii) An orbit is either a critical point, or, as $t \to \pm\infty$, it moves along a closed orbit, it approaches a closed orbit, or it leaves every bounded set.

A closed orbit corresponds to a periodic solution of (1.1). These facts severely limit the behavior of solutions in the phase plane. We discuss these matters further in the sequel.

Example 2.1

Consider the system

$$x' = y, \quad y' = -x. \tag{1.3}$$

Here $P(x, y) = y$ and $Q(x, y) = -x$ and $(0, 0)$ is the only critical point corresponding to the equilibrium solution $x(t) = 0$, $y(t) = 0$. One way find the orbits is to turn the system into a single linear equation

$$x'' + x - 0.$$

Then

$$x = x(t) = c_1 \cos t + c_2 \sin t, \quad y = y(t) = -c_1 \sin t + c_2 \cos t$$

give the solution, where c_1 and c_2 are constants. We can eliminate the time parameter t by squaring and adding to obtain

$$x^2 + y^2 = c_1^2 + c_2^2 = C.$$

Therefore the orbits are circles centered at the origin; the phase diagram is shown in Fig. 2.2. The critical point surrounded by periodic orbits is called a **center**. Alternatively, we may divide the two equations in the system (1.3) to obtain[2]

$$\frac{dy}{dx} = -\frac{x}{y}.$$

This can be solved directly to obtain the orbits

$$x^2 + y^2 = C.$$

---

[1] Henri Poincaré (1860–1934) was called the *last universalist*. His contributions to nearly all areas of mathematics made him one of the most influential mathematicians of all time. His work in dynamical systems paved the way to modern topological methods.

[2] This fact follows from an application of the chain rule, $\frac{dy}{dt} = \frac{dy}{dx}\frac{dx}{dt}$.

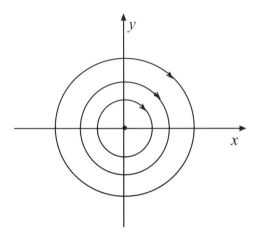

Figure 2.2 Circular orbits of (1.3) surrounding the critical point $(0,0)$, which is called a center. The orbits are counterclockwise because $x' > 0$ when $y > 0$, and $x' < 0$ when $y < 0$.

In this example all the paths are closed and each corresponds to a periodic solution of period 2π because $x(t+2\pi) = x(t)$, $y(t+2\pi) = y(t)$. Initial conditions fix a value of C and therefore a single orbit.   □

Example 2.2

The autonomous system

$$x' = 2x, \quad y' = 3y,$$

is *uncoupled* and each equation can be solved separately to obtain the solution

$$x = c_1 e^{2t}, \quad y = c_2 e^{3t}.$$

The origin $(0,0)$ is the only critical point. The orbits can be found either by eliminating t from the parametric equations, or by integrating the equation

$$\frac{dy}{dx} = \frac{3y}{2x}.$$

In any case we obtain $y^2 = cx^3$, where c is a constant, and the orbits are plotted in Fig. 2.3. A critical point having this local structure, where orbits emanate from the origin as $t \to -\infty$ and tend to infinity as $t \to \infty$, is called a **node**. Because the orbits are leaving the origin (the tangent vector field $(2x, 3y)$ points away from the origin), the origin is an unstable critical point, repeller, or source. (All of these terms are used.)   □

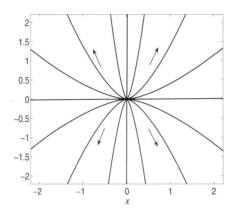

Figure 2.3 Phase diagram showing an unstable node at $(0,0)$.

Remark 2.3

It is a fact that any linear system, $x' = ax + by$, $y = cx + dy$, can be transformed into a second-order linear equation with constant coefficients. (The reader should be able to easily figure out how to do this.) The converse is also true. □

Example 2.4

(**Mechanics**) Consider a particle of mass m moving in one dimension x subject to a force $f(x, x')$. By Newton's second law the governing equation of motion is

$$mx'' = f(x, x').\qquad(1.4)$$

Equation (1.4) may be recast as a system of two first-order equations by introducing velocity y as a new variable: $y = x'$. Then

$$x' = y,$$
$$y' = \frac{1}{m}f(x, y).\qquad(1.5)$$

Critical points occur at $(x^*, 0)$, where $f(x^*, 0) = 0$; these correspond to points where the velocity and the acceleration vanish. Thus the particle is in equilibrium and no forces are acting on it. □

Example 2.5

(**Conservative forces**) If f depends only on position, $f = f(x)$, then the force is conservative and there is a potential function $V = V(x)$ defined by $\frac{dV}{dx}(x) = -f(x)$, or $V(x) = -\int f(x)\,dx$. In this special case the total energy (kinetic plus potential energy) is conserved, or

$$\frac{1}{2}my^2 + V(x) = E, \tag{1.6}$$

where E is the energy constant. This follows from

$$\frac{d}{dt}\left(\frac{1}{2}my^2 + V(x)\right) = myy' + \frac{dV}{dx}x' = myy' - f(x)y = 0.$$

For conservative systems we can draw the phase diagram by solving (1.6) for y (velocity) as a function of x (position) to get

$$y = \pm\sqrt{\frac{2}{m}(E - V(x))}. \tag{1.7}$$

Each total energy level E, determined by the particle's initial position and velocity, defines an orbit, or orbits, in intervals where $E > V(x)$. The orbit can be obtained quickly by plotting both the (generic) potential $V(x)$ and the constant energy level E on the same graph (the top plot in Fig. 2.4); then the positive right side of (1.7) can be plotted by subtracting and taking the square root graphically. Reflecting the positive plots through the x axis gives the lower plot. Figure 2.4 shows a generic potential energy function and one such orbit. We can get an entire phase diagram by carrying out this graphical procedure for several different energy levels E. Note that in spatial intervals where $E < V(x)$ there are no orbits; values of x for which $E = V(x)$ are called turning points.   □

We can also interpret the system (1.1) in the context of a fluid flow in the plane. Physically, the functions P and Q on the right side can be regarded as the components of a vector field $\mathbf{v}(x, y) = \langle P(x, y), Q(x, y)\rangle$ representing the velocity of a two-dimensional fluid motion, or flow. The orbits $x = x(t)$, $y = y(t)$ of the fluid particles are tangent to $\mathbf{v}$ and therefore satisfy (1.1). The critical points are points where $\mathbf{v} = \mathbf{0}$, or where the fluid particles are at rest. In fluid mechanics these points are called stagnation points.

In practice, to plot the orbital structure of the phase diagram, we often plot separately the locus $P(x, y) = 0$ (where the vector field is vertical) and the locus $Q(x, y) = 0$ (where the vector field is horizontal); these loci are called the x and y **nullclines**, respectively (also isoclines). The x and y nullclines intersect at the critical points. Sketching nullclines and critical points, and

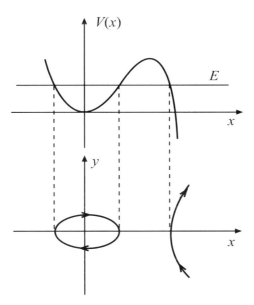

Figure 2.4 In a conservative mechanical system with generic potential energy $V(x)$, orbits showing constant energy curves (lower plot). Many such orbits can be drawn for several values of the energy to obtain the phase diagram. Equilibria occur at points where the potential energy has a maximum, minimum, or zero derivative.

plotting the directions of the vector field facilitate drawing the phase plane diagram. Typically, we usually indicate some key orbits that characterize the flow. Software can be useful to plot the vector field as well as the orbits.

In principle the critical points of (1.1) can be found by solving the simultaneous equations $P(x, y) = 0$ and $Q(x, y) = 0$. As observed in the examples, the orbits in the phase plane can sometimes be found by integrating the differential relationship,

$$\frac{dy}{dx} = \frac{Q(x, y)}{P(x, y)},$$

particularly when P and Q are fairly simple expressions.

A critical point of (1.1) is said to be **isolated** if there is an open neighborhood of the critical point that contains no other critical points. There are four types of isolated critical points that occur in linear system, where both P and Q are linear functions of x and y; these critical points are called **centers, nodes, saddles**, and *spirals*. In Fig. 2.5 these critical points are drawn schematically with the local orbital structure.

In nonlinear systems different types of critical points can occur. For ex-

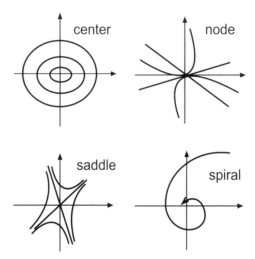

Figure 2.5 The four types of isolated critical points for a linear system. Centers are surrounded by closed, periodic orbits, either clockwise or counter-clockwise; at a node or spiral point, all orbits either approach the origin as $t \to +\infty$, or all orbits approach the origin as $t \to -\infty$. Near a saddle point, all orbits veer away from the origin except for special straight line orbits where one opposing pair enters the origin as $t \to +\infty$, and the other opposing pair approaches the origin $t \to -\infty$. The details of these constructions are discussed in the next section.

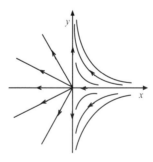

Figure 2.6 A possible orbital structure near an critical point in a nonlinear system.

ample, a critical point could have a nodal structure on one side and a saddle structure on the other, as shown in Fig. 2.6

Critical points represent equilibrium, or constant, solutions of (1.1). Another important issue is their stability. That is, does the equilibrium solution have

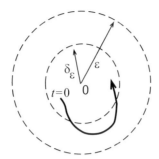

Figure 2.7 The definition of stability. Orbits that start at $t = 0$ close to a stable critical point remain close for all time $t > 0$.

some degree of permanence to it when it is subjected to small changes or perturbations? For example, we might expect a predator–prey system to have a state where the two species coexist. Roughly speaking, a critical point is stable if all paths that start nearby the critical point remain near the point for all time $t > 0$. All meaningful physical systems must have this property. We now formulate this concept in mathematical terms.

Definition 2.6

Suppose the origin $(0,0)$ is an isolated critical point of (1.1).

(i) The critical point $(0, 0)$ is **stable** if for each $\varepsilon > 0$ there exists a positive number δ_ε such that every path that starts inside the circle of radius δ_ε at some time t_0 (e.g., $t_0 = 0$) remains inside the circle of radius ε for all $t > t_0$ (see Fig. 2.7). This type of stability is sometimes called **Lyapunov stability**.

(ii) The critical point is **locally asymptotically stable** if it is stable and there exists a circle of radius δ_ε such that every path that is inside this circle at $t = t_0$ approaches $(0,0)$ as $t \to \infty$.

(iii) If $(0,0)$ is not stable, then it is **unstable** .

For a linear system, from Fig. 2.5, we note that a center is stable, but not asymptotically stable; a saddle is unstable. A spiral is either asymptotically stable or unstable, depending on the direction of the paths, and similarly for nodes.

To determine analytic criteria for stability we can proceed as we did for a single differential equation, namely to linearize near the critical point. Suppose (x_0, y_0) is an isolated critical point of (1.1). This means $x(t) = x_0$, $y(t) = y_0$, is an equilibrium, or constant, solution. Let

$$x(t) = x_0 + u(t),$$
$$y(t) = y_0 + v(t),$$

where $u(t)$ and $v(t)$ represent small perturbations from the equilibrium state. Substituting into (1.1) gives the perturbation equations (or, equations for the perturbations)

$$\frac{du}{dt} = P(x_0 + u, y_0 + v), \qquad \frac{dv}{dt} = Q(x_0 + u, y_0 + v).$$

By Taylor's theorem,

$$P(x_0 + u, y_0 + v) = P(x_0, y_0) + P_x(x_0, y_0)u + P_y(x_0, y_0)v + \cdots$$
$$= au + bv + \cdots,$$

where $a = P_x(x_0, y_0)$ and $b = P_y(x_0, y_0)$, and the dots represent higher order terms in u and v. Similarly,

$$Q(x_0 + u, y_0 + v) = cu + dv + \cdots,$$

where $c = Q_x(x_0, y_0)$ and $d = Q_y(x_0, y_0)$. Neglecting the higher-order terms, we obtain the linearized perturbation equations, or the **linearization** near the critical point:

$$\frac{du}{dt} = au + bv, \qquad\qquad\qquad (1.8)$$
$$\frac{dv}{dt} = cu + dv.$$

To emphasize the idea, the preceding equations express how the perturbations behave. The matrix associated with this linearization,

$$A(x_0, y_0) = \begin{pmatrix} a & b \\ c & d \end{pmatrix},$$

is called the **Jacobian matrix** of the system at the equilibrium (x_0, y_0). It seems reasonable that the stability of the equilibrium (x_0, y_0) of the nonlinear system (1.1) is indicated by the stability of $(0, 0)$ of the associated linearized system (1.8). Under suitable conditions, discussed in subsequent sections, this is true. In the next section we analyze linear systems.

In ecological contexts, where these ideas are central, the Jacobian is often called the **community matrix**.

EXERCISES

1. In the xy phase plane sketch several orbits of a conservative mechanical system governed by $mx'' = F(x)$ if the potential energy function is $V(x) = x^3(1-x)^2$. Take $m = 1$. For which values of the total energy E does the system oscillate?

2. Work Exercise 1 if the force is $F(x) = 3x^2 - 1$.

3. For the system

$$x' = y^2, \quad y' = -\frac{2}{3}x,$$

find the critical points, nullclines, and the direction of the orbits in the regions separated by the nullclines. Next, find the equations of the orbits and plot the phase plane diagram. Is the origin stable or unstable?

4. Repeat the questions in the last exercise for the system

$$x' = x^2, \quad y' = y.$$

5. Find the linearization of the system

$$
\begin{aligned}
x' &\quad - \quad y - x, \\
y' &\quad = \quad -y + \frac{5x^2}{4 + x^2},
\end{aligned}
$$

about the equilibrium point $(1, 1)$. Draw the nullclines and indicate the direction of the orbits in each region separated by the nullclines. Can you conclude anything about the stability of the critical point?

2.2 Linear Systems

In the preceding section we noted the role of the linear approximation near a critical point. Linear systems are also interesting in themselves and occur often in physics, engineering, and other contexts. Consider a two-dimensional linear system

$$
\begin{aligned}
x' &= ax + by, \\
y' &= cx + dy,
\end{aligned}
\tag{2.1}
$$

where a, b, c, and d are constants. We assume that the coefficient matrix is nonsingular, or

$$ad - bc \neq 0. \tag{2.2}$$

This guarantees that the algebraic system

$$ax + by = 0,$$
$$cx + dy = 0$$

has only the trivial solution and the linear system has only an isolated critical point at the origin. Otherwise, if $ad - bc = 0$, there is an entire line of non-isolated critical points.

It is easier to examine (2.1) using matrix notation. This is advantageous because the notation generalizes immediately to higher dimensions. Letting

$$\mathbf{x} = \begin{pmatrix} x \\ y \end{pmatrix}, \quad A = \begin{pmatrix} a & b \\ c & d \end{pmatrix},$$

(2.1) can be written

$$\mathbf{x}' = A\mathbf{x}. \tag{2.3}$$

Condition (2.2) is then

$$\det A \neq 0.$$

Following our experience with single linear equations with constant coefficients, we try a solution of (2.3) of the form

$$\mathbf{x} = \mathbf{v}e^{\lambda t}, \tag{2.4}$$

where $\mathbf{v}$ and λ are to be determined. Substituting (2.4) into (2.3) gives, after simplification,

$$A\mathbf{v} = \lambda\mathbf{v}, \tag{2.5}$$

which is the **algebraic eigenvalue problem**. The number λ is an **eigenvalue** of A and the nonzero vector $\mathbf{v}$ is an associated **eigenvector**. (Observe that any multiple of an eigenvector is also an eigenvector belonging to the same eigenvalue.) The eigenvalues of A are found as roots of the characteristic equation

$$\det(A - \lambda I) = 0, \tag{2.6}$$

and the corresponding eigenvector(s) are then determined by solving the homogeneous system

$$(A - \lambda I)\mathbf{v} = 0. \tag{2.7}$$

In terms of the elements of the matrix A, equation (2.6) is

$$\det \begin{pmatrix} a - \lambda & b \\ c & d - \lambda \end{pmatrix} = \lambda^2 - (\operatorname{tr} A)\lambda + \det A = 0,$$

where $\operatorname{tr} A = a + d$ is the trace of A. This quadratic equation is called the **characteristic equation** associated with A. It will have two roots: both real and unequal, both real and equal, or complex conjugate roots. Note that the assumption $\det A \neq 0$, so $\lambda = 0$ is not an eigenvalue in this case. In other contexts, zero can be an eigenvalue.

In summary, each eigenpair λ and $\mathbf{v}$ gives a solution $\mathbf{x} = \mathbf{v}e^{\lambda t}$ to the linear system. We summarize the solution structure in all of the cases.

Case I. If the eigenvalues λ_1, λ_2 are not equal, then the corresponding eigenvectors $\mathbf{v}_1$, $\mathbf{v}_2$ are linearly independent and the linear combination

$$\mathbf{x}(t) = c_1\mathbf{v}_1 e^{\lambda_1 t} + c_2\mathbf{v}_2 e^{\lambda_2 t} \tag{2.8}$$

is a solution to the linear system for all constants c_1 and c_2. This expression is the **general solution** to (2.3) because it contains all solutions. The initial value problem consisting of (2.3) and the initial condition $\mathbf{x}(0) = \mathbf{x}_0$ determine the two constants c_1 and c_2 uniquely.

Case II. When there is a single eigenvalue $\lambda \equiv \lambda_1 = \lambda_2$ of multiplicity two, the general solution depends on whether there are one or two linearly independent eigenvectors.

(a) If $\mathbf{v}_1$ and $\mathbf{v}_2$ are linearly independent eigenvectors corresponding to λ, then the general solution of (2.3) is

$$\mathbf{x}(t) = c_1\mathbf{v}_1 e^{\lambda t} + c_2\mathbf{v}_2 e^{\lambda t}$$
$$= (c_1\mathbf{v}_1 + c_2\mathbf{v}_2)e^{\lambda t}.$$

(b) If $\mathbf{v}$ is the only eigenvector corresponding to λ, then $\mathbf{v}e^{\lambda t}$ is a solution of (2.3). It is easily checked that a second linearly independent solution has the form $(\mathbf{w} + \mathbf{v}t)e^{\lambda t}$ for a vector $\mathbf{w}$ satisfying

$$(A - \lambda I)\mathbf{w} = \mathbf{v}.$$

Therefore the general solution is

$$\mathbf{x}(t) = (c_1\mathbf{v} + c_2\mathbf{w} + c_2\mathbf{v}t)e^{\lambda t}.$$

Case III. If the eigenvalues are complex, that is,

$$\lambda_1 = \alpha + i\beta, \quad \lambda_2 = \alpha - i\beta,$$

then the corresponding eigenvectors are complex,

$$\mathbf{w} + i\mathbf{v}, \quad \mathbf{w} - i\mathbf{v}.$$

Then a complex solution of (2.3) is

$$(\mathbf{w} + i\mathbf{v})e^{(\alpha + i\beta)t}.$$

Using Euler's formula, $\exp(i\theta) = \cos\theta + i\sin\theta$, we can expand this to

$$e^{\alpha t}(\mathbf{w}\cos\beta t - \mathbf{v}\sin\beta t) + ie^{\alpha t}(\mathbf{w}\sin\beta t + \mathbf{v}\cos\beta t).$$

The real and imaginary parts of a complex solution are independent real solutions, and therefore the general solution (2.3) is

$$\mathbf{x}(t) = c_1 e^{\alpha t}(\mathbf{w}\cos\beta t - \mathbf{v}\sin\beta t) + c_2 e^{\alpha t}(\mathbf{w}\sin\beta t + \mathbf{v}\cos\beta t).$$

The other eigenvalue gives the same two independent solutions. Notice that the two functions $\mathbf{w}\cos\beta t - \mathbf{v}\sin\beta t$ and $\mathbf{w}\sin\beta t + \mathbf{v}\cos\beta t$ are periodic functions with period $2\pi/\beta$, and the factor $e^{\alpha t}$ is an amplitude factor. If $\alpha = 0$, then the solution is periodic with period $2\pi/\beta$. Therefore the orbits are closed curves and $(0,0)$ is a **center**. If $\alpha < 0$, the amplitude of $\mathbf{x}$ decreases and the orbits form spirals that wind into the origin; these are decaying oscillations and the origin is an asymptotically stable **spiral point**. If $\alpha > 0$, the amplitude of $\mathbf{x}$ increases and the orbits are unstable spirals, representing growing oscillations, and the origin is an unstable spiral point.

The preceding discussion completely characterizes the solution structure near an isolated critical point $(0,0)$ of the linear system (2.3). We summarize the key stability result in the following theorem, whose proof easily follows from the form of the various solutions.

Theorem 2.7

The critical point $(0,0)$ of the linear system

$$\frac{d\mathbf{x}}{dt} = A\mathbf{x}, \quad \det A \neq 0,$$

is asymptotically stable if, and only if, the eigenvalues of A are negative or have negative real parts.   □

Remark 2.8

In the two-dimensional case the characteristic equation can be written simply as

$$\det(A - \lambda I) = \lambda^2 - \operatorname{tr} A\lambda + \det A = 0,$$

where $\operatorname{tr} A = a + d$ is the trace of the matrix A. Therefore, by the quadratic equation, the eigenvalues are given by

$$\lambda = \frac{1}{2}\left(p \pm \sqrt{p^2 - 4q}\right),$$

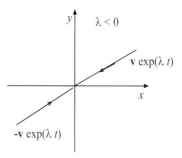

Figure 2.8 Opposing linear orbits corresponding to an eigenpair $\lambda < 0$, $\mathbf{v}$. These two linear orbits that approach the origin are the stable manifolds of $(0,0)$.

where $p = \operatorname{tr} A$ and $q = \det A$. It follows that the eigenvalues have negative real parts if, and only if,

$$p < 0 \quad \text{and} \quad q > 0,$$

or,

$$\operatorname{tr} A < 0 \quad \text{and} \quad \det A > 0.$$

These conditions on the trace and determinant are necessary and sufficient for asymptotic stability. □

In Case (III) above, when the eigenvalues are complex ($p^2 - 4q < 0$), we completely characterize the phase diagram as centers when $p = 0$ or spirals ($p \neq 0$). Now we examine the geometrical structure of the phase plane in Cases (I) and (II), when the eigenvalues are real.

If λ, $\mathbf{v}$ is an eigenpair, the solution $\mathbf{x}(t) = \mathbf{v}e^{\lambda t}$ is a straight line, or linear orbit. Note that $\mathbf{x}(0) = \mathbf{v}$, which is a nonzero point in the plane. Then for each time t, $\mathbf{v}e^{\lambda t}$ is a multiple of $\mathbf{v}$, and so it lies on the ray emanating from the origin and passing through $\mathbf{v}$. If $\lambda > 0$ then $|\mathbf{x}(t)| \to \infty$ as $t \to +\infty$, and $\mathbf{x}(t) \to \mathbf{0}$ as $t \to -\infty$. It is exactly opposite if $\lambda < 0$.

Further, when $\mathbf{v}$ is an eigenvector so is $-\mathbf{v}$. Therefore $-\mathbf{v}e^{\lambda t}$ is a straight line solution opposite to $\mathbf{v}e^{\lambda t}$. These two opposing linear orbits, shown in Fig. 2.8, either both point in ($\lambda < 0$), or both point out ($\lambda > 0$). The two opposing rays pointing inward are called the stable manifolds of $(0,0)$, and two pointing outward are called unstable manifolds of $(0,0)$.

In the case of a stable node with two negative unequal eigenvalues, there are two pairs of stable manifolds, and all other orbits approach the origin in the direction of the pair corresponding to the largest negative eigenvalue. For example, if $\lambda = -4, -1$, the general solution is

$$\mathbf{x}(t) = c_1 \mathbf{v}_1 e^{-4t} + c_2 \mathbf{v}_2 e^{-t} \approx c_2 \mathbf{v}_2 e^{-t}$$

as $t \to +\infty$. Similarly, as $t \to -\infty$, backward in time, the orbits approach the direction corresponding to the most negative eigenvalue.

For an unstable node, when the eigenvalues are both positive, and unequal, the phase plane diagram is just opposite.

For saddle points, the eigenvalues are real and have opposite signs. There is a pair of stable manifolds corresponding to the negative eigenvalue and a pair of unstable manifolds corresponding to the positive eigenvalue. All other orbits veer away from the origin and approach the unstable manifolds; in backward time they approach the stable manifolds.

Example 2.9

This example is a prototype of two real eigenvalues of opposite sign, which gives a saddle structure. Consider the linear system

$$x' = 3x - 2y,$$
$$y' = 2x - 2y.$$

The coefficient matrix

$$A = \begin{pmatrix} 3 & -2 \\ 2 & -2 \end{pmatrix}$$

has characteristic equation

$$\det \begin{pmatrix} 3 - \lambda & -2 \\ 2 & -2 - \lambda \end{pmatrix} = \lambda^2 - \lambda - 2 = 0,$$

and therefore the eigenvalues are $\lambda = -1, 2$. Therefore the origin is an unstable critical point. The eigenvectors are found by solving the linear homogeneous system

$$\begin{pmatrix} 3 - \lambda & -2 \\ 2 & -2 - \lambda \end{pmatrix} \begin{pmatrix} v_1 \\ v_2 \end{pmatrix} = \begin{pmatrix} 0 \\ 0 \end{pmatrix}.$$

When $\lambda = -1$,

$$4v_1 - 2v_2 = 0, \quad 2v_1 - v_2 = 0,$$

and an eigenvector corresponding to $\lambda = -1$ is $[v_1, v_2]^{\mathrm{T}} = [1, 2]^{\mathrm{T}}$. When $\lambda = 2$,

$$v_1 - 2v_2 = 0, \quad 2v_1 - 4v_2 = 0,$$

which gives $[v_1, v_2]^{\mathrm{T}} = [2, 1]^{\mathrm{T}}$. The general solution to the system is

$$\mathbf{x}(t) = c_1 \begin{pmatrix} 1 \\ 2 \end{pmatrix} e^{-t} + c_2 \begin{pmatrix} 2 \\ 1 \end{pmatrix} e^{2t}.$$

The two eigenvectors above define the directions of the linear orbits, or the pair of stable manifolds corresponding to negative eigenvalue and the pair of

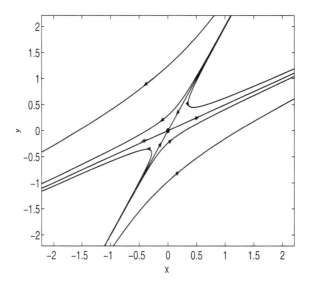

Figure 2.9 Saddle point structure.

unstable manifolds corresponding to the positive eigenvalue. The remaining
orbits, given as a linear combination of the two linear orbits, veer away from
the origin and approach the unstable manifolds as $t \to +\infty$. To get a more
accurate plot it is helpful to plot the nullclines, or loci of points where $x' = 0$
and where $y' = 0$. A phase diagram, computed in MATLAB, is shown in Fig.
2.9. □

EXERCISES

1. Find the general solution and sketch phase diagrams for the following systems; characterize the equilibria as to type (node, etc.) and stability.

   a) $x' = x - 3y, \; y' = -3x + y$.

   b) $x' = -x + y, \; y' = y$.

   c) $x' = 4y, \; y' = -9x$.

   d) $x' = x + y, \; y' = 4x - 2y$.

   e) $x' = 3x - 4y, \; y' = x - y$.

   f) $x' = -2x - 3y, \; y' = 3x - 2y$.

2. Determine the behavior of solutions near the origin for the system

$$\mathbf{x}' = \begin{pmatrix} 3 & b \\ 1 & 1 \end{pmatrix} \mathbf{x}$$

for different values of b.

3. Fill in the details of Remark 2.8.

4. For the linear system (2.3), letting $p = \operatorname{tr} A$ and $q = \det A$, shade regions in the pq plane representing values where the origin is a saddle point, a stable node, a stable spiral, an unstable node, an unstable spiral, or a center. What is the structure of the phase plane along the line $q = 0$.

5. (Environment) At time $t = 0$ a chemical herbicide is sprayed on a soil in a field of crops. Let x be the amount of herbicide in the crop, and let y be the amount of herbicide in the soil. Suppose herbicide is transferred from the soil to the crop at rate βy and transferred from the crop to the soil at rate αx; further, assume the chemical degrades in the soil at rate γy. Set up a model for the amounts of herbicide in the crop and in soil. Sketch a phase plane diagram indicating how the system evolves in time.

6. Consider the system

$$\mathbf{x}' = \begin{pmatrix} a & -1 \\ 1 & a \end{pmatrix} \mathbf{x}.$$

What are the possible behaviors of solutions to this system, depending upon the parameter a.

7. The equation for a damped spring-mass oscillator is

$$m\ddot{x} + a\dot{x} + kx = 0, \quad m, a, k > 0.$$

Write the equation as a system by introducing $y = \dot{x}$ and show that $(0, 0)$ is a critical point. Describe the nature and stability of the critical point in the following cases: $a = 0$; $a^2 - 4mk = 0$; $a^2 - 4km < 0$; $a^2 - 4km > 0$. Interpret the results physically.

2.3 Nonlinear Systems

We now examine nonlinear systems by first revisiting the question of whether the linearization

$$\mathbf{x}' = A\mathbf{x}, \quad A = \begin{pmatrix} P_x(x_0, y_0) & P_y(x_0, y_0) \\ Q_x(x_0, y_0) & Q_y(x_0, y_0) \end{pmatrix} \tag{3.1}$$

of a nonlinear system

$$x' = P(x, y), \quad y' = Q(x, y) \tag{3.2}$$

at a critical point (x_0, y_0) predicts its type (node, saddle, center, spiral point) and stability. The following result of Poincaré gives a partial answer to this question. The proof can be found in the references.

Theorem 2.10

Let (x_0, y_0) be an isolated critical point for the nonlinear system (3.2) and let $A = A(x_0, y_0)$ be the Jacobian matrix for the linearization (3.1), with $\det A \neq 0$. Then (x_0, y_0) is a critical point of the same type and stability as the origin $(0, 0)$ for the linearization in the following cases:

(i) The eigenvalues of A are real, either equal or distinct, and have the same sign (node).

(ii) The eigenvalues of A are real and have opposite signs (saddle).

(iii) The eigenvalues of A are complex but not purely imaginary (spiral).

The *exceptional case* is when the linearization has a center. In this case there are several interesting possibilities that can occur in the nonlinear system, and we will investigate some of them. The orbital structure for the nonlinear system near critical points mirrors that of the linearization in the nonexceptional cases indicated in the theorem, with a slight distortion in the phase plane diagram caused by the nonlinearity.

The question of stability is cataloged in the following theorem, which is an immediate corollary of the Poincaré theorem.

Theorem 2.11

If $(0, 0)$ is asymptotically stable for (3.1), then (x_0, y_0) is asymptotically stable for (3.2).   □

Example 2.12

Consider the nonlinear system

$$\dot{x} = 2x + 3y + xy,$$
$$\dot{y} = -x + y - 2xy^3,$$

which has an isolated critical point at $(0,0)$. The Jacobian matrix A at the origin is

$$A = \begin{pmatrix} -2 & 3 \\ -1 & 1 \end{pmatrix},$$

and it has eigenvalues $-\frac{1}{2} \pm (\sqrt{3}/2)i$. Thus the linearization has an asymptotically stable spiral at $(0,0)$, and thus the nonlinear system has an asymptotically stable spiral at $(0,0)$.   □

Example 2.13

The nonlinear system

$$x' = -y - x^3, \quad y' = x,$$

has a stable spiral at the origin. The linearization has Jacobian

$$A = \begin{pmatrix} 0 & -1 \\ 1 & 0 \end{pmatrix},$$

has purely imaginary eigenvalues $\pm i$, and hence $(0,0)$ is a center for the linearization. This is the exceptional case in the theorem and we are not guaranteed that $(0,0)$ is a center for the nonlinear system. In fact, we show later that there are no periodic solutions to the nonlinear system.   □

Example 2.14

(**Mechanics**) A particle of mass $m = 1$ moves on the x axis under the influence of a conservative force $f(x) = 3x^2 - 1$. The equations of motion in the phase plane are

$$
\begin{aligned}
x' &= y, \\
y' &= 3x^2 - 1,
\end{aligned}
$$

where the position x and the velocity y are functions of time t. The potential energy is $V(x) = -x^3 + x$, where the zero potential energy level is located at $x = 0$. The orbits are given by the conservation of energy law

$$\frac{1}{2}y^2 + (-x^3 + x) = E,$$

or

$$y = \sqrt{2}\sqrt{x^3 - x + E}, \quad y = -\sqrt{2}\sqrt{x^3 - x + E}.$$

They are plotted in Fig. 2.10. There are two equilibria, $x = \sqrt{\frac{1}{3}}, y = 0$ and $x = -\sqrt{\frac{1}{3}}, y = 0$, where the velocity is zero and the force is zero. The

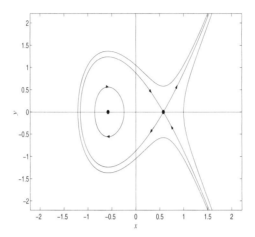

Figure 2.10 Orbits for the conservative system.

equilibrium solution $x = -\sqrt{\frac{1}{3}}$, $y = 0$ has the structure of a center, and for small initial values the system will oscillate. The other equilibrium $x = \sqrt{\frac{1}{3}}$, $y = 0$, has the structure of an unstable saddle point. Because $x' = y$, for $y > 0$ we have $x' > 0$, and the orbits are directed to the right in the upper half-plane. For $y < 0$ we have $x' < 0$, and the orbits are directed to the left in the lower half plane. The x nullcline is the y axis, where the vector field is vertical; the y nullclines are the lines $x = -\sqrt{\frac{1}{3}}$ and $x = \sqrt{\frac{1}{3}}$, where the vector field is horizontal. For large initial energies the system does not oscillate but rather goes to $x = +\infty$, $y = +\infty$; that is, the mass moves farther and farther to the right with faster speed. The Jacobian matrix is

$$A(x,y) = \begin{pmatrix} 0 & 1 \\ 6x & 0 \end{pmatrix}.$$

Then

$$A\left(-\sqrt{\frac{1}{3}},0\right) = \begin{pmatrix} 0 & 1 \\ -2\sqrt{3} & 0 \end{pmatrix}, \quad A\left(\sqrt{\frac{1}{3}},0\right) = \begin{pmatrix} 0 & 1 \\ 2\sqrt{3} & 0 \end{pmatrix}.$$

The eigenvalues of the first matrix are purely imaginary, so the linearization has a center at $\left(-\sqrt{\frac{1}{3}},0\right)$. This is the exceptional case and the theorem gives no information about the nonlinear system. However, for this problem we were able to find equations for the orbits, and we found that the nonlinear system does have a center at $\left(-\sqrt{\frac{1}{3}},0\right)$. The second matrix has real eigenvalues of

opposite sign, and therefore $\left(\sqrt{\frac{1}{3}}, 0\right)$ is a saddle point, confirming the analysis.
□

Example 2.15

(**Lotka–Volterra model**) Nonlinear equations play a central role in modeling population dynamics in ecology. Let $x = x(t)$ be the prey population and $y = y(t)$ be the predator population. We can think of lynx and hares, food fish and sharks, or any consumer–resource interaction, including herbivores and plants. If there is no predator we assume the prey dynamics is $x' = rx$, or exponential growth, where r is the *per capita* growth rate. In the absence of prey, we assume that the predator dies via $y' = -my$, where m is the *per capita* mortality rate. When there are interactions, we must include terms that decrease the prey population and increase the predator population. To determine the form of the predation term, let us assume that the rate of predation, or the number of prey consumed per unit time, per predator, is proportional to the number of prey. That is the rate of predation is ax. Thus, if there are y predators then the rate that prey is decreased is axy. This is **mass–action kinetics**. Note that the interaction term is proportional to xy, the product of the number of predators and the number of prey. For example, if there were 20 prey and 10 predators, there would be 200 possible interactions. Only a fraction of them, a, is assumed to result in a kill. The parameter a depends upon the fraction of encounters and the success of the encounters. The prey consumed cause a rate of increase in predators of εaxy, where ε is the conversion efficiency of the predator population. (There is not a one-to-one trade off of prey consumed to predators produced; it takes many small fish to create a shark.). Therefore, we obtain the simplest model of predator–prey interaction, called the **Lotka–Volterra** model:

$$\begin{aligned} x' &= rx - axy, \\ y' &= -my + bxy, \end{aligned}$$

where $b = \varepsilon a$. A. Lotka and V. Volterra developed this model in the 1920s to study fish populations in the Mediterranean Sea after the First World War. It was one of the first models of its kind developed in theoretical ecology, and it has a high historical interest despite some of its deficiencies.

To analyze the Lotka–Volterra model we factor the right sides of the equations to obtain

$$x' = x(r - ay), \quad y' = y(-m + bx). \tag{3.3}$$

Setting the right sides equal to zero gives two equilibria, $(0,0)$ and $(m/b, r/a)$. The origin represents extinction of both species, and the nonzero equilibrium

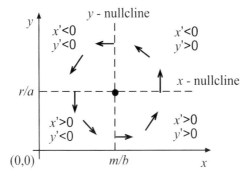

Figure 2.11 Nullclines and vector field for the Lotka–Volterra system.

represents a possible coexistent state. The x nullclines for (3.3), where $x' = 0$, are $x = 0$ and $y = r/a$. Orbits cross these two lines vertically. Orbits cross the y nullclines, where $y' = 0$, horizontally. The equilibria are the intersections of the x and y nullclines. See Fig. 2.11. Along each nullcline we can find the direction of the vector field. For example, on the ray to the right of the equilibrium we have $x > m/b$, $y = r/a$. We know the vector field is vertical so we need only check the sign of y'. We have $y' = y(-m + bx) = (r/a)(-m + bx) > 0$, so the vector field points upward. Similarly we can determine the directions along the other three rays. These are shown in the accompanying figure. Further, we can determine the direction of the vector field in regions between the nullclines either by selecting an arbitrary point in that region and calculating x' and y', or by just noting the sign of x' and y' in that region from information obtained from the system. For example, in the quadrant above and to the right of the nonzero equilibrium, it is easy to see that $x' < 0$ and $y' > 0$; so the vector field points upward and to the left. This task can be performed for each region to obtain the directions shown in the figure. Having the direction of the vector field along the nullclines and in the regions bounded by the nullclines tells us the directions of the orbits. Near $(0,0)$ the orbits appear to veer away and the equilibrium seems to have a saddle structure. The Jacobian matrix is

$$A(x, y) = \begin{pmatrix} r - ay & -ax \\ by & -m + bx \end{pmatrix}.$$

Clearly,

$$A(0,0) = \begin{pmatrix} r & 0 \\ 0 & -m \end{pmatrix}$$

has eigenvalues r and $-m$, confirming that $(0,0)$ is indeed a saddle. It appears that orbits circle around the nonzero equilibrium in a counterclockwise fashion.

The linearized matrix is

$$A\left(\frac{m}{b},\frac{r}{a}\right) = \left(\begin{array}{cc} 0 & -\frac{am}{b} \\ \frac{br}{a} & 0 \end{array}\right),$$

which has purely imaginary eigenvalues. So the linearization is inconclusive and additional work is required.

To obtain the equation of the orbits we divide the two equations in (3.3) to get

$$\frac{y'}{x'} = \frac{y(-m+bx)}{x(r-ay)}.$$

Rearranging and integrating gives

$$r\ln y - ay = bx - m\ln x + C,$$

which is the algebraic equation for the orbits. It remains obscure what these curves are because it is not possible to solve for either of the variables. But we can proceed as follows. If we exponentiate we get

$$y^r e^{-ay} = e^C e^{bx} x^{-m}.$$

Now consider the y nullcline where x is fixed at a value m/b, and fix a C value (i.e., an orbit). The right side of the last equation is a positive number γ, and so $y^r = \gamma e^{ay}$. If we plot both sides of this equation (a power function and a growing exponential) we observe that there can be at most two crossings; therefore, this equation can have at most two solutions for y. Hence, along the y nullcline there can be at most two crossings; an orbit cannot spiral into or out from the equilibrium point because that would mean many values of y would be possible. We conclude that the nonzero equilibrium is a center with closed, periodic orbits encircling it. A phase diagram is shown in Fig. 2.12. When the prey population is high the predators have a high food source and their numbers start to increase, thereby eventually driving down the prey population. Then the prey population gets low, ultimately reducing the number of predators because of lack of food. Then the process repeats, giving cycles. The Lotka–Volterra model is the simplest model in ecology showing how populations can cycle, and it was one of the first strategic models to explain qualitative observations in natural systems. Note that the nonzero equilibrium is stable. A small perturbation from equilibrium puts the populations on a periodic orbit that stays near the equilibrium. But the system does not return to equilibrium. □

Remark 2.16

From a historical point of view, there are many issues with the suitability of the Lotka–Volterra model as a predator–prey model. A classic data set obtained

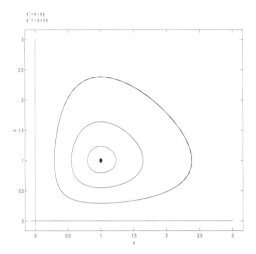

Figure 2.12 Counterclockwise periodic orbits of the Lotka–Volterra system.

in Canada during the period 1881-1940 on lynx–hare populations does not conform with the model. One problem, of many, is that the data indicate the cycles should go clockwise, not counterclockwise! And, the model lacks a certain structural stability in that if other terms are included, even if the effects are small, the behavior of the solution changes dramatically. Other problems exist in the data itself, e.g., its incompleteness, how it was collected, its omission of the role of the trappers, and so on. Many have written on this topic. A brief discussion with other references is in Brauer and Castillo-Chavez (2001), pp. 180–188.   □

A central problem in the theory of nonlinear systems is to determine whether the system admits any cycles, or closed orbits. These orbits represent periodic solutions of the system, or oscillations. In general, for nonlinear systems the existence of closed orbits may be difficult to decide. The following negative result, however, due to Bendixson and Dulac is easy to prove.

Theorem 2.17

For the system (3.2), if $P_x + Q_y$ is of one sign in a region of the phase plane, then system (3.2) cannot have a closed orbit in that region.   □

Proof

By way of contradiction assume the region contains a closed orbit Γ given by

$$x = x(t), \ y = y(t), \quad 0 \le t \le T,$$

and denote the interior of Γ by R. By Green's theorem in the plane,

$$\int_{\Gamma} P \, dy - Q \, dx = \int\int_{R} (P_x + Q_y) \, dx \, dy \ne 0.$$

On the other hand,

$$\int_{\Gamma} P \, dy - Q \, dx = \int_0^T (Py' - Qx') \, dt = \int_0^T (PQ - QP) \, dt = 0,$$

which is a contradiction. $\quad\square$

Another negative criterion is due to Poincaré.

Theorem 2.18

A closed orbit of the system (3.2) surrounds at least one critical point of the system. Thus, if no critical points exist in a region, then there can be no periodic orbits in that region. $\quad\square$

Positive criteria that ensure the existence of closed orbits are not abundant, particularly criteria that are practical and easy to apply. For certain special types of equations such criteria do exist. The following fundamental theorem, called the **Poincaré–Bendixson theorem**, is a general theoretical result.

Theorem 2.19

Let R be a closed bounded region in the plane containing no critical points of (3.2). If Γ is an orbit of (3.2) that lies in R for some t_0 and remains in R for all $t > t_0$, then Γ is either a closed orbit or it spirals toward a closed orbit as $t \to \infty$. $\quad\square$

The Poincaré–Bendixson theorem guarantees that, in the plane, either an orbit leaves every bounded set as $t \to \pm\infty$, it is a closed curve, it approaches a critical point or a closed curve as $t \to \pm\infty$, or it is a critical point. An orbit cannot just wander around chaotically forever. Thus, in the plane, the only *attractors* are closed orbits or critical points. Interestingly enough, the Poincaré–Bendixson theorem does not generalize directly to higher dimensions. In three dimensions, for example, there exist *strange attractors* that have neither the character of a point, a curve, or a surface, which act as attractors to all paths.

2.4 Bifurcations

Next we examine systems of equations of the form

$$x' = P(x, y, \mu), \quad y' = Q(x, y, \mu), \tag{4.1}$$

depending upon a real parameter μ. As μ varies, the nature of the solution often changes at special values of μ, giving fundamentally different critical point structures. For example, what was once a stable equilibrium can bifurcate into an unstable one. This type of phenomenon is ubiquitous in all physical problems involving differential equations. All systems contain parameters (often several), and it is essential to understand how the behavior of solutions changes as the parameters change. A simple example illustrates the idea.

Example 2.20

Consider the linear system

$$x' = x + \mu y, \quad y' = x - y,$$

where μ is a parameter. The eigenvalues λ are determined from

$$\det \begin{pmatrix} 1 - \lambda & \mu \\ 1 & -1 - \lambda \end{pmatrix} = 0,$$

or

$$\lambda^2 - (1 + \mu) = 0.$$

Hence

$$\lambda = \pm\sqrt{1 + \mu}.$$

If $\mu > -1$, then the eigenvalues are real and have opposite signs, and therefore the origin is a saddle point. If $\mu = -1$, then the system becomes

$$x' = x - y, \quad y' = x - y$$

and there is a line of equilibrium solutions along $y = x$. If $\mu < -1$, then the eigenvalues are purely imaginary and the origin is a center. Consequently, as μ decreases, the nature of the solution changes at $\mu = -1$; the equilibrium state $(0, 0)$ evolves from an unstable saddle to a stable center as μ passes through the critical value. We say a bifurcation occurs at $\mu = -1$. □

As noted above, these types of phenomena occur frequently in physical systems. For example, in a predator–prey interaction in ecology we may ask how coexistent states, or equilibrium populations, depend upon the carrying capacity K of the prey species. The parameter K then acts as a bifurcation parameter. For a given system, we may find a coexistent, asymptotically stable state for small carrying capacities. But, as the carrying capacity increases, the coexistent state can suddenly bifurcate to an unstable equilibrium and the dynamics could change dramatically. This type of behavior, if present, is an important characteristic of differential equations in physics, engineering, and the natural sciences.

Example 2.21

Consider the system

$$x' = -y - x(x^2 + y^2 - \mu), \tag{4.2}$$
$$y' = x - y(x^2 + y^2 - \mu).$$

The origin is an isolated equilibrium, and the linearization is

$$x' = \mu x - y, \quad y' = x + \mu y.$$

It is easy to see that the eigenvalues λ of the Jacobian matrix are given by

$$\lambda = \mu \pm i.$$

If $\mu < 0$, then $\operatorname{Re} \lambda < 0$, and $(0,0)$ is a stable spiral; if $\mu = 0$, then $\lambda = \pm i$, and $(0,0)$ is a center; if $\mu > 0$, then $\operatorname{Re} \lambda > 0$, and $(0,0)$ is an unstable spiral. From results in the preceding sections regarding the relationship between a nonlinear system and its linearization, we know $(0,0)$ is a stable spiral for the nonlinear system when $\mu < 0$ and an unstable spiral when $\mu > 0$. Therefore, there is a bifurcation at $\mu = 0$ where the equilibrium exchanges stability. □

The system in the last example deserves a more careful look. It is possible to solve the nonlinear problem (4.2) directly if we transform to polar coordinates $x = r \cos \theta, \quad y = r \sin \theta$. First, it is straightforward to show

$$xx' + yy' = rr', \quad xy' - yx' = r^2 \theta'.$$

If we multiply the first equation by x and the second by y and then add, we obtain

$$\dot{r} = r(\mu - r^2). \tag{4.3}$$

If we multiply the first equation by $-y$ and the second by x and then add, we obtain

$$\dot{\theta} = 1. \tag{4.4}$$

The equivalent system (4.3)–(4.4) in polar coordinates may be integrated directly. Clearly

$$\theta = t + t_0, \tag{4.5}$$

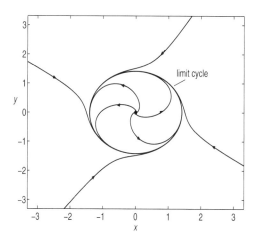

Figure 2.13 Limit cycle occurring in the case $\mu > 0$.

where t_0 is a constant. Therefore, the polar angle θ winds around counterclockwise as time increases.

Next we examine the r-equation when $\mu > 0$. By separating variables and using the partial fraction decomposition

$$\frac{1}{r(r^2 - \mu)} = \frac{-1/\mu}{r} + \frac{1/2\mu}{r - \sqrt{\mu}} + \frac{1/2\mu}{r + \sqrt{\mu}},$$

we obtain

$$r = \frac{\sqrt{\mu}}{\sqrt{1 + c\ \exp(-2\mu t)}}, \qquad (4.6)$$

where c is a constant of integration. When $c = 0$ we obtain the solution

$$r = \sqrt{\mu},$$
$$\theta = t + t_0,$$

which represents a periodic solution whose orbit is the circle $r = \sqrt{\mu}$ in the phase plane. If $c < 0$, then (4.6) represents an orbit that spirals counterclockwise toward the circle $r = \sqrt{\mu}$ from the outside, and if $c > 0$ the solution represents an orbit that spirals counterclockwise toward the circle from the inside. This means that the origin is an unstable spiral as indicated by the linear analysis, and the periodic solution $r = \sqrt{\mu}$ is approached by all other solutions. When a periodic orbit, or cycle, is approached by orbits from both the inside and outside, we say that the orbit is a stable **limit cycle**. See Fig. 2.13.

In the case $\mu = 0$ the r-equation becomes

$$\frac{dr}{r^3} = -\, dt,$$

and direct integration gives the solution

$$r = \frac{1}{\sqrt{2t + c}}, \qquad \theta = t + t_0.$$

This implies that the origin is a stable spiral. Thus, as μ changed from positive to zero, the limit cycle disappeared and the origin changed stability.

Finally, in the case $\mu < 0$, let $k^2 = -\mu > 0$. Then the radial r-equation is

$$\frac{dr}{r(r^2 + k^2)} = -\, dt. \tag{4.7}$$

The partial fraction decomposition is

$$\frac{1}{r(r^2 + k^2)} = \frac{1/k^2}{r} - \frac{(1/k^2)r}{r^2 + k^2},$$

and (4.7) integrates to

$$r^2 = \frac{ck^2 e^{-2k^2 t}}{1 - ce^{-2k^2 t}}.$$

It is easy to verify that this solution represents spirals that approach the origin at $t \to +\infty$.

Now let us review the results. As μ passes through the critical value $\mu = 0$ from negative to positive, the origin changes from a stable spiral to an unstable spiral and there is born a new time periodic solution, a limit cycle. This type of bifurcation is common and it is an example of a **Hopf bifurcation**. A graph in the complex plane of the eigenvalues $\lambda = \mu + i$ and $\lambda = \mu - i$ of the linearized system, as functions of μ, shows that the complex conjugate pair crosses the imaginary axis at $\mu = 0$. Generally this can signal a bifurcation to a time periodic solution, or limit cycle. In the general case (Fig. 2.14), the eigenvalues are $\lambda(\mu) = \text{Re}\lambda(\mu) \pm i\text{Im}\,\lambda(\mu)$ with $\lambda(\mu^*) = \pm i\text{Im}\lambda(\mu^*)$ purely imaginary, and $\text{Re}\lambda'(\mu) > 0$. Two cases can occur. There is an unstable limit cycle for $\mu < \mu^*$ with the critical point asymptotically stable, or there is a stable limit cycle for $\mu > \mu^*$ with the critical point unstable. This is essentially the Hopf bifurcation theorem. (The theorem was first mentioned by Poincaré; later, in the 1920s, A. Andronov proved the planar version, and E. Hopf generalized it to higher dimensional system in the early 1940s.)

There are several other theorems for special systems that guarantee the conditions for a Hopf bifurcation. One of these, due to A. Kolmogorov, is a result for systems of the form

$$x' = xf(x, y), \qquad y' = yg(x, y),$$

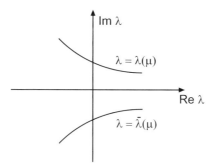

Figure 2.14 Schematic of a Hopf bifurcation in the complex λ plane showing how the eigenvalues at a critical point cross the imaginary axis from one side of the plane to the other, depending on the value of a parameter μ. Purely imaginary eigenvalues occur at $\mu = \mu^*$.

where f and g satisfy specialized conditions. Such systems are common in predator-prey models. We refer to Edelstein-Keshet (1988), p. 351, for a precise statement.

We mention one other important consequence of the Poincaré–Bendixson theorem that predicts the existence of limit cycles. Suppose there is a closed bounded region R in the plane where at each point (x, y) on the boundary of R the vector field $(P(x, y), Q(x, y))$ for the system points inward into R. We call R a **basin of attraction**. If R contains no critical points, then there must be a periodic orbit in R. This result applies equally well for an annular (donut-type) region R when the vector field points inward on both the interior and exterior boundaries. For example, if a rectangular region is attracting and contains a repeller in its interior, then we can form an annular region by deleting a small disc encircling the critical point; then we can apply the Poincaré–Bendixson theorem to show the annular region must contain a limit cycle.

EXERCISES

1. Determine the nature and stability properties of the critical points of the systems, and sketch the phase diagram:

   a) $x' = x + y - 2x^2$, $y' = -2x + y + 3y^2$.

   b) $x' = -x - y - 3x^2 y$, $y' = -2x - 4y + y \sin x$.

   c) $x' = y^2$, $y' = -\frac{2}{3}x$.

   d) $x' = x^2 - y^2$, $y' = x - y$.

   e) $x' = x^2 + y^2 - 4$, $y' = y - 2x$.

2. Consider the system

$$x' = 4x + 4y - x(x^2 + y^2),$$
$$y' = -4x + 4y - y(x^2 + y^2).$$

a) Show there is a closed orbit in the region $1 \le r \le 3$, where $r^2 = x^2 + y^2$.

b) Find the general solution.

3. Determine if the following systems admit periodic solutions.

a) $x' = y, \quad y' = (x^2 + 1)y - x^5$.

b) $x' = x + x^3 - 2y, \quad y' = -3x + y^5$.

c) $x' = y, \quad y' = y^2 + x^2 + 1$.

d) $x' = y, \quad y' = 3x^2 - y - y^5$.

e) $x' = 1 + x^2 + y^2, \quad y' = (x - 1)^2 + 4$.

4. Find the values of μ where solutions bifurcate and examine the stability of the origin in each case.

a) $x' = x + \mu y, \quad y' = \mu x + y$.

b) $x' = y, \quad y' = -2x + \mu y$.

c) $x' = x + y, \quad y' = \mu x + y$.

d) $x' = 2y, \quad y' = 2x - \mu y$.

e) $x' = y, \quad y' = \mu x + x^2$.

f) $x' = y, \quad y' = x^2 - x + \mu y$.

g) $x' = \mu y + xy, \quad y' = -\mu x + \mu y + x^2 + y^2$.

h) $x' = y, \quad y' = -y - \mu + x^2$.

i) $x' = \mu x - x^2 - 2xy, \quad y' = (\mu - \frac{5}{3})y + xy + y^2$.

5. Discuss the dependence on the sign of μ of the critical point at the origin of the system

$$x' = -y + \mu x^3, \quad y' = x + \mu y^3.$$

6. Consider the predator–prey model

$$x' = x(x(1 - x) - y), \quad y' = y(x - a), \quad a > 0,$$

where x and y are two species.

a) Sketch the nullclines and indicate the direction field.

b) Find the critical points and classify them.

c) Sketch a phase plane diagram in the case $a > 1$.

d) Explain what happens if $a = 1/2$.

e) Sketch the possible types of phase diagrams that can occur for $0 < a < 1$.

7. Prove a generalization of the Bendixson–Dulac theorem: For the system (3.2), if there is a smooth function $h(x, y)$ for which $(hP)_x + (hQ)_y$ is of one sign in a region R of the phase plane, then system (3.2) cannot have a closed orbit in R. Use this result to show that the system

$$x' = x(A - ax + by), \quad y' = y(B - cy + dx), \quad a, c > 0,$$

has no periodic orbits. (Hint: Take $h(x, y) = 1/xy$.)

8. If $p(0, 0)$ and $h(0)$ are positive, prove that the origin is an asymptotically stable critical point for the dynamical equation

$$x'' + p(x, x')x' + xh(x) = 0.$$

9. Examine the stability of the equilibrium solutions of the system

$$\frac{dx}{dt} = \gamma - y^2 - vx, \quad \frac{dy}{dt} = -vy + xy,$$

where γ and v are positive constants.

10. (Ecology) Let P denote the carbon biomass of plants in an ecosystem, H be the carbon biomass of herbivores, and ϕ the rate of primary carbon production in plants due to photosynthesis. A model of plant–herbivore dynamics is given by

$$
\begin{aligned}
P' &= \phi - aP - bHP, \\
H' &= \varepsilon bHP - cH,
\end{aligned}
$$

where a, b, c, and ε are positive parameters.

a) Explain the various terms and parameters in the model and determine the dimensions of each parameter.

b) Non-dimensionalize the model and find the equilibrium solutions.

c) Analyze the dynamics in the cases of high primary production ($\phi > ac/\varepsilon b$) and low primary production ($\phi < ac/\varepsilon b$). Explain what happens in the system if primary production is slowly increased from a low value to a high value.

11. A system

$$x' = f(x, y)$$
$$y' = g(x, y),$$

is called a **Hamiltonian system** if there is a function $H(x, y)$ for which $f = H_y$ and $g = -H_x$. The function H is called the Hamiltonian. Prove the following facts about Hamiltonian systems:

a) If $f_x + g_y = 0$, then the system is Hamiltonian. (Recall that $f_x + g_y$ is the divergence of the vector field (f, g).)

b) Prove that along any orbit, $H(x, y) = $ constant, and therefore all the orbits are given by $H(x, y) = $ constant.

c) Show that if a Hamiltonian system has an equilibrium, then it is not a source or sink (node or spiral).

d) Show that any conservative dynamical equation $x'' = f(x)$ leads to a Hamiltonian system, and show that the Hamiltonian coincides with the total energy.

e) Find the Hamiltonian for the system $x' = y$, $y' = x - x^2$, and plot the orbits.

12. In a Hamiltonian system the Hamiltonian given by $H(x, y) = x^2 + 4y^4$. Write down the system and determine the equilibria. Sketch the orbits.

13. A system

$$x' = f(x, y)$$
$$y' = g(x, y),$$

is called a **gradient system** if there is a function $G(x, y)$ for which $f = G_x$ and $g = G_y$.

a) If $f_y - g_x = 0$, prove that the system is a gradient system. (Recall that $f_y - g_x$ is the curl of the two-dimensional vector field (f, g); a zero curl ensures existence of a potential function on nice domains.)

b) Prove that along any orbit, $(d/dt)G(x, t) \geq 0$. Show that periodic orbits are impossible in gradient systems.

c) Show that if a gradient system has an equilibrium, then it is not a center or spiral.

d) Show that the system $x' = 9x^2 - 10xy^2$, $y' = 2y - 10x^2y$ is a gradient system.

e) Show that the system $x' = \sin y$, $y' = x \cos y$ has no periodic orbits.

14. Thoroughly analyze the system $x' = -2x(x-1)(2x-1)$, $y' = -2y$.

15. Analyze the system

$$x' = x^2 - 1, \quad y' = -xy + a(x^2 - 1),$$

in the cases $a = 0$, $a < 0$, and $a > 0$, and explain the bifurcations that occur.

16. (Circuits) An RCL circuit with a nonlinear resistance, where the voltage drop across the resistor is a nonlinear function of current, can be modeled by the Van der Pol equation

$$x'' + \rho(x^2 - 1)x' + x = 0,$$

where ρ is a positive constant, and $x(t)$ is the current.

a) In the phase plane, show that the origin is an unstable equilibrium.

b) Sketch the nullclines and the vector field. What are the possible dynamics? Is there a limit cycle?

c) Use software to sketch the phase diagram when $\rho = 1$, and plot $x = x(t)$ if $x(0) = 0.05$ and $x'(0) = 0$. Describe the dynamics in this case.

17. (Mechanics) Show that periodic orbits for the mechanical system

$$
\begin{aligned}
x' &\quad - \quad y, \\
y' &\quad = \quad -ky - V'(x),
\end{aligned}
$$

are possible only if $k = 0$.

18. (Biology) Analyze the following nonlinear system, which models cell differentiation.

$$x' = y - x, \quad y' = -y + \frac{5x^2}{4 + x^2}.$$

19. (Ecology) Consider the scaled predator–prey model

$$
\begin{aligned}
x' &= \frac{2}{3}x\left(1 - \frac{x}{4}\right) - \frac{xy}{1+x}, \\
y' &= ry\left(1 - \frac{y}{x}\right),
\end{aligned}
$$

where the carrying capacity of the predator depends upon the prey population. Examine the dynamics of the model as the predator growth rate r decreases from $\frac{1}{2}$ to a small value.

20. (Predation) In a simple model of predation a fraction of the prey take refuge and are not subject to predation. If $H = H(t)$ is the number of prey, and $P = P(t)$ is the number of predators, the model takes the form

$$\frac{dH}{dt} = rH - a(H - H_r)P, \quad \frac{dP}{dt} = -kp + b(H - H_r)P,$$

where r is the prey growth rate, k is the predator mortality rate, H_r is the number of prey in refuge, and a and b are predation rates. The model can be reduced to the dimensionless form

$$h' = h - (h - 1)p, \quad p' = -\alpha p + \beta(h - 1)p,$$

for appropriate choices of α and β. Find the equilibria, nullclines, and the community matrix. Determine the stability of the equilibria and sketch the phase diagram.

21. Present a thorough analysis of the model

$$\begin{aligned} p' &= mx(1 - p - q) - \mu pq, \\ q' &= \mu pq - \eta q, \end{aligned}$$

for $p, q \in [0, 1]$.

2.5 Reaction Kinetics

This section and the next contain two mathematically related applications involving differential equations: the basics of chemical kinetics and simple models of viruses. These topics are related in that they operate on the same scale, namely that of molecular biology and biochemical kinetics. There is an often-repeated cliche, that 'life is chemistry.' (Some might modify this and say 'electrochemistry.') It is hardly an understatement that chemistry and reaction kinetics pervade mathematical modeling of processes inside organisms. To drop a few reminders: metabolic pathways, transport of chemicals in organisms, photosynthesis, enzyme kinetics, nerve signal propagation, dynamics of viruses and immune system responses, chemical signalling, and so forth.

This section introduces some basics about chemical kinetics, the law of mass action, and enzyme kinetics. The next section introduces viral dynamics and the immune system response. These key ideas about kinetics help us better understand other dynamical issues on a larger scale, for example, population modeling and disease ecology.

2.5.1 The Law of Mass Action

The law of mass action is a rule that models the rate of chemical reactions and what concentrations of the constituents are involved. Elementary texts on reaction kinetics for chemical engineering students are good resources.

Example 2.22

(**Unitary reaction**) The simplest reaction is a single unimolecular (unitary) reaction where one molecule decays into a daughter particle and releases energy. For example, the unstable radioactive isotope $^3H_1$ (tritium) can decay into a stable isotope of helium $^3He_2$ while producing an electron and an antineutrino. We can think of these simple reactions as

$$X \to \text{products},$$

and we often model this reaction as radioactive decay by declaring that the rate of reaction is proportional to the amount present, or $r = kX$. (There should be no confusion in using the same letter X for both the molecule and its concentration; this is easier than using the common but unwieldy $[X]$ for concentration.) This means X is consumed at the rate kX, or

$$\frac{dX}{dt} = -kX.$$

The constant k is called the rate constant and it is usually written over the arrow in the chemical formula as

$$X \xrightarrow{k} \text{products}.$$

We easily find $X(t) = X_0 e^{-kt}$, where X_0 is the initial amount of X present. □

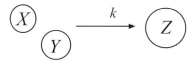

Figure 2.15 A binary reaction. Molecules X and Y combine to form the product Z.

Example 2.23

(**Binary reaction**) The next level of complication is to consider the bimolecular (binary) reaction

$$X + Y \to Z, \tag{5.1}$$

where one molecule of X and one molecule of Y (the reactants) combine to produce one molecule of Z (the product). Pictorially, Fig. 2.15 shows the mechanism. The question is: what is the reaction rate? Suppose the rate r is a function of the reactant concentrations X and Y, or $r = r(X, Y)$. Clearly we should have $r(X, 0) = r(0, Y) = 0$. Because the surface $r(X, Y)$ is identically zero along the X-axis, any order X partial derivative is zero, or $r_X(X, 0) = r_{XX}(X, 0) = r_{XXX}(X, 0) = \cdots$; similarly, $r_Y(0, Y) = r_{YY}(0, Y) = r_{YYY}(0, Y) = \cdots$. Therefore, by Taylor's theorem,

$$
\begin{aligned}
r(X, Y) &= r(0, 0) + r_X(0, 0)X + r_Y(0, 0)Y \\
&\quad + \frac{1}{2}r_{XX}(0, 0)X^2 + r_{XY}(0, 0)XY + \frac{1}{2}r_{YY}(0, 0)Y^2 + \cdots \\
&= r_{XY}(0, 0)XY + \cdots,
\end{aligned}
$$

where the dots denote higher-order terms. Letting $k = r_{XY}(0, 0)$, we have shown that the rate of the reaction (5.1) is given to leading order by

$$r = kXY. \tag{5.2}$$

This is called the **law of mass action**. Simply put, the rate of reaction is proportional to the product of the concentrations of the two reactants. The constant k is called the **rate constant** and it is written over the arrow in the reaction as

$$X + Y \overset{k}{\to} Z. \qquad \square$$

Remark 2.24

As it turns out, the rate constant is not constant, so *rate constant* is a bit of a misnomer. Chemical reactions almost always involve heat, and the rate of reaction depends strongly upon temperature. The usual assumption is that

$$k = k_0 e^{-E'/RT},$$

where R is the gas constant, T is temperature in degrees Kelvin, and E' is the activation energy which is related to the amount of energy required for the two reactants to 'get over the potential hump' to form the product. This rate is called the **Arrhenius rate**. In our discussion, however, we assume that the temperature is close to constant and the rate constant is independent of temperature. So, the reactions are isothermal. $\square$

Remark 2.25

We point out at this time the importance of the mass action assumption in other areas of dynamics. For example, in disease ecology we are interested in measuring the rate that susceptible individuals become infected. The simple argument is this. If there are S susceptible individuals and I infected individuals, then the number of possible contacts is SI. Only a fraction c of those contacts will occur, and of those only a fraction β will result in an infection. We conclude that the rate of infection is $r = bSI$, where $b = \beta c$. This elementary reasoning led to the mass action principle, and in an analog of the reaction kinetics, we have

$$S + I \xrightarrow{b} 2I,$$

or, a susceptible and an infective combine to make a product (in this case, two infectives) at rate $r = bSI$. The constant b is the transmission coefficient. A term of this form occurs often in modeling disease dynamics. Disease dynamics is discussed in a later section. □

Remark 2.26

In a different context, if X is the number of prey and Y is the number of predators, then XY is the total number of possible contacts. Only a fraction a of these will result in a kill, so the rate of predation is aXY, again mass action dynamics. This model was used earlier in describing predator-prey interactions in the Lotka–Volterra equations. □

Example 2.27

Now let's return to the main discussion of the binary chemical reaction

$$X + Y \xrightarrow{k} Z, \quad r = kXY.$$

In terms of the reaction rate we can write differential equations for the concentrations X, Y, and Z. The rates that X is consumed, Y is consumed, and Z is created are given by

$$\frac{dX}{dt} = -r, \quad \frac{dY}{dt} = -r, \quad \frac{dZ}{dt} = r, \quad r = kXY.$$

By adding the first two equations, it follows that

$$\frac{dX}{dt} - \frac{dY}{dt} = 0,$$

or $X - Y = \text{constant} = c$. This expression is called a **conservation law**. If at time $t = 0$ we have initial conditions

$$X(0) = X_0, \quad Y(0) = Y_0, \quad Z(0) = Z_0,$$

then $c = X_0 - Y_0$. Without loss of generality we can assume $c > 0$, or the initial concentration of X is larger than that of Y; this means Y will limit the reaction. The conservation law gives us a way to reduce the differential equations to a single equation in one unknown. Obviously we get

$$\frac{dX}{dt} = -kXY = -kX(X - c).$$

We can rewrite this equation in the familiar form

$$\frac{dX}{dt} = kcX\left(1 - \frac{X}{c}\right),$$

which is the logistic equation with intrinsic growth rate kc and carrying capacity c. So, the reaction evolves to the stable equilibrium $X = c$, and Y goes to zero. Notice also that $X + Z = \text{constant} = X_0$ (another conservation law), and so Z goes to $X_0 - c = Y_0$. This makes sense—all of Y is converted to Z during the course of the reaction.   □

Example 2.28

Chemists tell us that reactions also go the opposite way. That is, the product can break down into its reactant constituents. Therefore, instead of a single forward reaction, we write

$$X + Y \xrightarrow{k_1} Z, \quad r_1 \;\; = \;\; k_1 XY,$$
$$Z \xrightarrow{k_{-1}} X + Y, \quad r_{-1} \;\; = \;\; k_{-1}Z.$$

The **reversible reaction** (the second one) is a unimolecular reaction with a different rate and different rate constant than the first. Again, as in radioactive decay, the rate is proportional to Z, the concentration of the reactant. We are using subscripts on the reactions and the rates to denote which reaction. Usually we write these two reactions in shorthand as

$$X + Y \underset{k_{-1}}{\overset{k_1}{\rightleftharpoons}} Z.$$

The rate equations are now

$$\frac{dX}{dt} \;\; = \;\; -r_1 + r_{-1},$$
$$\frac{dY}{dt} \;\; = \;\; -r_1 + r_{-1},$$
$$\frac{dZ}{dt} \;\; = \;\; r_1 - r_{-1}.$$

We are using the property that for a system of reactions, the rates add. For example, in the first reaction X is consumed at rate r_1 and in the second X is recreated at rate r_{-1}. Immediately, by combining these rate equations, we get the conservation laws

$$X - Y = c_1, \quad X + Z = c_2.$$

The constants c_1 and c_2 are determined from the initial conditions, for example, $X - Y = c_1 = X(0) - Y(0)$. These expressions allow us to write a single differential equation for X,

$$
\begin{aligned}
\frac{dX}{dt} &= -k_1 XY + k_{-1} Z \\
&= -k_1 X(X - c_1) + k_{-1}(c_2 - X).
\end{aligned}
$$

This is a manageable equation and can be analyzed in the standard way with a phase line diagram (see Chapter 1). We make one important observation, that X is in equilibrium when the right side is zero, or

$$k_1 XY = k_{-1} Z.$$

Of course, from the dynamical system Y and Z will also be in equilibrium. This last expression can be written

$$\frac{XY}{Z} = \frac{k_{-1}}{k_1} = K,$$

where the ratio K of the backward to forward rate constants is called the **equilibrium constant**. (The X, Y, and Z values on the left are the equilibrium concentrations.) $\square$

A law of mass action applies to more general reactions than stated above. As a first step toward a generalization, consider a reaction of the form

$$mX + nY \xrightarrow{k} pW + qZ,$$

where m molecules of X react with n molecules of Y to produce p molecules of W and q molecules of Z. The law of mass action in this case states that the rate of the reaction is proportional to $X^m Y^n$, or

$$r = kX^m Y^n.$$

The differential equation for the chemical species X is then

$$\frac{dX}{dt} = -mr = -mkX^m Y^n,$$

because m molecules are consumed during the reaction. Similarly,

$$\frac{dY}{dt} = -nr, \quad \frac{dW}{dt} = pr, \quad \frac{dZ}{dt} = qr.$$

As we observed, conservation laws are important in reducing the number of variables. In this case, we see, for example, that

$$nX - mY = c = nX_0 - mY_0,$$

where the constant of motion is expressed in terms of the initial concentrations.

Consider the following variation of the previous reaction,

$$mX + nY \xrightarrow{k} pX + Z, \quad r = kX^mY^n.$$

Here, the species X appears as a reactant and as a product. The rate depends only on the reactant concentrations and is the same. But m molecules of X are consumed, while p are created. To fix the idea assume $p > m$. Then the net gain in X is $m - n$ molecules (which is negative). The rate law for X is then

$$\frac{dX}{dt} = (p - m)kX^mY^n.$$

Example 2.29

The reaction with accompanying rate law,

$$2X + Y \xrightarrow{k} 5X + Z, \quad r = kX^2Y, \tag{5.3}$$

has dynamics

$$\frac{dX}{dt} = -3r, \quad \frac{dY}{dt} = -r, \quad \frac{dZ}{dt} = r.$$

Independent conservation laws are

$$X - 3Y = c_1, \quad X + 3Z = c_2. \qquad \square$$

Remark 2.30

Now we make an extremely important observation. In the last reaction (5.3) we ask the likelihood of three molecules (2 X's and a Y) colliding simultaneously to react. The answer is that it is *very small*. In the chemical world, reactions are either unitary (unimolecular) or binary (bimolecular) involving either one or two reactants; such reactions are called **elementary reactions**; a higher-order, or nonelementary, reaction is unlikely. This means that a reaction like

$$2H + O \rightarrow H_2O$$

is not the correct **mechanism** for the formation of water. The actual mechanism involves a large number of intermediate, elementary reactions. Sometimes these intermediate reactions are not even known because intermediate species may be very short lived and cannot be observed. Determining the mechanism is a significant problem in chemistry. For nonchemical reactions, it is certainly possible to have tertiary reactions—e.g., in disease transmission two susceptible individuals can interact with a single infective at the same time.   □

Example 2.31

As a warm-up for a general statement of the law of mass action we consider two unitary reactions with their accompanying rates:

$$X_1 \xrightarrow{k_1} X_2, \quad r_1 = k_1 X_1,$$
$$X_2 \xrightarrow{k_2} X_1 + X_3, \quad r_2 = k_2 X_2.$$

Then the equations for the species (rate equations) are

$$\frac{dX_1}{dt} = -r_1 + r_2,$$
$$\frac{dX_2}{dt} = r_1 - r_2,$$
$$\frac{dX_3}{dt} = r_2.$$

This can be written in matrix form as

$$\begin{pmatrix} \frac{dX_1}{dt} \\ \frac{dX_2}{dt} \\ \frac{dX_3}{dt} \end{pmatrix} = \begin{pmatrix} -1 & 1 \\ 1 & -1 \\ 0 & 1 \end{pmatrix} \begin{pmatrix} r_1 \\ r_2 \end{pmatrix}.$$

In abbreviated vector notation

$$\frac{d\mathbf{X}}{dt} = S\mathbf{r},$$

where

$$\mathbf{X} = \begin{pmatrix} X_1 \\ X_2 \\ X_3 \end{pmatrix}, \quad \mathbf{r} = \begin{pmatrix} r_1 \\ r_2 \end{pmatrix}, \quad S = \begin{pmatrix} -1 & 1 \\ 1 & -1 \\ 0 & 1 \end{pmatrix}.$$

The matrix S is called the **stoichiometric matrix**, and it holds the coefficients of the reaction rates; $\mathbf{r}$ is the **rate vector**.   □

Now we generalize and state the general law of mass action.

Definition 2.32

Consider m different chemical species X_i, $i = 1, 2, ..., m$ and n reactions of the form

$$\sum_{i=1}^{m} a_{ij} X_i \overset{k_n}{\to} \sum_{i=1}^{m} b_{ij} X_i, \quad j = 1, 2, ..., n,$$

where the stoichiometric coefficients a_{ij} and b_{ij} are nonnegative integers. Notice that the a_{ij} is the stoichiometric coefficient of the reactant X_i in the jth reaction, and b_{ij} is the stoichiometric coefficient of the product X_i in the the jth equation. The reaction rates are given by

$$r_j = k_j \Pi_{i=1}^{m} X_i^{a_{ij}}, \quad j = 1, 2, ..., n.$$

Then the equation for the jth species is

$$\frac{dX_i}{dt} = \sum_{j=1}^{n} (b_{ij} - a_{ij}) r_j, \quad i = 1, 2, ..., m.$$

Letting $S = (s_{ij}) = (b_{ij} - a_{ij})$ denote the stoichiometric matrix, the rate equations are expressed by

$$\frac{d\mathbf{X}}{dt} = S\mathbf{r}. \quad \square$$

As we noted, conservation laws enable us to reduce the number of differential equations expressed by the rate laws. For a small number of equations and unknown species concentrations we can find conservation laws easily by inspection. For large systems it is harder. Therefore, let us derive a vector equation that gives the the conservation laws. A conservation law has the form

$$\frac{d}{dt} \sum_{i=1}^{m} c_i X_i = 0,$$

where the c_i is a set of constants to be determined. Then, in terms of the initial concentrations,

$$\sum_{i=1}^{m} c_i X_i = \sum_{i=1}^{m} c_i X_i(0).$$

Letting $\mathbf{c} = (c_1, ..., c_m)^{\mathrm{T}}$, this equation can be written in vector form as

$$\mathbf{c}^{\mathrm{T}} \mathbf{X} = \mathbf{c}^{\mathrm{T}} \mathbf{X}(0).$$

Multiplying the rate equation by $\mathbf{c}^{\mathrm{T}}$ we can write

$$\mathbf{c}^{\mathrm{T}} \frac{d\mathbf{X}}{dt} = \mathbf{c}^{\mathrm{T}} S\mathbf{r}.$$

But
$$\mathbf{c}^T \frac{d\mathbf{X}}{dt} = \frac{d}{dt}(\mathbf{c}^T \mathbf{X}) = \frac{d}{dt}(\mathbf{c}^T \mathbf{X}(0)) = 0.$$

Therefore,
$$\mathbf{c}^T S\mathbf{r} = (S^T \mathbf{c})^T \mathbf{r} = 0.$$

Because this holds for all rates, we have, by independence,

$$S^T \mathbf{c} = \mathbf{0}. \tag{5.4}$$

Consequently, the coefficients $\mathbf{c}$ in the conservation law are all the linearly independent solutions of the homogeneous system (5.4). In an exercise you are asked to show that this expression does indeed lead to a conservation law. For those who have studied linear algebra, $\mathbf{c} \in \ker S^T$, where ker denotes the kernel, or nullspace, of S^T.

Example 2.33

Returning to the last example, the coefficients c_i in a conservation law must satisfy

$$S^T \mathbf{c} = \begin{pmatrix} -1 & 1 & 0 \\ 1 & -1 & 1 \end{pmatrix} \begin{pmatrix} c_1 \\ c_2 \\ c_3 \end{pmatrix} = \begin{pmatrix} 0 \\ 0 \end{pmatrix}.$$

The coefficient can be reduced to row echelon form

$$\begin{pmatrix} -1 & 1 & 0 \\ 0 & 0 & 1 \end{pmatrix}.$$

Therefore, $-c_1 + c_2 = 0$ and $c_3 = 0$. Therefore, the solution is given by

$$\begin{pmatrix} c_1 \\ c_2 \\ c_3 \end{pmatrix} = c_2 \begin{pmatrix} 1 \\ 1 \\ 0 \end{pmatrix}, \qquad c_2 \text{ arbitrary.}$$

There is only one independent solution, or the kernel is one dimensional. The only conservation law is

$$\begin{pmatrix} 1 \\ 1 \\ 0 \end{pmatrix}^T \mathbf{X} = X_1 + X_2 = \text{ constant.} \qquad \square$$

EXERCISES

1. Make a generic plot of the Arrhenius rate k vs. temperature T,

$$k = k_0 e^{-E'/RT},$$

where E' is a large constant. Assume R and k_0 are positive constants.

2. For each of the following, find the rates of the reactions and the dynamical equations for the chemical constituents. Find all the independent conservation laws and reduce the dynamics to the fewest number of equations possible. Find the equilibria and sketch a phase line or phase plane, whichever is relevant. Sketch sample time series plots.

   a) $X \xrightarrow{k} 2Y$.

   b) $A \underset{k_{-1}}{\overset{k_1}{\rightleftharpoons}} C + D, \quad A + B \xrightarrow{k_2} 2A + C$.

   c) $A + X \underset{k_{-1}}{\overset{k_1}{\rightleftharpoons}} 2X$. Assume a surplus of A and its concentration is constant. (This is an autocatalytic reaction where X stimulates its own production.)

3. Consider the reaction

$$A + X \underset{k_{-1}}{\overset{k_1}{\rightleftharpoons}} 2X, \quad X + B \xrightarrow{k_2} C,$$

where X is used up in the production of C. Assume A and B are held constant. Find the reaction kinetics and determine conditions when the equilibrium $X^* = 0$.

4. Consider the hypothetical kinetics model

$$x' = a(1 - x) - xy^2, \quad y' = xy^2 - (a + b)y,$$

where a and b are positive constants. Show that a saddle-to-node bifurcation occurs when $b = -a \pm \sqrt{a}/2$.

5. Consider the reactions

$$X \underset{k_{-1}}{\overset{k_1}{\rightleftharpoons}} Y + Z, \quad X + W \xrightarrow{k_2} 2X + Y.$$

Write down the kinetic equations and reduce them to two equations for X and Y. Analyze the equations in the phase plane and state your conclusions.

6. Show that solutions of equation (5.4), $S^T \mathbf{c} = \mathbf{0}$, do indeed give a conservation law $\frac{d}{dt}(\mathbf{c}^T \mathbf{X}) = 0$.

7. (Chemical oscillator) For many years chemists thought that all reactions tended to an equilibrium state monotonically and could not oscillate. Then, in the 1950s, B. Belousov showed that certain reactions could oscillate for long time periods. This was signalled in a mixture by continual periodic changes in its color from yellow to clear. His initial work was not accepted readily by colleagues, but later it was noticed and he was awarded a Nobel Prize for his work. Now, many reactions have been found that exhibit oscillations. In this exercise we explore a system that closely approximates the concentrations of two chemical constituents in the real system:

$$x' = a - x - \frac{4xy}{1 + x^2}, \quad y' = bx\left(1 - \frac{y}{1 + x^2}\right).$$

a) Show that the equilibrium is $x^* = a/5$, $y^* = 1 + (a/5)^2$.

b) Sketch the nullclines and the the direction field.

c) Show that the equilibrium is a repeller if

$$b < \frac{3a}{5} - \frac{25}{a}.$$

d) Show that there is a closed orbit (limit cycle) bounded by the rectangular box bounded by $x = 0$, $y = 0$, $x = \bar{x}$, $y = \bar{y}$, where $\bar{x}$ is the point where the x-nullcline crosses the x axis, and $\bar{y} = 1 + \bar{x}^2$. (Use the Poincaré–Bendixson theorem.)

e) Perform numerical calculations and sketch the phase diagram in the cases $a = 10$, $b = 4$ and $a = 10$, $b = 2$.

2.5.2 Enzyme Kinetics

Many reactions in metabolic pathways are catalyzed by enzymes. Enzymes are proteins that can react with molecular substrates to break them down; they significantly speed the reaction and lower the activation energy. A model enzyme reaction that serves as a prototype to many other biological processes is the one shown in Fig. 2.16 and whose chemical equations are

$$S + E \underset{k_{-1}}{\overset{k_1}{\rightleftharpoons}} C \overset{k_2}{\rightarrow} E + P.$$

Here, one molecule of a substrate S reacts with an enzyme E to produce an intermediate complex C, and then the final product P is produced, along with the recovery of the enzyme. The formation of the complex C is a rapid

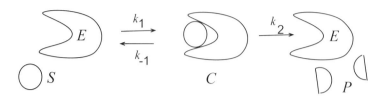

Figure 2.16 Schematic of an enzyme reaction.

reaction, and usually the initial concentration of the enzyme is small compared to the substrate. Initially, we assume $S(0) = S_0$, $E(0) = E_0$, $P(0) = 0$, and $C(0) = 0$. By the law of mass action, the rate equations are (we are using τ for real time)

$$\frac{dS}{d\tau} = -k_1 SE + k_{-1}C,$$

$$\frac{dE}{d\tau} = -k_1 SE + (k_{-1} + k_2)C,$$

$$\frac{dC}{d\tau} = k_1 SE - (k_{-1} + k_2)C,$$

$$\frac{dP}{d\tau} = k_2 C.$$

Notice that once C is determined, then P is determined by direct integration; P does not enter the first three equations. Thus, there are only three equations for S, E, and C. We can instantly find the conservation laws by observation; they are

$$E + C = E_0, \qquad S + C + P = S_0.$$

These equations allow us to eliminate E and reduce the rate equations to a system of two nonlinear equations for S and C. These equations are

$$\frac{dS}{d\tau} = -k_1 E_0 S + (k_{-1} + k_1 S)C,$$

$$\frac{dC}{d\tau} = k_1 E_0 S - (k_2 + k_{-1} + k_1 S)C.$$

Quasi-Steady State. These equations have exceedingly interesting behavior that we examine in Chapter 3. However, at present we can take the chemist's approach and gain a few insights to the behavior. The key idea is to use the fact that some reactions are very fast compared to others and therefore can equilibrate quickly, essentially being in steady state.

As we stated above, it is observed that the reaction $S + E \overset{k_1}{\rightarrow} C$, where the complex is formed, is very fast. For a typical enzyme reaction, the order of the rate constants may be

$$k_1 \approx 10^9, \quad k_{-1} \approx 10^5, \quad k_2 \approx 10^3, \tag{5.5}$$

where we are measuring concentrations in molarity M (1 M $= 6.02(10)^{23}$ molecules per liter), and time in seconds. These quantities confirm the dominant speed of complex formation. The **quasi-steady state assumption** is the assumption that the second reaction equilibrates so quickly that we can take, approximately, $dC/d\tau = 0$. Thus, in the quasi-steady state, the second equation reduces to an algebraic equation

$$C = \frac{k_1 E_0 S}{k_2 + k_{-1} + k_1 S}$$
$$= \frac{E_0 S}{K_m + S},$$

where

$$K_m = \frac{k_2 + k_{-1}}{k_1}$$

is called the Michaelis-Menten constant. We can now substitute this expression back into the S equation to obtain the simple equation

$$\frac{dS}{d\tau} = -\frac{V_m S}{K_m + S}, \quad V_m = k_2 E_0.$$

This equation for S can actually be solved implicitly (separate variables). More useful, however, the speed of the reaction, which can be measured, is defined to be the rate that the product is formed, or

$$V = \frac{dP}{d\tau} = k_2 C = \frac{V_m S}{K_m + S}.$$

This basic equation is the **Michaelis-Menten law**. Note that it is a type 2 functional response. See Fig. 2.17. Looking at the graph of V vs. S shows that V 'saturates' at the value V_m (or, $\lim V = V_m$); this is the limiting or maximum speed of the reaction. The value $S = V_m$ gives $V = \frac{1}{2}V_m$, half of the saturation value; K_m is called the half-saturation. Although many biochemists may be satisfied with this analysis, it lacks details. In Chapter 3 we give a more careful analysis of this important model using singular perturbation theory.

Example 2.34

(MATLAB code) With initial concentrations $S_0 = 10^{-3}$, $E_0 = 10^{-5}$, we invite the reader to run the following MATLAB code to obtain plots of the evolution of the Michaelis–Menten system. (Note $k_b \equiv k_{-1}$.)

```
function MichaelisMenten
global k1 kb k2 S0 E0 C0
k1=10^9; kb=10^5; k2=10^3; S0=10^(-3); E0=10^( 5); C0=0;
tspan=[0 200*10^(-7)];
```

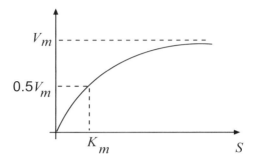

Figure 2.17 Velocity V of an enzyme reaction, which is the rate the product P is formed.

```
[T,Y] = ode15s(@kinetics,tspan,[S0 C0]);
plot(T,Y(:,1)/S0,'-',T,Y(:,2)/E0,'-.')
xlabel('time (sec)','Fontsize',14), ylabel('concentration','Fontsize',14),
ylim([0 1.1])
function dy = kinetics(t,y)
global k1 kb k2 S0 E0 C0
dy = zeros(2,1);
dy(1) = -k1*E0*y(1)+(kb+k1*y(1)).*y(2);
dy(2) = k1*E0*y(1)-(k2+kb+k1*y(1)).*y(2);
    □
```

EXERCISE

1. Sketch a generic phase plane diagram of the Michaelis-Menten system

$$\frac{dS}{d\tau} = -k_1 E_0 S + (k_{-1} + k_1 S)C,$$

$$\frac{dC}{d\tau} = k_1 E_0 S - (k_2 + k_{-1} + k_1 S)C.$$

What are the equilibria? What happens to the concentrations after a long time?

2.6 Pathogens

It takes only a small amount of reading to realize that viruses and other pathogens have been one of the great sources in human history of misery and death; disease from pathogens that enter the body more than rivals the effects of war. In this section we introduce models of viral infections, epidemics of common infections in populations, and macroparasite infections.

2.6.1 Virus Infections

A virus is a particle (*virion*) consisting of a center containing genetic material (e.g., RNA or DNA) surrounded by a protein coat. On the coating are chemical sites called *antigens*, which are macromolecules used for identification. Viruses have genetic material, but they cannot replicate by themselves. Thus, they hijack target cells in an organism and use the machinery of the cell to reproduce themselves.

As a beginning model we look at the primary phase of a viral infection; then we consider the effects on the immune system. Let V be the population of virions, X be the population of target, or healthy cells, and Y the population of infected cells. We can model the interactions using reaction kinetics and mass action as in previous sections. The model is

$$\frac{dV}{d\tau} = aY - d_v V, \tag{6.1}$$

$$\frac{dX}{d\tau} = c - d_x X - \beta XV, \tag{6.2}$$

$$\frac{dY}{d\tau} = \beta XV - d_y Y. \tag{6.3}$$

The parameters d_x, d_y, d_v are mortality rates; their inverses are average lifetimes. The parameter a is the rate that virus cells multiply, and β is the infection rate; finally, c is the rate that uninfected target cells are produced by the body.

The table shows the dimensions and ranges of the parameters. There are

parameter	a	d_v	c	d_x	d_y	β
dimensions	T^{-1}	T^{-1}	cells ml^{-1}T^{-1}	T^{-1}	T^{-1}	$(\text{cells/ml})^{-1}T^{-1}$
value range	100	5	10^5	0.1	0.5	$2 \cdot 10^{-7}$

six parameters and scaling is important. It is clear there are four possible time scales. We can scale X and Y by the virus-free steady-state value for X, namely,

$$x = \frac{X}{c/d_x}, \quad y = \frac{Y}{c/d_x}, \quad v = \frac{V}{ac/d_v d_y}.$$

We choose a dimensionless time as

$$t = \frac{\tau}{1/d_x}.$$

In dimensionless variables the equations become

$$\varepsilon \frac{dv}{dt} = \alpha y - v,$$
$$\frac{dx}{dt} = 1 - x - R_0 xv,$$
$$\frac{dy}{dt} = R_0 xv - \alpha y,$$

where

$$\alpha = \frac{d_y}{d_x}, \quad \varepsilon = \frac{d_x}{d_v}, \quad R_0 = \frac{\beta ac}{d_x d_v d_y}.$$

Using values in the table we have

$$\alpha = 5, \quad \varepsilon = 0.02, \quad R_0 = 4.$$

Note that ε is small compared to the other parameters.

In most infection models, we can identify a **basic reproduction number**, usually denoted by R_0, that is key in determining if an infection will be established or will die out. For simple epidemics, the value of R_0 is easily understood: it is the number of secondary infections caused by a single infection in the population. Hence, it is the rate of infection multiplied by the average time an infective has the disease. Here, for a virus infection, it is more complicated and we can regard R_0 as the expected number of virions that one virion gives rise to in a virus-free population. One virion infects target cells at the rate βX for a time of $1/d_v$, and each infected cell gives rise to virions at the rate a for an average time of $1/d_y$. Because $X = c/d_x$ is, on average, the number of target cells, we have

$$R_0 = \beta \frac{c}{d_x} \frac{a}{d_v d_x},$$

which is the basic reproduction number R_0. Notice that it turned up in the dimensionless formulation.

We can analyze this model in the usual way. There are two equilibria and they are easily found to be $(0, 1, 0)$ and $(1 - R_0^{-1}, R_0^{-1}, \alpha^{-1}(1 - R_0^{-1}))$. The former is the virus-free state, and the latter, a positive endemic state, exists only for $R_0 > 1$. A straightforward calculation shows that $(0, 1, 0)$ is unstable, and $(1 - R_0^{-1}, R_0^{-1}, \alpha^{-1}(1 - R_0^{-1}))$ is stable for $R_0 > 1$. We leave this as an exercise for the reader.

Observe that the parameter ε is small, and therefore we prefer to make the quasi-steady-state assumption and set $dv/dt = 0$ to get $v = \alpha y$. This reflects the fact that free virus populations turn over at a much faster rate than the populations of uninfected and infected cells. Under this assumption, the model becomes

$$\frac{dx}{dt} = 1 - x - R_0\alpha xy,$$

$$\frac{dy}{dt} = R_0\alpha xy - \alpha y.$$

This two-dimensional system is easily analyzed. The equilibria are

$$(1,0) \text{ and } \left(\frac{1}{R_0}, \frac{R_0 - 1}{\alpha R_0}\right),$$

and the nullclines are

$$y = \frac{1 - x}{R_0 \alpha x} \quad (x \text{ nullcline}) \tag{6.4}$$

$$y = 0; \; x = \frac{1}{R_0} \quad (y \text{ nullclines}) \tag{6.5}$$

There are two cases: $R_0 < 1$ and $R_0 > 1$. The two phase plane plots in Fig. 2.18 show the nullclines in each case, equilibria, and sample orbits. When $R_0 < 1$ there is only one equilibrium, located on the x axis, and it is an attractor. As R_0 increases the x nullcline crosses the vertical y nullcline at $R_0 = 1$, and the equilibrium bifurcates into two equilibria for $R_0 > 1$. The equilibrium on the x axis becomes unstable (a saddle) and the new positive equilibrium is an attractor.

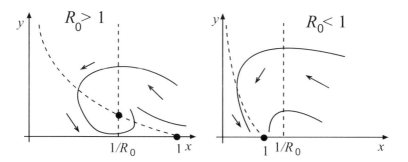

Figure 2.18 Phase plane diagrams for (6.5)–(6.5).

Now we confirm the stability of the equilibria. The Jacobian matrix is

$$J(x,y) = \begin{pmatrix} -1 - \alpha R_0 y & -\alpha R_0 x \\ \alpha R_0 y & \alpha R_0 x - \alpha \end{pmatrix}$$

At $(1,0)$,

$$J(1,0) = \begin{pmatrix} -1 & -\alpha R_0 \\ 0 & \alpha R_0 - \alpha \end{pmatrix}.$$

We have $\operatorname{tr} J = -1 + \alpha(R_0 - 1)$ and $\det = -\alpha(R_0 - 1)$. Therefore $(1,0)$ is stable for $R_0 < 1$ and unstable when $R_0 > 1$. At the other equilibrium,

$$J(1,0) = \begin{pmatrix} -R_0 & -\alpha \\ R_0 - 1 & 0 \end{pmatrix}.$$

Clearly, $\operatorname{tr} J < 0$ and $\det J = \alpha(R_0 - 1) > 0$ when $R_0 > 1$, and so the positive equilibrium is stable.

In summary, for $R_0 > 1$ the target cell population is reduced by the disease until each virion gives rise to exactly one new virion, or $x^* R_0 = 1$.

2.6.2 Immune System Response

The body's defense against a pathogen invasion is the immune system. You may have seen a speeded-up video of a dead animal going through decay; it is an impressive sight, and it shows what happens when there is no immune response—the dead animal is completely consumed. Our immune system prevents this by mounting a battle to fight invading organisms.

The immune system has three main tasks:

– To recognize the invading organism

– To mobilize weapons that intercept and kill the invader

– To remember the invader so that future infections of similar kind may be destroyed

The immune system response is by killer cells Z produced at a constant rate η with natural mortality d_z. Thus,

$$\frac{dZ}{d\tau} = \eta - d_z Z.$$

We assume they kill infected cells at rate $\rho Y Z$, so the Y equation (6.3) becomes

$$\frac{dY}{d\tau} = \beta X V - d_y Y - \delta Y Z.$$

These two equations for Y and Z couple with (6.2) and (6.3) to give the immune system model. We scale these equations as before, along with

$$z = \frac{Z}{\eta/d_z}.$$

Then we get, in summary,

$$\frac{dy}{dt} = R_0 x v - \alpha y - \rho Y Z, \tag{6.6}$$

$$\frac{dz}{dt} = \lambda(1 - z), \tag{6.7}$$

where

$$\lambda = \frac{d_z}{d_x}, \quad \rho = \frac{\delta \eta}{d_x d_z}.$$

With the addition of killer cells from the immune system response, the basic reproduction number R_0 should decrease. Motivated by its appearance in the steady state, we calculate the nonzero equilibrium for the scaled model (6.7)–(6.7). Setting the derivatives equal to zero, at equilibrium we have

$$v^* = \alpha y^*, \ z^* = 1, \ x^* = \frac{\alpha + \rho}{\alpha R_0}, \ y^* = \frac{1 - x^*}{R_0 \alpha x^*}.$$

Letting

$$R_0' = \frac{\alpha}{\alpha + \rho} R_0,$$

the positive equilibrium is

$$z^* = 1, x^* = \frac{1}{R_0'}, \ y^* = \frac{1}{\alpha + \rho}\left(1 - \frac{1}{R_0'}\right), \ v^* = \frac{\alpha}{\alpha + \rho}\left(1 - \frac{1}{R_0'}\right).$$

This shows that a positive endemic state occurs when $R_0' > 1$. Alternately, we could have shown that the other equilibrium, $v^* = 0, \ x^* = 0, \ y^* = 0, \ z^* = 1$, becomes unstable at $R_0' > 1$. Further, we observe that the killer cells can clear the virus when the immune system response rate satisfies

$$\rho > \alpha(R_0 - 1).$$

Example 2.35

(MATLAB code) We invite the reader to sketch solution profiles using the MATLAB code below along the same values of the parameters in the non-immune system response, and

$$dz = 5, \quad \eta = 75, \quad \rho = 0.10.$$

```
function killercells
global dx dy dv dz k gam beta rho eta
   dx=0.1; dy=0.5; dv=5; dz=5; k=100; gam=10^5; beta=2*10^(-7); rho=0.1;
eta=75; tend=75;
   tspan=[0 tend];
```

```
x0=gam/dx; y0=0; v0=1; z0=eta/dz; %actual initial values initial values
[T,Y] = ode15s(@killer,tspan,[x0 y0 v0 z0]);
subplot(1,2,1), plot(T,Y(:,2)/10^6,'−',T,Y(:,3)/10^6,'r')
xlabel('days','Fontsize',14), ylabel('concentrations x 10 ^ 6','Fontsize',14),
title('infected (dashed), virus (red)', 'Fontsize', 12)
subplot(1,2,2), plot(T,Y(:,1)/x0,'-',T,Y(:,4)/z0,'g')
ylim([0 1.1])
xlabel('days','Fontsize',14), ylabel('scaled cell concentrations','Fontsize',14),
title('killer (green), target (solid)', 'Fontsize', 12)
function der = killer(t,y)
global dx dy dv dz k gam beta rho eta
der = zeros(4,1);
der(1) = gam-dx*y(1)-beta*y(1)*y(3);
der(2) = beta*y(1)*y(3)-dy*y(2)-rho*y(2)*y(4);
der(3) = k*y(2)-dv*y(3);
der(4) = eta-dz*y(4);
```

□

EXERCISES

1. The simplest model of a virus is by a density-dependent logistic growth law,
$$v' = rv\left(1 - \frac{v}{T}\right) - av,$$
   where v is the virus population. The rate of growth is r, and the virus population clearly rate is a. The parameter T represents target cell limitation. Show that if $r < a$ the virus fails to establish an infection. What happens if $r > a$?

2. The simplest model of antigenic dynamics describes a virus that is opposed by a specific immune response. Let v be the population of virions and x the magnitude of the immune response. Then the model is
$$v' = av - bxv, \quad x' = cv - dx.$$
   Describe the dynamics of the model.

3. Verify the equilibria and their stability for the basic model (6.2), (6.3), and (6.3) as discussed in the text.

4. Use a numerical differential equation solver (e.g., MATLAB) to obtain a numerical solution to (6.2)–(6.3) with the values of the parameters given in the table. For initial conditions take $X(0) = 10^6$, $Y(0) = 0$, and $V(0) = 1$.

5. In an HIV infection, reverse transcriptase inhibitors are drugs that prevent the infection of new cells, or $\beta = 0$.

$$
\begin{aligned}
\frac{dX}{d\tau} &= \gamma - d_x X, \\
\frac{dY}{d\tau} &= -d_y Y, \\
\frac{dV}{d\tau} &= kY - d_v V.
\end{aligned}
$$

What are the long time dynamics of the virus?

6. Protease inhibitors for HIV are drugs that prevent infected cells from producing additional infected virions. Interpret and analyze the model

$$
\begin{aligned}
\frac{dX}{d\tau} &= \gamma - d_x X - \beta XV, \\
\frac{dY}{d\tau} &= \beta XV - d_y Y, \\
\frac{dV}{d\tau} &= -d_v V, \\
\frac{dW}{d\tau} &= kY - d_w W,
\end{aligned}
$$

where W is the population of uninfected virions produced.

The last two exercises are adapted from Britton (2003).

2.6.3 Epidemics in Populations

In the last section we presented a simple model of a virus infection and how the immune system attempts to clear it. A virus is one example of a **microparasitic** disease. Examples of viruses are common diseases like measles, chicken pox, the flu, whooping cough, and so forth. Other microparasitic diseases are bacterial based (TB) and protozoon infections such as malaria. These diseases are characterized by either having the disease, or not having it. Other infections are **macroparasitic**. That is, they are caused by multi-celled organisms, such as nematodes (schistosomiasis), tapeworms, or mites (ticks), and they have a more complex, longer life cycle in and out of their hosts. These diseases generally depend on the parasite load, or the degree of the infection.

Diseases like these can be studied on a population level in terms of susceptible, infected, and recovered individuals, and that is the topic of this section. Efforts to connect the microbiology of diseases and the population level effects is an active area of study in epidemiology.

We consider a simple epidemic model where, in a fixed population of size N, the function $I = I(t)$ represents the number of individuals that are infected with a contagious illness and $S = S(t)$ represents the number of individuals that are susceptible to the illness but not yet infected. We also introduce a removed, or recovered, class where $R = R(t)$ is the number who cannot get the illness because they have recovered permanently, are naturally immune, or have died. We assume $N = S(t) + I(t) + R(t)$, and each individual belongs to only one of the three classes. Observe that N includes the number who may have died. The evolution of the illness in the population can be described as follows. Infectives communicate the disease to susceptibles with a known infection rate; the susceptibles become infectives who have the disease a short time, recover (or die), and enter the removed class. Our goal is to set up a model that describes how the disease progresses with time. These models are called **SIR models**.

In this model we make several assumptions. First, we work in a time frame where we can ignore demographic effects such as births and immigration. We assume that the population mixes homogeneously, where all members of the population interact with one another to the same degree and each has the same risk of exposure to the disease. Think of measles, the flu, or chicken pox at an elementary school. We assume that individuals get over the disease quickly, so we are not modeling tuberculosis, AIDS, or other long-lasting or permanent diseases. More complicated models can be developed to account for these factors and all sorts of others, such as vaccination, sex, disease carriers (or vectors), the possibility of reinfection, and so on.

The disease spreads when a susceptible comes in contact with an infective. A reasonable measure of the number of contacts between susceptibles and infectives is $S(t)I(t)$. For example, if there are five infectives and twenty susceptibles, then one hundred contacts are possible. However, not every contact results in an infection. We use the letter a to denote the **transmission coefficient**, or the fraction of those contacts that usually result in infection. For example, a could be 0.02. The parameter a is the product of two effects, the fraction of the total possible number of encounters that occur, and the fraction of those that results in infection. The constant a has dimensions time^{-1} per individual. The quantity $aS(t)I(t)$ is the infection rate, or the rate at which members of the susceptible class become infected. Observe that this model is the same as the law of mass action in chemistry where the rate of chemical reaction between two reactants is proportional to the product of their concentrations. Therefore, if no other processes are included, we would have

$$\frac{dS}{dt} = -aSI, \quad \frac{dI}{dt} = aSI.$$

Remark 2.36

Mass action kinetics is only one possibility for selecting the effective encounter rate. More generally, we could model the infection process by

$$\frac{1}{S}\frac{dS}{dt} = -f(I),$$

where $f(I)$ is the force of infection. For example, f could be a saturating function of I, leveling off to a constant value as I gets large. For mass action kinetics, $f(I) = aI$.   □

However, as individuals get over the disease, they become part of the re-moved class R. The **recovery rate** r is the rate that infected individuals get over the disease; thus, the rate of removal is $rI(t)$. Importantly, the parameter r is measured in time^{-1} and $1/r$ can be interpreted as the **average time to recover**. Therefore, using the prime notation for derivative, we modify the last set of equations to obtain

$$S' = -aSI, \qquad (6.8)$$
$$I' = aSI - rI, \qquad (6.9)$$
$$R' = rI. \qquad (6.10)$$

This is the SIR model. We do not need equation (6.10) for R because R can be determined directly from $R = N - S - I$. At time $t = 0$ we assume there are I_0 infectives and S_0 susceptibles, but no one yet removed. Thus, initial conditions are

$$S(0) = S_0, \quad I(0) = I_0, \quad R(0) = 0, \qquad (6.11)$$

and $S_0 + I_0 = N$. SIR models are compartmental models and are commonly dia-grammed as boxes labeled S, I, and R with arrows indicating the rates at which individuals progress from one compartment to the other. An arrow entering a compartment represents a positive rate and an arrow leaving a compartment represents a negative rate.

Remark 2.37

Notice that this model can be deduced from simple chemical reaction kinetics

$$S + I \xrightarrow{a} 2I, \quad I \xrightarrow{r} R.$$

Other disease models can also be regarded as reaction kinetics as well.   □

As was the case with a virus, we can identify a priori a **basic reproduction number** that tells us if the disease will take hold or die out. If we introduce a single infective in a population N, then the rate that this individual infects others is approximately aN; the average time that this infected person has the disease is $1/r$. Thus, the expected number of individuals this person infects is

$$R_0 = aN\frac{1}{r}.$$

Hence, R_0 is the expected number of secondary infections produced by a single infective.

A qualitative analysis helps us understand how an SIR epidemic evolves. To carry out this analysis, we work in dimensionless variables. Instead of using the number of individual in each class, we use the fraction of individuals in each class. And we scale time by r^{-1}, the average time infected. Therefore, let

$$x = \frac{S}{N}, \quad yIN, \quad z = \frac{R}{N}, \quad \tau = \frac{t}{r^{-1}}.$$

The equations easily become

$$x' = -R_0 xy, \quad y' = R_0 xy - y, \quad z' = y. \tag{6.12}$$

Further, $x + y + z = 1$ and $x(0) + y(0) = 1$. The dynamics take place in a simplex in the xyz plane: x, y, $z \geq 0$, $x + y + z \leq 1$. In the xy plane, where we work, this is the triangle $x + y \leq 1$ in the first quadrant. The initial condition $x(0) = x_0$, $y(0) = y_0$ lies on the line $x + y = 1$. We track an orbit $x = x(t)$, $y = y(t)$ starting on the line $x + y = 1$. Note that x' is always negative so an orbit always moves to the left, decreasing x. Because $x' = y(R_0 x - 1)$, we see that infectives increase if $x > 1/R_0$, and infectives decrease if $x < 1/R_0$. The vertical line $x = 1/R_0$ is the y-nullcline. This information gives us the direction field. There are two cases, when $R_0 > 1$ and $R_0 < 1$, or when the y nullcline is to the left of $x = 1$ or to the right of $x = 1$. We show the case $R_0 > 1$ in Fig. 2.19. (The other case is an exercise for the reader.) If the initial condition is at point P, the orbit goes directly down and to the left until it hits $y = 0$, and the disease dies out. If the initial condition is at point Q, then the orbit increases to the left, reaching a maximum at $x = 1/R_0$. Then it decreases to the left and ends at a point x^* on $y = 0$.

We can obtain an equation for the orbits of (6.12) in a simple manner. If we divide the x and y equations in (6.12) we obtain

$$\frac{y'}{x'} = \frac{dy/d\tau}{dx/d\tau} = \frac{dy}{dx} = \frac{R_0 xy - y}{-R_0 xy} = -1 + \frac{1}{R_0 x}.$$

Integrating both sides with respect to S yields

$$y = -x + \frac{1}{R_0}\ln x + C,$$

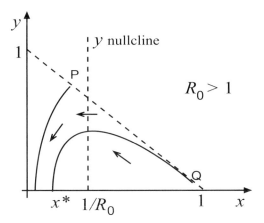

Figure 2.19 The xy phase plane showing two orbits in the case $R_0 > 1$. One starts at P and and one starts at Q, on the line $x + y = 1$. The second shows an epidemic where the number of infectives increases to a maximum value and then decreases to zero; x^* represents the number of individuals that do not get the disease.

where C is an arbitrary constant. Using the initial conditions to determine C, we obtain the orbit equation

$$y = 1 - x + \frac{1}{R_0} \ln \frac{x}{x_0}.$$

This orbit can be graphed with a calculator or computer algebra system, once parameter values are specified. Making such plots shows what the orbits looks like, as plotted in Fig. 2.19. Observe that an orbit cannot intersect the y axis, so it must intersect the x axis for $x > 0$, or at the root x^* of the nonlinear equation

$$-x + 1 + \frac{1}{R_0} \ln \frac{x}{x_0} = 0.$$

This root represents the fraction of individuals that do not get the disease. Once parameter values are specified, a numerical approximation of x^* can be obtained. In all cases, the disease dies out because of lack of infectives.

Finally, we can obtain an implicit differential equation for $z = z(\tau)$ by dividing the x and z equations in (6.12). We get

$$\frac{dx}{dz} = \frac{-R_0 xy}{y} = -R_0 x.$$

Therefore,

$$x = x_0 e^{-R_0 z},$$

and, again using (6.12),

$$\frac{dz}{d\tau} = 1 - x - x_0 e^{-R_0 z}. \tag{6.13}$$

This equation cannot be integrated in closed form, but we can obtain approximations in special cases.

EXERCISES

1. (SIS epidemic) In an SIS epidemic susceptibles get infected, and then infectives recover to become susceptible again. Using r and a as the recovery and infection rates, as in the SIR model, formulate equations for an SIS epidemic and find the basic reproduction number R_0. Assume a constant population N. Show that the disease dies out if $R_0 < 1$ and the disease is endemic, or continually present, if $R_0 > 1$. Discuss the dynamics in each case.

2. In a population of 200 individuals, 20 were initially infected with an influenza virus. After the flu ran its course, it was found that 100 individuals did not contract the flu. If it took about 3 days to recover, what was the transmission coefficient a? What was the average time that it might have taken for someone to get the flu?

3. In a population of 500 people, 25 have the contagious illness. On the average it takes about 2 days to contract the illness and 4 days to recover. How many in the population will not get the illness? What is the maximum number of infectives at a single time?

4. Beginning with the SIR model, assume that susceptible individuals are vaccinated at a constant rate ν. Formulate the model equations and describe the progress of the disease if, initially, there are a small number of infectives in a large population.

5. (SIRS disease) Beginning with the SIR model, assume that recovered individuals can lose their immunity and become susceptible again after an average recovery period of time μ. That is, the rate recovered individuals become susceptible is μR. Draw a compartmental diagram and formulate a two-dimensional system of model equations for S and I. Find the two equilibria. By sketching the nullclines and vector field, show that the disease-free equilibrium is unstable. Can you identify the type of equilibrium. Can you determine whether the nonzero equilibrium (the endemic state) is stable or unstable? What does it appear to be?

6. In an SIR epidemic with $R_0 > 1$, where $R_0 z$ is small for all time, use (6.13) to show, approximately,

$$\frac{dz}{d\tau} = (R_0 - 1)z \left(1 - \frac{z}{2(R_0 - 1)R_0^2} \right).$$

Note that this is the logistic equation. Sketch a graph of the rate of removal $dz/d\tau$ vs. τ. For example, in the plague, the removal rate closely approximates the death rate.

7. (SIR with Demographics) We can include demographics in the SIR model by assuming individuals are born into the susceptible class at the constant rate b and all classes die at the per capita rate μ. The model is

$$S' = b - aSI - \mu S, \quad I' = aSI - (r + \mu)I, \quad R' = rI - \mu I.$$

Assume the population is a constant N, and μ is much smaller that r.

   a) What constraint does the assumption of a constant population have on the parameters in the system?

   b) Nondimensionalize the model by taking $s = S/N$, $y = I/N$, $z = R/N$, and using the time scale r^{-1}.

   c) Working in the xy plane, find the nullclines and the equilibria and their stability. Consider all cases and write a few sentences discussing the results.

2.6.4 Macroparasitic Infections

Some diseases are not contagious or spread by contact, such as in the flu, measles, or sexually transmitted diseases. For example, in the case of malaria a mosquito is a carrier, or **vector**, in the transmission of the disease to different individuals. Malaria affects more individuals than any other disease, especially in tropical areas. Other vector diseases transmitted by mosquitos include West Nile virus, dengue and yellow fever, and filariasis.

The malaria carrier is the female *Anopheles* mosquito. The infectious agent is a protozoan parasite that is injected into the blood stream by a mosquito when she is taking a blood meal, which is necessary for the development of her eggs. The parasite develops inside the host and produces gametocytes which then can be taken up by another biting mosquito.

We present a simplified, classic model of R. Ross, who developed it in 1911, and which was modified by G. Macdonald in 1957. Sir Ronald Ross is given credit for first understanding and modeling the complex malarial cycle, for which he was awarded the Nobel Prize. (See R. M. Anderson & R. M. May, (1991), the standard reference for diseases, both micro- and macroparasitic.)

We assume that human victims have no immune system response and that they eventually recover from the disease without dying. We assume the mosquito and the human populations are approximately constant. Thus, the disease dynamics are fast compared to the dynamics of either hosts or mosquitos. Let H_T and M_T be the total number of hosts (humans) and total number of mosquitos, respectively, in a fixed region; both are assumed to be constant. Further, let

$$H(t) = \text{ number of infected hosts (humans)}$$
$$M(t) = \text{ number of infected mosquitos}$$

First we consider the hosts. The rate that a human gets infected depends on the number of mosquitos, the biting rate a (bites per time), and b, the fraction of bites that lead to an infection of a human, and the probability of the mosquito encountering a susceptible human. The fraction of susceptible humans is $(H_T - H)/H_T$. Finally, we assume that the per capita recovery rate of infected humans is r, where $1/r$ is the average time to recovery. Therefore, the rate equation for H is

$$\frac{dH}{dt} = abM\frac{H_T - H}{H_T} - rH,$$

which is the infection rate minus the recovery rate. Notice that the infection rate is proportional to the product of susceptible hosts and infected mosquitos, which should remind the reader of the simple SIR model studied earlier. The rate that mosquitos become infected from biting an infected host depends on a and c (the fraction of bites by an uninfected mosquito of an infected human that causes infection in the mosquito). If μ is the per capita death rate of infected mosquitos, then

$$\frac{dM}{dt} = ac(M_T - M)\frac{H}{H_T} - \mu M,$$

and H/H_T is the probability of encountering an infected human. Note that the infection rate is jointly proportional to $M_T - M$, the number of susceptible mosquitos, and the number of infected hosts, again a reminder of mass action kinetics. We can simplify these equations by introducing

$$h = \frac{H}{H_T}, \qquad m = \frac{M}{M_T},$$

which are the fractions of the populations that are infected. Then the governing equations become

$$\frac{dh}{dt} = ab \left(\frac{M_T}{H_T} \right) m(1 - h) - rh, \qquad (6.14)$$

$$\frac{dm}{dt} = ach(1 - m) - \mu m. \qquad (6.15)$$

For convenience, we define the parameters

$$\alpha = ab \left(\frac{M_T}{H_T} \right), \qquad \beta = ac.$$

Then

$$\frac{dh}{dt} = \alpha m(1 - h) - rh, \qquad (6.16)$$

$$\frac{dm}{dt} = \beta h(1 - m) - \mu m. \qquad (6.17)$$

We can analyze this geometrically in the phase plane in the usual way. Setting the right sides equal to zero gives the nullclines

$$m = \frac{rh}{\alpha(1 - h)}, \qquad (h \text{ nullcline}) \qquad (6.18)$$

$$m = \frac{\beta h}{\mu + \beta h}, \qquad (m \text{ nullcline}) \qquad (6.19)$$

Note that $h = m = 0$ is always an equilibrium. Also, the h nullcline is concave up with a vertical asymptote at $h = 1$; the m nullcline is concave down with a horizontal asymptote at $m = 1$. The two possibilities are shown in Fig. 2.20

There will be a nonzero equilibrium when these nullclines cross. In that case, the slope of the m nullcline must be steeper than the slope of the h nullcline at $h = 0$. Calculating these slopes from (6.18)–(6.19), respectively, we get

$$m'(0) = \frac{r}{\alpha}, \qquad (h \text{ nullcline})$$

$$m'(0) = \frac{\beta}{\mu}. \qquad (m \text{ nullcline})$$

Therefore, for a nonzero equilibrium, we must have

$$\frac{\beta}{\mu} > \frac{r}{\alpha}. \qquad (6.20)$$

We show that this nonzero equilibrium is asymptotically stable, which means the infectious populations approach a nonzero endemic state. First, however, let's interpret this result (6.20) in terms of the actual parameter values. We can rewrite (6.20) as

$$\frac{ac}{\mu} \frac{ab \frac{M_T}{H_T}}{r} > 1.$$

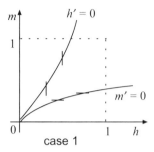
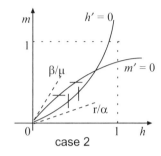

Figure 2.20 Two cases in the malaria model: the nullclines cross only at the origin, and the nullclines cross at the origin and at a nonzero state. The second case occurs only when the slope of the mosquito nullcline exceeds the slope of the host nullcline at the origin, or $\beta/\mu > r/\alpha$.

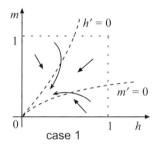
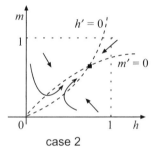

Figure 2.21 Orbits in the two cases. In case 1 the origin is a stable node and the infection dies out. In case 2, the origin is unstable and the disease becomes endemic.

The first factor is the rate of infection of mosquitos (ac) times their average lifetime $(1/\mu)$. The second factor is the rate of infection of human hosts (abM_T/H_T) times the average length of infection $(1/r)$.

We can easily check the stability of the nonzero equilibrium by sketching the direction field. Or, we can approach this analytically by finding the equilibrium and checking the Jacobian matrix. Setting (6.18) equal to (6.19) and solving for h gives

$$h^* = \frac{\alpha\beta - \mu r}{\beta(r + \alpha)}.$$

Then,

$$m^* = \frac{\alpha\beta - \mu r}{\alpha(\mu + \beta)}.$$

Notice that this is a viable equilibrium only if the numerator is positive, which is the same as the condition (6.20). Otherwise it is not viable and the origin, $(0,0)$, is the only equilibrium. The Jacobian matrix at an arbitrary (h, m) is easily

$$J(h, m) = \begin{pmatrix} -\alpha m - r & \alpha(1 - h) \\ \beta(1 - m) & -\beta h - \mu \end{pmatrix}.$$

Clearly

$$J(0,0) = \begin{pmatrix} -r & \alpha \\ \beta & -\mu \end{pmatrix}.$$

The trace is negative in both cases. The determinant $\mu r - \alpha\beta$ is positive when condition (6.20) holds and negative when it does not hold. Therefore, the origin (extinction of the disease) is asymptotically stable when it is the only equilibrium, and it is unstable when a nonzero equilibrium exists.

For the nonzero equilibrium

$$J(h^*, m^*) = \begin{pmatrix} -\alpha m^* - r & \alpha(1 - h^*) \\ \beta(1 - m^*) & -\beta h^* - \mu \end{pmatrix}.$$

The trace is negative and

$$\det J(h^*, m^*) = (\alpha m^* + r)(\beta h^* + \mu) - \alpha\beta(1 - m^*)(1 - h^*)$$
$$= \alpha\beta - \mu r > 0,$$

by condition (6.20), and after considerable simplification. Thus, (h^*, m^*) is asymptotically stable.

EXERCISES

1. Using data in the table below, calculate solution curves for the malaria model. Demonstrate both cases (extinction and coexistence).

Parameter	Name	Sample Value
M_T/H_T	population ratio	2
a	biting rate	0.2–0.5 per day
b	effective bites infecting humans	0.5
c	effective bites infecting mosquitos	0.5
r	recovery rate	0.01–0.05 per day
μ	mortality rate	0.05–0.5 per day

Table 2.1 Sample malaria parameter values

2. (Malaria) In this exercise develop and analyze a simplified version of the malaria model under the condition that r is much less than μ.

a) Beginning with (6.14)–(6.15), nondimensionalize these equations by rescaling time by taking $\tau = \mu t$. Obtain

$$\frac{dh}{d\tau} = \lambda m(1 - h) - \varepsilon h,$$

$$\frac{dm}{d\tau} = \eta h(1 - m) - m,$$

where

$$\varepsilon = \frac{r}{\mu}, \quad \lambda = \frac{ab}{\mu}\frac{M_T}{H_T}, \quad \eta = \frac{ac}{\mu}.$$

b) Assuming ε is very small, neglect the εh term in the host equation and draw the phase portrait. Include the equilibria, nullclines, direction field, and a local stability analysis for the equilibria.

c) For the simplified dimensionless model in part (b), with the values given in Table 1, specifically, $a = 0.5$, $r = 0.01$, and $\mu = 0.5$, use a numerical method to draw time series plots of h and m for various initial initial conditions.

3. (Schistosomiasis) Schistosomiasis is a macroparasitic disease of humans caused by trematode worms, or blood flukes. Trematodes form a class of flatworms in the phylum *Platyhelminthes*, or helminths. Schistosomiasis is highly prevalent in tropical areas, and it is estimated that hundreds of millions of people suffer from it. The life cycle of the parasite is complicated and involves a definitive host (humans), where maturity and reproduction occur, and a secondary host (e.g., snails), in which the intermediate larval stage develop into infectious larva (cercaria) that are shed and then penetrate, or are ingested, by the definitive host, completing the cycle. Figure 2.22 is a diagrammatic flow chart summarizing the principle processes. In this exercise we formulate a simplified model for the number of infected snails I and the average worm burden m in the host (the total number of mature worms divided by the constant number of hosts).

The dynamics for parasites in a host is

$$\frac{dm}{dt} = -\mu m + a\frac{I}{N}, \tag{6.21}$$

where a is the rate that infected snails produce the free-living stage larva that infects the host through ingestion or skin penetration. The factor I/N is the fraction of snails infected, and μ is the per capita mortality rate. The dynamics for the number of infected snails is

$$\frac{dI}{dt} = -\delta I + C(m)(N - I), \tag{6.22}$$

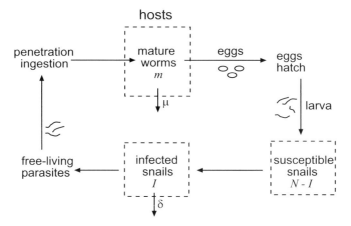

hosts

penetration
ingestion → mature worms m

eggs → eggs hatch

larva

free-living parasites ← infected snails I ← susceptible snails $N-I$

μ

δ

Figure 2.22   Diagrammatic life cycle of schistosome parasites in humans.

where δ represents the per capita mortality rate of the infected snails, $N-I$, is the number of susceptible snails, and $C(m)$ is proportional to the rate of production of eggs by (*paired*) female adult worms; the latter includes the rate of hatching of the eggs that eventually produce the infecting, free-living larva. Thus, C contains several rates in the life cycle and perhaps complicated dependence on the fraction of paired females. The function $C(m)$ is a type 3 functional response (a sigmoid or S-shaped curve).

a) Find the nullclines and sketch them in the mI plane. In terms of the bifurcation parameter a, show three cases: $a < a^*$, $a = a^*$, $a > a^*$, for some critical value a^* to be determined.

b) For each case in (a), indicate graphically the equilibria and the direction field. Can you conclude anything about stability?

c) Find an equation, or equations, that determine a^*.

d) Sketch a bifurcation diagram of the equilibria m^* vs. the parameter a. Indicate the stability of each branch.

e) Interpret the results in a biological sense.

4. Gregarines (Ph. Apicomplexa) are parasites that infect various insects, such as flour beetles, damselflies, and dragonflies. They are among the most evolutionary long-lived, widespread parasite species. In a freshwater environment, for example, the adult stage of the parasite (the trophont) infects the intestine of the larval stage of the host insect species; adults pair and produce a gametocyst, which is shed back into the water environment. The gametocyst undergoes gametogenesis and eventually releases a large num-

ber of infective oocysts that are ingested by the insects, and the life cycle proceeds again. Assuming a constant number of hosts and letting P and C be the average number (per host values) of adults and oocysts, respectively, the governing dynamics for the parasite life cycle is

$$\frac{dC}{dt} = bP - dC - \alpha \frac{C^2}{\gamma^2 + C^2}, \quad \frac{dP}{dt} = -\mu P + \alpha \frac{C^2}{\gamma^2 + C^2},$$

where d and μ are mortalities, b is the production rate of adults, and α and γ are parameters in a type 3 contact rate. (Details can be found in J. D. Logan, J. Janovy, & B. E. Bunker, 2012. *Ecological Modelling* 233, 31–40.)

a) In terms of the *fitness* parameter R defined by

$$R = \frac{\alpha}{2\gamma d} \left(\frac{b}{\mu} - 1 \right),$$

show that the equilibria are given by $P = C = 0$ (extinction) and

$$C = \gamma \left(R \pm \sqrt{R^2 - 1} \right), \quad P = \frac{\alpha}{\mu} \frac{C^2}{\gamma^2 + C^2}.$$

b) Sketch phase plane diagrams for $R < 1$, $R = 1$, and $R > 1$.

c) Describe the bifurcation and sketch a bifurcation diagrams of the equilibrium C vs. the parameter R.

REFERENCES AND NOTES

An elementary account of systems of differential equations in the spirit of this book can be found in the author's text, Logan (2010); most elementary texts have a good treatment of systems. At the next level we refer to notable treatment by Strogatz (1994), which discusses applications to physics, biology, engineering, and chemistry. A readable theoretical text is Kelley and Peterson (2010).

Models of epidemiology and chemical kinetics can be found in many books. Chemical engineering texts are especially good for reactions. Epidemiology is an extensive area of study. The tome by Anderson and May (1989) is a definitive treatment. Additional discussions, less extensive, can be found in Edelstein-Keshet (1987), Brauer and Castillo-Chavez (2001), Murray (2002), Britton (2003), Nowak (2006), Wodarz (2007), Brauer et al. (2008), and Logan and Wolesensky (2009).

Brauer, F. & Castillo-Chavez, C. 2001. *Mathematical Models in Population Biology and Epidemiology*, Springer, New York.

Brauer, F., van den Driessche, P., & Wu, J., eds. 2008. *Mathematical Epidemiology*, Lecture Notes in Mathematics, Vol. 1945, Springer, New York.

Britton, N. J. 2003. *Essential Mathematical Biology*, Springer-Verlag, New York.

Edelstein-Keshet, L. 1987. *Mathematical Models in Biology*, Random House, New York (reprinted by SIAM, Philadelphia 2000).

Kelley, W. G. & Peterson, A. C. 2010. *The Theory of Differential Equations*, 2nd ed., Springer, NY.

Logan, J. D. 2010. *A First Course in Differential Equations*, 2nd ed., Springer, NY.

Logan, J. D. & Wolesensky, W. R. 2009. *Mathematical Methods in Biology*, John Wiley & Sons, New York.

Murray, J. D. 2002. *Mathematical Biology*, 3rd ed., Springer-Verlag, New York.

Nowak, M. A. 2006. *Evolutionary Dynamics*, Belnap Press of Harvard University, Cambridge, MA.

Strogatz, S. H. 1994. *Nonlinear Dynamics and Chaos*, Addison-Wesley, Reading, MA.

Wodarz, D. 2007. *Killer Cell Dynamics*, Springer-Verlag, New York.

3

Perturbation Methods and Asymptotic Expansions

Equations arising from mathematical models usually cannot be solved in exact form. Therefore, we often resort to approximation and numerical methods. Foremost among approximation techniques are perturbation methods. Perturbation methods lead to an approximate solution to a problem when the model equations have terms that are small. These terms arise because the underlying physical process has small effects. For example, in a fluid flow problem the viscosity may be small compared to advection, or in the motion of a projectile the force caused by air resistance may be small compared to gravity. These low-order effects are represented by terms in the model equations that, when compared to the other terms, are negligible. When scaled properly, the order of magnitude of these terms is represented by a small coefficient parameter, say ε. By a perturbation solution we mean an approximate solution that is the first few terms of a Taylor-like expansion in the small parameter ε.

Perturbation methods apply to differential equations (ordinary and partial), algebraic equations, integral equations, and all of the other types of equations we encounter in applied mathematics. Their importance cannot be overstated.

Applied Mathematics 4th ed.,
By J. David Logan
Copyright © 2013 John Wiley & Sons, Inc.

3.1 Regular Perturbation

To fix the idea, consider a differential equation

$$F(t, y, y', y'', \varepsilon) = 0, \quad t \in I, \tag{1.1}$$

where t is the independent variable, I is the time interval, and y is the dependent variable. The appearance of a small parameter ε is shown explicitly. In general, initial or boundary conditions may accompany the equation, but for the present we ignore auxiliary conditions. To denote that ε is a small parameter we write

$$\varepsilon \ll 1.$$

This means that ε is *small compared to unity*, a glib phrase in which no specific cutoff is explicitly defined; certainly $0.001 \ll 1$, but 0.75 is not small compared to unity. The previous remarks also apply to equations with a large parameter λ, since in that case we may introduce $\varepsilon = 1/\lambda$, which is small.

By a **perturbation series** we understand a power series in ε of the form

$$y_0(t) + \varepsilon y_1(t) + \varepsilon^2 y_2(t) + \cdots. \tag{1.2}$$

The basis of the **regular perturbation** method is to assume a solution of the differential equation of this form, where the functions $y_0, y_1, y_2, \ldots$ are found by substitution into the differential equation. The first few terms of such a series form an approximate solution, called a **perturbation solution** or **approximation**; usually no more than two or three terms are taken. Generally, the method will be successful if the approximation is uniform; that is, the difference between the approximate solution and the exact solution converges to zero at some well-defined rate as ε approaches zero, uniformly on I. These ideas are made precise later. We emphasize in this context that there is no advantage in taking *a priori* a specific value for ε. Rather, we consider ε to be an arbitrary small number, so that the analysis will be valid for any choice of ε. Of particular interest in many problems is the behavior of the solutions as $\varepsilon \to 0$. Generally, the perturbation series is a valid representation for $\varepsilon < \varepsilon_0$ for some undetermined ε_0.

The term y_0 in perturbation series is called the *leading order* term. The terms $\varepsilon y_1, \varepsilon^2 y_2, \ldots$ are regarded as higher-order correction terms that are expected to be small. If the method is successful, y_0 will be the solution of the **unperturbed problem**

$$F(t, y, y', y'', 0) = 0, \quad t \in I,$$

in which ε is set to zero. In this context (1.1) is called the **perturbed problem**. Therefore, when a model equation is obtained in which a small parameter appears, it is often regarded as a perturbed equation where the term(s) containing the small parameter represent small perturbations or changes from some basic unperturbed problem. It is implicit that the unperturbed equation should always be solvable so that the leading-order behavior of the solution can be found. One frequently encounters problems of this sort whose solutions are facilitated by the fact that the governing equation, the boundary condition, or the shape of the region in higher-dimensional problems is not too different from those of a simpler problem.

For a variety of reasons, sometimes the naive perturbation approach described above fails. This failure often occurs when the leading order problem is not well posed, or its solution is invalid on all parts of the domain, for example, when there are multiple time or spatial scales. Then we must modify the approach. These modifications, which are myriad, belong to the class of **singular perturbation** methods.

In preparation for the discussion in this chapter and the exercises, we list the Taylor series expansions of several common functions, along with the interval of convergence, where restricted.

$$f(x) = f(a) + f'(a)(x-a) + \frac{1}{2!}f''(a)(x-a)^2 + \frac{1}{3!}f'''(a)(x-a)^3 + \cdots$$

$$e^x = 1 + x + \frac{1}{2!}x^2 + \frac{1}{3!}x^3 + \cdots$$

$$\sin x = x - \frac{1}{3!}x^3 + \frac{1}{5!}x^5 - \frac{1}{7!}x^7 + \cdots$$

$$\cos x = 1 - \frac{1}{2!}x^2 + \frac{1}{4!}x^4 - \frac{1}{6!}x^6 + \cdots$$

$$(1+x)^p = 1 + px + \frac{p(p-1)}{2!}x^2 + \frac{p(p-1)(p-2)}{3!}x^3 + \cdots, \quad |x| < 1$$

$$\ln(1+x) = x - \frac{1}{2}x^2 + \frac{1}{3}x^3 - \frac{1}{4}x^4 + \cdots, \quad |x| < 1$$

$$(a+x)^p = a^p + pa^{p-1}x + \frac{p(p-1)}{2!}a^{p-2}x^2 + \frac{p(p-1)(p-2)}{3!}a^{p-3}x^3 + \cdots,$$
$$\left|\frac{x}{a}\right| < 1$$

The basic idea of the perturbation method is easily illustrated by investigating a simple algebraic equation for which we know the solution.

Example 3.1

Consider the quadratic equation

$$x^2 + 2\varepsilon x - 3 = 0,$$

where ε is a small positive parameter. We are not assigning a value to the parameter; all we know is that it is small in some sense. Of course we can solve this equation exactly. But here we illustrate a different method. Assume a solution in the form of a perturbation series $x = x_0 + x_1\varepsilon + x_2\varepsilon^2 + \cdots$, where the dots represent higher-order terms and the x_k are to be determined. Substituting into the equation gives

$$(x_0 + x_1\varepsilon + x_2\varepsilon^2 + \cdots)^2 + 2\varepsilon(x_0 + x_1\varepsilon + x_2\varepsilon^2 + \cdots) - 3 = 0.$$

Expanding out and collecting terms in powers of ε gives

$$x_0^2 - 3 + 2x_0(x_1 + 1)\varepsilon + (x_1^2 + 2x_0x_2 + 2x_1)\varepsilon^2 + \cdots = 0.$$

Equating the coefficients to zero yields

$$x_0^2 = 3, \ x_1 = -1, \ x_1^2 + 2x_0x_2 + 2x_1 = 0, \dots$$

Solving,

$$x_0 = \pm\sqrt{3}, \ x_1 = -1, \ x_2 = \pm\frac{1}{2\sqrt{3}}, \dots$$

Therefore, we obtain two approximate solutions

$$
\begin{aligned}
x &= \sqrt{3} - \varepsilon + \frac{1}{2\sqrt{3}}\varepsilon^2 + \cdots, \\
x &= -\sqrt{3} - \varepsilon - \frac{1}{2\sqrt{3}}\varepsilon^2 + \cdots.
\end{aligned}
$$

Each is a three term perturbation approximation. We can calculate as many terms as we wish, and often computer algebra systems can be quite useful in performing all the tedious algebra. The approximate solution can be compared to the exact solution obtained by the quadratic formula. We have

$$x = \frac{1}{2}\left(-2\varepsilon \pm \sqrt{4\varepsilon^2 + 12}\right) = -\varepsilon \pm \sqrt{3 + \varepsilon^2}.$$

The radical can be expanded by the binomial formula to get

$$\sqrt{3 + \varepsilon^2} = \sqrt{3}\left(1 + \frac{\varepsilon^2}{3}\right)^{1/2} = \sqrt{3}\left(1 + \frac{\varepsilon^2}{6} + \cdots\right).$$

Therefore, the quadratic formula leads to

$$x = \pm\sqrt{3} - \varepsilon \pm \frac{1}{2\sqrt{3}}\varepsilon^2 + \cdots,$$

which is the same as the three-term perturbation approximation. The point is that we can carry out this procedure to obtain approximate solutions even when we cannot solve a problem exactly. $\square$

3.1.1 Motion in a Resistive Medium

Next consider a differential equation. Suppose a body of mass m, initially with velocity V_0, moves in a straight line in a medium that offers a resistive force of magnitude $av - bv^2$, where $v = v(\tau)$ is the velocity of the object as a function of time τ, and a and b are positive constants with $b \ll a$ (meaning b is much smaller than a). Therefore, the nonlinear part of the force is assumed to be small compared to the linear part. The constants a and b have units of force per velocity and force per velocity-squared, respectively. By Newton's second law, the equation of motion is

$$m\frac{dv}{d\tau} = -av + bv^2, \quad v(0) = V_0.$$

We first non-dimensionalize the problem. Later we see why we must do this. A velocity scale is the maximum velocity, which is clearly V_0, because the object must slow down in the resistive medium. If the small nonlinear term bv^2 were not present, then the velocity would decay like $e^{-a\tau/m}$; therefore, the characteristic time is m/a. Introducing dimensionless variables

$$y = \frac{v}{V_0}, \quad t = \frac{\tau}{m/a},$$

the problem becomes

$$\frac{dy}{dt} = -y + \varepsilon y^2, \quad t > 0, \tag{1.3}$$

$$y(0) = 1, \tag{1.4}$$

where

$$\varepsilon \equiv \frac{bV_0}{a} \ll 1.$$

Here $\varepsilon \ll 1$ means that the dimensionless parameter ε is small compared to unity. We do not specify exactly what that means; certainly $\varepsilon = 0.01$ satisfies the condition, but $\varepsilon = 0.8$ probably does not. But the relationship defines precisely what we mean when we say glibly that $b \ll a$ (a and b do not have the same dimensions and really cannot be compared); we mean bV_0 is much smaller than a.

Equation (1.3) is a Bernoulli equation and can be solved exactly by making the substitution $w = y^{-1}$, and then integrating the resulting linear equation to obtain

$$y_{\text{ex}} = \frac{e^{-t}}{1 + \varepsilon(e^{-t} - 1)}.$$

Equation (1.3) is a slightly altered or perturbed form of the linear equation

$$\frac{dy}{dt} = -y, \quad y(0) = 1, \tag{1.5}$$

which is easily solved by $y = e^{-t}$. Because the nonlinear term εy^2 is small, the function e^{-t} appears to be a good approximate solution to the problem, provided the scaling was performed correctly. The exact solution can be expanded in a Taylor series in powers of ε as

$$y_{ex} = e^{-t} + \varepsilon(e^{-t} - e^{-2t}) + \varepsilon^2(e^{-t} - 2e^{-2t} + e^{-3t}) + \cdots. \tag{1.6}$$

If ε is small, the leading order term $y_0 = e^{-t}$ is a reasonable approximation, but it does not include any effects of the nonlinear term in the original equation.

To obtain an approximate solution by a perturbation method we employ a perturbation series. That is, we assume that the solution of (1.3) is representable as

$$y = y_0(t) + \varepsilon y_1(t) + \varepsilon^2 y_2(t) + \cdots, \tag{1.7}$$

which is a series in powers of ε. The functions $y_0, y_1, y_2, \ldots$ are determined by substituting (1.7) into both the differential equation (1.3) and initial condition, and then equating coefficients of like powers of ε. Thus,

$$y_0' + \varepsilon y_1' + \varepsilon^2 y_2' + \cdots = -(y_0 + \varepsilon y_1 + \varepsilon^2 y_2 + \cdots) + \varepsilon(y_0 + \varepsilon y_1 + \varepsilon^2 y_2 + \cdots)^2,$$

which, when coefficients are collected, gives a sequence of linear differential equations

$$\begin{aligned}
y_0' &= -y_0, \\
y_1' &= -y_1 + y_0^2, \\
y_2' &= -y_2 + 2y_0 y_1, \ldots.
\end{aligned}$$

The initial condition leads to

$$y_0(0) + \varepsilon y_1(0) + \varepsilon^2 y_2(0) + \cdots = 1,$$

or the sequence of initial conditions

$$y_0(0) = 1, \quad y_1(0) = y_2(0) = \cdots = 0.$$

Therefore we have obtained a recursive set of linear initial value problems for $y_0, y_1, y_2, \ldots$. These are easily solved in sequence to obtain

$$\begin{aligned}
y_0 &= e^{-t}, \\
y_1 &= e^{-t} - e^{-2t}, \\
y_2 &= e^{-t} - 2e^{-t} + e^{-3t}, \ldots.
\end{aligned}$$

Notice that y_1 and y_2 are the first- and second-order correction terms to the leading-order approximation $y_0 = e^{-t}$, which is in agreement with (1.6). Therefore we have obtained a three-term perturbation solution

$$y_a = e^{-t} + \varepsilon(e^{-t} - e^{-2t}) + \varepsilon^2(e^{-t} - 2e^{-2t} + e^{-3t}), \tag{1.8}$$

which is an approximation of y_{ex} and includes nonlinear effects due to the term εy^2 in the original differential equation. The approximate solution is the first three terms of the Taylor expansion of the exact solution. In this problem the exact solution is known (this is rare), and we can compare it to the approximate solution. The error in the approximation (1.8) is given by the difference

$$y_{ex} - y_a = m_1(t)\varepsilon^3 + m_2(t)\varepsilon^4 + \cdots, \quad t > 0,$$

for bounded functions $m_1, m_2, \ldots$. For a fixed positive t the error approaches zero as $\varepsilon \to 0$ as the same rate as ε^3 goes to zero. In fact, one can show that the convergence is uniform as $\varepsilon \to 0$ in the interval $0 \le t < \infty$. Convergence notions are explored further after an additional example.

3.1.2 Nonlinear Oscillations

In the last example the regular perturbation method led to a satisfactory result, consistent with intuition and that compared favorably with the exact solution. In the next example the procedure is the same, but the result does not turn out favorably; it is the first signal of the need to modify the regular perturbation method and develop a *singular* perturbation method. In this example, the perturbation approximation will be valid only if certain restrictions are placed on the interval of time that the solution evolves.

Consider a spring–mass oscillator where a mass m is connected to a spring whose restoring force has magnitude $ky + ay^3$, where y is the displacement of the mass, measured positively from equilibrium, and k and a are positive constants characterizing the stiffness properties of the spring. Assume that the nonlinear portion of the restoring force is small in magnitude compared to the linear part, or $a \ll k$. If, initially, the mass is released from a positive displacement A, then the motion is given by a function $y = y(\tau)$ of time τ, which satisfies Newton's second law

$$m\frac{d^2y}{d\tau^2} = -ky - ay^3, \quad \tau > 0, \tag{1.9}$$

and the initial conditions

$$y(0) = A, \quad \frac{dy}{d\tau}(0) = 0. \tag{1.10}$$

Because of the presence of the nonlinear term ay^3, this problem cannot be solved exactly. Because $a \ll k$, however, a perturbation method is suggested.

From the projectile problem in Chapter 1 the reader should have gained enough skepticism not to just neglect the term ay^3 and proceed without scaling. Therefore, to properly analyze the problem we seek appropriate time and length scales that reduce the problem to dimensionless form. The constants in the problem are k, a, m, and A have dimensions

$$[k] = \frac{\text{mass}}{\text{time}^2}, \quad [a] = \frac{\text{mass}}{\text{length}^2 \cdot \text{time}^2}, \quad [m] = \text{mass}, \quad [A] = \text{length}.$$

An obvious choice is to scale y by the amplitude A of the initial displacement. To scale time τ we argue as follows. If the small term ay^3 is neglected, then the differential equation is $my'' = -ky$, which has periodic solutions of the form $\cos \sqrt{k/mt}$ with a period proportional to $\sqrt{m/k}$. Hence, we choose $\sqrt{m/k}$ to be the characteristic time and introduce dimensionless variables t and u via

$$t = \frac{\tau}{\sqrt{m/k}}, \quad u = \frac{y}{A}.$$

Under this change of variables the differential equation (1.9) and initial conditions become (*overdots* denote derivatives with respect to t)

$$\ddot{u} + u + \varepsilon u^3 = 0, \quad t > 0,$$
$$u(0) = 1, \quad \dot{u}(0) = 0, \tag{1.11}$$

where

$$\varepsilon \equiv \frac{aA^2}{k} \ll 1$$

is a dimensionless small parameter. It now becomes clear exactly what the assumption of a small nonlinear restoring force means; the precise assumption is $aA^2 \ll k$. Equation (1.11) is the *Duffing equation*.

We attempt a solution of the form

$$u(t) = u_0(t) + \varepsilon u_1(t) + \varepsilon^2 u_2(t) + \cdots,$$

where $u_0, u_1, \dots$ are to be determined. Substituting into the differential equation and initial conditions, and equating coefficients of like powers of ε, gives a sequence of linear initial value problems

$$\ddot{u}_0 + u_0 = 0, \quad u_0(0) = 1, \quad \dot{u}_0(0) = 0, \tag{1.12a}$$
$$\ddot{u}_1 + u_1 = -u_0^3, \quad u_1(0) = 0, \quad \dot{u}_0(0) = 0, \dots \tag{1.12b}$$

The leading-order solution to (1.12a) is easily found to be

$$u_0(t) = \cos t.$$

Then the problem (1.12b) for the first correction term becomes

$$\ddot{u}_1 + u_1 = -\cos^3 t, \quad u_1(0) = \dot{u}_1(0) = 0.$$

We employ the trigonometric identity $\cos 3t = 4\cos^3 t - 3\cos t$ to write the differential equation as

$$\ddot{u}_1 + u_1 = -\frac{1}{4}(3\cos t + \cos 3t), \qquad (1.13)$$

which can be solved by standard methods. The general solution of the homogeneous equation $\ddot{u}_1 + u_1 = 0$ is $c_1 \cos t + c_2 \sin t$. A particular solution can be found by the method of undetermined coefficients and is of the form

$$u_p = C\cos 3t + Dt\cos t + Et\sin t.$$

Substituting u_p into the differential equation and equating like terms gives $C = \frac{1}{32}, D = 0$, and $E = -\frac{3}{8}$. Thus the general solution of (1.13) is

$$u_1 = c_1 \cos t + c_2 \sin t + \frac{1}{32}\cos 3t - \frac{3}{8}t\sin t.$$

Applying the initial conditions on u_1 gives

$$u_1 = \frac{1}{32}(\cos 3t - \cos t) - \frac{3}{8}t\sin t.$$

Therefore, in scaled variables, a two-term approximate solution takes the form

$$u_a = \cos t + \varepsilon\left[\frac{1}{32}(\cos 3t - \cos t) - \frac{3}{8}t\sin t\right]. \qquad (1.14)$$

The leading-order behavior of the approximate solution is $\cos t$, which is an oscillation. The second term, or the correction term, however, is not necessarily small. For a fixed value of time t the term goes to zero as $\varepsilon \to 0$, but if t itself is of the order of ε^{-1} or larger as $\varepsilon \to 0$, then the term $-\frac{3}{8}t\sin t$ will be large. Such a term is called a **secular term**. Therefore, we expect that the amplitude of the approximate solution will grow with time, which is not consistent with the physical situation or with the exact solution. Indeed, it is easy to show that the exact solution is bounded for all $t > 0$. In this approximation, the correction term cannot be made arbitrarily small for t in $(0, \infty)$ by choosing ε small enough. And, it is not possible to improve on (1.14) by calculating additional higher-order terms; they too will contain secular terms that will not cancel the effects of the lower-order terms. One may legitimately ask, therefore, if there is any value at all in the approximate solution (1.14). In a limited sense the answer is yes. If we restrict the independent variable t to a *finite* interval $[0, T]$, then the correction term $\varepsilon\left[\frac{1}{2}(\cos 3t - \cos t) - \frac{3}{8}t\sin t\right]$ can be made arbitrarily small by choosing ε sufficiently small, for any $t \in [0, T]$. So, as long as the coefficient $3\varepsilon t/8$ is kept small by limiting t and taking ε small, the leading-order term $\cos t$ is a reasonable approximate solution.

3.1.3 The Poincaré–Lindstedt Method

A straightforward application of the regular perturbation method on the initial value problem (1.11) for the Duffing equation led to a secular term in the first correction that ruined the approximation unless t was restricted. In this section we present a method to remedy this type of singular behavior, which occurs in all cases of periodic motion. The key to the analysis is to recognize that not only does the correction term grow in amplitude, but it also does not correct for the difference between the exact period of oscillation and the approximate period 2π of the leading order term $\cos t$. Over several oscillations the error in the period increases until the approximation and the actual solution are completely out of phase. To confirm this, the reader should use a software system to solve (1.11) numerically, and then plot the numerical solution vs. $\cos t$.

The idea of the Poincaré–Lindstedt method is to introduce a distorted time scale in the perturbation series. In particular, we let

$$u(\tau) = u_0(\tau) + \varepsilon u_1(\tau) + \varepsilon^2 u_2(\tau) + \cdots, \qquad (1.15)$$

where

$$\tau = \omega t, \qquad (1.16)$$

with

$$\omega = 1 + \varepsilon \omega_1 + \varepsilon^2 \omega_2 + \cdots. \qquad (1.17)$$

Here ω_0 has been chosen to be unity, the frequency of the solution of the unperturbed problem. Under the scale transformation (1.16) the initial value problem (1.11) becomes

$$\omega^2 u'' + u + \varepsilon u^3 = 0, \quad \tau > 0, \qquad (1.18)$$
$$u(0) = 1, \quad u'(0) = 0, \qquad (1.19)$$

where $u = u(\tau)$ and prime denotes differentiation with respect to τ. Substituting (1.15) and (1.17) into (1.18) and (1.19) gives

$$(1 + 2\varepsilon\omega_1\omega_0 + \cdots)(u_0'' + \varepsilon u_1'' + \cdots) + (u_0 + \varepsilon u_1 + \cdots) + \varepsilon(u_0^3 + 3\varepsilon u_0^2 u_1 + \cdots) = 0$$

and

$$u_0(0) + \varepsilon u_1(0) + \cdots = 1, \quad u_0'(0) + \varepsilon u_1'(0) + \cdots = 0.$$

Collecting coefficients of powers of ε gives

$$u_0'' + u_0 = 0, \quad u_0(0) = 1, \quad u_0'(0) = 0, ..., \qquad (1.20)$$
$$u_1'' + u_1 = -2\omega_1 u_0'' - u_0^3, \quad u_1(0) = u_1'(0) = 0, ... \qquad (1.21)$$

The solution of (1.20) is

$$u_0(\tau) = \cos \tau.$$

Then the differential equation in (1.21) becomes

$$u_1'' + u_1 = 2\omega_1 \cos \tau - \cos^3 \tau = \left(2\omega_1 - \frac{3}{4}\right)\cos\tau - \frac{1}{4}\cos 3\tau. \qquad (1.22)$$

Because $\cos \tau$ is a solution to the homogeneous equation, the $\cos \tau$ term on the right side of the equation leads to a particular solution with a term $\tau \cos \tau$, which is a secular term. The key observation is that we can avoid a secular term if we choose

$$\omega_1 = \frac{3}{8}.$$

Then (1.22) becomes

$$u_1'' + u_1 = -\frac{1}{4}\cos 3\tau,$$

which has the general solution

$$u_1(\tau) = c_1 \cos \tau + c_2 \sin \tau + \frac{1}{32}\cos 3\tau.$$

The initial conditions on u_1 lead to

$$u_1(\tau) = \frac{1}{32}(\cos 3\tau - \cos \tau).$$

Therefore, a first-order, uniformly valid perturbation solution of (1.11) is

$$u(\tau) = \cos\tau + \frac{1}{32}\varepsilon(\cos 3\tau - \cos \tau) + \cdots,$$

where

$$\tau = t + \frac{3}{8}\varepsilon t + \cdots.$$

The Poincaré–Lindstedt method is successful on a number of similar problems, and it is one of a general class of multiple scale methods. Generally, it works successfully on some (not all) equations of the form

$$u'' + \omega_0^2 u = \varepsilon F(t, u, u'), \quad 0 < \varepsilon \ll 1.$$

These are problems whose leading order is oscillatory with frequency ω_0. The basic technique is to change variables to one with a different frequency, $\tau = (\omega_0 + \omega_1 \varepsilon + \ldots)t$, and then assume $u = u(\tau)$ is a perturbation series in ε. The constants $\omega_0, \omega_1, \ldots$ are chosen at each step to avoid the presence of secular terms in the expansion.

3.1.4 Asymptotic Analysis

With insight from the previous examples we define some notions regarding convergence and uniformity. We observed that substitution of a perturbation series into a differential equation does not always lead to a valid approximate solution. Ideally we would like to say that a few terms in a truncated perturbation series provides, for a given ε, an approximate solution for the entire range of the independent variable t. Unfortunately, as we have seen, this is not always the case. Failure of this regular perturbation method is the rule rather than the exception.

To aid the analysis of approximate solutions we introduce some basic notation and terminology that permits the comparison of two functions as their common argument approaches some fixed value. These comparisons are called order relations.

Let $f(\varepsilon)$ and $g(\varepsilon)$ be defined in some neighborhood (or punctured neighborhood) of $\varepsilon = 0$. We write

$$f(\varepsilon) = o(g(\varepsilon)) \quad \text{as} \quad \varepsilon \to 0, \tag{1.23}$$

if

$$\lim_{\varepsilon \to 0} \left| \frac{f(\varepsilon)}{g(\varepsilon)} \right| = 0,$$

and we write

$$f(\varepsilon) = O(g(\varepsilon)) \quad \text{as} \quad \varepsilon \to 0, \tag{1.24}$$

if there exists a positive constant M such that

$$|f(\varepsilon)| \leq M|g(\varepsilon)|$$

for all ε in some neighborhood (punctured neighborhood) of zero. The comparison function g is called a **gauge function**.

In this definition, $\varepsilon \to 0$ may be replaced by a one-sided limit or by $\varepsilon \to \varepsilon_0$, where ε_0 is any finite or infinite number, with the domain of f and g defined appropriately. If (1.23) holds, we say f is **little oh** of g as $\varepsilon \to 0$, and if (1.24) holds we say f is **big oh** of g as $\varepsilon \to 0$. Common gauge functions are $g(\varepsilon) = \varepsilon^n$ for some exponent n, and $g(\varepsilon) = \varepsilon^n (\ln \varepsilon)^m$ for exponents m and n. The statement $f(\varepsilon) = O(1)$ means f is bounded in a neighborhood of $\varepsilon = 0$, and $f(\varepsilon) = o(1)$ means $f(\varepsilon) \to 0$ as $\varepsilon \to 0$. If $f = o(g)$, then f goes to zero faster than g goes to zero as $\varepsilon \to 0$, and we write $f(\varepsilon) \ll g(\varepsilon)$. The following examples illustrate a few methods to prove order relations.

Example 3.2

Verify $\varepsilon^2 \ln \varepsilon = o(\varepsilon)$ as $\varepsilon \to 0^+$. By L'Hôpital's rule,

$$\lim_{\varepsilon \to 0^+} \frac{\varepsilon^2 \ln \varepsilon}{\varepsilon} = \lim_{\varepsilon \to 0^+} \frac{\ln \varepsilon}{(1/\varepsilon)} = \lim_{\varepsilon \to 0^+} \frac{(1/\varepsilon)}{(-1/\varepsilon^2)} = 0. \quad \square$$

Example 3.3

Verify $\sin \varepsilon = O(\varepsilon)$ as $\varepsilon \to 0^+$. By the mean value theorem there is a number c between 0 and ε such that

$$\frac{\sin \varepsilon - \sin 0}{\varepsilon - 0} = \cos c.$$

Hence $|\sin \varepsilon| = |\varepsilon \cos c| \leq |\varepsilon|$, because $|\cos c| \leq 1$. An alternate argument is to note that $(\sin \varepsilon)/\varepsilon \to 1$ as $\varepsilon \to 0^+$. Because the limit exists, the function $(\sin \varepsilon)/\varepsilon$ must be bounded for $0 < \varepsilon < \varepsilon_0$, for some ε_0. Therefore $|(\sin \varepsilon)/\varepsilon| \leq M$, for some constant M and $\sin \varepsilon = O(\varepsilon)$. $\quad \square$

The order definitions may be extended to functions of ε and another variable t lying in an interval I. First we review the notion of uniform convergence. Let $h(t, \varepsilon)$ be a function defined for ε in a neighborhood of $\varepsilon = 0$, possibly not including the value $\varepsilon = 0$ itself, and for t in some interval I, either finite or infinite. We say

$$\lim_{\varepsilon \to 0} h(t, \varepsilon) = 0 \text{ uniformly on } I,$$

if the convergence to zero is at the same rate for each $t \in I$; that is, if for any positive number η there can be chosen a positive number ε_0, independent of t, such that $|h(t, \varepsilon)| < \eta$ for all $t \in I$, whenever $|\varepsilon| < \varepsilon_0$. In other words, if $h(t, \varepsilon)$ can be made arbitrarily small over the entire interval I by choosing ε small enough, then the convergence is uniform. If merely $\lim_{\varepsilon \to 0} h(t_0, \varepsilon) = 0$ for each fixed $t_0 \in I$, then we say that the convergence is *pointwise* on I.

To prove $\lim_{\varepsilon \to 0} h(t, \varepsilon) = 0$ uniformly on I it is sufficient to find a function $H(\varepsilon)$ such that $|h(t, \varepsilon)| \leq H(\varepsilon)$ holds for all $t \in I$, with $H(\varepsilon) \to 0$ as $\varepsilon \to 0$. To prove that convergence is not uniform on I, it is sufficient to produce a $\bar{t} \in I$ such that $|h(\bar{t}, \varepsilon)| \geq \eta$ for some positive η, regardless of how small ε is chosen.

Let $f(t, \varepsilon)$ and $g(t, \varepsilon)$ be defined for all $t \in I$ and all ε in a (punctured) neighborhood of $\varepsilon = 0$. We write

$$f(t, \varepsilon) = o(g(t, \varepsilon)) \quad \text{as} \quad \varepsilon \to 0,$$

if

$$\lim_{\varepsilon \to 0} \left| \frac{f(t, \varepsilon)}{g(t, \varepsilon)} \right| = 0 \tag{1.25}$$

pointwise on I. If the limit is uniform on I, we write $f(t, \varepsilon) = o(g(t, \varepsilon))$ as $\varepsilon \to 0$, uniformly on I. If there exists a positive function $M(t)$ on I such that

$$|f(t, \varepsilon)| \leq M(t)|g(t, \varepsilon)|$$

for all $t \in I$ and ε in some neighborhood of zero, then we write

$$f(t, \varepsilon) = O(g(t, \varepsilon)) \quad \text{as} \quad \varepsilon \to 0, \quad t \in I.$$

If $M(t)$ is a bounded function on I, we write

$$f(t, \varepsilon) = O(g(t, \varepsilon)) \quad \text{as} \quad \varepsilon \to 0, \quad \text{uniformly on } I.$$

The big oh and little oh notations permit us to make quantitative statements about the error in a given approximation. We can make the following definition. A function $y_a(t, \varepsilon)$ is a *uniformly valid asymptotic approximation* to a function $y(t, \varepsilon)$ on an interval I as $\varepsilon \to 0$ if the error $E(t, \varepsilon) \equiv y(t, \varepsilon) - y_a(t, \varepsilon)$ converges to zero as $\varepsilon \to 0$ uniformly for $t \in I$. We often express the fact that $E(t, \varepsilon)$ is little oh or big oh of ε^n (for some n) as $\varepsilon \to 0$ to make an explicit statement regarding the rate at which the error goes to zero, and whether or not the convergence is uniform.

Example 3.4

Let

$$y(t, \varepsilon) = e^{-t\varepsilon}, \quad t > 0, \quad \varepsilon \ll 1.$$

The first three terms of the Taylor expansion in powers of ε provide an approximation

$$y_a(t, \varepsilon) = 1 - t\varepsilon + \tfrac{1}{2}t^2\varepsilon^2.$$

The error is

$$E(t, \varepsilon) = e^{-t\varepsilon} - 1 + t\varepsilon - \tfrac{1}{2}t^2\varepsilon^2 = -\frac{1}{3!}t^3\varepsilon^3 + \cdots.$$

For a fixed t the error can be made as small as desired by choosing ε sufficiently small. Thus $E(t, \varepsilon) = O(\varepsilon^2)$ as $\varepsilon \to 0$. If ε is fixed, however, regardless of how small, t may be chosen large enough so that the approximation is totally invalid. Thus, the approximation is not uniform on $I = [0, \infty)$. Clearly, by choosing $t = 1/\varepsilon$ we have $E(1/\varepsilon, \varepsilon) = e^{-1} - \tfrac{1}{2}$, which is not small. We may *not* write $E(t, \varepsilon) = O(\varepsilon^2)$ as $\varepsilon \to 0$ uniformly on $[0, \infty)$.   □

Example 3.5

(**Transcendentally small**) Some functions cannot be gauged against polynomials. Consider $f(\varepsilon) = e^{-1/\varepsilon}$ for small, positive ε. From calculus, this function does not have a Taylor expansion about ε; further,

$$\lim_{\varepsilon \to 0^+} \frac{e^{-1/\varepsilon}}{\varepsilon^n} = \lim_{\alpha \to +\infty} \alpha^n e^{-\alpha} = 0,$$

because exponentials decay faster than polynomials grow. This means

$$e^{-1/\varepsilon} = o(\varepsilon^n) \quad \text{as} \quad \varepsilon \to 0^+$$

for any $n = 0, 1, 2, \dots$. Thus, $e^{-1/\varepsilon}$ goes to zero faster than any positive power of ε. In this case we say the function $f(\varepsilon)$ is transcendentally small.   □

The difficulty of these definitions with regard to differential equations is that the exact solution to the equation is seldom known, and therefore a direct error estimate cannot be made. Consequently, we require some notion of how well an approximate solution satisfies the differential equation and the auxiliary conditions. For definiteness consider the differential equation in (1.1). We say that an approximate solution $y_a(t, \varepsilon)$ *satisfies the differential equation* (1.1) *uniformly for* $t \in I$ *as* $\varepsilon \to 0$ if

$$r(t, \varepsilon) \equiv F(t, y_a(t, \varepsilon), \dot{y}_a(t, \varepsilon), \ddot{y}_a(t, \varepsilon), \varepsilon) \to 0$$

uniformly on I as $\varepsilon \to 0$. The quantity $r(t, \varepsilon)$ is the residual error, which measures how well the approximate solution $y_a(t, \varepsilon)$ satisfies the equation.

Example 3.6

Consider the initial value problem

$$\ddot{y} + \dot{y}^2 + \varepsilon y = 0, \quad t > 0, \quad 0 < \varepsilon \ll 1,$$
$$y(0) = 0, \quad \dot{y}(0) = 1.$$

Substituting the perturbation series $y = y_0 + \varepsilon y_1 + \cdots$ gives the initial value problem

$$\ddot{y}_0 + \dot{y}_0^2 = 0, \quad t > 0,$$
$$y_0(0) = 0, \quad \dot{y}_0(0) = 1,$$

for the leading-order term y_0. It is easily found that $y_0(t) = \ln(t+1)$, and hence

$$r(t, \varepsilon) \equiv \ddot{y}_0 + \dot{y}_0^2 + \varepsilon y_0 = \varepsilon \ln(t + 1).$$

Thus $r(t, \varepsilon) = O(\varepsilon)$ as $\varepsilon \to 0$, but not uniformly on $[0, \infty)$. On any finite interval $[0, T]$, however, we have $|\varepsilon \ln(t+1)| \leq \varepsilon \ln(T+1)$, and so $r(t, \varepsilon) = O(\varepsilon)$ as $\varepsilon \to 0$ uniformly on $[0, T]$. $\quad\square$

Specifically, the regular perturbation method produces an asymptotic expansion

$$y_0(t) + y_1(t)\varepsilon + y_2(t)\varepsilon^2 + \cdots,$$

for which an approximate solution can be obtained by taking the first few terms. Such an expansion in the integral powers of ε, that is, $1, \varepsilon, \varepsilon^2, \ldots$, is called an asymptotic power series. In some problems the expansion may take the form

$$y_0(t) + y_1(t)\sqrt{\varepsilon} + y_2(t)\varepsilon + y_3(t)\varepsilon^{3/2} + \cdots,$$

in terms of the sequence $1, \varepsilon^{1/2}, \varepsilon, \varepsilon^{3/2}, \ldots$ In yet other problems the required expansion must have the form

$$y_0(t) \ln \varepsilon + y_1(t)\varepsilon \ln \varepsilon + y_2(t)\varepsilon + y_3(t)\varepsilon^2 \ln^2 \varepsilon + y_4(t)\varepsilon^2 \ln \varepsilon + y_5(t)\varepsilon^2 + \cdots.$$

The type of expansion depends on the problem. In general, we say a sequence of gauge functions $\{g_n(t, \varepsilon)\}$ is an **asymptotic sequence** as $\varepsilon \to 0$, $t \in I$, if

$$g_{n+1}(t, \varepsilon) = o(g_n(t, \varepsilon)) \quad \text{as} \quad \varepsilon \to 0,$$

for $n = 0, 1, 2, \ldots$. That is, each term in the sequence tends to zero faster than its predecessor, as $\varepsilon \to 0$. Given a function $y(t, \varepsilon)$ and an asymptotic sequence $\{g_n(t, \varepsilon)\}$ as $\varepsilon \to 0$, the formal series

$$\sum_{n=0}^{\infty} a_n g_n(t, \varepsilon) \qquad (1.26)$$

is said to be an **asymptotic expansion** of $y(t, \varepsilon)$ as $\varepsilon \to 0$, if

$$y(t, \varepsilon) - \sum_{n=0}^{N} a_n g_n(t, \varepsilon) = o(g_N(t, \varepsilon)), \quad \text{as} \quad \varepsilon \to 0,$$

for every N. In other words, for any partial sum the remainder is little oh of the last term. If the limits just cited are uniform for $t \in I$, then we speak of a uniform asymptotic sequence and uniform asymptotic expansion. In most cases the sequence $\{g_n(t, \varepsilon)\}$ is of the form of a product, $g_n(t, \varepsilon) = y_n(t)\phi_n(\varepsilon)$, as in the previous examples. The notation

$$y \sim \sum_{n=0}^{\infty} a_n g_n(t, \varepsilon)$$

often denotes an asymptotic expansion.

The formal series (1.26) need not converge to be valuable. The value of such expansions, although perhaps divergent, is that often only a few terms are required to obtain an accurate approximation, whereas a convergent Taylor series may yield an accurate approximation only if many terms are calculated.

A rather obvious question arises. If a given approximate solution $y_a(t, \varepsilon)$ satisfies the differential equation uniformly on $t \in I$, is it in fact a uniformly valid approximation to the exact solution $y(t, \varepsilon)$? A complete discussion of this question is beyond our scope, but a few remarks are appropriate to caution the reader regarding the nature of this problem. Probably more familiar is the situation in linear algebra where we consider a linear system of equations $A\mathbf{x} = \mathbf{b}$. Let $\mathbf{x}_a$ be an approximate solution. A measure of how well it satisfies the system is the magnitude $|\mathbf{r}|$ of the *residual vector* $\mathbf{r}$ defined by $\mathbf{r} = A\mathbf{x}_a - \mathbf{b}$. If $\mathbf{r} = \mathbf{0}$, then $\mathbf{x}_a$ must be the exact solution $\bar{\mathbf{x}}$. But if $|\mathbf{r}|$ is small, it does not necessarily follow that the magnitude $|\mathbf{e}|$ is small, where $\mathbf{e} = \bar{\mathbf{x}} - \mathbf{x}_a$ is the error. In *ill-conditioned systems*, where $\det A$ is close to zero, a small residual may not imply a small error. A similar state of affairs exists for differential equations. Therefore one must proceed cautiously in interpreting the validity of a perturbation solution. Numerical calculations or the computation of additional correction terms may aid in the interpretation. Often a favorable comparison with experiment leads one to conclude that an approximation is valid.

EXERCISES

1. In a spring–mass problem assume that the restoring force is $-ky$ and that there is a resistive force numerically equal to $a\dot{y}^2$, where k and a are constants with appropriate units. With initial conditions $y(0) = A$, $\dot{y}(0) = 0$, determine the correct time and displacement scales for small damping and show that the problem can be written in dimensionless form as

    $$\bar{y}'' + \varepsilon(\bar{y}')^2 + \bar{y} = 0,$$
    $$\bar{y}(0) = 1, \quad \bar{y}'(0) = 0,$$

    where $\varepsilon \equiv aA/m$ is a dimensionless parameter and prime denotes the derivative with respect to the scaled time $\bar{t}$.

2. Consider the initial value problem

    $$u'' - u = \varepsilon t u, \quad t > 0, \quad u(0) = 1, \quad u'(0) = -1.$$

    Find a two-term perturbation approximation for $0 < \varepsilon \ll 1$ and compare it graphically to a six-term Taylor series approximation (centered at $t = 0$) where $\varepsilon = 0.04$. Use a numerical differential equation solver to find an accurate numerical solution and compare.

3. Write down an order relation (big oh or little oh) that expresses the fact that e^{-t} decays to zero faster that $1/t^2$ as $t \to \infty$, and prove your assertion.

4. Let
$$f(y, \varepsilon) = \frac{1}{(1 + \varepsilon y)^{3/2}}, \quad y = y_0 + y_1 \varepsilon + \cdots .$$

Expand f in powers of ε up to $O(\varepsilon^2)$.

5. Verify the following order relations:

   a) $t^2 \tanh t = O(t^2)$ as $t \to \infty$.

   b) $\exp(-t) = o(1)$ as $t \to \infty$.

   c) $\sqrt{\varepsilon(1 - \varepsilon)} = O(\sqrt{\varepsilon})$ as $\varepsilon \to 0^+$.

   d) $\frac{\sqrt{\varepsilon}}{1 - \cos \varepsilon} = O(\varepsilon^{-3/2})$ as $\varepsilon \to 0^+$.

   e) $t = O(t^2)$ as $t \to \infty$.

   f) $\exp(\varepsilon) - 1 = O(\varepsilon)$ as $\varepsilon \to 0$.

   g) $\int_0^\varepsilon \exp(-x^2)dx = O(\varepsilon)$ as $\varepsilon \to 0^+$.

   h) $\exp(\tan \varepsilon) = O(1)$ as $\varepsilon \to 0$.

   i) $e^{-\varepsilon} = O(\varepsilon^{-p})$ as $\varepsilon \to \infty$ for all $p > 0$.

   j) $\ln \varepsilon = o(\varepsilon^{-p})$ as $\varepsilon \to 0^+$ for all $p > 0$.

6. Consider the algebraic equation $\phi(x, \varepsilon) = 0$, $0 < \varepsilon \ll 1$, where ϕ is a function having derivatives of all order. Assuming $\phi(x, 0) = 0$ is solvable to obtain x_0, show how to find a three-term perturbation approximation of the form $x = x_0 + x_1 \varepsilon + x_2 \varepsilon^2$. What condition on ϕ is required to determine x_1 and x_2? Find a three-term approximation to the roots of $\exp(\varepsilon x) = x^2 - 1$.

7. Find a three-term approximation to the real solution of $(x + 1)^3 = \varepsilon x$.

8. Use the Poincaré-Lindstedt method to obtain a two-term perturbation approximation to the following problems:

   a) $y'' + y = \varepsilon y y'^2$, $y(0) = 1$, $y'(0) = 0$.

   b) $y'' + 9y = 3\varepsilon y^3$, $y(0) = 0$, $y'(0) = 1$.

   c) $y'' + y = \varepsilon y(1 - y'^2)$, $y(0) = 1$, $y'(0) = 0$.

9. To find approximations to the roots of the cubic equation
$$x^3 - 4.001x + 0.002 = 0,$$

why is it easier to examine the equation

$$x^3 - (4 + \varepsilon)x + 2\varepsilon = 0?$$

Find a two-term approximation to this equation.

10. Consider the algebraic system

$$0.01x + y = 0.1, \qquad x + 101y = 11.$$

In the first equation ignore the apparently small first term $0.01x$ and obtain an approximate solution for the system. Is the approximation a good one? Analyze what went wrong by examining the solution of the system

$$\varepsilon x + y = 0.1, \qquad x + 101y = 11.$$

11. In elementary physics courses it is often stated that for the pendulum equation

$$\theta'' + \frac{g}{l} \sin \theta = 0,$$

if the oscillations $\theta = \theta(t)$ are small, then $\sin \theta \approx \theta$ and the equation can be approximated by

$$\theta'' + \frac{g}{l} \theta = 0.$$

In what sense is this true? In what sense is this false?

12. In Section 1.3 we obtained the initial value problem

$$\frac{d^2 h}{dt^2} = -\frac{1}{(1 + \varepsilon h)^2}, \qquad h(0) = 0, \quad h'(0) = 1, \quad 0 < \varepsilon \ll 1,$$

governing the motion of a projectile. Use regular perturbation theory to obtain a three-term perturbation approximation. Up to the accuracy of ε^2 terms, determine the value t_m when h is maximum. Find $h_{\max} \equiv h(t_m)$ up through order ε^2 terms.

13. A mass m is attached to an aging spring that exerts a force $F_s = -kye^{-rt}$, where k and r are positive constants, and $y = y(t)$ is the displacement from equilibrium. The system is also immersed in a medium that exerts a drag force $F_d = -ay'$, where a is a positive constant. Initially, the mass is displaced to position y_0 and released.

   a) Formulate an initial value problem that governs the motion of the mass on the spring.

   b) Make a table of the parameters in the problem and state their dimensions.

c) Non-dimensionalize the problem in the simplest way in the case that the constant k is very small. Use y_0 as the characteristic length. Denote the small dimensionless parameter by ε.

d) Use a perturbation method to obtain a two-term approximation (up to $O(\varepsilon)$) to the scaled problem.

e) In mks units, take $m = 0.01$, $a = 0.01$, $r = 0.07$, $k = 0.0001$. Plot an accurate numerical approximation and the two-term approximation on the same set of axes, and comment on the accuracy of the approximation.

14. Consider the boundary value problem

$$y'' = 3y \cos\left(\frac{\pi x}{2}\right), \quad y(0) = 0, \ y(1) = 1.$$

Obtain an approximation by considering the equation

$$y'' = \varepsilon y \cos\left(\frac{\pi x}{2}\right)$$

for small ε and then setting $\varepsilon = 3$. How accurate is the approximation?

15. Find the exact solution to the initial value problem

$$\varepsilon y' = y - 1, \quad y(0) = 0, \ 0 < \varepsilon \ll 1.$$

Next, use perturbation to find an approximate solution. What went wrong? Explain why regular perturbation doesn't work.

16. Consider the problem

$$\frac{dy}{dt} = e^{-\varepsilon/y}, \quad y(0) = 1.$$

Find the leading order approximation and derive initial value problems for the $O(\varepsilon)$ and $O(\varepsilon^2)$ terms. (Do not solve.) Write down the exact solution of the problem.

17. Consider the scaled pendulum problem

$$\frac{d^2\theta}{d\tau^2} + \frac{\sin \varepsilon\theta}{\varepsilon} = 0, \quad \tau > 0, \quad 0 < \varepsilon \ll 1,$$

$$\theta(0) = 1, \quad \frac{d\theta}{d\tau}(0) = 0.$$

a) Apply the regular perturbation method to find a two-term expansion. Show that the correction term is secular and comment on the validity of the approximation.

b) Show that the exact period of oscillation is

$$P = \frac{4\varepsilon}{\sqrt{2}} \int_0^1 \frac{dx}{\sqrt{\cos(\varepsilon x) - \cos \varepsilon}}.$$

c) Show that $P = 2\pi + \frac{\pi^2}{8}\varepsilon^2 + \cdots$. Find an expansion for the frequency ω.

18. Consider the initial value problem $y'' = \varepsilon t y$, $0 < \varepsilon \ll 1$, $y(0) = 0$, $y'(0) = 1$. Using regular perturbation theory obtain a three-term approximate solution on $t \geq 0$. Does the approximation satisfy the differential equation uniformly on $t \geq 0$ as $\varepsilon \to 0^+$?

19. Consider the boundary value problem

$$Ly \equiv t^2 y'' + \varepsilon t^2 y' + \frac{1}{4}y = 0, \quad 1 \leq t \leq e, \quad 0 < \varepsilon \ll 1,$$

$$y(1) = 1, \quad y(e) = 0.$$

a) Use regular perturbation to find the leading-order behavior $y_0(t)$.

b) Compute an upper bound for $|Ly_0|$ on $1 \leq t \leq e$ when $\varepsilon = 0.01$. Can you conclude that y_0 is a good approximation to the exact solution?

20. Find a two-term perturbation solution of

$$u' + u = \frac{1}{1 + \varepsilon u}, \quad u(0) = 0, \quad 0 < \varepsilon \ll 1.$$

21. A wildebeest starts at $t = 0$ from a point (a, b) in the xy-plane and moves along the line $x = a$. A lion, starting from the origin at $t = 0$ begins the chase, always running in a direction pointing at the wildebeest. The differential equation for the lion's path $y(x)$ is

$$(a - x)y'' = \varepsilon\sqrt{1 + y'^2},$$

where ε is a small parameter giving the ratio of the speed of the wildebeest to the speed of the lion.

a) Find a leading order perturbation approximation. Based upon your answer, do you expect the lion to catch the wildebeest?

b) Find the equation for a correction term in the perturbation approximation, but do not solve.

3.2 Singular Perturbation

3.2.1 Algebraic Equations

As we observed in Section 2.1 for the Duffing equation, a straightforward application of the regular perturbation method failed because secular terms appeared in the expansion, making the approximation nonuniform for large times. We remedied this by adjusting the frequency of oscillations using a scale transformation. Now we indicate a different type of singular behavior, first for a simple algebra problem.

Example 3.7

Solve the quadratic equation

$$\varepsilon x^2 + 2x + 1 = 0, \quad 0 < \varepsilon \ll 1.$$

This equation can be solved exactly, of course, but our goal is to illustrate the failure of the regular perturbation method. Observe that the equation is a slight alteration of the unperturbed equation

$$2x + 1 = 0,$$

which has solution $x = -\frac{1}{2}$, and the unperturbed problem (linear) is fundamentally different from the original problem (quadratic). If we attempt regular perturbation by substituting the perturbation series

$$x = x_0 + x_1\varepsilon + x_2\varepsilon^2 + \cdots,$$

then we obtain, after setting the coefficients of like powers of ε equal to zero, the sequence of equations

$$2x_0 + 1 = 0,$$
$$x_0^2 + 2x_1 = 0,$$
$$2x_1x_0 + 2x_2 = 0. \dots$$

Hence, $x_0 = -\frac{1}{2}, x_1 = -\frac{1}{8}, x_2 = -\frac{1}{16}, \dots$, and we obtain a single perturbation solution

$$x = -\frac{1}{2} - \frac{1}{8}\varepsilon - \frac{1}{16}\varepsilon^2 - \cdots.$$

What happened to the other solution? Regular perturbation assumed a leading term of order unity, and it is not surprising that it recovered only one root, of order unity. The other root could be different order, either large or small. To find the second root we examine the three terms, εx^2, $2x$, and 1, of the equation more closely. Discarding the εx^2 gave the root close to $x = -\frac{1}{2}$, and in that case, the term εx^2 is small compared to $2x$ and 1. So it is reasonable to ignore εx^2; the apparent small term is indeed small. For the second root, εx^2 may not be small (because x may be large). Two cases are possible if one term is to be neglected from the equation to make a simplification. (i) εx^2 and 1 are the same order and $2x \ll 1$, or (ii) εx^2 and $2x$ are the same order and both are large compared to 1.

In case (i) we have $x = O(1/\sqrt{\varepsilon})$; but then $2x \ll 1$ could not hold because $2x$ would be large. This in inconsistent. In case (ii) we have $x = O(1/\varepsilon)$; hence εx^2 and $2x$ both are of order $1/\varepsilon$ and both are large compared to 1. Therefore case (ii) is consistent and the second root is of order $1/\varepsilon$, which is large. This provides a clue to a scaling. Let us choose a new variable y of order 1 defined by

$$y = \frac{x}{1/\varepsilon} = \varepsilon x.$$

With this change of variables the original equation becomes

$$y^2 + 2y + \varepsilon = 0.$$

Now the principle of proper scaling holds—each term has a magnitude defined by its coefficient. Assuming a regular perturbation series

$$y = y_0 + y_1 \varepsilon + y_2 \varepsilon^2 + \cdots.$$

and substituting into the last equation, we obtain

$$y_0^2 + 2y_0 = 0,$$
$$2y_0 y_1 + 2y_1 + 1 = 0, \ldots$$

Hence $y_0 = -2$, $y_1 = \frac{1}{2}, \ldots$, giving

$$y = -2 + \frac{1}{2}\varepsilon + \cdots,$$

or

$$x = -\frac{2}{\varepsilon} + \frac{1}{2} + \cdots,$$

as the second root. In summary, the roots are different order, and one expansion does not reveal both.   □

The reasoning we used in this example is called **dominant balancing**. We examine each term carefully and determine which ones combine to give a dominant balance. In the next section we show type of balancing argument applies equally well to differential equations.

First, however, before continuing with additional examples, we review the procedure for finding roots of complex numbers. This is useful in solving the exercises.

Example 3.8

(**Roots of complex numbers**) Let z_0 be a fixed, given, complex number. We are interested in finding its nth roots, or solving the equation

$$z^n = z_0 \qquad (2.1)$$

for z. Here, $n \geq 2$ is a positive integer. From the fundamental theorem of algebra we know that (2.1) has n roots. First, write z and z_0 in polar form,

$$z = re^{i\theta}, \quad z_0 = r_0 e^{i\theta_0},$$

where r and r_0 are the moduli (distances from the origin in the complex plane), and θ and θ_0 are the arguments (the angles measured counterclockwise from the positive real axis between 0 and 2π.) Then,

$$r^n e^{in\theta} = r_0 e^{i\theta_0}.$$

Thus, $r = r_0^{1/n}$ and $in\theta = i\theta_0 + 2k\pi i$, $k = 0, \pm 1, \pm 2, \ldots$. The latter follows from the $2\pi i$ periodicity of $e^{i\theta}$. Thus, the n roots of z_0 are

$$z = z_0^{1/n} = r_0^{1/n} e^{i\left(\frac{\theta_0}{n} + \frac{2k\pi}{n}\right)}, \qquad (2.2)$$

where $k = 0, 1, 2, \ldots, n-1$. (Only n of the roots in (2.2) are different.)

If $z_0 = 1$, then the nth roots of 1 are called the **roots of unity**. These are spaced uniformly around the unit circle in the complex plane. □

Example 3.9

The three roots of the equation $z^3 = 1$ are given by (2.2) as

$$1^{1/3} = e^{i\left(\frac{0}{3} + \frac{2k\pi}{3}\right)}, \quad k = 0, 1, 2.$$

Thus, the roots are

$$1, \ e^{2\pi i/3}, \ e^{4\pi i/3}, \ \text{or } 1, \ e^{2\pi i/3}, \ e^{-2\pi i/3}. □$$

Example 3.10

Find a leading order approximation of the four roots to the equation

$$\varepsilon x^4 - x - 1 = 0,$$

where $0 < \varepsilon \ll 1$. When $\varepsilon = 0$ we get only a single root $x = -1$, which is order 1. To determine the leading order of the other roots we use dominant balancing. If the first and third terms balance then $x = O(\varepsilon^{-1/4})$, which is large; this is inconsistent. If the first and second terms balance, then $x = O(\varepsilon^{-1/3})$, which is large compared to 1; this case is consistent. Therefore, we re-scale according to $y = \varepsilon^{1/3}x$, which gives

$$y^4 - y - \varepsilon^{1/3} = 0.$$

To leading order $y^4 - y = 0$, giving $y = 1$, $e^{2\pi i/3}$, $e^{-2\pi i/3}$. We discarded $y = 0$ because that corresponds to $x = 0$, which is clearly not a valid approximation to the equation. Consequently, the leading order behavior of the four roots is

$$x = -1, \ \varepsilon^{-1/3}, \ \varepsilon^{-1/3}e^{2\pi i/3}, \ \varepsilon^{-1/3}e^{-2\pi i/3}.$$

Three of them are large. We can obtain higher-order correction terms using the perturbation series

$$y = y_0 + \varepsilon^{1/3}y_1 + \varepsilon^{2/3} + \cdots. \quad \square$$

3.2.2 Differential Equations

Now we encounter a different type of singular behavior from that caused by the appearance of secular terms in the perturbation series. Often, with a perturbation series, some problems do not even permit the calculation of the leading-order term because the perturbed problem is of totally different character from the unperturbed problem.

Consider the boundary problem

$$\varepsilon y'' + (1 + \varepsilon)y' + y = 0, \quad 0 < x < 1, \quad 0 < \varepsilon \ll 1,$$
$$y(0) = 0, \quad y(1) = 1, \tag{2.3}$$

and assume a naive perturbation series of the form

$$y = y_0(x) + \varepsilon y_1(x) + \varepsilon^2 y_2(x) + \cdots.$$

Substitution into the differential equation gives

$$\varepsilon(y_0'' + \varepsilon y_1'' + \varepsilon^2 y_2'' + \cdots) + (y_0' + \varepsilon y_1' + \varepsilon^2 y_2' + \cdots)$$
$$+ (\varepsilon y_0' + \varepsilon^2 y_1' + \varepsilon^3 y_2' + \cdots) + (y_0 + \varepsilon y_1 + \varepsilon^2 y_2 + \cdots) = 0.$$

Equating the coefficients of like powers of ε to zero gives a sequence of problems

$$y_0' + y_0 = 0,$$
$$y_1' + y_1 = -y_0'' - y_0', \ldots.$$

The boundary conditions force

$$y_0(0) = 0, \quad y_0(1) = 1,$$
$$y_1(0) = 0, \quad y_1(1) = 0, \ldots$$

Therefore, we have obtained a sequence of boundary value problems for y_0, $y_1, \ldots$. The leading-order problem is

$$y_0' + y_0 = 0, \quad y_0(0) = 0, \quad y_0(1) = 1. \tag{2.4}$$

Already we see a difficulty. The differential equation is first-order, yet there are two conditions to be satisfied. The general solution of the equation is

$$y_0(t) = ce^{-x}. \tag{2.5}$$

Application of the boundary condition $y_0(0) = 0$ gives $c = 0$, and so $y_0(x) = 0$. This function cannot satisfy the boundary condition at $x = 1$. Conversely, application of the boundary condition $y_0(1) = 1$ in (2.5) gives $c = e$, and so

$$y_0(t) = e^{1-x}. \tag{2.6}$$

This function cannot satisfy the condition at $x = 0$. Therefore we are at an impasse; regular perturbation fails at the first step.

3.2.3 Boundary Layers

Careful examination of the preceding problem shows what went wrong and points the way toward a correct, systematic method to obtain an approximate solution. First of all, we should have been suspicious from the beginning. The unperturbed problem, found by setting $\varepsilon = 0$, is

$$y' + y = 0, \quad y(0) = 0, \quad y(1) = 1. \tag{2.7}$$

This problem is of a different character than the perturbed problem (2.3) in that it is first order rather than second. Because the small parameter multiplied the highest derivative in (2.3), the second derivative term disappeared when ε was set to zero. This type of phenomenon signals in almost all cases the failure of the regular perturbation method.

Equation (2.3), a linear equation with constant coefficients, can be solved exactly to get

$$y(x) = \frac{1}{e^{-1} - e^{-1/\varepsilon}}(e^{-x} - e^{-x/\varepsilon}). \tag{2.8}$$

A look at the graph of the solution of (2.3) reveals the reason for the difficulty. See Fig. 3.1. We observe that $y(x)$ is changing very rapidly in a narrow interval,

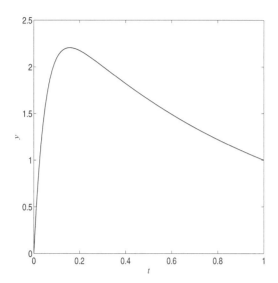

Figure 3.1 The exact solution (2.8) showing a narrow layer near $x = 0$ where rapid changes are occurring.

called the **boundary layer**, near the origin, and more slowly in the larger interval, called the **outer layer** away from the origin. One spatial scale does not describe the variations in both layers. Rather, two spatial scales are indicated, one for each zone. It is instructive and revealing to compute the derivatives of y and estimate the size of the terms in the equation. An easy calculation shows

$$y' = \frac{1}{e^{-1} - e^{-1/\varepsilon}} \left(-e^{-x} + \frac{1}{\varepsilon} e^{-x/\varepsilon} \right),$$

$$y'' = \frac{1}{e^{-1} - e^{-1/\varepsilon}} \left(e^{-x} - \frac{1}{\varepsilon^2} e^{-x/\varepsilon} \right).$$

First we examine the second derivative. Suppose ε is small and x is in the narrow boundary layer near $x = 0$. For definiteness assume $x = \varepsilon$. Then

$$y''(\varepsilon) = \frac{1}{e^{-1} - e^{-1/\varepsilon}} \left(e^{-\varepsilon} - \frac{1}{\varepsilon^2} e^{-1} \right) = O(\varepsilon^{-2})$$

Thus y'' is very large inside this narrow band, and therefore the term $\varepsilon y''$ is not small, as would be anticipated in a regular perturbation calculation; in fact $\varepsilon y'' = O(\varepsilon^{-1})$. We see again that terms that appear small in a differential equation are not always necessarily small. The problem is that the differential equation (2.3), as it stands, is not scaled properly for small x, and a rescaling must occur in the boundary layer if correct conclusions are to be drawn. For values of x away from the boundary layer, the term $\varepsilon y''$ is indeed small. For example, if $x = \frac{1}{2}$, then

$$y''\left(\tfrac{1}{2}\right) = \frac{1}{e^{-1} - e^{-1/\varepsilon}}\left(e^{-1/2} - \frac{1}{\varepsilon^2}e^{-1/2\varepsilon}\right) = \mathrm{O}(1).$$

Consequently, in the outer layer $\varepsilon y''$ is small and may be safely neglected. The reader may deduce similar conclusions regarding the first derivative term.

This calculation suggests that in the outer region, the region away from the boundary layer, the leading-order problem (2.7) obtained by setting $\varepsilon = 0$ in the original problem is a valid approximation provided we take only the right $(x = 1)$ boundary condition. Thus,

$$y_0(x) = e^{1-x}. \tag{2.9}$$

This is consistent with the exact solution (2.8). For, if ε is small, then $e^{-1} - e^{-1/\varepsilon} \approx e^{-1}$ and y may be approximated by

$$y(x) \approx e^{1-x} - e^{1-x/\varepsilon}. \tag{2.10}$$

For x of order 1 we have $e^{1-x/\varepsilon} \approx 0$, and so

$$y(x) \approx e^{1-x}, \quad x = \mathrm{O}(1).$$

The approximate solution (2.9), which is valid in the outer region, is called the **outer approximation** and is plotted in Fig. 3.2.

What about an approximate solution in the narrow boundary layer near $x = 0$? From (2.10), if x is small, then

$$y(x) \approx e - e^{1-x/\varepsilon}, \quad x \text{ small}. \tag{2.11}$$

This approximate solution, which is valid only in the boundary layer where rapid changes are taking place, is called the **inner approximation** and is denoted by $y_i(x)$; it is also plotted in Fig. 3.2.

In this problem we have the advantage of knowing the exact solution, and as yet we have little clue how to determine the inner solution if the exact solution is unknown. The key to the analysis in the boundary layer is rescaling, as in the polynomial equations examined earlier. In the layer, the term $\varepsilon y''$ in the differential equation is not small, as it appears. Consequently, we must rescale the independent variable x in the boundary layer by selecting a small spatial scale that will reflect rapid and abrupt changes and will force each term in the equation into its proper form in the rescaled variables, namely as the product of a coefficient representing the magnitude of the term and a term of order unity. In the next section we develop a general procedure for dealing with boundary layers.

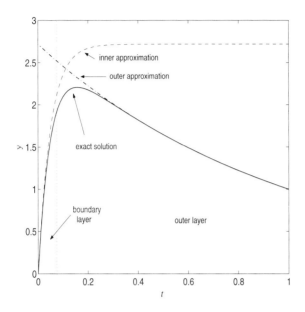

Figure 3.2 The exact solution (2.9) and inner and outer approximations.

Before embarking on an analysis of the boundary layer phenomena, we summarize our observations and make some general remarks. We have noted that a naive regular perturbation expansion does not always produce an approximate solution. In fact, there are several indicators that often suggest its failure.

1. When the small parameter multiplies the highest derivative in the problem.

2. When setting the small parameter equal to zero changes the character of the problem, as in the case of a partial differential equation changing type (elliptic to parabolic, for example), or an algebraic equation changing degree. In other words, the solution for $\varepsilon = 0$ is fundamentally different in character from the solutions for ε close to zero.

3. When problems occur on infinite domains, giving, for example, secular terms.

4. When singular points are present in the interval of interest.

5. When the equations that model physical processes have multiple time or spatial scales.

Such perturbation problems fall in the general category of singular perturbation problems. For ordinary differential equations, problems involving boundary layers are common. The procedure is to determine whether there is a boundary layer and where it is located. If there is a boundary layer, then the leading-order perturbation term found by setting $\varepsilon = 0$ in the equation often provides a valid approximation in a large outer region. The inner approximation in the boundary layer is found by rescaling, which we discuss below. We show that the inner and outer approximations can be *matched* to obtain a uniformly valid approximation over the entire interval of interest. The singular perturbation method applied in this context is also called the method of *matched asymptotic expansions* or *boundary layer theory*. The latter arises from its inception in the study of boundary layer phenomena in air flow over a wing.

EXERCISES

1. Use a dominant balancing method to determine the leading-order behavior of the roots of the following algebraic equations. In all cases $0 < \varepsilon \ll 1$.

   a) $\varepsilon x^4 + \varepsilon x^3 - x^2 + 2x - 1 = 0$.

   b) $\varepsilon x^3 + x - 2 = 0$.

   c) $\varepsilon^2 x^6 - \varepsilon x^4 - x^3 + 8 = 0$.

   d) $\varepsilon x^5 - x^3 - 1 = 0$.

   e) $\varepsilon^2 x^5 - x^3 + 1 = 0$.

2. In (1b) find a first-order correction for the leading behavior (Hint: for the scaled equation assume $\bar{x} = x_0 + x_1 \sqrt{\varepsilon} + x_2 \varepsilon + \cdots .)$

3. Find a two-term approximation to the three roots $x^3 + \varepsilon x + 2\varepsilon^2 = 0$, $\quad 0 < \varepsilon \ll 1$.

4. Show that regular perturbation fails on the boundary value problem

$$\varepsilon y'' + y' + y = 0, \quad 0 < x < 1, \quad 0 < \varepsilon \ll 1,$$
$$y(0) = 0, \quad y(1) = 1.$$

Find the exact solution and sketch it for $\varepsilon = 0.05$ and $\varepsilon = 0.005$. If $x = O(\varepsilon)$, show that $\varepsilon y''(x)$ is large; if $x = O(1)$, show that $\varepsilon y''(x) = O(1)$. Find the inner and outer approximations from the exact solution.

3.3 Boundary Layer Analysis

3.3.1 Inner and Outer Approximations

We return to the boundary value problem

$$\varepsilon y'' + (1+\varepsilon)y' + y = 0, \quad 0 < x < 1, \tag{3.1}$$
$$y(0) = 0, \quad y(1) = 1,$$

where $0 < \varepsilon \ll 1$. By examining the exact solution near $x = 0$ we found that rapid changes were occurring in y, y', and y'', suggesting a small characteristic length; the term $\varepsilon y''$ was not small as it appears to be in the equation. Away from the boundary layer, in the region where $x = O(1)$, it was noted that $\varepsilon y''$ and $\varepsilon y'$ are small, and so the solution could be approximated accurately by setting $\varepsilon = 0$ in the equation to obtain

$$y' + y = 0,$$

and selecting the boundary condition $y(1) = 1$. This gives the outer approximation

$$y_0(t) = e^{1-x}. \tag{3.2}$$

To analyze the behavior in the boundary layer, we notice that significant changes in y take place on a very short spatial interval, which suggests a length scale on the order of a function of ε, say $\delta(\varepsilon)$. If we change variables via

$$\xi = \frac{x}{\delta(\varepsilon)}, \quad Y(\xi) = y(\delta(\varepsilon)\xi) \tag{3.3}$$

and use the chain rule, the differential equation (3.1) becomes

$$\frac{\varepsilon}{\delta(\varepsilon)^2} Y''(\xi) + \frac{(1+\varepsilon)}{\delta(\varepsilon)} Y'(\xi) + Y(\zeta) - 0, \tag{3.4}$$

where prime denotes derivatives with respect to ξ. Another way of looking at the rescaling is to regard (3.3) as a scale transformation that permits examination of the boundary layer close up, as under a microscope.

The coefficients of the four terms in the differential equation are

$$\frac{\varepsilon}{\delta(\varepsilon)^2}, \quad \frac{1}{\delta(\varepsilon)}, \quad \frac{\varepsilon}{\delta(\varepsilon)}, \quad 1. \tag{3.5}$$

If the scaling is correct, each will reflect the order of magnitude of the term in which it appears. To determine the scale factor $\delta(\varepsilon)$ we estimate these magnitudes by considering all possible dominant balances between pairs of terms in

(3.5). In the pairs we include the first term because it was ignored in the outer layer, and it is known that it plays a significant role in the boundary layer. Because the goal is to make a simplification in the problem, we do not consider dominant balancing of three terms. If all four terms are equally important, no simplification can be made at all. Therefore there are three cases to consider. For notation we will sometimes use the symbol $\sim$ to denote "of the same order."

(i) The terms $\varepsilon/\delta(\varepsilon)^2$ and $1/\delta(\varepsilon)$ are of the same order and $\varepsilon/\delta(\varepsilon)$ and 1 are small in comparison.

(ii) The terms $\varepsilon/\delta(\varepsilon)^2$ and 1 are of the same order and $1/\delta(\varepsilon)$ and $\varepsilon/\delta(\varepsilon)$ are small in comparison.

(iii) The terms $\varepsilon/\delta(\varepsilon)^2$ and $\varepsilon/\delta(\varepsilon)$ are of the same order and $1/\delta(\varepsilon)$ and 1 are small in comparison.

Only case (i) is possible. For, in case (ii), $\varepsilon/\delta(\varepsilon)^2 \sim 1$ implies $\delta(\varepsilon) = O(\sqrt{\varepsilon})$; but then $1/\delta(\varepsilon)$ is not small compared to 1. This case is inconsistent. In case (iii), $\varepsilon/\delta(\varepsilon)^2 \sim \varepsilon/\delta(\varepsilon)$ implies $\delta(\varepsilon) = O(1)$, which leads to the outer approximation. In case (i), $\varepsilon/\delta(\varepsilon)^2 \sim 1/\delta(\varepsilon)$ forces $\delta(\varepsilon) = O(\varepsilon)$; then $\varepsilon/\delta(\varepsilon)^2$ and $1/\delta(\varepsilon)$ are both order $1/\varepsilon$, which is large compared to $\varepsilon/\delta(\varepsilon)$ and 1. Therefore, a consistent scaling is possible if we select $\delta(\varepsilon) = O(\varepsilon)$; hence, we take

$$\delta(\varepsilon) = \varepsilon. \tag{3.6}$$

Therefore, the scaled differential equation (3.4) becomes

$$Y'' + Y' + \varepsilon Y' + \varepsilon Y = 0. \tag{3.7}$$

Now, (3.7) is amenable to regular perturbation. Because we are interested only in the leading-order approximation, which we denote by Y_i, we set $\varepsilon = 0$ in (3.7) to obtain

$$Y_i'' + Y_i' = 0. \tag{3.8}$$

The general solution is

$$Y_i(\tau) = C_1 + C_2 e^{-\xi}.$$

Because the boundary layer is located near $x = 0$, we apply the boundary condition $y(0) = 0$, or $Y_i(0) = 0$. This yields $C_2 = -C_1$, and so

$$Y_i(\xi) = C_1(1 - e^{-\xi}). \tag{3.9}$$

In terms of y and x,

$$y_i(t) = C_1(1 - e^{-x/\varepsilon}). \tag{3.10}$$

This is the inner approximation for $x = O(\varepsilon)$.

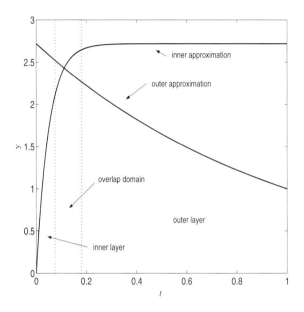

Figure 3.3 The overlap domain where inner and outer approximations are matched.

In summary, we have the approximate solution

$$y_0(x) = e^{1-x}, \quad x = O(1)$$
$$y_i(x) = C_1(1 - e^{-x/\varepsilon}), \quad x = O(\varepsilon), \tag{3.11}$$

each valid for an appropriate range of x. There remains to determine the constant C_1, which is accomplished by the process of matching.

3.3.2 Matching

By matching we do *not* mean selecting a specific value of ε, say ε_0, and requiring that $y_0(\varepsilon_0) = y_i(\varepsilon_0)$ to obtain the constant C_1. This process is nothing more than patching the two approximations together so as to be continuous at some fixed $x = \varepsilon_0$. Rather, the goal is to construct a single composite expansion in ε that is uniformly valid on the entire interval $[0, 1]$ as $\varepsilon \to 0$. Thus, there is little gain in pinpointing the edge of the boundary layer, since it becomes narrower as $\varepsilon \to 0$. We stated figuratively that the boundary layer for this problem has width ε. In general, it is the scaling factor $\delta(\varepsilon)$ that defines the **width** of the boundary layer. One can proceed as follows, however. It seems reasonable that the inner and outer expansions should agree to some order in an overlap domain, intermediate between the boundary layer and outer region (see Fig. 3.3). If $x = O(\varepsilon)$, then x is in the boundary layer, and if $x = O(1)$, then x is in the outer region; therefore, this overlap domain could be characterized as values of x for which $x = O(\sqrt{\varepsilon})$, for example, because $\sqrt{\varepsilon}$ is between ε and 1 ($\sqrt{\varepsilon}$ goes to zero slower than ε does, or $\varepsilon \ll \sqrt{\varepsilon}$). This intermediate scale suggests a new scaled independent variable η in the overlap domain defined by

$$\eta = \frac{x}{\sqrt{\varepsilon}}. \tag{3.12}$$

The condition for matching is that the inner approximation, written in terms of the **intermediate variable** η, should agree with the outer approximation, written in terms of the intermediate variable η, in the limit as $\varepsilon \to 0^+$. In symbols, for matching we require that for fixed η

$$\lim_{\varepsilon \to 0^+} y_0(\sqrt{\varepsilon}\eta) = \lim_{\varepsilon \to 0^+} y_i(\sqrt{\varepsilon}\eta). \tag{3.13}$$

For the present problem,

$$\lim_{\varepsilon \to 0^+} y_0(\sqrt{\varepsilon}\eta) = \lim_{\varepsilon \to 0^+} e^{1-\sqrt{\varepsilon}\eta} = e$$

and

$$\lim_{\varepsilon \to 0^+} y_i(\sqrt{\varepsilon}\eta) = \lim_{\varepsilon \to 0^+} C_1(1 - e^{-\eta/\sqrt{\varepsilon}}) = C_1.$$

Therefore, matching requires $C_1 = e$ and the inner approximation becomes

$$y_i(t) = e(1 - e^{-x/\varepsilon}). \tag{3.14}$$

Because we are seeking only a leading order approximation, we can avoid introducing an intermediate variable and just write the **matching condition** (3.13) simply as

$$\lim_{x \to 0^+} y_0(x) = \lim_{\xi \to \infty} Y_i(\xi). \tag{3.15}$$

This states that the outer approximation, as the outer variable moves into the inner region, must equal the inner approximation, as the inner variable moves into the outer region. Finally, we point out that we have only introduced approximations of leading order. Higher-order approximations can be obtained by more elaborate matching schemes, and we refer to the references for discussions of these methods. For example, see Bender and Orszag (1978). In the sequel we usually use (3.15) instead of introducing an intermediate variable.

3.3.3 Uniform Approximations

To obtain a composite expansion that is uniformly valid throughout $[0, 1]$, we note the sum of the outer and inner approximations is

$$y_0(x) + y_i(x) = e^{1-x} + e - e^{1-x/\varepsilon}$$

$$\approx \begin{cases} e^{1-x} + e, & x = O(1) \\ 2e - e^{1-x/\varepsilon}, & x = O(\varepsilon). \end{cases}$$

By subtracting the common limit (3.13), which is e, we have

$$y_u(x) \equiv y_0(x) + y_i(x) - e = e^{1-x} - e^{1-x/\varepsilon}. \tag{3.16}$$

When x is in the outer region the second term is small and $y_u(x)$ is approximately e^{1-x}, which is the outer approximation. When x is in the boundary layer the first term is nearly e and $y_u(x)$ is approximately $e(1 - e^{-x/\varepsilon})$, which is the inner approximation. In the intermediate or overlap region, both the inner and outer approximations are approximately equal to e. Therefore, in the overlap domain the sum of $y_0(x)$ and $y_i(x)$ gives $2e$, or twice the contribution. This is why we must subtract the common limit from the sum. In summary, $y_u(x)$ provides a uniform approximate solution throughout the interval $[0, 1]$.

Briefly summarizing, the **matching condition** is:

uniform approximation $=$ outer $+$ inner $-$ common limit.

Substituting $y_u(x)$ into the differential equation shows that

$$\varepsilon y_u'' + (1 + \varepsilon)y_u' + y_u = 0,$$

so $y_u(x)$ satisfies the differential equation exactly on $(0, 1)$. Checking the boundary conditions,

$$y_u(0) = 0, \quad y_u(1) = 1 - e^{1-1/\varepsilon}.$$

The left boundary condition is satisfied exactly and the right boundary condition holds up to $O(\varepsilon^n)$, for any $n > 0$, because

$$\lim_{\varepsilon \to 0^+} \frac{e^{1-1/\varepsilon}}{\varepsilon^n} = 0,$$

for any $n > 0$. Consequently, y_u is a uniformly valid approximation on $[0, 1]$.

Example 3.11

We work through another simple example without the detailed exposition of the preceding paragraphs. We determine an approximate solution of the boundary value problem

$$\varepsilon y'' + y' = 2x, \quad 0 < x < 1, \quad 0 < \varepsilon \ll 1,$$
$$y(0) = 1, \quad y(1) = 1,$$

using singular perturbation methods. Clearly, regular perturbation will fail since the unperturbed problem is

$$y' = 2x,$$

which has the general solution

$$y(x) = x^2 + C.$$

Such a function cannot satisfy both boundary conditions. Consequently we assume a boundary layer at $x = 0$ and impose the boundary condition $y(1) = 1$ (since $x = 1$ is in the outer layer) to get the outer approximation

$$y_0(x) = x^2, \quad x = O(1).$$

To determine the width $\delta(\varepsilon)$ of the boundary layer we rescale near $x = 0$ via

$$\xi = \frac{x}{\delta(\varepsilon)}, \quad Y(\xi) = y(x).$$

In scaled variables the differential equation becomes

$$\frac{\varepsilon}{\delta(\varepsilon)^2} Y'' + \frac{1}{\delta(\varepsilon)} Y' = 2\delta(\varepsilon)\xi.$$

If $\varepsilon/\delta(\varepsilon)^2 \sim 2\delta(\varepsilon)$ is the dominant balance, then $\delta(\varepsilon) = O(\varepsilon^{1/3})$ and the second term $1/\delta(\varepsilon)$ would be $O(\varepsilon^{-1/3})$, which is not small compared to the assumed dominant terms. Therefore assume $\varepsilon/\delta(\varepsilon)^2 \sim 1/\delta(\varepsilon)$ is the dominant balance. In that case $\delta(\varepsilon) = O(\varepsilon)$ and the term $2\delta(\varepsilon)$ has order $O(\varepsilon)$, which is small compared to $\varepsilon/\delta(\varepsilon)^2$ and $1/\delta(\varepsilon)$, both of which are $O(\varepsilon^{-1})$. Therefore $\delta(\varepsilon) = \varepsilon$ is consistent and the scaled differential equation is

$$Y'' + Y' = 2\varepsilon^2\xi.$$

The inner approximation to first order satisfies

$$Y_i'' + Y_i' = 0,$$

whose general solution is

$$Y(\xi) = C_1 + C_2 e^{-\xi}.$$

In terms of y and x

$$y(x) = C_1 + C_2 e^{-x/\varepsilon}.$$

Applying the boundary condition $y(0) = 1$ in the boundary layer gives $C_1 = 1 - C_2$, and therefore the inner approximation is

$$y_i(x) = (1 - C_2) + C_2 e^{-x/\varepsilon}.$$

To find C_2 we introduce an overlap domain of order $\sqrt{\varepsilon}$ and an appropriate intermediate scaled variable

$$\eta = \frac{x}{\sqrt{\varepsilon}}.$$

Then $x = \sqrt{\varepsilon}\eta$ and the matching condition becomes (with η fixed)

$$\lim_{\varepsilon \to 0^+} y_0(\sqrt{\varepsilon}\eta) = \lim_{\varepsilon \to 0^+} y_i(\sqrt{\varepsilon}\eta),$$

or

$$\lim_{\varepsilon \to 0^+} \varepsilon\eta^2 = \lim_{\varepsilon \to 0^+} [(1 - C_2) + C_2 e^{-\eta/\sqrt{\varepsilon}}].$$

This gives $C_2 = 1$, and thus the inner approximation is

$$y_i(x) = e^{-x/\varepsilon}.$$

A uniform composite approximation $y_u(x)$ is found by adding the inner and outer approximations and subtracting the common limit in the overlap domain, which is zero in this case. Consequently,

$$y_u(x) = x^2 + e^{-x/\varepsilon}.$$

As an exercise, the reader should plot this approximation for different values of ε. Again, as a shortcut, the matching condition may be written

$$\lim_{x \to 0} y_0(x) = \lim_{\xi \to \infty} Y_i(\xi),$$

which gives the same result. □

3.3.4 General Procedures

In the preceding examples the boundary layer occurred at $x = 0$, or at the left endpoint. In the general case boundary layers can occur at any point in the interval, at the right endpoint or an interior point; in fact, multiple boundary layers can occur in the same problem. Boundary layers also appear in initial value problems, which we discuss in the next section. The Exercises point out different boundary layer phenomena. When solving a problem one should assume a boundary layer at $x = 0$ (or the left endpoint) and then proceed. If this assumption is in error, then the procedure will break down when trying to match the inner and outer approximations. At that time make an assumption of a boundary layer at the right endpoint. The analysis is exactly the same, but the scale transformation to define the inner variable in the boundary layer becomes

$$\xi = \frac{x_0 - x}{\delta(\varepsilon)},$$

where $x = x_0$ is the right endpoint, and $Y(\xi) = y(x_0 - \delta(\varepsilon)\xi)$. To compute transformed derivatives in this case, the chain rule shows

$$\frac{dy}{dx} = -\frac{1}{\delta(\varepsilon)}\frac{dY}{d\xi}, \quad \frac{d^2y}{dx^2} = \frac{1}{\delta(\varepsilon)^2}\frac{d^2Y}{d\xi^2}.$$

Observe the negative sign on the first derivative. The matching condition at a right boundary layer is

$$\lim_{\xi \to \infty} Y_i(\xi) = \lim_{x \to x_0} y_0(x).$$

In words, the limit of the inner solution as the inner variable moves out of the layer equals the limit of the outer solution as the outer variable moves into the layer.

To reemphasize, boundary layers can occur at the left endpoint, right endpoint, at both endpoints, or neither. There can also be internal layers, called *shock layers*. The latter often leads to difficult computations.

Further, even though preceding examples exhibit a boundary layer width of $\delta(\varepsilon) = \varepsilon$, this is not the rule. Also, we have only matched the leading-order behavior of the inner and outer approximations. Refined matching procedures can include matching the higher order terms in expansions. (For reference, the reader should search on Van Dyke's matching principle.) Finally we point out that this method is not a universal technique. For certain classes of problems it works well, but for other problems significant modifications must be made. Singular perturbation theory is an active area of research in applied mathematics and a well-developed rigorous theory is only available for restricted classes of differential equations.

We conclude with a theorem for linear equations with variable coefficients. For this class of problems, the boundary layer can be completely characterized, provided suitable restrictions are placed on the coefficients.

Theorem 3.12

Consider the boundary value problem

$$\varepsilon y'' + p(x)y' + q(x)y = 0, \quad 0 < x < 1, \quad 0 < \varepsilon \ll 1, \tag{3.17}$$
$$y(0) = a, \quad y(1) = b,$$

where p and q are continuous functions on $0 \le x \le 1$, and $p(x) > 0$ for $0 \le x \le 1$. Then there exists a boundary layer at $x = 0$ with inner and outer approximations given by

$$y_i(x) = C_1 + (a - C_1)e^{-p(0)x/\varepsilon}, \tag{3.18}$$

$$y_0(x) = b \exp \left(\int_x^1 \frac{q(s)}{p(s)} \, ds \right), \tag{3.19}$$

where

$$C_1 - b \exp \left(\int_0^1 \frac{q(s)}{p(s)} \, ds \right). \quad \square \tag{3.20}$$

Proof

To demonstrate the theorem, we show that the assumption of a boundary layer at $x = 0$ is consistent and leads to the approximations given above. If the boundary layer is at $x = 0$, then the outer solution $y_0(x)$ will satisfy

$$p(x)y_0' + q(x)y_0 = 0$$

and the condition $y_0(1) = b$. Separating variables and applying the condition at $x = 1$ gives (3.19). In the boundary layer we introduce a scaled variable ξ defined by $\xi = x/\delta(\varepsilon)$, where $\delta(\varepsilon)$ is to be determined. If $Y(\xi) = y(\delta(\varepsilon)\xi)$, then the differential equation becomes

$$\frac{\varepsilon}{\delta(\varepsilon)^2}Y'' + \frac{p(\delta(\varepsilon)\xi)}{\delta(\varepsilon)}Y' + q(\delta(\varepsilon)\xi)Y = 0. \tag{3.21}$$

As $\varepsilon \to 0^+$ the coefficients behave like

$$\frac{\varepsilon}{\delta(\varepsilon)^2}, \quad \frac{p(0)}{\delta(\varepsilon)}, \quad q(0).$$

It is easy to see that the dominant balance is $\varepsilon/\delta(\varepsilon)^2 \sim p(0)/\delta(\varepsilon)$, and therefore the boundary layer has thickness $\delta(\varepsilon) = O(\varepsilon)$. For definiteness take $\delta(\varepsilon) = \varepsilon$. Equation (3.21) then becomes

$$Y'' + p(\varepsilon\xi)Y' + \varepsilon q(\varepsilon\xi)Y = 0,$$

which to leading order is

$$Y_i'' + p(0)Y_i' = 0.$$

The general solution is

$$Y(\xi) = C_1 + C_2 e^{-p(0)\xi}.$$

Applying the boundary condition $Y_i(0) = a$ yields $C_2 = a - C_1$, and thus the inner approximation is

$$y_i(x) = C_1 + (a - C_1)e^{-p(0)x/\varepsilon}. \tag{3.22}$$

To match we introduce the intermediate variable $\eta = x/\sqrt{\varepsilon}$ and require that for fixed η,

$$\lim_{\varepsilon \to 0^+} y_i(\sqrt{\varepsilon}\eta) = \lim_{\varepsilon \to 0^+} y_0(\sqrt{\varepsilon}\eta).$$

In this case the matching condition becomes

$$\lim_{\varepsilon \to 0^+} \{C_1 + (a - C_1)e^{-p(0)\eta/\sqrt{\varepsilon}}\} = \lim_{\varepsilon \to 0^+} \left\{ b \exp\left(\int_{\sqrt{\varepsilon}\eta}^1 \frac{q(s)}{p(s)} \, ds \right) \right\},$$

which forces

$$C_1 = b \exp\left(\int_0^1 \frac{q(s)}{p(s)} \, ds \right).$$

Consequently, the inner approximation is given as stated in (3.18) and (3.20), and the proof is complete. A uniform composite approximation $y_u(x)$ is given by

$$y_u(x) = y_0(x) + y_i(x) - C_1.$$

It can be shown that $y_u(x) - y(x) = O(\varepsilon)$ as $\varepsilon \to 0^+$, uniformly on $[0, 1]$, where $y(x)$ is the exact solution to (3.17). $\square$

Remark 3.13

If $p(x) < 0$ on $0 \le x \le 1$, then no match is possible because $y_i(x)$ would grow exponentially, unless $C_1 = a$. On the other hand, a match is possible if $p(x) < 0$ if the boundary layer is at $x = 1$. In summary, one can show that the boundary layer is at $x = 0$ if $p(x) > 0$ and at $x = 1$ if $p(x) < 0$. As a final observation, there can be no boundary layer at an interior point $0 < x < 1$ in either case. If p changes sign in the interval, then interior layers are possible; these problems are called **turning point** problems. $\square$

EXERCISES

1. Use singular perturbation methods to obtain a uniform approximate solution to the following problems. In each case assume $0 < \varepsilon \ll 1$ and
$0 < x < 1$.

   a) $\varepsilon y'' + 2y' + y = 0$, $y(0) = 0$, $y(1) = 1$.

   b) $\varepsilon y'' + y' + y^2 = 0$, $y(0) = \frac{1}{4}$, $y(1) = \frac{1}{2}$.

   c) $\varepsilon y'' + (1 + x)y' = 1$, $y(0) = 0, y(1) = 1 + \ln 2$.

   d) $\varepsilon y'' + (x + 1)y' + y = 0$, $y(0) = 0$, $y(1) = 1$

   e) $\varepsilon y'' + x^{1/3}y' + y = 0$, $y(0) = 0$, $y(1) = \exp\left(-\frac{3}{2}\right)$.

   f) $\varepsilon y'' + xy' - xy = 0$, $y(0) = 0$, $y(1) = e$.

   g) $\varepsilon y'' + 2y' + e^y = 0$, $y(0) = y(1) = 0$.

   h) $\varepsilon y'' - (2 - x^2)y = -1$, $y'(0) = 0$, $y(1) = 1$.

   i) $\varepsilon y'' - b(x)y' = 0$, $y(0) = \alpha$, $y(1) = \beta$, where $\alpha \neq \beta$, $b(x) > 0$.

   j) $\varepsilon y'' - 4(\pi - x^2)y = \cos x$, $y(0) = 0$, $y(\pi/2) = 1$.

2. Examine the exact solution to show why singular perturbation methods
fail on the boundary value problem

$$\varepsilon u'' + u = 0, \quad 0 < x < 1,$$
$$u(0) = 1, \quad u(1) = 2.$$

3. Determine values of a for which the problem

$$\varepsilon y'' + y' + ae^y = 0, \quad y(0) = 0, \, y(1) = 0$$

has a solution with a boundary layer structure.

4. Obtain a uniform approximation to the problem

$$\varepsilon u'' - (2x + 1)u' + 2u = 0, \quad 0 < x < 1,$$
$$u(0) = 1, \quad u(1) = 0.$$

5. Is the problem

$$\varepsilon y'' + \frac{1}{x} y' + y = 0, \quad x > 0,$$
$$y(0) = 1, \quad y'(0) = 0$$

a singular perturbation problem? Discuss.

6. Attempt singular perturbation methods on the boundary value problem

$$\varepsilon y'' + \left(x - \tfrac{1}{2}\right) y = 0, \quad 0 < x < 1,$$
$$y(0) = 1, \quad y(1) = 2.$$

Give a complete discussion of your results.

7. Find a leading order approximation to

$$\varepsilon y'' + e^x y' = 1, \quad y(0) = 1, \ y(\infty) = 0.$$

8. Let f be a smooth function, monotonic on $[0, 1]$, and let $f'(0) = b \neq 0$. Either find a uniformly valid approximation to the problem

$$\varepsilon y'' - \frac{1}{f'(x)} y' = -1, \quad y'(0) = 1, \ y(1) = 0,$$

or show that such an approximation does not exist. Does the answer depend upon the value b?

9. Find a uniformly valid approximation to

$$-\varepsilon u'' + \cos(b - x)u = 1, \quad 0 < x < 1,$$
$$u(0) = 0, \quad u'(b) = 0,$$

where $0 < \varepsilon \ll 1$ and $b > 0$. For what values of b is the approximation valid?

10. Find a uniformly valid approximation to

$$\varepsilon u'' - a(x)u = f(t), \quad 0 < x < 1,$$
$$u(0) = 0, \quad u(b) = -f(1)/a(1),$$

where $0 < \varepsilon \ll 1$ and $a > 0$, and a and f have infinitely many derivatives on $\mathbb{R}$.

11. Find a uniform approximation to the boundary value problem

$$\varepsilon y'' + \frac{1}{1 + x} y' + \varepsilon y \ = \ 0, \quad 0 < x < 1, \quad 0 < \varepsilon \ll 1,$$
$$y(0) \ = \ 0, \quad y(1) = 1.$$

Draw a graph of the approximation.

12. Consider the boundary value problem on $0 < x < 1$:

$$\varepsilon^2 u'' + \varepsilon x u' - u = e^{-x}, \quad u(0) = 2, \ u(1) = 1.$$

Find a uniform approximation. (Hint: try a multiple layer.)

13. Find a uniform approximation:

$$\varepsilon y'' - 2y' + 2y = 0, \quad y(0) = 0, /, /, y(1) = 1.$$

14. Find a uniform approximation:

$$\varepsilon y'' - \varepsilon y y' - y = x, \quad y(0) = 2, /, /, y(1) = 0.$$

15. Find a uniform approximation to the boundary value problem

$$\varepsilon^2 u'' - x u' - x = 0, \quad u(0) = 2, \ u(1) = -2.$$

(Hint: try an interior, or *shock*, layer.)

16. Investigate the problem

$$\varepsilon y'' - y y' - y = 0, \quad y(0) = 2, \ y(1) = -2.$$

17. Find uniform asymptotic approximations to the solutions of the following problems on $0 \le x \le 1$ with $0 < \varepsilon \ll 1$.

a) $\varepsilon y'' + (x^2 + 1)y' - x^3 y = 0, \quad y(0) = y(1) = 1.$

b) $\varepsilon y'' + (\cosh x)y' - y = 0, \quad y(0) = y(1) = 1.$

c) $\varepsilon y'' + 2\varepsilon x^{-1} y' - y = 0, \quad y(0) = 0, \quad y'(1) = 1.$

d) $\varepsilon(y'' + y') - y = 0, \quad y(0) = y(1) = 0.$

e) $\varepsilon y'' + (1 + x)^{-1} y' + \varepsilon y = 0, \quad y(0) = 0, \quad y(1) = 1.$

3.4 Initial Layers

3.4.1 Damped Spring–Mass System

Initial value problems containing a small parameter can have a layer near $t = 0$. For evolution problems these layers are called initial layers. They are a zones in which there are rapid temporal changes. We will illustrate the idea on a simple, solvable initial value problem. Consider a damped spring-mass system with y denoting the positive displacement of the mass m from equilibrium, and with the forces given by

$$F_{\text{spring}} = -ky, \quad F_{\text{damping}} = -a\dot{y}, \quad (k, a > 0),$$

where k and a are the spring constant and damping constant, respectively. By Newton's second law the governing differential equation is

$$m\ddot{y} + a\dot{y} + ky = 0, \quad t > 0. \tag{4.1}$$

Initially, we assume the displacement is zero and that the mass is put into motion by imparting to it (say, with a hammer blow) a positive impulse I. Therefore, the initial conditions are given by

$$y(0) = 0, \quad m\dot{y}(0) = I. \tag{4.2}$$

We want to determine the leading-order behavior of the system in the case that the mass has very small magnitude.

The first step is to recast the problem in dimensionless form. The independent and dependent variables are t and y having dimensions of time T and length L, respectively. The constants m, a, k, and I have dimensions

$$[m] = M, \quad [a] = MT^{-1}, \quad [k] = MT^{-2}, \quad [I] = MLT^{-1},$$

where M is a mass dimension. We can identify three possible time scales,

$$\frac{m}{a}, \quad \sqrt{\frac{m}{k}}, \quad \frac{a}{k}. \tag{4.3}$$

These correspond to balancing inertia and damping terms, the inertia and the spring terms, and the damping and the spring terms, respectively. Possible length scales are

$$\frac{I}{a}, \quad \frac{I}{\sqrt{km}}, \quad \frac{aI}{km}. \tag{4.4}$$

By assumption $m \ll 1$, and so $m/a \ll 1$, $\sqrt{m/k} \ll 1$, $I/\sqrt{km} \gg 1$, and $aI/km \gg 1$. Such relations are often important in determining appropriate time and length scales, because scales should be chosen so that the dimensionless dependent and independent variables are of order 1.

Before proceeding further, we use our intuition regarding the motion of the mass. The positive impulse given to the mass will cause a rapid displacement to some maximum value, at which time the force due to the spring will attempt to restore it to its equilibrium position. Because the mass is small there will be very little inertia, and therefore it will probably not oscillate about equilibrium; the system is strongly over-damped. Consequently we expect the graph of $y = y(t)$ to exhibit the features shown in Fig. 3.4—a quick rise and then a gradual decay. Such a function is indicative of a multiple time-scale description. In the region near $t = 0$, where there is an abrupt change, a short time scale seems appropriate; we call this the **initial layer**. Away from the initial layer an order

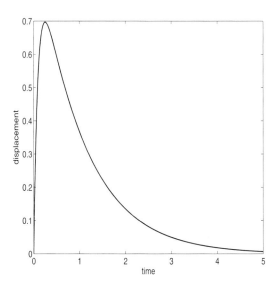

Figure 3.4 Displacement vs. time.

1 time scale seems appropriate. In all, the problem has the characteristics of a singular perturbation problem, namely multiple time scales and an apparent small quantity m multiplying the highest derivative term. Of the length scales given in (4.4), it appears that only I/a is suitable. The remaining two are large, which violates our intuition regarding the maximum displacement. Of the time scales, only a/k is first order. The remaining two are small, and one of them may be suitable in the assumed initial layer near $t = 0$. As a guess, since m/a depends on the mass and the damping, and $\sqrt{m/k}$ depends on the mass and the spring, we predict the former. The reason lies in intuition concerning the dominant processes during the early stages of motion. The high initial velocity should influence the damping force more than the force due to the spring. Therefore, the terms $m\ddot{y}$ and $a\dot{y}$ should dominate during the initial phase, whereas $m\ddot{y}$ should be low-order in the later phase.

With these remarks as motivation, we introduce scaled variables

$$\bar{t} = \frac{t}{a/k}, \quad \bar{y} = \frac{y}{I/a}, \tag{4.5}$$

which seem appropriate for the outer layer, away from the initial layer at $t = 0$. Note that both $\bar{t}$ and $\bar{y}$ are of order 1. In terms of these variables the initial value problem becomes

$$\varepsilon\bar{y}'' + \bar{y}' + \bar{y} = 0,$$
$$\bar{y}(0) = 0, \quad \varepsilon\bar{y}'(0) = 1, \tag{4.6}$$

where prime denotes the derivative with respect to $\bar{t}$, and the dimensionless constant ε is given by

$$\varepsilon = \frac{mk}{a^2} \ll 1. \tag{4.7}$$

Notice, by proper scaling, the small term occurs where we expect it, namely in the inertia term. Using singular perturbation techniques we can obtain an approximate solution to (4.5). The leading order approximation in the outer layer is

$$\bar{y}' + \bar{y} = 0,$$

where we have set $\varepsilon = 0$. Then,

$$\bar{y}_0(\bar{t}) = Ce^{-\bar{t}} \tag{4.8}$$

is the outer approximation. Neither initial condition is appropriate to apply because both are at $\bar{t} = 0$, which is in the assumed initial layer. To obtain an inner approximation we rescale according to

$$\tau = \frac{\bar{t}}{\delta(\varepsilon)}, \quad Y = \bar{y}. \tag{4.9}$$

Then (4.6) becomes

$$\frac{\varepsilon}{\delta(\varepsilon)^2} Y'' + \frac{1}{\delta(\varepsilon)} Y' + Y = 0. \tag{4.10}$$

The dominant balance is $\varepsilon/\delta(\varepsilon)^2 \sim 1/\delta(\varepsilon)$, which gives the width of the initial layer, $\delta(\varepsilon) = \varepsilon$. (The balance $\varepsilon/\delta(\varepsilon)^2 \sim 1$ implies $\delta(\varepsilon) = O(\sqrt{\varepsilon})$, but then $1/\delta(\varepsilon)$ is not small, giving inconsistency.) Consequently, the initial layer has width ε, or $O(m)$. The scale transformation (4.9) is then

$$\tau = \frac{\bar{t}}{\varepsilon}, \quad Y = \bar{y},$$

and the differential equation (4.10) becomes

$$Y'' + Y' + \varepsilon Y = 0.$$

Setting $\varepsilon = 0$ and solving the resulting equation gives the inner approximation

$$Y_i(\tau) = A + Be^{-\tau},$$

or

$$\bar{y}_i(\bar{t}) = A + Be^{-\bar{t}/\varepsilon}.$$

The initial condition $\bar{y}_i(0) = 0$ forces $B = -A$, and the condition $\varepsilon \bar{y}_i'(0) = 1$ forces $A = 1$. Therefore, the inner approximation is completely determined by

$$\bar{y}_i(\bar{t}) = 1 - e^{-\bar{t}/\varepsilon}. \tag{4.11}$$

Next we use matching to determine the constant C. We can introduce an overlap domain with time scale $\sqrt{\varepsilon}$ and scaled dimensionless variable $\eta = \frac{\bar{t}}{\sqrt{\varepsilon}}$, but we prefer to apply the simple condition (3.15), or

$$\lim_{\bar{t} \to 0} \bar{y}_0(\bar{t}) = \lim_{\tau \to \infty} Y_i(\tau).$$

Both limits give $C = 1$. A uniformly valid approximation is therefore

$$\bar{y}_u(\bar{t}) = \bar{y}_0(\bar{t}) + \bar{y}_i(\bar{t}) - \lim_{\varepsilon \to 0^+} y_0(\bar{t}),$$

or

$$\bar{y}_u(\bar{t}) = e^{-\bar{t}} - e^{-\bar{t}/\varepsilon}.$$

Here the limits are taken with η fixed. In terms of the original dimensioned variables t and y,

$$y_u(t) = \frac{I}{a}(e^{-kt/a} - e^{-at/m}). \tag{4.12}$$

Intuition proved to be correct; for small t we have

$$y_u(t) \cong \frac{I}{a}(1 - e^{-at/m}),$$

which describes a rapidly rising function with time scale m/a. For larger t the first term in (4.12) dominates and

$$y_u(t) \cong \frac{I}{a}e^{-kt/a},$$

which represents a function that is decaying exponentially on a time scale a/k.

This example shows how physical reasoning can complement mathematical analysis in solving problems arising out of empirics. The constant interplay between the physics and the mathematics leads to a good understanding of a problem and gives insight into its structure, as well as strategies for obtaining solutions.

3.4.2 Enzyme Kinetics

In chemical processes it is important to know the concentrations of the chemical species when reactions take place. The differential equations of reaction kinetics provide a rich source of singular perturbation phenomena because reactions in a chain often occur on different time scales. For example, some intermediary chemical species may be short-lived, and reactions creating those species may be able to be ignored, thereby leading to simplification of the system.

In Section 2.6 of Chapter 2 we introduced mass action dynamics for chemical kinetics. We review some of those ideas and we recommend readers review that material as well, especially enzyme kinetics.

The basic enzyme reaction is

$$S + E \rightleftarrows C \rightarrow P + E, \tag{4.13}$$

where reactant molecules S (a substrate) and E (an enzyme) combine to form a complex molecule C. The molecule C breaks up to form a product molecule P and the original enzyme E, or the reverse reaction, C disassociating into the original reactants S and E can occur. There are three reactions in (4.13), namely $S + E \overset{k_1}{\rightarrow} C$, $C \overset{k_{-1}}{\rightarrow} S + E$, and $C \overset{k_2}{\rightarrow} P + E$. Here, one molecule of a substrate S reacts with an enzyme E to produce an intermediate complex C, and then the final product P is produced, along with the recovery of the enzyme. The formation of the complex C is a rapid reaction, and usually the initial concentration of the enzyme is small compared to the substrate. Initially, we assume $S(0) = S_0$, $E(0) = E_0$, $P(0) = 0$, and $C(0) = 0$. By the law of mass action, the rate equations are (we are using τ for real time)

$$\frac{dS}{d\tau} = -k_1 SE + k_{-1}C,$$
$$\frac{dE}{d\tau} = -k_1 SE + (k_{-1} + k_2)C,$$
$$\frac{dC}{d\tau} = k_1 SE - (k_{-1} + k_2)C,$$
$$\frac{dP}{d\tau} = k_2 C.$$

Notice that once C is determined, then P is determined by direct integration; P does not enter the first three equations. Thus, there are only three equations for S, E, and C. We can instantly find the conservation laws by observation; they are

$$E + C = E_0, \qquad S + C + P = S_0.$$

These equations permit elimination of E and the reduction to a system of two nonlinear equations for S and C,

$$\frac{dS}{d\tau} = -k_1 E_0 S + (k_{-1} + k_1 S)C, \tag{4.14}$$
$$\frac{dC}{d\tau} = k_1 E_0 S - (k_2 + k_{-1} + k_1 S)C. \tag{4.15}$$

Singular Perturbation. Let us rescale

$$x = \frac{S}{S_0}, \quad y = \frac{C}{E_0}, \quad t = \frac{\tau}{T},$$

where T is a yet unchosen time scale. Then the system becomes

$$\frac{dx}{dt} = -k_1 E_0 T x + (k_{-1} + k_1 S_0 x) T \frac{E_0}{S_0} y,$$

$$\frac{dy}{dt} = k_1 S_0 T x - (k_2 + k_{-1} + k_1 S_0 x) T y.$$

From these equations we can note that there are two obvious possible time scales T,

$$T_s = \frac{1}{k_1 E_0}, \quad T_f = \frac{1}{k_1 S_0},$$

where the subscripts s and f denote slow and fast; typically, E_0 is much smaller than S_0. To understand the longer time behavior we use the slow time scale T_s. Substituting into the equations, while letting

$$\mu = \frac{k_{-1}}{k_1 S_0}, \quad \lambda = \frac{k_{-1} + k_2}{k_1 S_0}, \quad \varepsilon = \frac{E_0}{S_0},$$

we get

$$\frac{dx}{dt} = -x + (\mu + x)y,$$

$$\varepsilon \frac{dy}{dt} = x - (\lambda + x)y.$$

It is usually the case that μ and λ are order 1 quantities, but $\varepsilon \ll 1$.

Outer approximation. The equations are

$$\frac{dx_0}{dt} = -x_0 + (\mu + x_0)y,$$

$$0 = x_0 - (\lambda + x_0)y_0.$$

This gives

$$y_0 = \frac{x_0}{\lambda + x_0},$$

$$\frac{dx_0}{dt} = -\frac{(\mu - \lambda)x_0}{\lambda + x_0}.$$

Solving the second equation gives

$$x_0 + \ln x_0 = -(\mu - \lambda)t + A,$$

where A is a constant of integration.

Inner approximation. Let

$$\bar{t} = \frac{t}{\varepsilon} = \frac{\tau}{T_f},$$

which is the fast time scale. Letting $X = x$ and $Y = y$ we get

$$\frac{dX}{d\bar{t}} = \varepsilon\left(-X + (\mu + X)Y\right),$$

$$\frac{dY}{d\bar{t}} = X - (\lambda + X)Y.$$

We may apply the initial conditions $X(0) = 1$ and $Y(0) = 0$. Then to leading order,

$$\frac{dX_0}{d\bar{t}} = 0,$$

$$\frac{dY_0}{d\bar{t}} = X_0 - (\lambda + X_0)Y_0.$$

This means $X_0 = D = $ const.$= 1$. Thus

$$\frac{dY_0}{d\bar{t}} = 1 - (\lambda + 1)Y_0, \quad Y_0(0) = 0.$$

The general solution of the differential equation is

$$Y_0(\bar{t}) = Me^{-(\lambda+1)\bar{t}} + \frac{1}{\lambda+1}.$$

Applying the initial condition gives $M = -1/(\lambda + 1)$. Therefore,

$$Y_0(\bar{t}) = \frac{1}{\lambda+1}\left(1 - e^{-(\lambda+1)\bar{t}}\right).$$

Matching. We must have

$$\lim_{t \to 0} x_0(t) = \lim_{\bar{t} \to \infty} X_0(\bar{t}), \quad \lim_{t \to 0} y_0(t) = \lim_{\bar{t} \to \infty} Y_0(\bar{t}),$$

or

$$\lim_{t \to 0} x_0(t) = 1, \quad \lim_{t \to 0} \frac{x_0}{\lambda + x_0} = \frac{1}{\lambda+1}.$$

This gives consistency. Therefore,

$$\lim_{t \to 0}[x_0 + \ln x_0 + (\mu - \lambda)t + A] = 1 + A = 1,$$

or $A = 1$. In summary, the outer and inner solutions are

$$x_0 + \ln x_0 = -(\mu - \lambda)t, \quad y_0 = \frac{x_0}{\lambda + x_0},$$

$$X_0(\bar{t}) = 1, \quad Y_0(\bar{t}) = Me^{-(\lambda+1)\bar{t}} + \frac{1}{\lambda+1}.$$

The uniform approximation is the outer solution plus the inner solution, minus the common limit. Thus,

$$
\begin{aligned}
x_{\text{unif}}(t) &= x_0 + 1 - 1 = x_0 \\
y_{\text{unif}}(t) &= \frac{x_0}{\lambda + x_0} + \frac{1}{\lambda + 1}\left(1 - e^{-(\lambda+1)t/\varepsilon}\right) - \frac{1}{\lambda + 1} \\
&= \frac{x_0}{\lambda + x_0} - \frac{1}{\lambda + 1}e^{-(\lambda+1)t/\varepsilon}.
\end{aligned}
$$

Figures 3.5 and 3.6 show the fast time solution that approximates the initial layer where the complex is rapidly formed and the slow time solution that approximates the longer behavior.

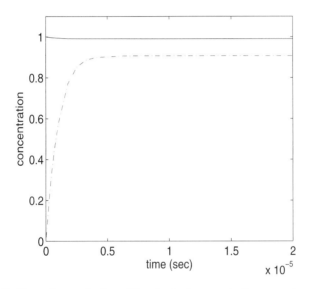

Figure 3.5 Fast time solution. The dashed curve is the complex and the solid curve is the substrate.

EXERCISES

1. Use a singular perturbation method to find a leading-order approximation to the initial value problem

$$
\varepsilon y' + y = e^{-t}, \quad y(0) = 2,
$$

on $t \geq 0$. Calculate the residual and show that it converges to zero uniformly on $t \geq 0$ as $\varepsilon \to 0$.

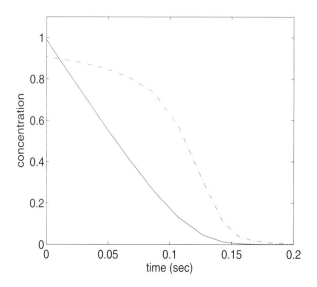

Figure 3.6 Slow time solution. The dashed curve is the complex and the solid curve is the substrate.

2. Find a uniformly valid approximation to the problem

$$\varepsilon y'' + b(t)y' + y = 0, \quad 0 < t < T,$$
$$y(0) = 1, \quad y'(0) = \beta/\varepsilon + \gamma,$$

as $\varepsilon \to 0$. Here, b is smooth and $b(t) > 0$, and T, β, $\gamma > 0$.

3. Find a uniformly valid approximation to the problem

$$\varepsilon y'' + (t+1)^2 y' = 1, \quad t > 0, \ 0 < \varepsilon \ll 1,$$
$$y(0) = 1, \quad \varepsilon y'(0) = 1.$$

4. Consider a nonlinear damped oscillator whose dynamics is governed by the initial value problem

$$m\ddot{y} + a\dot{y} + ky^3 = 0, \quad t > 0,$$
$$y(0) = 0, \quad m\dot{y}(0) = I,$$

where the mass m is small.

a) Non-dimensionalize the problem with a correct scaling.

b) Find a leading-order perturbation approximation.

5. Find approximations to the following systems where $0 < \varepsilon \ll 1$:

   a) $\frac{dx}{dt} = y - \varepsilon \sin x$, $\quad \varepsilon \frac{dy}{dt} = x^2 y + \varepsilon y^3$, $\quad x(0) = k$, $\quad y(0) = 0$.

   b) $\frac{du}{dt} = v$, $\quad \varepsilon \frac{dv}{dt} = -u^2 - v$, $\quad u(0) = 1$, $\quad v(0) = 0$.

6. Find a uniform approximation to the initial value problem

$$\varepsilon y'' + y' + y = f(t), \quad y(0) = a, \ y'(0) = b/\varepsilon.$$

7. Consider an object of mass m falling through a fluid. It experiences the force of gravity and a resistive force proportional to its velocity. At time $\tau = 0$ it is located at height H and at some time T it is located at height 0.

   a) Set up a boundary value problem for the height $h = h(\tau)$. (Orient $h > 0$ upward.)

   b) Assume the mass is very small and nondimensionalize the problem by the obvious scales. (Use y for dimensionless height and t for dimensionless time.) Obtain

$$\varepsilon y'' = -\varepsilon Q - y', \quad y(0) = 1, \quad y(1) = 0,$$

   for appropriate choices of ε and Q.

   c) Obtain a uniform approximation using perturbation methods and plot it for generic small ε. Finally, compute the residual.

8. Consider an exothermic reaction $\mathbf{X} \rightarrow$ products that releases heat. Initially, the concentration and temperature are $x(0) = x_0$ and $T(0) = T_0$, respectively. The rate $k = k(T)$ is a function of temperature and is given by the Arrhenius form

$$k(T) = re^{E/RT_0} e^{-E/RT},$$

   where E is the activation energy, R is the gas constant, and r is a multiplicative constant. As the reaction proceeds, T is related to x via the constitutive equation $T = T_0 + h(x_0 - x)$, where h is the increase in temperature when one gram of x is consumed by reaction.

   a) If $y = (x_0 - x)/x_0$ is a progress variable that measures the extent of the reaction, find a differential equation for y.

   b) Define a scaled temperature and time by $\theta = T/T_0$ and $\tau = rt$, respectively, and show that the model can be written

$$\frac{dy}{d\tau} = e^A e^{-A/\theta}(1 - y), \quad \theta = 1 + \beta y,$$

   for appropriately chosen constants A and β.

   c) Find an initial value problem for the scaled temperature θ.

   d) Assume the parameter A is small and obtain approximations for θ and y.

   e) Does a perturbation method work for A large? Perform some numerical experiments to support your answer to this question.

9. A thermochemical model for the scaled concentration u and temperature q in a reacting fluid is given by

$$\frac{du}{dt} = 1 - ue^{\varepsilon(q-1)}, \qquad \frac{dq}{dt} = -q + ue^{\varepsilon(q-1)},$$

   where $u(0) = q(0) = 0$ and $\varepsilon \ll 1$. Find a two-term perturbation solution.

10. Consider the system

$$x' = ky - x + \varepsilon axy,$$
$$\varepsilon y' = x - \varepsilon axy - y,$$
$$x(0) = 1, \quad y(0) = 0$$

   where $\varepsilon \ll 1$. Find the outer and inner approximations and use matching to find a uniform composite approximation.

11. Modify the Michaelis–Menten model by including the reverse reaction $P + E \overset{k_{-2}}{\to} C$ where the enzyme and product reform into the complex.

   a) Make the quasi-steady-state assumption and find $C = C(S)$.

   b) Find the velocity $V = dP/dt$ and compare it to the original model.

3.5 The WKB Approximation

The WKB method (Wentzel–Kramers–Brillouin) is a perturbation method that applies to a variety of problems—in particular, to linear differential equations of the form

$$\varepsilon^2 y'' + q(x)y = 0, \quad 0 < \varepsilon \ll 1, \tag{5.1}$$
$$y'' + (\lambda^2 p(x) - q(x))y = 0, \quad \lambda \gg 1, \tag{5.2}$$
$$y'' + q(\varepsilon x)^2 y = 0, \quad 0 < \varepsilon \ll 1. \tag{5.3}$$

We note that (5.1) and (5.3) have a small parameter, while (5.2) has a large parameter; the latter may be changed to the former by letting $\varepsilon = \lambda^{-1}$.

It is not possible, in general, to obtain a closed-form solution, in terms of elementary functions, to differential equations with variable coefficients like equations (5.1)–(5.3). So approximation methods are often useful. The WKB method was developed by several individuals. The first work was by Liouville[1] in the mid 1800s; in the early 1900s Liouville's method was extended by others, for example, Rayleigh (in the context of sound wave reflection) and Jeffrey (Mathieu's equation); the method now bears the initials of Wentzel, Kramers, and Brillouin because of their important application of the method in quantum mechanics. In the next few paragraphs we briefly develop some of the basic concepts of quantum mechanics, which gives a physical context in which to examine the WKB approximation.

For simplicity, consider a particle of mass m moving on the x axis under the influence of a conservative force given by $F(x)$. The potential $V(x)$ is defined by the equation $F(x) = -V'(x)$. According to the canon of classical particle mechanics, the motion $x = x(t)$ is governed by the dynamical equation

$$m\frac{d^2x}{dt^2} = F(x), \tag{5.4}$$

which is Newton's second law of motion. If the initial position $x(0)$ and velocity $dx/dt(0)$ are specified, then one can, in theory, determine the state of the particle (position and velocity) for all times $t > 0$. In this sense, classical mechanics is a *deterministic* model. The dynamics can be easily analyzed using the conservation of energy law

$$\frac{1}{2}m(y')^2 + V(x) = E,$$

where $y = x'$ is the velocity and E is the constant energy in the system, determined by the initial state. In xy (position-velocity) phase space the orbits are given by

$$y = \pm\sqrt{\frac{2}{m}(E - V(x))}.$$

These orbits are valid in the domain where $E > V(x)$, which is the classical region. The particle cannot occupy the domain $E < V(x)$, which is the nonclassical region. Values of x for which $E = V(x)$ are called *turning points*.

In the early 1900s it was found that this classic model of motion fails on the atomic scale. Out of this revolution quantum mechanics was developed. Quantum theory dictates that the particle has no definite position or velocity; rather, it postulates a statistical or probabilistic interpretation of the state of the particle in terms of a **wave function** $\Psi(x,t)$, which is complex. The

---

[1] J. Liouville (1809–1882) was one of the great mathematicians of the 19th century. His name is attached to key results in differential equations, complex analysis, and many other areas of mathematics.

square of the wave function is the probability density for the position, which is a random variable X. Therefore,

$$\Pr(a < X \le b) = \int_a^b |\Psi(x,t)|^2\, dx$$

is the probability of the particle being in the interval $a < x \le b$ at time t. Also,

$$\int_{-\infty}^{\infty} |\Psi(x,t)|^2\, dx = 1,$$

because the particle is located somewhere on the x axis. From elementary probability theory it is known that the probability density contains all of the statistical information for a given problem (for example, the mean and variance of position).

The question is how to find the wave function. The equation that governs the evolution of a quantum mechanical system (the analog of Newton's law for the classic system) is the **Schrödinger**[2] **equation**, a second-order partial differential equation having the form

$$i\hbar\Psi_t = -\frac{\hbar^2}{2m}\Psi_{xx} + V(x)\Psi, \tag{5.5}$$

where V is the potential energy, m is the mass, and $\hbar = h/2\pi$, where $h = 6.625 \cdot 10^{-34}$ kg m^2/sec is Planck's constant.

Assuming solutions have the form of a product $\Psi(x,t) = y(x)\phi(t)$, we obtain, upon substitution into (5.5) and after rearrangement,

$$\frac{i\hbar\frac{d\phi}{dt}}{\phi} = \frac{\frac{-\hbar^2}{2m}y'' + V(x)y}{y}.$$

The left side of this equation depends only on t, and the right side depends only on x. The only way equality can occur for all t and x is if both sides are equal to the same constant, which we shall call E. Therefore we obtain two equations, one for $\phi(t)$,

$$\frac{d\phi}{dt} = (-iE/\hbar)\phi,$$

and one for $y(x)$, namely

$$-\frac{\hbar^2}{2m}y'' + (V(x) - E)y = 0. \tag{5.6}$$

The solution of the time equation is the periodic function $\phi = C\exp(-iEt/\hbar)$, where C is a constant. Equation (5.6), whose solution $y(x)$ gives the spatial (steady-state) part of the wave function, is called the **time-independent**

---

[2] E. Schrödinger was one of the major scientists in the early 20th century who developed the theory underlying quantum mechanics.

Schrödinger equation, and it is one of the fundamental equations of mathematical physics. The condition

$$\int_{\infty}^{\infty} |y(x)|^2 dx = 1$$

is a normalizing condition.

Because $\hbar$ is extremely small, we may set $\varepsilon = \hbar/\sqrt{2m} \ll 1$ and obtain a problem of the form (5.1), where $q(x) = E - V(x)$. From our knowledge of constant coefficient equations, if $q(x) = E - V(x) > 0$ we expect the equation

$$\varepsilon y'' + q(x)y = 0$$

has rapidly varying oscillatory solutions (this is the classical region), and if $q(x) = E - V(x) < 0$ we expect exponentially growing and decaying solutions (nonoscillatory). The latter region is forbidden in classical physics, but we shall observe later that in quantum mechanics there can be nonzero probability of the particle existing in this region.

3.5.1 The Nonoscillatory Case

We first consider equation (5.1) in the case that $q(x)$ is strictly negative on the interval of interest. For definiteness, let us write $q(x) = -k(x)^2$, where $k(x) > 0$, and consider the equation

$$\varepsilon^2 y'' - k(x)^2 y = 0. \tag{5.7}$$

If k were a constant, say $k = k_0$, then (5.7) would have real, rapidly increasing or decreasing exponential solutions of the form $\exp(\pm k_0 x/\varepsilon)$. This suggests making in (5.7) the substitution

$$y = e^{u(x)/\varepsilon}.$$

Then (5.7) becomes

$$\varepsilon v' + v^2 - k(x)^2 = 0, \quad v = u'. \tag{5.8}$$

Now we try a regular perturbation expansion

$$v(x) = v_0(x) + \varepsilon v_1(x) + \mathrm{O}(\varepsilon^2).$$

Substituting this expansion into (5.8) gives

$$v_0 = \pm k(x), \quad 2v_0 v_1 = -v_0',$$

at $O(1)$ and $O(\varepsilon)$, respectively. Therefore,

$$v_0 = \pm k(x), \quad v_1 = -\frac{k'(x)}{2k(x)},$$

and the expansion for v becomes

$$v(x) = \pm k(x) - \varepsilon \frac{k'(x)}{2k(x)} + O(\varepsilon^2).$$

Consequently, u is given by

$$u(x) = \pm \int_a^x k(\xi)\, d\xi - \frac{\varepsilon}{2} \ln k(x) + O(\varepsilon^2),$$

where a is an arbitrary constant; the other constants are incorporated into the indefinite integral. Finally, the expansion for $y(x)$ is

$$y(x) = e^{u(x)/\varepsilon} = \frac{1}{\sqrt{k(x)}} \exp\left(\pm\frac{1}{\varepsilon}\int_a^x k(\xi)\, d\xi\right)(1 + O(\varepsilon)), \quad \varepsilon \ll 1.$$

This equation defines two linearly independent approximations (one with the plus sign and one with the minus sign) to the linear equation (5.7); taking the linear combination of these gives the WKB approximation to (5.7) in the nonoscillatory case:

$$y_{\mathrm{WKB}}(x) = \frac{c_1}{\sqrt{k(x)}} \exp\left(\frac{1}{\varepsilon}\int_a^x k(\xi)\, d\xi\right) + \frac{c_2}{\sqrt{k(x)}} \exp\left(-\frac{1}{\varepsilon}\int_a^x k(\xi)\, d\xi\right).$$
$$(5.9)$$

The arbitrary constants c_1 and c_2 can be determined by boundary data; the arbitrary lower limit of integration a should be taken at the point where the data is given. We note that the solutions may also be expressed in terms of the hyperbolic functions cosh and sinh.

Example 3.14

Find the WKB approximation to the equation

$$\varepsilon^2 y'' - (1+x)^2 y = 0, \quad x > 0.$$

Here, $k(x) = 1 + x$ and $\int_0^x k(\xi)\, d\xi = x + x^2/2$. Therefore the WKB approximation is

$$y_{\mathrm{WKB}}(x) = \frac{c_1}{\sqrt{1+x}} e^{(x+x^2/2)/\varepsilon} + \frac{c_2}{\sqrt{1+x}} e^{-(x+x^2/2)/\varepsilon}.$$

If we append boundary conditions $y(0) = 0$ and $\lim_{x\to\infty} y(x) = 0$, then $c_1 = 0$ to ensure the latter condition at infinity. Then the condition at $x = 0$ forces $c_2 = 1$. So

$$y_{\mathrm{WKB}}(x) = \frac{1}{\sqrt{1+x}} e^{-(x+x^2/2)/\varepsilon}. \quad \square$$

The quantum mechanical interpretation of exponential solutions is discussed in Chapter 8. See also the highly readable text by D. J. Griffiths (2005).

3.5.2 The Oscillatory Case

In the case that $q(x)$ is strictly positive, we expect oscillatory solutions of (5.1). In quantum mechanics, this is the classical region. So, we consider the equation

$$\varepsilon^2 y'' + k(x)^2 y = 0, \tag{5.10}$$

where $k(x) > 0$. Nearly the same calculation can be made as for the nonoscillatory case. Now we make the substitution $y(x) = \exp(iu(x)/\varepsilon)$ into equation (5.10) and proceed exactly as before to obtain the WKB approximation in the oscillatory case as

$$y_{\text{WKB}}(x) = \frac{c_1}{\sqrt{k(x)}} \exp\left(\frac{i}{\varepsilon} \int_a^x k(\xi)\, d\xi\right) + \frac{c_2}{\sqrt{k(x)}} \exp\left(-\frac{i}{\varepsilon} \int_a^x k(\xi)\, d\xi\right).$$

This is the same result except for the presence of the complex number i in the exponential. Using Euler's formula, $e^{i\theta} = \cos\theta + i\sin\theta$, we can rewrite the approximation to (5.10) in terms of sines and cosines as

$$y_{\text{WKB}}(x) = \frac{c_1}{\sqrt{k(x)}} \sin\left(\frac{1}{\varepsilon} \int_a^x k(\xi)\, d\xi\right) + \frac{c_2}{\sqrt{k(x)}} \cos\left(-\frac{1}{\varepsilon} \int_a^x k(\xi)\, d\xi\right). \tag{5.11}$$

These are rapidly oscillating solutions.

Example 3.15

(**Schrödinger equation**) In the case that $E > V(x)$, which is the oscillatory case, the WKB approximation (5.11) to the Schrödinger equation (5.6) is often written in terms of an amplitude A and phase ϕ (which play the role of arbitrary constants) as

$$y_{\text{WKB}} = \frac{A}{(E - V(x))^{1/4}} \cos\left(\frac{\sqrt{2m}}{\hbar} \int_a^x \sqrt{E - V(\xi)}\, d\xi + \phi\right). \quad \square$$

Example 3.16

The WKB approximation can be used to determine the large eigenvalues for simple differential operators. (A detailed discussion of eigenvalue problems is presented in Chapter 5.) For the present, consider the boundary value problem

$$y'' + \lambda q(x)y = 0, \quad 0 < x < \pi; \; y(0) = y(\pi) = 0. \tag{5.12}$$

We assume $q(x) > 0$. A number λ is called an **eigenvalue** of the boundary value problem (5.12) if there exists a nontrivial solution of (5.12) for that particular value of λ; the corresponding nontrivial solutions are called **eigenfunctions**. It is an important problem to determine all the eigenvalues and eigenfunctions for a given boundary value problem. In the present case, if we define $\varepsilon = 1/\sqrt{\lambda}$ and $k(x) = \sqrt{q(x)}$, then the differential equation in (5.12) has the form

$$\varepsilon^2 y'' + k(x)^2 y = 0,$$

for which the WKB method applies, provided ε is small or, equivalently, if λ is large in (5.12). The WKB approximation is given by

$$y_{\text{WKB}}(x) = \frac{1}{q(x)^{1/4}} \left(c_1 \sin\left(\sqrt{\lambda} \int_0^x \sqrt{q(\xi)} \, d\xi \right) + c_2 \cos\left(\sqrt{\lambda} \int_0^x \sqrt{q(\xi)} \, d\xi \right) \right).$$

But $y(0) = 0$ forces $c_2 = 0$, and $y(\pi) = 0$ forces

$$\frac{c_1}{q(\pi)^{1/4}} \sin\left(\sqrt{\lambda} \int_0^\pi \sqrt{q(\xi)} \, d\xi \right) = 0.$$

Choosing $c_1 = 0$ gives the trivial solution, which we are trying to avoid; so we force

$$\sin\left(\sqrt{\lambda} \int_0^\pi \sqrt{q(\xi)} \, d\xi \right) = 0,$$

which can be satisfied by choosing λ such that

$$\sqrt{\lambda} \int_0^\pi \sqrt{q(\xi)} \, d\xi = n\pi,$$

where n is a large integer (recall, the approximation is valid only for large values of λ). Thus, the large eigenvalues of problem (5.12) are given by

$$\lambda = \lambda_n = \left(\frac{n\pi}{\int_0^\pi \sqrt{q(\xi)} \, d\xi} \right)^2,$$

for large n. The corresponding eigenfunctions are

$$y_{\text{WKB}}(x) = \frac{c_1}{q(x)^{1/4}} \sin\left(\frac{n\pi \int_0^x \sqrt{q(\xi)} \, d\xi}{\int_0^\pi \sqrt{q(\xi)} \, d\xi} \right),$$

for large n. $\square$

If the function $q(x)$ has zeros in the interval of interest, then the problem of determining asymptotic approximations of (5.1) is significantly more difficult. On one side of a simple zero of q the solution is oscillatory, and on the other side it is exponential; therefore these two solutions must be connected, or matched, in a layer located at that root. As we remarked earlier, the points x where $q(x) = 0$ are called **turning points**; in terms of quantum mechanics and the Schrödinger equation (5.6), the turning points are the values of x where $E = V(x)$, or where the total energy equals the potential energy. We refer the reader to Holmes (1995) for an elementary treatment of turning point problems. Murray (1984) shows how to apply the WKB method to the more general Liouville equation

$$y'' + (\lambda^2 q_0(x) + \lambda q_1(x) + q_2(x) + \cdots)y = 0, \quad \lambda \gg 1.$$

Interestingly enough, even though the WKB approximation is valid for $\lambda \gg 1$ or $\varepsilon \ll 1$, it often provides a good approximation even when λ and ε are O(1). Several numerical examples are discussed in Bender and Orszag (1978).

EXERCISES

1. Use the WKB method to find an approximate solution to the initial value problem
$$y'' - \lambda(1 + x^2)^2 y = 0, \quad y(0) = 0, \quad y'(0) = 1,$$
   for large λ.

2. Show that the large eigenvalues of the problem
$$y'' + \lambda(x + \pi)^4 y = 0, \quad y(0) = y(\pi) = 0,$$
   are given approximately by
$$\lambda = \frac{9n^2}{49\pi^4}$$
   for large integers n, and find the corresponding eigenfunctions.

3. Consider the eigenvalue problem
$$-\frac{1}{x}y'' = \lambda y, \quad 1 < x < 4,$$
$$y(1) = y(4) = 0.$$
   Find an approximation for the large eigenvalues and eigenfunctions.

4. Show that the differential equation $y'' + p(x)y' + q(x)y = 0$ can be transformed into the equation $u'' + r(x)u = 0$, $r = q - p'/2 - p^2/4$, without a first derivative, by the Liouville transformation $u = y \exp(\frac{1}{2}\int_a^x p(\xi)\, d\xi)$.

5. Find a formula in terms of the sinh function for the WKB approximation to the initial value problem

$$\varepsilon^2 y'' - q(x)y = 0, \quad y(0) = 0, \quad y'(0) = 1; \quad q(x) > 0.$$

6. The *slowly varying oscillator* equation is

$$y'' + q(\varepsilon x)^2 y = 0,$$

where $\varepsilon \ll 1$ and q is strictly positive. Find an approximate solution. Why is the equation called "slowly varying"?

7. Find an approximation for the large eigenvalues of

$$y'' + \lambda e^{4x} y = 0, \quad 0 < x < 1, \quad y(0) = y(1) = 1.$$

8. Find an approximation for the general solution of

$$y'' + (\lambda^2 x^2 + x)y = 0, \quad x > 0, \quad \lambda \gg 1.$$

Hint: let $y = \exp(iu/\varepsilon)$, $\varepsilon = 1/\lambda$.

3.6 Asymptotic Expansion of Integrals

The solution of differential equations often leads to formulas involving complicated integrals that cannot be evaluated in closed form. For example, the initial value problem

$$y'' + 2\lambda t y' = 0, \quad y(0) = 0, \quad y'(0) = 1,$$

has solution

$$y = y(t; \lambda) = \int_0^t e^{-\lambda s^2}\, ds,$$

but we cannot calculate the integral because there is no antiderivative in terms of simple functions. For some problems, however, we may want to know the behavior of y for large t (with λ fixed), or we may want to determine the value of y for a fixed value of t in the case that λ is a large parameter. Problems like this are common in applied mathematics, and in this section we introduce standard techniques to approximate certain kinds of integrals.

3.6.1 Laplace Integrals

One type of integral we wish to study has the form

$$I(\lambda) = \int_a^b f(t)e^{-\lambda g(t)}\,dt, \quad \lambda \gg 1, \tag{6.1}$$

where g is a strictly increasing function on $[a, b]$ and the derivative g' is continuous. Here, $a < b \leq +\infty$, and $\lambda \gg 1$ is another notation for the asymptotic statement "as $\lambda \to \infty$." An example is the Laplace transform

$$U(\lambda) = \int_0^\infty f(t)e^{-\lambda t}\,dt.$$

Actually, it is sufficient to examine integrals of the form

$$I(\lambda) = \int_0^b f(t)e^{-\lambda t}\,dt, \quad \lambda \gg 1. \tag{6.2}$$

To observe this, we can make a change of variables $s = g(t) - g(a)$ in (6.1) to obtain

$$I(\lambda) = e^{-\lambda g(a)} \int_0^{g(b)-g(a)} \frac{f(t(s))}{g'(t(s))} e^{-\lambda s}\,ds,$$

where $t = t(s)$ is the solution of the equation $s = g(t) - g(a)$ for t. The integral on the right has the form of (6.2).

The fundamental idea in obtaining an approximation for (6.2) is to determine what subinterval gives the dominant contribution. The function $\exp(-\lambda t)$ is a rapidly decaying exponential; thus, if f does not grow too fast at infinity and if it is reasonably well behaved at $t = 0$, then it appears that the main contribution comes from a neighborhood of $t = 0$. A detailed example will illustrate our reasoning; the method is called **Laplace's method**.

Example 3.17

Consider the integral

$$I(\lambda) = \int_0^\infty \frac{\sin t}{t} e^{\lambda t}\,dt, \quad \lambda \gg 1. \tag{6.3}$$

Figure 2.8, which shows a graph of the integrand for different values of λ, confirms that the main contribution to the integral occurs near $t = 0$. To proceed we partition the interval of integration into two parts, $[0, T]$ and $[T, \infty]$, to obtain

$$I(\lambda) - \int_0^T \frac{\sin t}{t} e^{-\lambda t}\,dt + \int_T^\infty \frac{\sin t}{t} e^{-\lambda t}\,dt, \tag{6.4}$$

where T is any positive real number. For any T, the second integral is an exponentially small term (EST); that is, it has order $\lambda^{-1} \exp(-\lambda T)$ as $\lambda \to \infty$. A simple calculation confirms this fact:

$$|I(\lambda)| \leq \int_T^\infty \left| \frac{\sin t}{t} \right| e^{-\lambda t}\, dt \leq \int_T^\infty e^{-\lambda t}\, dt = \frac{1}{\lambda} e^{-\lambda T}.$$

We note that an exponentially small term is $o(\lambda^{-m})$, for any positive integer m, which means that it decays much faster than any power of λ. Thus,

$$I(\lambda) = \int_0^T \frac{\sin t}{t} e^{-\lambda t}\, dt + \text{EST}.$$

In the finite interval $[0, T]$ we can replace $\sin t/t$ by its Taylor series about $t = 0$,

$$\frac{\sin t}{t} = 1 - \frac{t^2}{3!} + \mathrm{O}(t^4).$$

Thus

$$I(\lambda) = \int_0^T \left(1 - \frac{t^2}{3!} + \mathrm{O}(t^4) \right) e^{-\lambda t}\, dt + \text{EST}.$$

Now change variables via $u = \lambda t$ to obtain

$$I(\lambda) = \frac{1}{\lambda} \int_0^{\lambda T} \left(1 - \frac{u^2}{3!\lambda^2} + \mathrm{O}\left(\frac{1}{\lambda^4} \right) \right) e^{-u}\, du + \text{EST}.$$

But now the upper limit can be replaced by ∞ since the error incurred by doing this is exponentially small as $\lambda \to \infty$. (Since we are already discarding algebraically small terms, there is no worry about adding exponentially small terms to the integral.) Consequently, we obtain

$$I(\lambda) = \frac{1}{\lambda} \int_0^\infty \left(1 - \frac{u^2}{3!\lambda^2} + \mathrm{O}\left(\frac{1}{\lambda^4} \right) \right) e^{-u}\, du.$$

We can now use the integration formula

$$\int_0^\infty u^m e^{-u}\, du = m!, \quad m = 0, 1, 2, \ldots, \tag{6.5}$$

to obtain the asymptotic approximation of (6.3) given by

$$I(\lambda) \sim \frac{1}{\lambda} - \frac{2!}{3!\lambda^3} + \mathrm{O}\left(\frac{1}{\lambda^5} \right), \quad \lambda \gg 1. \quad \square$$

Before discussing general problems, we pause to define a special function, the *gamma function*, which plays an important role in subsequent calculations. It is defined by

$$\Gamma(x) = \int_0^\infty u^{x-1} e^{-u}\, du, \quad x > 0.$$

Because the **gamma function** has the property

$$\Gamma(x+1) = x\Gamma(x),$$

it generalizes the idea of a factorial function to non-integer values of the argument; note that $\Gamma(n) = (n-1)!$ for positive integers n, and $\Gamma(\frac{1}{2}) = \sqrt{\pi}$. Other values of the gamma function are tabulated in handbooks on special functions.

We now formulate a general result known as **Watson's lemma**, which gives the asymptotic behavior of a large number of integrals. We shall only outline the proof, and those interested in a more detailed discussion should consult, for example, Murray (1984).

Theorem 3.18

(**Watson's Lemma**) Consider the integral

$$I(\lambda) = \int_0^b t^\alpha h(t) e^{-\lambda t}\, dt, \tag{6.6}$$

where $\alpha > -1$, where $h(t)$ has a Taylor series expansion about $t = 0$, with $h(0) \neq 0$, and where $|h(t)| < ke^{ct}$, $0 < t < b$, for some positive constants k and c. Then

$$I(\lambda) \sim \sum_{n=0}^{\infty} \frac{h^{(n)}(0)\Gamma(\alpha + n + 1)}{n!\lambda^{\alpha+n+1}}, \quad \lambda \gg 1. \quad \square \tag{6.7}$$

Proof

The condition on α will guarantee the convergence of the improper integral at $t = 0$ and the exponential boundedness of h is required to ensure that the integral converges as t goes to infinity. The argument is similar to the one in the previous example. We split up the integral (6.6) for any $T > 0$, where T lies in the radius of convergence of the power series for h. Then

$$I(\lambda) = \int_0^T t^\alpha h(t) e^{-\lambda t}\, dt + \int_T^b t^\alpha h(t) e^{-\lambda t}\, dt.$$

The second integral is exponentially small. Now we substitute the Taylor series expansion for h about $t = 0$ in the first integral to obtain

$$I(\lambda) = \int_0^T t^\alpha \left(h(0) + h'(0)t + \frac{1}{2!}h''(0)t^2 + \cdots \right) e^{-\lambda t}\, dt + \text{EST}.$$

Making the substitution $u = \lambda t$ and then replacing the upper limit of integration by ∞ gives

$$I(\lambda) \sim \frac{1}{\lambda} \int_0^\infty \left(h(0) \left(\frac{u}{\lambda}\right)^\alpha + h'(0) \left(\frac{u}{\lambda}\right)^{\alpha+1} \right.$$
$$\left. + \frac{1}{2!} h''(0) \left(\frac{u}{\lambda}\right)^{\alpha+2} + \cdots \right) e^{-\lambda t} \, dt.$$

The result (6.7) now follows from the definition of the gamma function.   □

Example 3.19

Some integrals that do not have the form (6.6) can be transformed into an integral that does. For example, consider the complementary error function, erfc defined by

$$\text{erfc}(\lambda) = \frac{2}{\sqrt{\pi}} \int_\lambda^\infty e^{-s^2} \, ds. \tag{6.8}$$

Letting $t = s - \lambda$ gives

$$\text{erfc}(\lambda) = \frac{2}{\sqrt{\pi}} e^{-\lambda^2} \int_0^\infty e^{-t^2} e^{-2\lambda t} \, dt = \frac{1}{\sqrt{\pi}} e^{-\lambda^2} \int_0^\infty e^{-\tau^2/4} e^{-\lambda \tau} \, d\tau.$$

Watson's lemma applies, and we leave it as an exercise to show that

$$\text{erfc}(\lambda) \sim \frac{2}{\sqrt{\pi}} e^{-\lambda^2} \left(\frac{1}{2\lambda} - \frac{\Gamma(3)}{(2\lambda)^3} + \frac{\Gamma(5)}{2!(2\lambda)^5} - \frac{\Gamma(7)}{3!(2\lambda)^7} + \cdots \right). \quad □ \tag{6.9}$$

3.6.2 Integration by Parts

Sometimes an asymptotic expansion can be found by a simple, successive integration by parts procedure. However, this method is not often as applicable as the Laplace method, and it usually involves more work. Let us consider the erfc function. Using integration by parts, we have

$$\text{erfc}(\lambda) = \frac{2}{\sqrt{\pi}} \int_\lambda^\infty e^{-t^2} \, dt$$
$$= \frac{2}{\sqrt{\pi}} \int_\lambda^\infty \frac{-2te^{-t^2}}{-2t} \, dt$$
$$= \frac{2}{\sqrt{\pi}} \left(\left. \frac{e^{-t^2}}{-2t} \right|_\lambda^\infty - \int_\lambda^\infty \frac{e^{-t^2}}{2t^2} \, dt \right)$$
$$= \frac{2}{\sqrt{\pi}} \left(\frac{e^{-\lambda^2}}{2\lambda} - \int_\lambda^\infty \frac{e^{-t^2}}{2t^2} \, dt \right).$$

The integral on the right may be integrated by parts to get

$$\int_\lambda^\infty \frac{e^{-t^2}}{2t^2}\, dt = \frac{1}{2}\int_\lambda^\infty \frac{-2te^{-t^2}}{-2t^3}\, dt$$
$$= \frac{1}{2}\left(\frac{e^{-\lambda^2}}{2\lambda^3} - \int_\lambda^\infty \frac{3e^{-t^2}}{2t^4}\, dt\right).$$

Thus,

$$\mathrm{erfc}(\lambda) = \frac{2}{\sqrt{\pi}}e^{-\lambda^2}\left(\frac{1}{2\lambda} - \frac{1}{4\lambda^3}\right) + \frac{1}{\sqrt{\pi}}\int_\lambda^\infty \frac{3e^{-t^2}}{2t^4}\, dt.$$

By repeated integration by parts we can eventually reproduce (6.9). It is important in this method to ensure that each succeeding term generated in the expansion is asymptotically smaller than the preceding term; this is required by the definition of asymptotic expansion in Section 2.1.

Using the ratio test, we can easily observe that for fixed λ the series in (6.9) for erfc is not convergent. One may legitimately ask, therefore, how a divergent series can be useful in approximating values of $\mathrm{erfc}(\lambda)$; for fixed λ we cannot obtain a good approximation by taking more and more terms. For calculations, the idea is to *fix* the number of terms n and apply the resulting approximation for large values of λ. The error in this approximation is about the magnitude of the first discarded term. Practically, then, to approximate $\mathrm{erfc}(\lambda)$, we compute the magnitude of the first few terms of the series and then terminate the series when the magnitude begins to increase.

To illustrate this strategy, we approximate $\mathrm{erfc}(2)$. We have

$$\mathrm{erfc}(2) = \frac{e^{-4}}{\sqrt{\pi}}\left(\frac{1}{2} - \frac{1}{2^4} + \frac{1\cdot 3}{2^7} - \frac{1\cdot 3\cdot 5}{2^{10}} + \frac{1\cdot 3\cdot 5\cdot 7}{2^{13}}\right.$$
$$\left. - \frac{1\cdot 3\cdot 5\cdot 7\cdot 9}{2^{16}} + \cdots\right).$$

After the fifth term, the magnitude of the terms begins to increase. So we terminate after five terms and obtain $\mathrm{erfc}(2) \approx 0.004744$; the tabulated value is 0.004678 and thus our approximation has an error of 0.000066.

The reader might inquire why one shouldn't just write

$$\mathrm{erfc}(\lambda) = \frac{2}{\sqrt{\pi}}\int_0^\infty e^{-t^2}\, dt - \frac{2}{\sqrt{\pi}}\int_0^\lambda e^{-t^2}\, dt$$
$$= 1 - \frac{2}{\sqrt{\pi}}\int_0^\lambda e^{-t^2}\, dt,$$

and then replace $\exp(-t^2)$ by its Taylor series expansion and integrate term-by-term. This procedure would yield a convergent series. So, adopting this strategy we get

$$\operatorname{erfc}(\lambda) = 1 - \frac{2}{\sqrt{\pi}} \int_0^\lambda \left(1 - \frac{t^2}{2} + \frac{1}{2!} \left(\frac{t^2}{2} \right)^2 - \cdots \right) dt$$

$$= 1 - \frac{2}{\sqrt{\pi}} \left(\lambda - \frac{\lambda^3}{3} + \frac{\lambda^5}{5 \cdot 2!} - \frac{\lambda^7}{7 \cdot 3!} + \cdots \right).$$

The reader should check that it takes about 20 terms to get the same accuracy that the divergent, asymptotic series gave in only five terms. Therefore, asymptotic expansions can be significantly more efficient than traditional Taylor series expansions. For small λ, however, the Taylor series in preferable.

3.6.3 Other Integrals

We can extend the previous ideas to approximate integrals of the form

$$I(\lambda) = \int_a^b f(t) e^{\lambda g(t)} \, dt, \quad \lambda \gg 1, \tag{6.10}$$

where f is continuous and g is sufficiently smooth and has a unique maximum at the point $t = c$ in (a, b). Specifically, $g'(c) = 0$, $g''(c) < 0$. Again we expect the main contribution of (6.10) to come from where g assumes its maximum. To obtain a leading order approximation we replace $f(t)$ by $f(c)$, and we replace g by the first three terms of its Taylor series (the g' term is zero) to obtain

$$I(\lambda) \sim \int_a^b f(c) e^{\lambda(g(c) + g''(c)(t-c)^2/2)} \, dt \tag{6.11}$$

$$\sim f(c) e^{\lambda g(c)} \int_a^b e^{\lambda/2 g''(c)(t-c)^2} \, dt. \tag{6.12}$$

Letting $s = (t - c)\sqrt{-\lambda g''(c)/2}$, we get

$$I(\lambda) \sim f(c) e^{\lambda g(c)} \sqrt{\frac{-2}{\lambda g''(c)}} \int_{(a-c)\sqrt{-\lambda g''(c)/2}}^{(b-c)\sqrt{-\lambda g''(c)/2}} e^{-s^2} \, ds. \tag{6.13}$$

We may now replace the limits of integration by $-\infty$ and $+\infty$ and use $\int_{-\infty}^{+\infty} \exp(-s^2) ds = \sqrt{\pi}$ to obtain

$$I(\lambda) \sim f(c) e^{\lambda g(c)} \sqrt{\frac{-2\pi}{\lambda g''(c)}}, \quad \lambda \gg 1. \tag{6.14}$$

Remark 3.20

If the maximum of g occurs at an endpoint, then $c = a$ or $c = b$ and one of the limits of integration in (6.11) is zero; thus we will obtain one-half of the value indicated in (6.12). □

Example 3.21

Approximate the integral

$$I(\lambda) = \int_0^\pi e^{\lambda \sin t} \, dt$$

for large λ. Here $f(t) = 1$, and $g(t) = \sin t$ has its maximum at $t = \pi/2$ with $g''(\pi/2) = -1$. Thus, by (6.12), $I(\lambda) \sim e^\lambda \sqrt{2\pi/\lambda}$. □

EXERCISES

1. Verify (6.9).

2. Find a two-term approximation for large λ for

$$I(\lambda) = \int_0^{\pi/2} e^{-\lambda \tan^2 \theta} \, d\theta.$$

3. Consider the integral in (6.10), where the maximum of g occurs at an endpoint of the interval with a nonzero derivative. To fix the idea assume that the maximum occurs at $t = b$ with $g'(t) > 0$ for $a \le t \le b$. Derive the formula

$$I(\lambda) \sim \frac{f(b)e^{\lambda g(b)}}{\lambda g'(b)}, \quad \lambda \gg 1.$$

4. Verify the following approximations for large λ.

   a) $\int_0^\infty e^{-\lambda t} \ln(1 + t^2) \, dt \sim \frac{2!}{\lambda^3} - \frac{1}{2} \frac{4!}{\lambda^5} + \cdots$.

   b) $\int_0^1 \sqrt{1 + t} e^{\lambda(2t - t^2)} \, dt \sim \sqrt{\pi/2\lambda} e^\lambda$.

   c) $\int_1^2 \sqrt{3 + t} e^{\lambda/(t+1)} \, dt \sim \frac{8}{\lambda} e^{\lambda/2}$.

5. Use integration by parts to show

$$\Gamma(x + 1) = x\Gamma(x).$$

6. Find a three-term asymptotic expansion of

$$\int_0^\infty e^{-\lambda t^2} \sin t \, dt$$

   for large λ.

7. Find a leading-order asymptotic approximation for the following integrals, $\lambda \gg 1$.

   a) $\int_0^\pi e^{-\lambda t} \frac{\cos t}{\sqrt{t}} \, dt$.

   b) $\int_0^1 e^{-\lambda t} \frac{1}{\sqrt{t}} \frac{1}{\sqrt{1-t/2}} \, dt$.

   c) $\int_0^2 e^{\lambda(2-t)t} \ln(t+1) \, dt$.

   d) $\int_0^1 e^t e^{\lambda(3t^2 + 2t^3)} \, dt$.

8. The modified *Bessel* function $I_n(x)$ has integral representation

$$I_n(x) = \frac{1}{\pi} \int_0^\pi e^{x \cos \theta} \cos n\theta \, d\theta, \quad n = 1, 2, 3, \dots$$

   Find the leading-order asymptotic approximation of $I_n(x)$ as $x \to +\infty$.

9. Find the first two terms in the asymptotic expansion of the integral

$$I(\lambda) = \int_0^1 e^{-\lambda(1+s)} \frac{1}{\sqrt{2 + 3s + 3s^2 + s^3}} \, ds, \quad \text{as} \quad \lambda \to \infty.$$

10. Show that

$$\int_0^{\pi/2} e^{-x \sin^2 t} dt \sim \frac{\sqrt{\pi}}{2\sqrt{x}} \left(1 + \frac{1}{4x} + \cdots \right), \quad \lambda \gg 1.$$

11. Use integration by parts to find a two-term asymptotic approximation of the function
$$\text{Ci}(\lambda) = \int_\lambda^\infty \frac{\cos x}{x} \, dx$$
   for large λ.

12. Consider the real exponential integral
$$\text{Ei}(\lambda) = \int_\lambda^\infty \frac{e^{-t}}{t} \, dt$$
   for large values of λ.

   a) Use integration by parts to show that
$$\text{Ei}(\lambda) = e^{-\lambda} \left(\frac{1}{\lambda} - \frac{1}{\lambda^2} + \frac{2!}{\lambda^3} - \frac{3!}{\lambda^4} + \cdots + \frac{(-1)^{n-1}(n-1)!}{\lambda^n} \right) + r_n(\lambda),$$

      where
$$r_n(\lambda) = (-1)^n n! \int_\lambda^\infty \frac{e^{-t}}{t^{n+1}} \, dt.$$

3.6 Asymptotic Expansion of Integrals   219

b) Show that for fixed n, $|r_n(\lambda)| \to 0$ as $\lambda \to \infty$.

c) Show that for fixed n, $r_n(\lambda) = \mathrm{o}\left(\frac{(n-1)!e^{-\lambda}}{\lambda^n}\right)$ as $\lambda \to \infty$, and conclude that $\mathrm{Ei}(\lambda)$ has asymptotic expansion

$$\mathrm{Ei}(\lambda) \sim e^{-\lambda}\left(\frac{1}{\lambda} - \frac{1}{\lambda^2} + \frac{2!}{\lambda^3} - \frac{3!}{\lambda^4} + \cdots\right).$$

d) Show that the asymptotic series in part (c) diverges for fixed λ.

e) Approximate $\mathrm{Ei}(10)$.

13. Use integration by parts to obtain an asymptotic expansion for large λ of

$$I(\lambda) = \int_0^\infty \frac{e^{-t}}{(t+\lambda)^2}\, dt.$$

REFERENCES AND NOTES

The four books listed below cover a large amount of material on perturbation methods and asymptotic analysis. Bender and Orszag (1978) is a definitive treatment with many challenging exercises.

Bender, C. M. & and Orszag, S. A. 1978. *Advanced Mathematical Methods for Scientists and Engineers*, McGraw-Hill, New York (reprinted by Springer–Verlag, New York 1998).

Griffiths, D. J. 2005. *Introduction to Quantum Mechanics*, 2nd ed., Pearson Prentice-Hall, Upper Saddle River, NJ.

Holmes, M. 1995. *Introduction to Perturbation Methods*, Springer-Verlag, New York (1995).

Lin, C. C. & Segel, L. A. 1974. *Mathematics Applied to Deterministic Problems in the Natural Sciences*, Macmillan, New York (reprinted by SIAM, Philadelphia 1988).

Murray, J. D. 1984. *Asymptotic Analysis*, Springer–Verlag, New York (1984).

4

Calculus of Variations

Generally, the calculus of variations is concerned with the optimization of variable quantities called functionals over some admissible class of competing objects. Many of its methods were developed over two hundred years ago by L. Euler[1], L. Lagrange[2], and others. It continues to the present day to bring important techniques to many branches of engineering, physics, and optimal control theory, and it introduces many important methods to applied analysis.

This chapter is more or less independent from the remainder of the text. One concept, that of a normed linear space, appears in later chapters; at that time the reader may refer here, to Section 4.2.1, for the basic definitions.

4.1 Variational Problems

4.1.1 Functionals

To motivate the basic concepts of the calculus of variations we review some simple notions in elementary differential calculus. One of the central problems

---

[1] Leonard Euler (1701–1783), a Swiss mathematician who spent much of his professional life in St. Petersburg, was perhaps one of the most prolific contributors to mathematics and science of all time. Some may rank him at the top of all mathematicians. His name is attached to major results in nearly every area of study in mathematics.

[2] Joseph Louis Lagrange (1736–1813) was one of the great mathematicians in the 18th century, whose work had a deep influence on subsequent research.

Applied Mathematics 4th ed.,
By J. David Logan
Copyright © 2013 John Wiley & Sons, Inc.

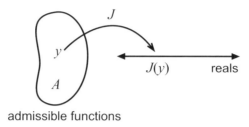

admissible functions

Figure 4.1 Schematic of a functional $J : A \to$ reals mapping a function space A into the real numbers.

in the calculus is to maximize or minimize a given real valued function of a single variable. If f is a given function defined in an open interval I, then f has a **local** (or *relative*) **minimum** at a point x_0 in I if $f(x_0) \le f(x)$ for all x, satisfying $|x - x_0| < \delta$ for some δ. If f has a local minimum at x_0 in I and f is differentiable in I, then it is well known that

$$f'(x_0) = 0, \tag{1.1}$$

where the prime denotes the ordinary derivative of f. Similar statements can be made if f has a local maximum at x_0. Condition (1.1) is called a **necessary condition** for a local minimum; that is, if f has a local minimum at x_0, then (1.1) necessarily follows. However, equation (1.1) is not sufficient for a local minimum; that is, if (1.1) holds, it does not guarantee that x_0 provides an actual minimum. The following conditions are **sufficient conditions** for f to have a local minimum at x_0,

$$f'(x_0) = 0 \quad \text{and} \quad f''(x_0) > 0,$$

provided f'' exists. Again, similar conditions can be formulated for local maxima. If (1.1) holds, we say f is **stationary** at x_0 and that x_0 is an critical point of f.

The calculus of variations deals with generalizations of this problem from the calculus. Rather than find conditions under which functions have extreme values, the calculus of variations deals with extremizing general quantities called functionals. A **functional** is a rule that assigns a real number to each function $y(t)$ in a well-defined class. Like a function, a functional is a rule or association, but its domain is a set of functions rather than a subset of real numbers. To be more precise let A be a set of functions $y, z, \ldots$; then a functional J on A is a rule that associates to each $y \in A$ a real number denoted by $J(y)$. Figure 4.1 illustrates this notion and it is symbolically represented as $J : A \to \mathbb{R}$. We say J *maps* A into $\mathbb{R}$.

A fundamental problem of the calculus of variations can be stated as follows: Given a functional J and a well-defined set of functions A, determine which functions in A afford a minimum (or maximum) value to J. The word *minimum* can be interpreted as a local minimum (if A is equipped with some device to measure closeness of its elements) or an absolute minimum, that is, a minimum relative to all elements in A. The set A is called the set of **admissible functions**, and those in A are the competing functions for extremizing J. For example, the set of admissible functions might be the set of all continuous functions on an interval $[a, b]$, the set of all continuously differentiable functions on $[a, b]$ satisfying the condition $f(a) = 0$, or whatever, as long as the set is well defined. Later we impose some general conditions on A. For the most part the classical calculus of variations restricts itself to functionals that are defined by certain integrals and to the determination of both necessary and sufficient conditions for extrema. The problem of extremizing a functional J over the set A is also called a variational problem. Several examples are presented in the next section.

In summary, the calculus of variations may be characterized as the 'calculus of functionals.' In this chapter we restrict ourselves to an analysis of necessary conditions for extrema. Applications of the calculus of variations are found in geometry, physics, engineering, numerical analysis, ordinary and partial differential equations, and optimal control theory.

4.1.2 Examples

We now formulate a few classical variational problems to illustrate the notions of a functional and the corresponding admissible functions.

Example 4.1

(**Arclength**) Let A be the set of all continuously differentiable functions on an interval $a \leq x \leq b$ that satisfy the boundary conditions $y(a) = y_0$, $y(b) = y_1$. Let J be the arclength functional on A defined by

$$J(y) = \int_a^b \sqrt{1 + y'(x)^2} \, dx.$$

To each y in A the functional J associates a real number that is the arclength of the curve $y = y(x)$ between the two fixed points $P\colon (a, y_0)$ and $Q\colon (b, y_1)$. The associated problem in the calculus of variations is to minimize J, that is, the arclength, over the set A. Clearly the minimizing arc is the straight line between P and Q given by

$$y(x) = \frac{y_1 - y_0}{b - a}(x - a) + y_0.$$

The general methods of the calculus of variations lead to a proof of this fact: the shortest distance between two points is a straight line. □

Example 4.2

(**Brachistochrone problem**) A bead of mass m with initial velocity zero slides with no friction under the force of gravity g from a point $(0, b)$ to a point $(a, 0)$ along a wire defined by a curve $y = y(x)$ in the xy plane. Which curve leads to the fastest time of descent? Historically this problem was important in the development of the calculus of variations. J. Bernoulli posed it in 1696 and several well-known mathematicians of that day proposed solutions; Euler's solution led to general methods that were useful in solving a wide variety of such problems. The solution curve itself is an arc of a cycloid, the curve traced out by a point on the rim of a rolling wheel; so the *brachistochrone*, or the curve of shortest time, is a cycloid. To formulate this problem analytically we compute the time of descent T for a fixed curve $y = y(x)$ connecting the points $(0, b)$ and $(a, 0)$ (see Fig. 4.2). Letting s denote the arc-length along the curve measured from the initial point $(0, b)$, we have

$$T = \int_0^T dt = \int_0^S \frac{dt}{ds}\, ds = \int_0^S \frac{1}{v}\, ds,$$

where S is the total arc-length of the curve and v denotes the velocity. Because $ds = (1 + y'(x)^2)^{1/2}\, dx$ we have

$$T = \int_0^a \frac{\sqrt{1 + y'(x)^2}}{v}\, dx.$$

To obtain an expression for v in terms of x we use the fact that energy is conserved energy conservation; that is,

(Kinetic energy at $t > 0$) + (potential energy at $t > 0$)
$$= \text{(kinetic energy at } t = 0) + \text{(potential energy at } t = 0).$$

In terms of our notation,

$$\frac{1}{2}mv^2 + mgy = 0 + mgb.$$

Solving for v gives

$$v = \sqrt{2g(b - y(x))}.$$

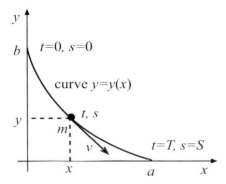

Figure 4.2 The brachistochrone problem.

Therefore the time required for the bead to descend is

$$T(y) = \int_0^a \frac{\sqrt{1 + y'(x)^2}}{\sqrt{2g(b - y(x))}} \, dx, \qquad (1.2)$$

where we have explicitly noted that T depends on the curve y. Here, the set of admissible functions A is the set of all continuously differentiable functions y defined on $0 \leq x \leq a$ that satisfy the boundary conditions $y(0) = b$, $y(a) = 0$, and the conditions

$$y \leq b, \quad \int_0^a (b - y)^{-1/2} \, dx < +\infty.$$

Equation (1.2) defines a functional on A. A real number is associated to each curve, namely the time required for the bead to descend the curve. The associated variational problem is to minimize $T(y)$, where $y \in A$, that is, to find which curve y in A minimizes the time. Because of the fixed boundary conditions, this problem, like Example 4.1, is called a **fixed-endpoint problem**.
◻

Example 4.3

Let A be the set of all nonnegative continuous functions on an interval $x_1 \leq x \leq x$, and define the functional J on A by

$$J(y) = \int_{x_1}^{x_2} y(x) \, dx,$$

which gives the area under the curve $y = y(x)$. Clearly $y(x) = 0$ provides an absolute minimum for J and there is no maximum.   ◻

The types of functionals that are of interest in the classic calculus of variations are ones defined by integral expressions of the form

$$J(y) = \int_a^b L(x, y(x), y'(x)) \, dx, \quad y \in A, \tag{1.3}$$

where $L = L(x, y, y')$ is some given function and A is a well-defined class of functions. In the last three examples the functionals are each of the form (1.3) with

$$
\begin{align}
L &= \sqrt{1 + (y')^2}, &\tag{1.4}\\
L &= \sqrt{1 + (y')^2}/\sqrt{2g(y_1 - y)}, &\tag{1.5}\\
L &= y, &\tag{1.6}
\end{align}
$$

respectively. The function L is called the **Lagrangian** (after Lagrange). We assume that L is twice continuously differentiable in each of its three arguments. It is customary to drop the independent variable x in parts of the integrand in (1.3) and just write

$$J(y) = \int_a^b L(x, y, y') \, dx, \quad y \in A.$$

As is customary practice in physics, in problems where time is the independent variable we often write

$$J(y) = \int_a^b L(t, y, \dot{y}) \, dt, \quad y \in A,$$

where the overdot on y denotes differentiation with respect to t.

EXERCISES

1. Show that the functional

$$J(y) = \int_0^1 \{\dot{y} \sin \pi y - (t + y)^2\} \, dt,$$

   where y is continuously differentiable on $0 \le t \le 1$, actually assumes its maximum value $2/\pi$ for the function $y(t) = -t$.

2. Consider the functional $J(y) = \int_0^1 (1 + x)(y')^2 \, dx$, where y is twice continuously differentiable and $y(0) = 0$, $y(1) = 1$. Of all functions of the form $y(x) = x + c_1 x(1 - x) + c_2 x^2 (1 - x)$, where c_1 and c_2 are constants, find the one that minimizes J.

4.2 Necessary Conditions for Extrema

4.2.1 Normed Linear Spaces

As noted in Section 4.1, a necessary condition for a function f to have a local minimum at some point x_0 is $f'(x_0) = 0$. A similar condition holds for functionals, provided the concept of derivative can be defined. In elementary calculus the derivative of a function f at x_0 is defined by a limit process, that is,

$$f'(x_0) = \lim_{\Delta x \to 0} \frac{f(x_0 + \Delta x) - f(x_0)}{\Delta x},$$

if the limit exists. For a functional $J : A \to \mathbb{R}$, this formula makes no sense if f is replaced by J, and x_0 and Δx are replaced by functions in A. However, the essence of the derivative, namely the limit process, can be carried over in a natural way to functionals.

To define the derivative of a functional J at some function y_0 in its domain A, we require a notion of closeness in the set A so that the value $J(y_0)$ can be compared to $J(y)$ for a function y *close to* y_0. To carry out this idea in a general manner we assume that the set of admissible functions A has a linear algebraic structure imposed on it. In particular, we assume A is a subset of a **normed linear space** where addition of functions and multiplication by scalars (real numbers) are defined, and there is some measuring device that gives definiteness to the idea of the *size* of a function. Actually, the objects in the normed linear space may be vectors (in a geometric sense), functions, numbers, tensors, matrices, and so on. Normed linear spaces whose elements are functions are frequently called **function spaces**. We make a precise definition in two parts—first we define a linear space, and then we define a norm.

A collection V of objects $u, v, w, \ldots$ is called a **real linear space** if the following conditions are satisfied:

1. There is a binary operation $+$ on V such that $u + v \in V$ whenever $u, v \in V$. That is, V is *closed* under the operation $+$, which is called *addition*.

2. Addition is *commutative* and *associative*; that is, $u + v = v + u$ and $u + (v + w) = (u + v) + w$.

3. There is *zero element* 0 in V that has the property that $u + 0 = u$ for all $u \in V$. The zero element is also called the identity under addition and it is unique.

4. For each $u \in V$ there is a $v \in V$ such that $u + v = 0$. The element v is called the *inverse* and u and is denoted by $-u$.

5. For each $u \in V$ and $\alpha \in \mathbb{R}$, *scalar multiplication* αu is defined and $\alpha u \in V$.

6. Scalar multiplication is associative and distributive; that is,

$$\alpha(\beta u) = (\alpha\beta)u,$$
$$(\alpha + \beta)u = \alpha u + \beta u,$$
$$\alpha(u + v) = \alpha u + \alpha v,$$

for all $\alpha, \beta \in \mathbb{R}$, and $u, v \in V$.

7. $1u = u$ for all $u \in V$.

These conditions, or axioms, put algebraic structure on the set V. The first four conditions state that there is an addition defined on V that satisfies the usual rules of arithmetic; the last three conditions state that the objects of V can be multiplied by real numbers and that reasonable rules governing that multiplication hold true. The reader is already familiar with some examples of linear spaces. In linear algebra courses, linear spaces are usually called vector spaces.

Example 4.4

(The set $\mathbb{R}^n$) Here an object is an n-tuple of real numbers $(x_1, x_2, \dots, x_n)$, where $x_i \in \mathbb{R}$. Addition and scalar multiplication are defined by

$$(x_1, x_2, \dots, x_n) + (y_1, y_2, \dots, y_n) = (x_1 + y_1, x_2 + y_2, \dots, x_n + y_n)$$

and

$$\alpha(x_1, x_2, \dots, x_n) = (\alpha x_1, \alpha x_2, \dots, \alpha x_n).$$

The zero element is $(0, 0, \dots, 0)$ and the inverse of $(x_1, \dots, x_n)$ is $(-x_1, \dots, -x_n)$. It is easily checked that the remaining properties hold true. The objects of $\mathbb{R}^n$ are called vectors and they are represented by directed arrows in two and three dimensions. $\square$

Example 4.5

(**Continuous functions**) The function space $C[a, b]$ represents the set of all continuous functions $f, g, h, \dots$ defined on an interval $a \leq x \leq b$. Addition and scalar multiplication are defined pointwise by

$$(f + g)(x) = f(x) + g(x),$$
$$(\alpha f)(x) = \alpha f(x).$$

From calculus we know that the sum of two continuous functions is a continuous function, and a continuous function multiplied by a constant is also a continuous function. The zero element is the 0 function, that is, the function that is constantly zero on $[a, b]$. The inverse of f is $-f$, which is the reflection of $f(x)$ through the x axis. Again it is easily checked that the other conditions are satisfied. Geometrically the sum of the two functions f and g is the graphical sum, and multiplication of f by α stretches the graph of f vertically by a factor α. □

Example 4.6

The function space $C^1[a, b]$ is the set of all continuous functions on an interval $[a, b]$ whose derivative is also continuous. Addition and scalar multiplication are defined as in the last example. It is clear that $C^1[a, b]$ satisfies the other axioms for a linear space. This linear space is commonly called the set of continuously differentiable, or smooth, functions on the interval $[a, b]$. In a similar way we define $C^2[a, b]$ as the set of continuous functions on $[a, b]$ whose second derivatives are also continuous. Similarly, we define $C^n[a, b]$ as the set of continuous functions whose nth derivatives are continuous. □

The linear spaces, or function spaces, in the last two examples are typical of those occurring in the calculus of variations. Later, in Chapter 5, we define the L^p spaces, which play a key role in modern applied analysis.

A linear space is said to have a norm if there is a rule that determines the size of a given element in the space. In particular if V is a linear space, then a **norm** on V is a mapping that associates to each $y \in V$ a nonnegative real number denoted by $||y||$, called the norm of y, and that satisfies the following conditions:

1. $||y|| = 0$ if, and only if, $y = 0$.

2. $||\alpha y|| = |\alpha| ||y||$ for all $\alpha \in \mathbb{R}$, $y \in V$.

3. $||y_1 + y_2|| \leq ||y_1|| + ||y_2||$ for all $y_1, y_2 \subset V$.

A **normed linear space** is a linear space V on which there is defined a norm $||\cdot||$. The number $||y||$ is interpreted as the magnitude or size of y. Thus, a norm puts geometric structure on V. The examples show that there are several ways to define a norm on a given space.

Example 4.7

The linear space $\mathbb{R}^n$ is a normed linear space if we define a norm by

$$||\mathbf{y}|| = \sqrt{y_1^2 + \cdots + y_n^2}, \quad \mathbf{y} = (y_1, y_2, \ldots, y_n).$$

It is not difficult to show that this (Euclidean) norm satisfies the conditions 1 through 3. A norm on a linear space is not unique. In $\mathbb{R}^n$ the quantity

$$||\mathbf{y}|| = \max\{|y_1|, |y_2|, \ldots, |y_n|\}$$

also defines a norm. $\square$

Example 4.8

$C[a, b]$ becomes a normed linear space if we define a norm of a function by

$$||y||_M = \max_{a \leq x \leq b} |y(x)|. \tag{2.1}$$

This norm is called the **strong norm**. The quantity

$$||y||_1 = \int_a^b |y(x)| \, dx \tag{2.2}$$

also defines a norm on $C[a, b]$, called the L^1–norm. Both these norms provide a way to define the magnitude of a continuous function on $[a, b]$, and both easily satisfy the Axioms 1 through 3. $\square$

Example 4.9

In the function space $C^1[a, b]$ the norm defined by

$$||y||_w = \max_{a \leq x \leq b} |y(x)| + \max_{a \leq x \leq b} |y'(x)| \tag{2.3}$$

is called the **weak norm** and it plays an important role in the calculus of variations. The same norm could also be defined in $C^n[a, b]$, $n \geq 2$. $\square$

If V is a normed linear space with norm $|| \cdot ||$, then the **distance** between the elements y_1 and y_2 in V is defined by

$$||y_1 - y_2||.$$

The motivation for this definition comes from the set of real numbers $\mathbb{R}$, where the absolute value $|x|$ is a norm. The distance between two real numbers a and b is $|a - b|$. Another example is the distance between two functions y_1 and y_2 in $C^1[a, b]$ measured in the weak norm (2.3),

$$||y_1 - y_2||_w = \max_{a \leq x \leq b} |y_1(x) - y_2(x)| + \max_{a \leq x \leq b} |y_1'(x) - y_2'(x)|. \tag{2.4}$$

Having the notion of distance allows us to talk about nearness. Two elements of a normed linear space are close if the distance between them is small. Two

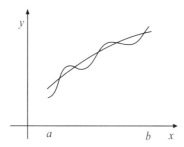

Figure 4.3 Two functions in $C^1[a, b]$ close in the strong norm but not close in the weak norm.

functions y_1 and y_2 in $C^1[a, b]$ are close in the weak norm (2.3) if the quantity $||y_1 - y_2||_w$ is small; this means that y_1 and y_2 have values that do not differ greatly, and their derivatives do not differ greatly. The two functions y_1 and y_2 shown in Fig. 4.3 are not close in the weak norm, but they are close in the strong norm because $\max |y_1(x) - y_2(x)|$ is small. Hence, closeness in one norm does not imply closeness in another.

4.2.2 Derivatives of Functionals

For a variational problem we assume the set of admissible functions A is a subset of a normed linear space V. Then the set A can inherit the norm and algebraic properties of the space V. Therefore the notions of local maxima and minima make sense. If $J\colon A \to \mathbb{R}$ is a functional on A, where $A \subseteq V$ and V is a normed linear space with norm $||\cdot||$, then J has a **local minimum** at $y_0 \in A$, provided $J(y_0) \leq J(y)$ for all $y \in A$ with $||y - y_0|| < d$, for some positive number d. We call y_0 an **extremal** of J and say J is **stationary** at y_0. In the special case that A is a set of functions and $A \subseteq C^1[a, b]$ with the weak norm, we say J has a **weak relative minimum** at y_0. If the norm on $C^1[a, b]$ is the strong norm, then we say J has a **strong relative minimum** at y_0.

In the calculus of variations the norm comes into play only in the theory of sufficient conditions for extrema. In this chapter we focus only on necessary conditions.

To motivate the definition of a derivative of a functional, let us rewrite the limit definition of derivative of a function f at x_0 as

$$f(x_0 + \Delta x) - f(x_0) = f'(x_0)\Delta x + \mathrm{o}(\Delta x).$$

The **differential** of f at x_0, defined by $df(x_0, \Delta x) = f'(x_0)\Delta x$, is the linear part in the increment Δx of the total change $\Delta f \equiv f(x_0 + \Delta x) - f(x_0)$.

Therefore,

$$\Delta f = df(x_0, \Delta x) + \mathrm{o}(\Delta x).$$

Now consider a functional $J\colon A \to \mathbb{R}$, where A is a subset of a normed linear space V. Specifically, V may be a function space. Let $y_0 \in A$. To increment y_0 we fix an element $h \in V$ such that $y_0 + \varepsilon h$ is in A for all real numbers ε sufficiently small. The increment εh is called the **variation** of the function y_0 and is often denoted by δy_0; that is, $\delta y_0 \equiv \varepsilon h$. Then we define the total change in the functional J due to the change εh in y_0 by

$$\Delta J = J(y_0 + \varepsilon h) - J(y_0).$$

Our goal is to calculate the linear part of this change. To this end, we define the real-valued function

$$\mathcal{J}(\varepsilon) \equiv J(y_0 + \varepsilon h),$$

which is a function of a real variable ε defined by evaluating the functional J on the one parameter family of functions $y_0 + \varepsilon h$. Assuming $\mathcal{J}$ is sufficiently differentiable in ε, we have

$$\Delta J = \mathcal{J}(\varepsilon) - \mathcal{J}(0) = \mathcal{J}'(0)\varepsilon + \mathrm{o}(\varepsilon).$$

Therefore, $\mathcal{J}'(0)\varepsilon$ is like a differential for the functional J, being the linear part, or lowest order, of the increment ΔJ Thus, we are led to the following definition.

Definition 4.10

Let $J\colon A \to \mathbb{R}$ be a functional on A, where $A \subseteq V$, and V a normed linear space. Let $y_0 \in A$ and $h \in V$ such that $y_0 + \varepsilon h \in A$ for all ε sufficiently small. Then the **first variation** (also called the **Gâteaux derivative**) of J at y_0 in the direction of h is defined by

$$\delta J(y_0, h) \equiv \mathcal{J}'(0) \equiv \frac{d}{d\varepsilon} J(y_0 + \varepsilon h)|_{\varepsilon=0}, \qquad (2.5)$$

provided the derivative exists. Such a direction h for which (2.5) exists is called an **admissible variation** at y_0. $\square$

There is an analogy of δJ with a directional derivative. Let $f(\mathbf{X})$ be a real valued function defined for $\mathbf{X} = (x, y)$ in $\mathbb{R}^2$, and let $\mathbf{n} = (n_1, n_2)$ be a unit vector. We recall from calculus that the directional derivative of f at $\mathbf{X}_0 = (x_0, y_0)$ in the direction $\mathbf{n}$ is defined by

$$D_{\mathbf{n}} f(\mathbf{X}_0) \equiv \lim_{\varepsilon \to 0} \frac{f(\mathbf{X}_0 + \varepsilon \mathbf{n}) - f(\mathbf{X}_0)}{\varepsilon}.$$

When (2.5) is written out in limit form, it becomes

$$\delta J(y_0, h) = \lim_{\varepsilon \to 0} \frac{J(y_0 + \varepsilon h) - J(y_0)}{\varepsilon},$$

and the analogy stands out clearly.

The recipe for calculating the first variation $\delta J(y_0, h)$ is to use (2.5); that is, evaluate J at $y_0 + \varepsilon h$, where h is chosen such that $y_0 + \varepsilon h \in A$; then take the derivative with respect to ε, afterwards setting $\varepsilon = 0$. Examples are in the next section.

4.2.3 Necessary Conditions

Let $J \colon A \to \mathbb{R}$ be a functional on A, where $A \subseteq V$, and let $y_0 \in A$ be a local minimum for J relative to a norm $||\cdot||$ in V. Guided by results from the calculus, we expect that the first variation δJ of J should vanish at y_0. It is not difficult to see that this is indeed the case. Let $h \in V$ be such that $y_0 + \varepsilon h$ is in A for sufficiently small values of ε. Then the function $\mathcal{J}(\varepsilon)$ defined by

$$\mathcal{J}(\varepsilon) = J(y_0 + \varepsilon h)$$

has a local minimum at $\varepsilon = 0$. Hence from ordinary calculus we must have $\mathcal{J}'(0) = 0$ or $\delta J(y_0, h) = 0$. This follows regardless of which $h \in V$ is chosen. Hence we have proved the following theorem that gives a necessary condition for a local minimum.

Theorem 4.11

Let $J \colon A \to \mathbb{R}$ be a functional, $A \subseteq V$. If $y_0 \in A$ provides a local minimum for J relative to the norm $|| \cdot ||$, then

$$\delta J(y_0, h) = 0 \tag{2.6}$$

for all admissible variations h. □

The fact that (2.6) holds for all *admissible variations* h often allows us to eliminate h from the condition and obtain an equation just in terms of y_0, which can then be solved for y_0. In the calculus of variations, the equation for y_0 is a differential equation. Since (2.6) is a necessary condition we are not guaranteed that solutions y_0 actually will provide a minimum. Therefore, all we know is that y_0 satisfying (2.6) are extremals, which are the candidates for maxima and minima. In other words, $J(y)$ is *stationary* at y_0 in the direction h.

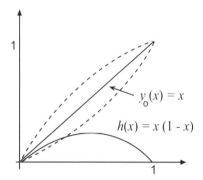

Figure 4.4 Two members of the one-parameter family of curves $y_0(x) +$ $\varepsilon h(x) = x + \varepsilon x(1 - x)$ (dashed) for two different values of the parameter ε.

Example 4.12

Consider the functional $J(y) = \int_0^1 (1 + y'(x)^2)dx$, where $y \in C^1[0, 1]$ and $y(0) =$ 0, $y(1) = 1$. Let $y_0(x) = x$ and $h(x) = x(1 - x)$. Two members of the family of curves $y_0 + \varepsilon h = x + \varepsilon x(1 - x)$ are sketched in Fig. 4.4. These are the variations in $y_0(x) = x$. We evaluate J on the family $y_0 + \varepsilon h$ to get

$$\mathcal{J}(\varepsilon) = J(y_0 + \varepsilon h) = \int_0^1 \left[1 + (y_0'(x) + \varepsilon h'(x))^2\right] dx$$

$$= \int_0^1 \left[1 + (1 + \varepsilon(1 - 2x))^2\right] dx = 2 + \frac{\varepsilon^2}{3}.$$

Then $\mathcal{J}'(\varepsilon) = \frac{2}{3}\varepsilon$ and $\mathcal{J}'(0) = 0$. Hence we conclude that $\delta J(y_0, h) = 0$ and $y_0 = x$ is an extremal; J is stationary at $y_0 = x$ in the direction $h = x(1 - x)$.
□

EXERCISES

1. Determine whether the given set constitutes a real linear space under the usual operations associated with elements of the set.

   a) The set of all polynomials of degree ≤ 2.

   b) The set of all continuous functions on $[0, 1]$ satisfying $f(0) = 0$.

   c) The set of all continuous functions on $[0, 1]$ satisfying $f(1) = 1$.

2. If $J(y) = \int_0^1 (x^2 - y^2 + (y')^2)dx$, $y \in C^2[0, 1]$, calculate ΔJ and $\delta J(y, h)$ when $y(x) = x$ and $h(x) = x^2$.

3. Consider the functional $J(y) = \int_0^1 (1+x)(y')^2 dx$ where $y \in C^2$ and $y(0) = 0$ and $y(1) = 1$. Show that $\delta J(y_0, h) = 0$ for all $h \in C^2$ and $h(0) = h(1) = 0$, where $y_0(x) = \ln(1+x)/\ln 2$.

4. Consider the functional $J(y) = \int_0^{2\pi} (y')^2 \, dx$. Sketch the function $y_0(x) = x$ and the family $y_0(x) + \varepsilon h(x)$, where $h(x) = \sin x$. Compute $\mathcal{J}(\varepsilon) \equiv J(y_0 + \varepsilon h)$ and show that $\mathcal{J}'(0) = 0$. Deduce that J is stationary at x in the direction $\sin x$.

5. Let $J(y) = \int_0^1 (3y^2 + x) dx + \frac{1}{2} y(0)^2$, where y is a continuous function on $0 \le x \le 1$. Let $y = x$ and $h = x + 1$, $0 \le x \le 1$. Compute $\delta J(x, h)$.

6. Prove that (2.1) and (2.2) both satisfy properties for a norm on $C[a, b]$.

7. In $C^1[0, 1]$ compute the distances between the two functions $y_1(x) = 0$ and $y_2(x) = \frac{1}{100} \sin(1000x)$ in both the weak and strong norms.

8. Show that the first variation $\delta J(y_0, h)$ satisfies the homogeneity condition

$$\delta J(y_0, \alpha h) = \alpha \delta J(y_0, h), \quad \alpha \in \mathbb{R}.$$

9. A functional $J: V \to \mathbb{R}$ is *linear* if

$$J(y_1 + y_2) = J(y_1) + J(y_2), \quad y_1, y_2 \in V$$

and

$$J(\alpha y_1) = \alpha J(y_1), \quad \alpha \in \mathbb{R}, \quad y_1 \in V$$

Which of the following functionals on $C^1[a, b]$ are linear?

a) $J(y) = \int_a^b yy' \, dx$.

b) $J(y) = \int_a^b ((y')^2 + 2y) \, dx$.

c) $J(y) = e^{y(a)}$.

d) $J(y) = \int_a^b \int_a^b K(x, t) y(x) y(t) \, dx \, dt$.

e) $J(y) = \int_a^b y \sin x \, dx$.

f) $J(y) = \int_a^b (y')^2 dx + G(y(b))$, where G is a given differentiable function

10. A functional $J: V \to \mathbb{R}$, where V is a normed linear space, is continuous at $y_0 \in V$ if for any $\varepsilon > 0$ there is a $\delta > 0$ such that $|J(y) - J(y_0)| < \varepsilon$ whenever $||y - y_0|| < \delta$.

a) Let $V = C^1[a, b]$. Prove that if J is continuous in the strong norm, then it is continuous in the weak norm.

    b) Give a specific example to show that continuity in the weak norm does
       not imply continuity in the strong norm.

11. Compute the first variation of the functionals on $C^1[a, b]$ given in Exercise
    5(a)–(f).

12. Let $J: V \to \mathbb{R}$ be a linear functional and V a normed linear space. Prove
    that if J is continuous at $y_0 = 0$, then it is continuous at each $y \in V$.

13. The second variation of a functional $J: A \to \mathbb{R}$ at $y_0 \in A$ in the direction
    h is defined by

$$\delta^2 J(y_0, h) \equiv \frac{d^2}{d\varepsilon^2} J(y_0 + \varepsilon h)|_{\varepsilon=0}.$$

Find the second variation of the functional

$$J(y) = \int_0^1 (xy'^2 + y \sin y') \, dx,$$

where $y \in C^2[0, 1]$.

4.3 The Simplest Problem

4.3.1 The Euler Equation

The *simplest problem* in the calculus of variations is to find a local minimum
for the functional

$$J(y) = \int_a^b L(x, y, y') \, dx, \tag{3.1}$$

where $y \in C^2[a, b]$ and $y(a) = y_0$, $y(b) = y_1$. Here L is a given, twice continu-
ously differentiable function on $[a, b] \times \mathbb{R}^2$, and $|| \cdot ||$ is some norm in $C^2[a, b]$.
Actually, our assumption that y is of class C^2 could be weakened to $y \in C^1[a, b]$.
The increased smoothness assumption, however, will make the work a little easier. In this discussion we assume that the competing functions are class C^2. The
references at the end of the chapter can be consulted to see how the analysis
changes if smoothness conditions are relaxed.

    We seek a necessary condition. Guided by the last two sections we compute
the first variation of J. Let y be a local minimum and h a twice continuously
differentiable function satisfying $h(a) = h(b) = 0$. Then $y + \varepsilon h$ is an admissible
function and

$$J(y + \varepsilon h) = \int_a^b L(x, y + \varepsilon h, y' + \varepsilon h') \, dx.$$

Therefore,

$$\frac{d}{d\varepsilon} J(y + \varepsilon h) = \int_a^b \frac{\partial}{\partial \varepsilon} L(x, y + \varepsilon h, y' + \varepsilon h') \, dx$$

$$= \int_a^b \left[L_y(x, y + \varepsilon h, y' + \varepsilon h')h + L_{y'}(x, y + \varepsilon h, y' + \varepsilon h')h' \right] dx,$$

where L_y denotes $\partial L/\partial y$ and $L_{y'}$ denotes $\partial L/\partial y'$. Hence,

$$\delta J(y, h) = \frac{d}{d\varepsilon} J(y + \varepsilon h)|_{\varepsilon = 0} = \int_a^b \{ L_y(x, y, y')h + L_{y'}(x, y, y')h' \} \, dx.$$

Therefore, a necessary condition for $y(x)$ to be a local minimum is

$$\int_a^b \{ L_y(x, y, y')h + L_{y'}(x, y, y')h' \} \, dx = 0 \tag{3.2}$$

for all $h \in C^2[a, b]$ with $h(a) = h(b) = 0$.

Condition (3.2) is not useful as it stands for determining $y(x)$. Using the fact that it must hold for all h, however, we can eliminate h and thereby obtain a condition for y alone. First we integrate the second term in (3.2) by parts to obtain

$$\int_a^b \left(L_y(x, y, y') - \frac{d}{dx} L_{y'}(x, y, y') \right) h \, dx + L_{y'}(x, y, y')h|_{x=a}^{x=b} = 0. \tag{3.3}$$

Because h vanishes at a and b, the boundary terms vanish and the necessary condition becomes

$$\int_a^b \left(L_y(x, y, y') - \frac{d}{dx} L_{y'}(x, y, y') \right) h \, dx = 0 \tag{3.4}$$

for all $h \in C^2[a, b]$ with $h(a) = h(b) = 0$. The following lemma, due to Lagrange, provides the final step in our calculation by effectively allowing us to eliminate h from condition (3.4). This lemma is one version of what is called the *fundamental lemma* of the calculus of variations.

Lemma 4.13

If $f(x)$ is continuous on $[a, b]$ and if

$$\int_a^b f(x)h(x) \, dx = 0 \tag{3.5}$$

for every twice continuously differentiable function h with $h(a) = h(b) = 0$, then $f(x) = 0$ for $x \in [a, b]$. □

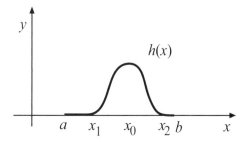

Figure 4.5 Plot of the C^2 function $h(x)$.

Proof

The proof is by contradiction. Assume for some x_0 in (a, b) that $f(x_0) > 0$. Because f is continuous, $f(x) > 0$ for all x in some interval (x_1, x_2) containing x_0. For $h(x)$ choose

$$h(x) = \begin{cases} (x - x_1)^3(x_2 - x)^3, & x_1 \leq x \leq x_2 \\ 0, & \text{otherwise} \end{cases}$$

a graph of which is shown in Fig. 4.5. The cubic factors $(x - x_1)^3$ and $(x_2 - x)^3$ appear so that h is smoothed out at x_1 and x_2 and is, therefore, of class C^2. Then

$$\int_a^b f(x)h(x)\, dx = \int_{x_1}^{x_2} f(x)(x - x_1)^3(x_2 - x)^3\, dx > 0,$$

because f is positive on (x_1, x_2). This contradicts (3.5) and the lemma is proved.
□

Applying the lemma to (3.4) with $f(x) = L_y(x, y, y') - (d/dx)L_{y'}(x, y, y')$, which is continuous because L is twice continuously differentiable, we obtain the following important theorem.

Theorem 4.14

If a function y provides a local minimum for the functional

$$J(y) = \int_a^b L(x, y, y')\, dx,$$

where $y \in C^2[a, b]$ and

$$y(a) = y_0, \quad y(b) = y_1,$$

then y must satisfy the differential equation

$$L_y(x, y, y') - \frac{d}{dx} L_{y'}(x, y, y') = 0, \quad x \in [a, b]. \; \square \tag{3.6}$$

Equation (3.6) is called the **Euler equation** or **Euler–Lagrange equation**. It is a second-order ordinary differential equation that is, in general, nonlinear. It represents a *necessary condition* for a local minimum and it is analogous to the derivative condition $f'(x) = 0$ in differential calculus. Therefore its solutions are not necessarily local minima. The solutions of (3.6) are (local) **extremals**. If y is an extremal, then $\delta J(y, h) = 0$ for all h, and J is stationary at y.

By writing out the total derivative in (3.6) using the chain rule, the Euler equation becomes

$$L_y - L_{y'x}(x, y, y') - L_{y'y}(x, y, y')y' - L_{y'y'}(x, y, y')y'' = 0.$$

Thus it is evident that the Euler equation is second order, provided $L_{y'y'} \neq 0$.

4.3.2 Solved Examples

We now determine the extremals for some specific variational problems.

Example 4.15

Find the extremals of the functional

$$J(y) = \int_0^1 ((y')^2 + 3y + 2x)\, dx, \quad y(0) = 0, \quad y(1) = 1.$$

The Lagrangian is $L = (y')^2 + 3y + 2x$. Then $L_y = 3$, $L_{y'} = 2y'$ and the Euler equation is

$$3 - \frac{d}{dx}(2y') = 0.$$

Integrating twice gives the extremals

$$y = \frac{3}{4}x^2 + C_1 x + C_2.$$

The boundary conditions readily give $C_2 = 0$ and $C_1 = \frac{1}{4}$. Therefore the extremal of J satisfying the boundary conditions is

$$y = \frac{3}{4}x^2 + \frac{1}{4}x.$$

Further calculations would be required to determine whether this maximizes or minimizes J. $\square$

Example 4.16

The arclength functional is

$$J(y) = \int_a^b \sqrt{1 + (y')^2} \, dx,$$

where $y \in C^2[a,b]$, $y(a) = y_0$ and $y(b) = y_1$. A necessary condition for J to have a local minimum at y is that y satisfy the Euler equation

$$L_y - \frac{d}{dx} L_{y'} = 0,$$

or

$$\frac{d}{dx} \left(\frac{y'}{\sqrt{1 + (y')^2}} \right) = 0.$$

Hence,

$$\frac{y'}{\sqrt{1 + (y')^2}} = C,$$

where C is a constant. Solving for y' gives

$$y' = K,$$

where K is a constant. Therefore,

$$y = Kx + M,$$

and the extremals are straight lines. Applying the boundary conditions determines the constants K and M,

$$K = \frac{y_1 - y_0}{b - a}, \quad M = \frac{by_0 - ay_1}{b - a}.$$

Therefore the unique extremal is $y = \frac{y_1 - y_0}{b-a} x + \frac{by_0 - ay_1}{b-a}$, which is the equation of the straight line connecting (a, y_0) and (b, y_1). $\square$

4.3.3 First Integrals

The Euler equation is a second-order ordinary differential equation and the solution will generally contain two arbitrary constants. In special cases, when the Lagrangian does not depend explicitly on one of its variables x, y, or y', it is possible to make an immediate simplification of the Euler equation. We consider three cases:

1. If $L = L(x, y)$, then the Euler equation is $L_y(x, y) = 0$, which is an algebraic equation.

2. If $L = L(x, y')$, then the Euler equation is

$$L_{y'}(x, y') = C, \tag{3.7}$$

where C is any constant.

3. If $L = L(y, y')$, then the Euler equation is

$$L(y, y') - y' L_{y'}(y, y') = C, \tag{3.8}$$

where C is any constant.

Propositions (1) and (2) follow obviously from the Euler equation. To prove (3) we note that

$$\frac{d}{dx}[L - y' L_{y'}] = \frac{dL}{dx} - y' \frac{d}{dx} L_{y'} - L_{y'} y''$$

$$= L_y y' + L_{y'} y'' - y' \frac{d}{dx} L_{y'} - L_{y'} y''$$

$$= y' \left(L_y - \frac{d}{dx} L_{y'} \right) = 0.$$

Equations (3.7) and (3.8) are first integrals of the Euler equation. A **first integral** of a second-order differential equation $F(x, y, y', y'') = 0$ is an expression of the form $g(x, y, y')$, involving one lower derivative, which is constant whenever y is a solution of the original equation $F(x, y, y', y'') = 0$. Hence

$$g(x, y, y') = C$$

represents an integration of the second-order equation. In the calculus of variations, if L is independent of x, then $L - y' L_{y'}$ is a first integral of the Euler equation and we say $L - y' L_{y'}$ is constant on the extremals. In mechanics, first integrals are called **conservation laws**.

Example 4.17

(Brachistochrone) We determine the extremals of the brachistochrone problem. The functional is

$$J(y) = \int_{x_1}^{x_2} \frac{\sqrt{1 + (y')^2}}{\sqrt{2g(y_1 - y)}} \, dx,$$

subject to the boundary conditions $y(0) = b$, $y(a) = 0$. The Lagrangian is independent of x, and therefore a first integral of the Euler equation is given by (3.8). Then

$$\frac{\sqrt{1 + (y')^2}}{\sqrt{b - y}} - (y')^2 \frac{(1 + (y')^2)^{-1/2}}{\sqrt{b - y}} = C,$$

which simplifies to

$$\left(\frac{dy}{dx}\right)^2 = \frac{1 - C^2(b - y)}{C^2(b - y)}.$$

Taking the square root of both sides and separating variables gives

$$dx = -\frac{\sqrt{b - y}}{\sqrt{C_1 - (b - y)}} \, dy, \quad C_1 = C^{-2},$$

where the minus sign is taken because $dy/dx < 0$. The last equation can be integrated by making the trigonometric substitution

$$b - y = C_1 \sin^2 \frac{\phi}{2}. \tag{3.9}$$

One obtains

$$dx = C_1 \sin^2 \frac{\phi}{2} \, d\phi = \frac{C_1}{2}(1 - \cos\phi) \, d\phi,$$

which yields

$$x = \frac{C_1}{2}(\phi - \sin\phi) + C_2. \tag{3.10}$$

Equations (3.9) and (3.10) are parametric equations for a cycloid. Here, in contrast to the problem of finding the curve of shortest length between two points, it is not clear that the cycloids just obtained actually minimize the given functional. Further calculations would be required for confirmation. □

Example 4.18

(**Fermat's principle**) In the limit of geometric optics the principle of Fermat (1601–1665) states that the time elapsed in the passage of light between two fixed points in a medium is an extremum with respect to all possible paths connecting the points. For simplicity we consider only light rays that lie in the xy plane. Let $c = c(x, y)$ be a positive continuously differentiable function representing the velocity of light in the medium. Its reciprocal $n = c^{-1}$ is called the index of refraction of the medium. If $P: (x_1, y_1)$ and $Q: (x_2, y_2)$ are two fixed points in the plane, then the time required for light to travel along a given path $y = y(x)$ connecting the two points is

$$T(y) = \int_P^Q \frac{ds}{c} = \int_{x_1}^{x_2} n(x, y)\sqrt{1 + (y')^2} \, dx.$$

Therefore, the actual light path connecting P and Q is the one that extremizes the integral $T(y)$. The differential equation for the extremal path is the Euler equation

$$n_y(x, y)\sqrt{1 + (y')^2} - \frac{d}{dx}\left(\frac{n(x, y)y'}{\sqrt{1 + (y')^2}}\right) = 0. \quad \square$$

EXERCISES

1. Find the extremal(s).

   a) $J(y) = \int_0^1 y' \, dx, \quad y \in C^2[0, 1], \quad y(0) = 0, \quad y(1) = 1.$

   b) $J(y) = \int_0^1 yy' \, dx, \quad y \in C^2[0, 1], \quad y(0) = 0, \quad y(1) = 1.$

   c) $J(y) = \int_0^1 xyy' \, dx, \quad y \in C^2[0, 1], \quad y(0) = 0, \quad y(1) = 1.$

2. Find extremals for the following functionals:

   a) $\int_a^b \frac{(y')^2}{x^3} \, dx.$

   b) $\int_a^b (y^2 + (y')^2 + 2ye^x) \, dx.$

3. Show that the Euler equation for the functional

$$J(y) = \int_a^b f(x, y) \sqrt{1 + (y')^2} \, dx$$

   has the form

$$f_y - f_x y' - \frac{f y''}{1 + (y')^2} = 0.$$

4. Find an extremal for

$$J(y) = \int_1^2 \frac{\sqrt{1 + (y')^2}}{x} \, dx, \quad y(1) = 0, \quad y(2) = 1.$$

5. Prove that the functional $J(y) = \int_a^b (p(x)(y')^2 + q(x)y^2) dx$, $y \in C^2[a, b]$, $y(a) = y_a$, $y(b) = y_b$, where p and q are positive, assumes its absolute minimum value for the function $y = Y(x)$, where Y is the solution to the Euler equation.

6. Obtain a necessary condition for a function $y \in C[a, b]$ to be a local minimum of the functional

$$J(y) = \int \int_R K(s, t) y(s) y(t) \, ds \, dt + \int_a^b y(t)^2 \, dt - 2 \int_a^b y(t) f(t) \, dt,$$

   where $K(s, t)$ is a given continuous function of s and t on the square $R: \{(s, t) \mid a \le s \le b, a \le t \le b\}$, $K(s, t) = K(t, s)$, and $f \in C[a, b]$. (The Euler equation is an integral equation.)

7. Find the extremal for $J(y) = \int_0^1 (1 + x)(y')^2 dx$, $y \in C^2$ and $y(0) = 0$, $y(1) = 1$. What is the extremal if the boundary condition at $x = 1$ is changed to $y'(1) = 0$?

8. Describe the paths of light rays in the plane where the medium has index of refraction given by

   a) $n = kx$.

   b) $n = ky^{-1}$.

   c) $n = ky^{1/2}$.

   d) $n = ky^{-1/2}$, where $k > 0$.

9. Show that the Euler equation for the problem

$$J(y) = \int_a^b L(t, y, \dot{y})\, dt$$

   can be written in the form

$$L_t - \frac{d}{dt}(L - \dot{y}L_{\dot{y}}) = 0.$$

10. Show that the minimal area of a surface of revolution is a catenoid, that is, the surface found by revolving a catenary

$$y = c_1 \cosh\left(\frac{x + c_2}{c_1}\right)$$

   about the x axis.

11. Find the extremals for

$$J(y) = \int_a^b (x^2 y'^2 + y^2)\, dx, \quad y \in C^2[a, b].$$

12. A Lagrangian has the form

$$L(x, y, y') = \frac{a^2}{12} y'^4 + ay'^2 G(y) - G(y)^2,$$

   where G is a given differentiable function. Find Euler's equation and a first integral.

13. Find extremals of the functional

$$J(y) = \int_0^1 (y^2 + yy' + (y' - 2)^2)\, dx$$

   over the domain $A = \{y \in C^2[0, 1] : y(0) = y(1) = 0\}$. Show that J does not assume a maximum value at these extremals. Explain why it does, or does not, follow that the extremals yield minimum values of J.

14. (Economics) Let $y = y(t)$ be an individual's total capital at time t and let $r = r(t)$ be the rate that capital is spent. If $U = U(r)$ is the rate of enjoyment, then his total enjoyment over a lifetime $0 \leq t \leq T$ is

$$E = \int_0^T e^{-\beta t} U(r(t)) \, dt,$$

where the exponential factor reflects the fact that future enjoyment is discounted over time. Initially, his capital is Y, and he desires $y(T) = 0$. Because his capital gains interest at rate α,

$$y' = \alpha y - r(t).$$

Assume $\alpha < 2\beta < 2\alpha$. Determine $r(t)$ and $y(t)$ for which the individual's total enjoyment is maximized if his enjoyment function is $U(r) = 2\sqrt{r}$.

4.4 Generalizations

4.4.1 Higher Derivatives

In the simplest variational problem the Lagrangian depends on x, y, and y'. An obvious generalization is to include higher derivatives in the Lagrangian. Thus we consider the *second-order problem*

$$J(y) = \int_a^b L(x, y, y', y'') \, dx, \quad y \in A, \tag{4.1}$$

where A is the set of all $y \in C^4[a, b]$ that satisfy the boundary conditions

$$y(a) = A_1, \quad y'(a) = A_2, \quad y(b) = B_1, \quad y'(b) = B_2. \tag{4.2}$$

The function L is assumed to be twice continuously differentiable in each of its four arguments.

Again we seek a necessary condition. Let $y \in A$ provide a local minimum for J with respect to some norm in $C^4[a, b]$, and choose $h \in C^4[a, b]$ satisfying

$$h(a) = h'(a) = h(b) = h'(b) = 0. \tag{4.3}$$

To compute the first variation, we form the function $J(y + \varepsilon h)$. Then

$$\begin{aligned}
\delta J(y, h) &= \frac{d}{d\varepsilon} J(y + \varepsilon h)|_{\varepsilon=0} \\
&= \frac{d}{d\varepsilon} \int_a^b L(x, y + \varepsilon h, y' + \varepsilon h', y'' + \varepsilon h'') \, dx|_{\varepsilon=0} \\
&= \int_a^b [L_y h + L_{y'} h' + L_{y''} h''] \, dx,
\end{aligned}$$

where L_y, and $L_{y'}$, and $L_{y''}$ are evaluated at (x, y, y', y''). In this calculation two integrations by parts are required on the term $L_{y''}h''$. Proceeding,

$$\delta J(y, h) = \int_a^b \left[L_y - \frac{d}{dx}L_{y'} + \frac{d^2}{dx^2}L_{y''} \right] h\, dx$$

$$+ \left[L_{y''}h' - \frac{d}{dx}L_{y''}h + hL_{y'} \right]_{x=a}^{x=b}. \qquad (4.4)$$

By (4.3) the terms on the boundary at $x = a$ and $x = b$ vanish. Therefore,

$$\int_a^b \left[L_y - \frac{d}{dx}L_{y'} + \frac{d^2}{dx^2}L_{y''} \right] h\, dx = 0 \qquad (4.5)$$

for all $h \in C^4[a, b]$ satisfying (4.3). To complete the argument, another version of the fundamental lemma is needed.

Lemma 4.19

If f is a continuous function on $[a, b]$, and if

$$\int_a^b f(x)h(x)\, dx = 0$$

for all $h \in C^4[a, b]$ satisfying (4.3), then $f(x) \equiv 0$ for $x \in [a, b]$.   □

The proof is nearly the same as before and is left as an exercise. Applying this lemma to (4.5) gives a necessary condition for y to be a local minimum, that is,

$$L_y - \frac{d}{dx}L_{y'} + \frac{d^2}{dx^2}L_{y''} = 0, \quad a \leq x \leq b. \qquad (4.6)$$

This is the Euler equation for the second-order problem defined by (4.1). Equation (4.6) is a fourth-order ordinary differential equation for y, and its general solution involves four arbitrary constants that can be determined from the boundary conditions.

The *nth-order variational problem* is defined by the functional

$$J(y) = \int_a^b L(x, y, y', y'', \ldots, y^{(n)})\, dx,$$

$y \in C^{2n}[a, b]$, with

$$y(a) = A_1, \quad y'(a) = A_2, \ldots, y^{(n-1)}(a) = A_n,$$
$$y(b) = B_1, \quad y'(b) = B_2, \ldots, y^{(n-1)}(b) = B_n.$$

The Euler equation in this case is

$$L_y - \frac{d}{dx}L_{y'} + \frac{d^2}{dx^2}L_{y''} - \cdots + (-1)^n \frac{d^n}{dn^n}L_{y^{(n)}} = 0,$$

which is an ordinary differential equation of order $2n$.

4.4.2 Several Functions

The functional J may depend on several functions $y_1, \ldots, y_n$. To fix the notion, let $n = 2$ and consider the functional

$$J(y_1, y_2) = \int_a^b L(x, y_1, y_2, y_1', y_2') \, dx, \tag{4.7}$$

where $y_1, y_2 \in C^2[a, b]$ with boundary conditions given by

$$y_1(a) = A_1, \quad y_1(b) = B_1,$$
$$y_2(a) = A_2, \quad y_2(b) = B_2. \tag{4.8}$$

In this case suppose the pair y_1 and y_2 provide a local minimum for J. We vary each independently by choosing h_1 and h_2 in $C^2[a, b]$, satisfying

$$h_1(a) = h_2(a) = h_1(b) = h_2(b) = 0, \tag{4.9}$$

and forming a one-parameter admissible pair of functions $y_1 + \varepsilon h_1$ and $y_2 + \varepsilon h_2$. Then

$$\mathcal{J}(\varepsilon) = \int_a^b L(x, y_1 + \varepsilon h_1, y_2 + \varepsilon h_2, y_1' + \varepsilon h_1', y_2' + \varepsilon h_2') \, dx,$$

and $\mathcal{J}$ has a local minimum at $\varepsilon = 0$. Now

$$\mathcal{J}'(0) = \int_a^b (L_{y_1} h_1 + L_{y_2} h_2 + L_{y_1'} h' + L_{y_2'} h_2') \, dx.$$

Integrating the last two terms in the integrand by parts and applying conditions (4.9) gives

$$\int_a^b \left\{ \left(L_{y_1} - \frac{d}{dx}L_{y_1'} \right) h_1 + \left(L_{y_2} - \frac{d}{dx}L_{y_2'} \right) h_2 \right\} dx = 0 \tag{4.10}$$

for all $h_1, h_2 \in C^2[a, b]$. Picking $h_2 = 0$ on $[a, b]$ gives

$$\int_a^b \left(L_{y_1} - \frac{d}{dx}L_{y_1'} \right) h_1 \, dx = 0$$

for all $h_1 \in C^2[a,b]$. By the fundamental lemma,

$$L_{y_1} - \frac{d}{dx} L_{y_1'} = 0. \tag{4.11}$$

Now in (4.10) pick $h_1 = 0$ on $[a,b]$. Then

$$\int_a^b \left(L_{y_2} - \frac{d}{dx} L_{y_2'} \right) h_2 \, dx = 0$$

for all $h_2 \in C^2[a,b]$. Applying the fundamental lemma again yields

$$L_{y_2} - \frac{d}{dx} L_{y_2'} = 0. \tag{4.12}$$

Consequently, if the pair y_1, y_2 provides a local minimum for J, then y_1 and y_2 must satisfy the system of two ordinary differential equations (4.11) and (4.12). These are the Euler equations for the problem. The four arbitrary constants appearing in the general solution can be determined from the boundary data.

In general, if J depends on n functions,

$$J(y_1, \ldots, y_n) = \int_a^b L(x, y_1, \ldots, y_n, y_1', \ldots, y_n') \, dx, \tag{4.13}$$

where $y_i \in C^2[a,b]$, and

$$y_i(a) = A_i, \quad y_i(b) = B_i, \quad i = 1, \ldots, n,$$

then a necessary condition for $y_1, \ldots, y_n$ to provide a local minimum for J is that $y_1, \ldots, y_n$ satisfy the Euler system of n ordinary differential equations

$$L_{y_i} - \frac{d}{dx} L_{y_i'} = 0, \quad i = 1, \ldots, n. \tag{4.14}$$

If the Lagrangian L is independent of explicit dependence on x, that is, $L_x = 0$, then it is left as an exercise to show that

$$L - \sum_{i=1}^n y_i' L_{y_i'} = C$$

is a first integral of (4.14).

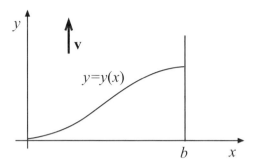

Figure 4.6 Path $y = y(x)$ across a river with velocity field $\mathbf{v}(x, y) = v(x)\mathbf{j}$.

4.4.3 Natural Boundary Conditions

Let us consider the following problem: A river with parallel straight banks b units apart has stream velocity given by $\mathbf{v}(x, y) = v(x)\mathbf{j}$, where $\mathbf{j}$ is the unit vector in the y direction (see Fig. 4.6). Assuming that one of the banks is the y axis and that the point $(0, 0)$ is the point of departure, what route should a boat take to reach the opposite bank in the shortest possible time? Assume that the speed of the boat in still water is c, where $c > v$.

This problem differs from those in earlier sections in that the right-hand endpoint, the point of arrival on the line $x = b$, is not specified; it must be determined as part of the solution to the problem. It can be shown (see the Exercises) that the time required for the boat to cross the river along a given path $y = y(x)$ is

$$J(y) = \int_0^b \frac{\sqrt{c^2(1 + (y')^2) - v^2} - vy'}{c^2 - v^2} \, dx. \tag{4.15}$$

The variational problem is to minimize $J(y)$ subject to the conditions

$$y(0) = 0, \quad y(b) \quad \text{free.}$$

Such a problem is called a **free endpoint problem**, and if $y(x)$ is an extremal then a certain condition must hold at $x = b$. Conditions of these types, called **natural boundary conditions**, are the subject of this section. Just as common are problems where both endpoints are unspecified.

To fix the notion we consider the problem

$$J(y) = \int_a^b L(x, y, y') \, dx, \tag{4.16}$$

where $y \in C^2[a, b]$ and

$$y(a) = y_0, \quad y(b) \quad \text{free.} \tag{4.17}$$

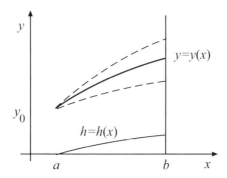

Figure 4.7 Variations $y = y(x) + \varepsilon h(x)$ of $y(x)$ satisfying $h(a) = 0$.

Let y be a local minimum. Then the variations h must be C^2 functions that satisfy the single condition

$$h(a) = 0. \tag{4.18}$$

Figure 4.7 shows several variations $y + \varepsilon h$ of the extremal y. No condition on h at $x = b$ is required, because the admissible functions are unspecified at the right endpoint.

The first variation is

$$\delta J(y, h) = \frac{d}{d\varepsilon} \int_a^b L(x, y + \varepsilon h, y' + \varepsilon h') \, dx|_{\varepsilon=0}$$

$$= \int_a^b \left(L_y - \frac{d}{dx} L_{y'} \right) h \, dx + L_{y'} h|_{x=a}^{x=b}.$$

Because $h(a) = 0$, a necessary condition for y to be a local minimum is

$$\int_a^b \left(L_y - \frac{d}{dx} L_{y'} \right) h \, dx + L_{y'}(b, y(b), y'(b)) h(b) = 0 \tag{4.19}$$

for every $h \in C^2(a, b)$ satisfying $h(a) = 0$. Because (4.19) holds for these h, it must hold for h also satisfying the condition $h(b) = 0$. Hence

$$\int_a^b \left(L_y - \frac{d}{dx} L_{y'} \right) h \, dx = 0, \tag{4.20}$$

and by the fundamental lemma, y must satisfy the Euler equation

$$L_y - \frac{d}{dx} L_{y'} = 0. \tag{4.21}$$

Then, substituting (4.21) into (4.19) gives

$$L_{y'}(b, y(b), y'(b)) h(b) = 0. \tag{4.22}$$

Now, because (4.22) holds for all choices of $h(b)$, we must have

$$L_{y'}(b, y(b), y'(b)) = 0, \qquad (4.23)$$

which is a condition on the extremal y at $x = b$. Equation (4.23) is called a **natural boundary condition**. The Euler equation (4.21), the fixed boundary condition $y(a) = A$, and the natural boundary condition (4.23) are enough to determine the extremal for the variational problem (4.16) and (4.17). By similar arguments if the left endpoint $y(a)$ is unspecified, then the natural boundary condition on an extremal y at $x = a$ is $L_{y'}(a, y(a), y'(a)) = 0$.

Example 4.20

We find the natural boundary condition at $x = b$ for the river crossing problem (4.15). In this case

$$L_{y'} = \frac{1}{c^2 - v^2} \left(\frac{c^2 y'}{(c^2(1 + y'^2) - v^2)^{1/2}} - v \right).$$

The boundary condition (4.23) becomes

$$\frac{c^2 y'(b)}{(c^2(1 + y'(b)^2) - v(b)^2)^{1/2}} - v(b) = 0,$$

which can be simplified to

$$y'(b) = \frac{v(b)}{c}.$$

Thus, the slope that the boat enters the bank at $x = b$ is the ratio of the water speed at the bank to the boat velocity in still water.   □

Example 4.21

Find the differential equation and boundary conditions for the extremal of the variational problem

$$J(y) = \int_0^1 (p(x)y'^2 - q(x)y^2)\, dx,$$

$$y(0) = 0, \quad y(1) \text{ free},$$

where p and q are positive smooth functions on $[0, 1]$. In this case $L_y = -2q(x)y$, $L_{y'} = 2p(x)y'$, and the Euler equation is

$$\frac{d}{dx}(p(x)y') + q(x)y = 0, \quad 0 < x < 1,$$

which is a Sturm–Liouville type equation . The natural boundary condition at $x = 1$ is given by (4.23), or

$$2p(1)y'(1) = 0.$$

Because $p > 0$, the boundary conditions are

$$y(0) = 0, \quad y'(1) = 0. \quad \square$$

EXERCISES

1. Find extremals for the following functionals:

   a) $J(y_1, y_2) = \int_0^{\pi/4} [4y_1^2 + y_2^2 + y_1' y_2'] \, dx, \quad y_1(0) = 1, \; y_1\left(\frac{\pi}{4}\right) = 0, \; y_2(0) = 0, \; y_2\left(\frac{\pi}{4}\right) = 1.$

   b) $J(y) = \int_0^1 (1 + (y'')^2) \, dx, \quad y(0) = 0, \; y'(0) = 1, \; y(1) = 1, \; y'(1) = 1.$

   c) $J(y) = \int_a^b (y^2 + 2\dot{y}^2 + (y'')^2) \, dt.$

   d) $J(y) = \int_0^1 (yy' + (y'')^2) \, dx, \; y(0) = 0, \; y'(0) = 1, \; y(1) = 2, \; y'(1) = 4.$

   e) $J(y) = \int_0^\infty [y^2 + y'^2 + (y'' + y')^2] \, dx, \quad y(0) = 1, \; y'(0) = 2, \; y(\infty) = 0, \; y'(\infty) = 0.$

2. Prove that the Euler equation for $J(y) = \int_a^b L(x, y, y', y'') \, dx$ has first integral $L_{y'} - (d/dx)L_{y''} = C$ when L is independent of y. If L is independent of x, show that

$$L - y' \left(L_{y'} - \frac{d}{dx} L_{y''} \right) - y'' L_{y''} = C$$

   is a first integral.

3. One version of the Ramsey growth model in economics involves minimizing the *total product*

$$J(M) = \int_0^T (aM - M' - b)^2 dt, \quad a, b > 0,$$

   over a fixed planning period $[0, T]$, where $M = M(t)$ is the capital at time t and $M(0) = M_0$ is the initial capital. If $M(t)$ minimizes J, find the capital $M(T)$ at the end of the planning period.

4. Find the extremals for a functional of the form

$$J(y, z) = \int_a^b L(\dot{y}, \dot{z}) \, dt,$$

   given that L satisfies the condition $L_{\dot{y}\dot{y}} L_{\dot{z}\dot{z}} - L_{\dot{y}\dot{z}}^2 \neq 0.$

5. Find the extremals for

   a) $J(y) = \int_0^1 (y'^2 + y^2) \, dx, \quad y(0) = 1, \ y(1)$ free.

   b) $J(y) = \int_0^3 e^{2x}(y'^2 - y^2) \, dx, \quad y(0) = 1, \ y(3)$ free.

   c) $J(y) = \int_0^1 (\frac{1}{2}y'^2 + y'y + y' + y) \, dx, \quad y(0) = \frac{1}{2}, \ y(1)$ free.

   d) $J(y) = \int_1^e (\frac{1}{2}x^2y'^2 - \frac{1}{8}y^2) \, dx, \quad y(1) = 1, \ y(e)$ free.

   e) $J(y) = \int_0^1 y'^2 \, dx + y(1)^2, \quad y(0) = 1, \ y(1)$ free.

6. Determine the natural boundary condition at $x = b$ for the variational problem defined by

$$J(y) = \int_a^b L(x, y, y') \, dx + G(y(b)), \quad y \in C^2[a, b], \ y(a) = y_0,$$

   where G is a given differentiable function on $\mathbb{R}$.

7. Find the extremal(s) of the functional

$$J(y) = \int_0^b (\sqrt{1 - k^2 + y'^2} - ky') \, dx,$$

   in the class of all $C^2[0, b]$ functions with $y(0) = 0$ and $y(b)$ free. Take $0 < k < 1$.

8. Find the natural boundary condition at $x = 2$ associated with the functional

$$J(y) = (1 - 3y(2))^2 + \int_0^2 [(x + 1)^2 y'^2 + xy' - y^2] \, dx.$$

4.5 Hamilton's Principle

According to the doctrine of classical dynamics, one associates with the system being described a set of quantities or dynamical variables, each of which has a well–defined value at each instant of time and which defines the state of the dynamical system at that instant. Further, it is assumed that the time evolution of the system is completely determined if its state is known at some given instant. Analytically this doctrine is expressed by the fact that the dynamical variables satisfy a set of differential equations (the equations of motion of the system) as functions of time, along with initial conditions. The program of classical dynamics consists of listing the dynamical variables and formulating the equations of motion that predict the system's evolution in time. Newton's second law of motion describes the dynamics of a mechanical system.

Another method of obtaining the equations of motion is from a variational principle. This method is based on the idea that a system should evolve along a path of least resistance. Principles of this sort have a long history in physical theories dating back to antiquity when Hero of Alexandria stated a minimum principle concerning the path of reflected light rays. In the 17th century, Fermat's principle, that light rays travel along the path of shortest time, was put forth. For mechanical systems, Maupertuis's principle of least action stated that a system should evolve from one state to another in such a way that the *action* (a vaguely defined term with the energy × time) is smallest. Lagrange and Gauss were advocates of similar principles. In the early part of the 19th century, however, W. R. Hamilton (1805–1865) stated what has become an encompassing, aesthetic principle that can be generalized to embrace many areas of physics.

Hamilton's principle states that the time evolution of a mechanical system occurs in such a manner that the integral of the difference between kinetic and potential energy is stationary. To be more precise, let $y_1, \ldots, y_n$ denote a set of *generalized coordinates* of a given dynamical system. That is, regarded as functions of time, we assume that $y_1, \ldots, y_n$ completely specify the state of the system at any instant. Further, we assume that there are no relations among the y_i, so that they may be regarded as independent. In general, the y_i may be lengths, angles, or whatever. The time derivatives $\dot{y}_1, \ldots, \dot{y}_n$ are called the *generalized velocities*. The kinetic energy T is, in the most general case, a quadratic form in the $\dot{y}_i$, that is,

$$T = \sum_{i=1}^{n} \sum_{j=1}^{n} a_{ij}(y_1, \ldots, y_n) \dot{y}_i \dot{y}_j, \tag{5.1}$$

where the a_{ij} are known functions of the coordinates $y_1, \ldots, y_n$. The potential energy V is a scalar function

$$V = V(t, y_1, \ldots, y_n). \tag{5.2}$$

We define the *Lagrangian* of the system by

$$L = L(t, y_1, \ldots, y_n, \dot{y}_1, \ldots, \dot{y}_n) \equiv T - V. \tag{5.3}$$

Hamilton's principle for these systems may then be stated as follows: Consider a mechanical system described by generalized coordinates $y_1, \ldots, y_n$ with Lagrangian given by (5.3). Then the motion of the system from time t_0 to t_1 is such that the functional

$$J(y_1, \ldots, y_n) = \int_{t_0}^{t_1} L(t, y_1, \ldots, y_n, \dot{y}_1, \ldots, \dot{y}_n) \, dt \tag{5.4}$$

is stationary for the functions $y_1(t), \ldots, y_n(t)$, which describe the actual time evolution of the system. If we regard the set of coordinates $y_1, \ldots, y_n$ as coordinates in n dimensional space, then the equations

$$y_i = y_i(t), \quad i = 1, \ldots, n, \quad (t_0 \le t \le t_1)$$

are parametric equations of a curve C that joins two states $S_0: (y_1(t_0), \ldots, y_n(t_0))$ and $S_1: (y_1(t_1), \ldots, y_n(t_1))$. Hamilton's principle then states that among all paths in configuration space connecting the initial state S_0 to the final state S_1, the actual motion takes place along the path that affords an extreme value to the integral (5.4). The actual path is an extremal. In physics and engineering, Hamilton's principle is often stated concisely as

$$\delta \int_{t_0}^{t_1} L(t, y_1, \ldots, y_n, \dot{y}_1, \ldots, \dot{y}_n) \, dt = 0.$$

The functional $\int L \, dt$ is called the **action integral** for the system.

Because the curve $y_i = y_i(t)$, $i = 1, \ldots, n$, along which the motion occurs, makes the functional J stationary, it follows from the calculus of variations that the $y_i(t)$ must satisfy the Euler equations

$$L_{y_i} - \frac{d}{dt} L_{\dot{y}_i} = 0, \quad i = 1, \ldots, n. \tag{5.5}$$

In mechanics, the Euler equations (5.5) are called **Lagrange's equations**. They form the equations of motion, or governing equations, for the system. We say that the governing equations follow from a *variational principle* if we can find an L such that $\delta \int L \, dt = 0$ gives those governing equations as necessary conditions for an extremum.

If the Lagrangian L is independent of time t, that is, $L_t = 0$, then a first integral is

$$L - \sum_{i=1}^{n} \dot{y}_i L_{\dot{y}_i} = \text{const.}$$

This equation is a **conservation law**. The quantity $-L + \sum_{i=1}^{n} \dot{y}_i L_{\dot{y}_i}$ is called the **Hamiltonian** of the system, and it frequently represents the total energy. Thus, if L is independent of time, then energy is conserved.

Example 4.22

(**Harmonic oscillator**) Consider the motion of a mass m attached to a spring obeying Hooke's law, where the restoring force is $F = -ky$. The variable y is a generalized coordinate representing the positive displacement of the mass from equilibrium, and $k > 0$ is the spring constant. The kinetic and potential energies are

$$T = \tfrac{1}{2}m\dot{y}^2, \quad V = \tfrac{1}{2}ky^2.$$

The Lagrangian is therefore

$$L = T - V = \tfrac{1}{2}m\dot{y}^2 - \tfrac{1}{2}ky^2,$$

and Hamilton's principle states that the motion takes place in such a way that

$$J(y) = \int_{t_0}^{t_1} \left(\tfrac{1}{2}m\dot{y}^2 - \tfrac{1}{2}ky^2\right) dt$$

is stationary. Lagrange's equation is

$$L_y - \frac{d}{dt}L_{\dot{y}} = -ky - \frac{d}{dt}(m\dot{y}) = 0,$$

or

$$\ddot{y} + \frac{k}{m}y = 0,$$

which expresses Newton's second law that force equals mass times acceleration. Its solution gives the extremals

$$y(t) = C_1 \cos\sqrt{\frac{k}{m}}t + C_2 \sin\sqrt{\frac{k}{m}}t,$$

where C_1 and C_2 are constants that can be determined from boundary conditions.   □

Example 4.23

(**Pendulum**) Consider a simple pendulum of length l and bob mass m suspended from a frictionless support. To describe the state at any time t we choose the generalized coordinate θ measuring the angle displaced from the vertical equilibrium position. If s denotes the actual displacement of the bob along a circular arc measured from equilibrium, then the kinetic energy is

$$T = \tfrac{1}{2}m\dot{s}^2 = \tfrac{1}{2}ml^2\dot{\theta}^2,$$

where we have used $s = l\theta$. The potential energy of the bob is mg times the height above its equilibrium position, or

$$V = mg(l - l\cos\theta).$$

By Hamilton's principle the motion takes place so that

$$J(\theta) = \int_{t_0}^{t_1} \left(\tfrac{1}{2}ml^2\dot{\theta}^2 - mg(l - l\cos\theta)\right) dt$$

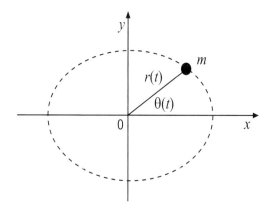

Figure 4.8 The motion of a mass in an inverse-square, central force field.

is stationary. Lagrange's equation is

$$L_\theta - \frac{d}{dt} L_{\dot\theta} = -mgl \sin\theta - \frac{d}{dt}(ml^2 \dot\theta) = 0,$$

or

$$\ddot\theta + \frac{g}{l} \sin\theta - 0,$$

which describes the evolution the system. For small displacements $\sin\theta \approx \theta$ and the equation becomes $\ddot\theta + (g/l)\theta = 0$, whose solution exhibits simple harmonic motion. $\square$

Example 4.24

(**Central force field**) Consider the planar motion of a mass m that is attracted to the origin with a force inversely proportional to the square of the distance from the origin (see Fig. 4.8). For generalized coordinates we take the polar coordinates r and θ of the position of the mass. The kinetic energy is

$$T = \tfrac{1}{2}mv^2 = \tfrac{1}{2}m(\dot x^2 + \dot y^2)$$

$$= \tfrac{1}{2}m\left[\left(\frac{d}{dt}(r\cos\theta)\right)^2 + \left(\frac{d}{dt}(r\sin\theta)\right)^2\right]$$

$$= \tfrac{1}{2}m[(\dot r\cos\theta - r\sin\theta\dot\theta)^2 + (\dot r\sin\theta + r\cos\theta\dot\theta)^2]$$

$$= \tfrac{1}{2}m(\dot r^2 + r^2\dot\theta^2).$$

The force is $-k/r^2$ for some constant k, and thus the potential energy is

$$V = -\int -\frac{k}{r^2}\, dr = -\frac{k}{r}.$$

Hamilton's principle requires that

$$J(r, \theta) = \int_{t_0}^{t_1} \left(\frac{m}{2}(\dot{r}^2 + r^2\dot{\theta}^2) + \frac{k}{r} \right) dt$$

is stationary. Lagrange's equations are

$$L_\theta - \frac{d}{dt}L_{\dot{\theta}} = 0, \quad L_r - \frac{d}{dt}L_{\dot{r}} = 0,$$

or

$$-\frac{d}{dt}(mr^2\dot{\theta}) = 0, \quad mr\dot{\theta}^2 - \frac{k}{r^2} - \frac{d}{dt}(m\dot{r}) = 0.$$

These can be rewritten as

$$mr^2\dot{\theta} = \text{constant}, \quad m\ddot{r} - mr\dot{\theta}^2 + \frac{k}{r^2} = 0,$$

which is a coupled system of ordinary differential equations. The Exercises request a solution. □

In summary, Hamilton's principle gives us a procedure for finding the equations of motion of a system if we can write down the kinetic and potential energies. This offers an alternative approach to writing down Newton's second law for a system, which requires that we know the forces. Because Hamilton's principle only results in writing the equations of motion, why not just directly determine the governing equations and forgo the variational principle altogether? Actually, this may be a legitimate objection, particularly in view of the fact that the variational principle is usually derived *a posteriori*, that is, from the known equations of motion and not conversely, as would be relevant from the point of view of the calculus of variations. Moreover, if a variational principle is given as the basic principle for the system, then there are complicated sufficient conditions for extrema that must be considered, and they seem to have little or no role in physical problems. Finally, although variational principles do to some extent represent a unifying concept for physical theories, the extent is by no means universal; it is impossible to state such principles for some systems with constraints or dissipative forces.

On the other hand, aside from the aesthetic view, the *ab initio* formulation of the governing law by a variational principle has arguments on its side. The action integral plays a fundamental role in the development of numerical methods for solving differential equations (Rayleigh–Ritz method, Galerkin methods, etc.); it also plays a decisive role in the definition of Hamilton's characteristic function, the basis for the Hamilton–Jacobi theory. Furthermore, many variational problems occur in geometry and other areas apart from physics; in these problems the action or fundamental integral *is* an *a priori* notion. In summary, the calculus of variations provides a general context in which to study wide classes of problems of interest in many areas of science, engineering, and mathematics.

4.5.1 Hamilton's Equations

The Euler equations for the variational problem

$$J(y_1, \ldots, y_n) = \int_{t_0}^{t_1} L(t, y_1, \ldots, y_n, \dot{y}_1, \ldots, \dot{y}_n) \, dt \tag{5.6}$$

form a system of n second-order ordinary differential equations. We now introduce a canonical method for reducing these equations to a system of $2n$ first-order equations. For simplicity we examine the case $n = 1$,

$$J(y) = \int_{t_0}^{t_1} L(t, y, \dot{y}) \, dt. \tag{5.7}$$

The Euler equation is

$$L_y - \frac{d}{dt} L_{\dot{y}} = 0. \tag{5.8}$$

First we define a new variable p, called the **canonical momentum** by

$$p \equiv L_{\dot{y}}(t, y, \dot{y}). \tag{5.9}$$

If $L_{\dot{y}\dot{y}} \neq 0$, then the implicit function theorem guarantees that (5.9) can be solved for $\dot{y}$ in terms of t, y, and p to get

$$\dot{y} = \phi(t, y, p). \tag{5.10}$$

Now we define the **Hamiltonian** H by

$$H(t, y, p) \equiv -L(t, y, \phi(t, y, p)) + \phi(t, y, p)p. \tag{5.11}$$

Notice that H is just the expression $-L + \dot{y}L_{\dot{y}}$, where $\dot{y}$ is given by (5.10). In many systems H is the total energy.

Example 4.25

For a particle of mass m moving in one dimension with potential energy $V(y)$, the Lagrangian is given by $L = \frac{1}{2}m\dot{y}^2 - V(y)$. Hence $p = m\dot{y}$, which is the momentum of the particle, and

$$H = -\frac{1}{2}m \left(\frac{p}{m}\right)^2 + V(y) + \frac{p^2}{m} = \frac{1}{2}\frac{p^2}{m} + V(y),$$

which is the total energy of the particle written in terms of position and momentum. □

Now, from (5.11),

$$\frac{\partial H}{\partial p} = -\frac{\partial L}{\partial \dot{y}}\frac{\partial \phi}{\partial p} + \phi + p\frac{\partial \phi}{\partial p} = \phi = \dot{y},$$

and

$$\frac{\partial H}{\partial y} = -\frac{\partial L}{\partial y} - \frac{\partial L}{\partial \dot{y}}\frac{\partial \phi}{\partial y} + p\frac{\partial \phi}{\partial y} = -L_y = -\frac{d}{dt}L_{\dot{y}} = -\dot{p}.$$

Therefore, we have shown that the Euler equation (5.8) can be written as an equivalent system of equations

$$\dot{y} = \frac{\partial H}{\partial p}(t, y, p), \quad \dot{p} = -\frac{\partial H}{\partial y}(t, y, p). \tag{5.12}$$

Equations (5.12) are called **Hamilton's equations**. They form a system of first-order differential equations for y and p.

Example 4.26

Consider the harmonic oscillator whose Lagrangian is

$$L(t, y, \dot{y}) = \tfrac{1}{2}m\dot{y}^2 - \tfrac{1}{2}ky^2. \tag{5.13}$$

The canonical momentum is

$$p = L_{\dot{y}} = m\dot{y}.$$

Solving for $\dot{y}$ gives

$$\dot{y} = \frac{p}{m}, \quad (= \phi(t, y, p)),$$

and therefore the Hamiltonian is

$$H = -L + \dot{y}L_{\dot{y}} = -\left(\tfrac{1}{2}m\left(\frac{p}{m}\right)^2 - \tfrac{1}{2}ky^2\right) + \left(\frac{p}{m}\right)p$$

$$= \tfrac{1}{2}\frac{p^2}{m} + \tfrac{1}{2}ky^2,$$

which is the sum of the kinetic and potential energy, or the total energy of the system. Now

$$\frac{\partial H}{\partial y} = ky, \quad \frac{\partial H}{\partial p} = \frac{p}{m},$$

and so Hamilton's equations are

$$\dot{y} = \frac{p}{m}, \quad \dot{p} = -ky. \tag{5.14}$$

One way to solve these equations in the yp (phase) plane is to divide them to obtain

$$\frac{dp}{dy} = \frac{-ky}{p/m}.$$

Separating variables and integrating yields

$$p^2 + kmy^2 = C,$$

which is a family of ellipses in the yp plane. $\square$

For the general action integral

$$J(y_1, \ldots, y_n) = \int_{t_0}^{t_1} L(t, y_1, \ldots, y_n, \dot{y}_1, \ldots, \dot{y}_n) \, dt$$

depending on n functions $y_1, \ldots, y_n$, the canonical momenta are

$$p_i = L_{\dot{y}_i}(t, y_1, \ldots, y_n, \dot{y}_1, \ldots, \dot{y}_n), \quad i = 1, \ldots, n. \tag{5.15}$$

Assuming

$$\det(L_{\dot{y}_i \dot{y}_j}) \neq 0,$$

we can solve the system of n equations (5.15) for $\dot{y}_1, \ldots, \dot{y}_n$ to obtain

$$\dot{y}_i = \phi_i(t, y_1, \ldots, y_n, p_1, \ldots, p_n), \quad i = 1, \ldots, n.$$

The Hamiltonian H is defined by

$$H(t, y_1, \ldots, y_n, p_1, \ldots, p_n) = -L + \sum_{i=1}^{n} \dot{y}_i L_{\dot{y}_i}$$

$$= -L(t, y_1, \ldots, y_n, \phi_1, \ldots, \phi_n)$$

$$+ \sum_{i=1}^{n} \phi_i(t, y_1, \ldots, y_n, p_1, \ldots, p_n) p_i.$$

Using an argument exactly like the one used for the case $n = 1$, we obtain the canonical form of the Euler equations,

$$\dot{y}_i = \frac{\partial H}{\partial p_i}(t, y_1, \ldots, y_n, p_1, \ldots, p_n), \tag{5.16a}$$

$$\dot{p}_i = -\frac{\partial H}{\partial y_i}(t, y_1, \ldots, y_n, p_1, \ldots, p_n), \tag{5.16b}$$

$i = 1, \ldots, n$, which are Hamilton's equations. They represent $2n$ first-order ordinary differential equations for the $2n$ functions $y_1, \ldots, y_n, p_1, \ldots, p_n$. A complete discussion of the role of the canonical formalism in the calculus of variations, geometry, and physics can be found in Rund (1966).

4.5.2 The Inverse Problem

In general, a variational principle exists for a given physical system if there is a Lagrangian L such that the Euler equations

$$L_{y_i} - \frac{d}{dt} L_{\dot{y}_i} = 0, \quad i = 1, \dots, n,$$

corresponding to the action integral $\int L \, dt$ coincide with the governing equations of the system. For conservative mechanical systems Hamilton's principle tells us that $L = T - V$. How can we determine the Lagrangian L for other systems if we know the equations of motion? This problem is known as the **inverse problem** of the calculus of variations. Let us formulate the inverse problem for $n = 1$. Given a second-order ordinary differential equation

$$\ddot{y} = F(t, y, \dot{y}), \tag{5.17}$$

find a Lagrangian $L(t, y, \dot{y})$ such that (5.17) is the Euler equation

$$L_y - \frac{d}{dt} L_{\dot{y}} = 0. \tag{5.18}$$

Generally, the problem has infinitely many solutions. To determine L we write the Euler equation as

$$L_y - L_{\dot{y}t} - L_{\dot{y}y} \dot{y} - L_{\dot{y}\dot{y}} \ddot{y} = 0,$$

or

$$L_y - L_{\dot{y}t} - L_{\dot{y}y} \dot{y} - F L_{\dot{y}\dot{y}} = 0.$$

There is a general procedure for characterizing all such Lagrangians, but it is often simpler to proceed directly by matching terms in the Euler equation to terms in the given differential equation.

Example 4.27

Consider a damped spring–mass oscillator where a mass m is suspended from a spring with spring constant k and there is a dashpot offering a resistive force numerically equal to $a\dot{y}$, where a is the damping constant, and y is the displacement of the mass from equilibrium. By Newton's law

$$m\ddot{y} = -ky - a\dot{y},$$

or

$$m\ddot{y} + a\dot{y} + ky = 0. \tag{5.19}$$

Equation (5.19) is the damped harmonic oscillator equation. The system is not conservative because of the damping, and there is not a scalar potential. We cannot apply Hamilton's principle directly. We seek a Lagrangian $L(t, y, \dot{y})$ such that (5.19) is the Euler equation. Multiplying (5.19) by a nonnegative function $f(t)$ we obtain

$$mf(t)\ddot{y} + af(t)\dot{y} + kf(t)y = 0. \tag{5.20}$$

Equation (5.20) must coincide with the Euler equation

$$-L_y + L_{\dot{y}t} + L_{\dot{y}y}\dot{y} + L_{\dot{y}\dot{y}}\ddot{y} = 0.$$

An obvious choice is to take $L_{\dot{y}\dot{y}} = mf(t)$, so that

$$L_{\dot{y}} = mf(t)\dot{y} + M(t, y),$$

and

$$L = mf(t)\frac{\dot{y}^2}{2} + M(t, y)\dot{y} + N(t, y),$$

where M and N are arbitrary functions. It follows that

$$-L_y + L_{\dot{y}t} + L_{\dot{y}y}\dot{y} = af(t)\dot{y} + kf(t)y,$$

or

$$-(M_y\dot{y} + N_y) + mf'(t)\dot{y} + M_t + M_y\dot{y} = af(t)\dot{y} + kf(t)y.$$

Thus $mf'(t) = af(t)$, or $f(t) = e^{at/m}$. Hence

$$-N_y + M_t = kye^{at/m}.$$

Selecting $M = 0$ gives $N = -(k/2)y^2 e^{at/m}$, and therefore a Lagrangian for the damped harmonic oscillator is

$$L(t, y, \dot{y}) = \left(\frac{m}{2}\dot{y}^2 - \frac{k}{2}y^2\right)e^{at/m}.$$

This expression is the Lagrangian for the undamped oscillator times a time-dependent factor. □

EXERCISES

1. Consider the functional $J(y) = \int_a^b [r(t)\dot{y}^2 + q(t)y^2]\, dt$. Find the Hamiltonian $H(t, y, p)$ and write down Hamilton's equations for the problem.

2. Write down Hamilton's equations for the functional

$$J(y) = \int_a^b \sqrt{(t^2 + y^2)(1 + \dot{y}^2)}\, dt.$$

Solve the equations and sketch the solution curves in the yp plane.

3. Derive Hamilton's equations for the functional

$$J(\theta) = \int_{t_1}^{t_2} \left(\frac{m}{2} l^2 \dot{\theta}^2 + mgl \cos\theta - mgl \right) dt,$$

representing the motion of a pendulum in a plane.

4. A particle of unit mass moves along a y axis under the influence of a potential given by

$$f(y) = -\omega^2 y + ay^2,$$

where ω and a are positive constants.

a) What is the potential energy $V(y)$? Determine the Lagrangian and write down the equation of motion.

b) Find the Hamiltonian $H(y, p)$ and show it coincides with the total energy. Write down Hamilton's equations. Is energy conserved? Is momentum conserved?

c) If the total energy E is $\omega^2/10$, and at time $t = 0$ the particle is at $y = 0$, what is the initial velocity?

d) Sketch the possible phase trajectories in phase space when the total energy in the system is given by $E = \omega^6/12a^2$. [Hint: note that $p = \pm\sqrt{2}\sqrt{E - V(y)}$.) What is the value of E above which oscillatory motion is not possible?

5. Give an alternate derivation of Hamilton's equations by rewriting the action integral

$$\int_{t_0}^{t_1} L(t, y, \dot{y}) \, dt$$

as

$$J(p, y) = \int_{t_0}^{t_1} F(t, y, \dot{y}, p, \dot{p}) \, dt,$$

where $F = \dot{y}p - H(t, y, p)$, treating y and p as independent and finding the Euler equations.

6. A particle of mass m moves in one dimension under the influence of a force $F(y, t) = ky^{-2}e^t$, where y is position, t is time, and k is a constant. Formulate Hamilton's principle for this system and derive the equation of motion. Determine the Hamiltonian and compare it with the total energy. Is energy conserved?

7. A particle of mass m is falling under the action of gravity with resistive forces neglected. If y denotes the distance measured downward, find the Lagrangian for the system and determine the equation of motion from a variational principle.

8. Consider a system of n particles where m_i is the mass of the ith particle and (x_i, y_i, z_i) is its position in space. The kinetic energy of the system is

$$T = \frac{1}{2} \sum_{i=1}^{n} m_i(\dot{x}_i^2 + \dot{y}_i^2 + \dot{z}_i^2),$$

and assume that the system has a potential energy

$$V = V(t, x_1, \ldots, x_n, y_1, \ldots, y_n, z_1, \ldots, z_n)$$

such that the force acting on the ith particle has components

$$F_i = -\frac{\partial V}{\partial x_i}, \quad G_i = -\frac{\partial V}{\partial y_i}, \quad H_i = -\frac{\partial V}{\partial z_i}.$$

Show that Hamilton's principle applied to this system yields

$$m_i\ddot{x}_i = F_i, \quad m_i\ddot{y}_i = G_i, \quad m_i\ddot{z}_i = H_i,$$

which are Newton's equations for a system of n particles.

9. Consider a simple plane pendulum with a bob of mass m attached to a string of length l. After the pendulum is set in motion the string is shortened by a constant rate $dl/dt = -\alpha = \text{const}$. Formulate Hamilton's principle and determine the equation of motion. Compare the Hamiltonian to the total energy. Is energy conserved?

10. Consider the differential equations

$$r^2\dot{\theta} = C, \quad \ddot{r} - r\dot{\theta}^2 + \frac{k}{m}r^{-2} = 0,$$

governing the motion of a mass in an inverse square central force field.

a) Using the chain rule show that

$$\dot{r} = Cr^{-2}\frac{dr}{d\theta}, \quad \ddot{r} = C^2r^{-4}\frac{d^2r}{d\theta^2} - 2C^2r^{-5}\left(\frac{dr}{d\theta}\right)^2,$$

and therefore the equations of motion may be written

$$\frac{d^2r}{d\theta^2} - 2r^{-1}\left(\frac{dr}{d\theta}\right)^2 - r + \frac{k}{C^2m}r^2 = 0.$$

b) Let $r = u^{-1}$ and show that

$$\frac{d^2u}{d\theta^2} + u = \frac{k}{C^2m}.$$

c) Solve the differential equation in part (b) to obtain

$$u = r^{-1} = \frac{k}{C^2 m}(1 + \varepsilon \cos(\theta - \theta_0)),$$

where ε and θ_0 are constants of integration.

d) Show that elliptical orbits are obtained when $\varepsilon < 1$.

11. Find a Lagrangian $L(t, y, \dot{y})$ such that the Euler equation coincides with the Emden–Fowler equation

$$\ddot{y} + \frac{2}{t}\dot{y} + y^5 = 0.$$

12. A particle moving in one dimension in a constant external force field with frictional force proportional to its velocity has equation of motion

$$\ddot{y} + a\dot{y} + b = 0, \qquad a, b > 0.$$

Find a Lagrangian $L(t, y, \dot{y})$.

13. Using the electrical–mechanical analogy and the result for the damped harmonic oscillator, determine a Lagrangian for an RCL electrical circuit whose differential equation is

$$L\ddot{I} + R\dot{I} + \frac{1}{C}I = 0,$$

where $I(t)$ is the current, L the inductance, C the capacitance, and R the resistance.

14. Consider the nonlinear equation

$$y'' + p(t)y' + f(y) = 0,$$

where p and f are continuous functions with antiderivatives $P(t)$ and $F(y)$, respectively. Find a Lagrangian $L = L(t, y, y')$ such that the Euler equation associated with the functional $J(y) = \int_a^b L\, dt$ is equivalent to the given equation. (Hint: Multiply the equation by $e^{P(t)}$.)

4.6 Isoperimetric Problems

In differential calculus the problem of minimizing a given function $f(x, y)$ subject to a subsidiary condition $g(x, y) = $ constant is a common one. Now we consider classes of variational problems in which the competing functions are required to conform to certain restrictions, in addition to the normal endpoint conditions. Such restrictions are called *constraints* and they may take the form of integral relations, algebraic equations, or differential equations. In this section we limit the discussion to integral constraints.

One method for solving the constraint problem in calculus is the **Lagrange multiplier rule**. We state this necessary condition precisely because it is used in the subsequent discussion.

Theorem 4.28

Let f and g be differentiable functions with $g_x(x_0, y_0)$ and $g_y(x_0, y_0)$ not both zero. If (x_0, y_0) provides an extreme value to f subject to the constraint $g = C =$ constant, then there exists a constant λ such that

$$f_x^*(x_0, y_0) = 0, \tag{6.1}$$
$$f_y^*(x_0, y_0) = 0, \tag{6.2}$$

and

$$g(x_0, y_0) = C, \tag{6.3}$$

where

$$f^* \equiv f + \lambda g. \quad \square$$

The proof can be found in most calculus books. For practical calculations the three conditions in the theorem can be used to determine x_0, y_0, and the Lagrange multiplier λ.

A variational problem having an integral constraint is known as an **isoperimetric problem**. In particular, consider the problem of minimizing the functional

$$J(y) = \int_a^b L(x, y, y') \, dx \tag{6.4}$$

subject to

$$W(y) \equiv \int_a^b G(x, y, y') \, dx = k, \tag{6.5}$$

where $y \in C^2[a, b]$ and

$$y(a) = y_0, \quad y(b) = y_1, \tag{6.6}$$

and k is a fixed constant. The given functions L and G are assumed to be twice continuously differentiable. The subsidiary condition (6.5) is called an **isoperimetric constraint**.

In essence, we follow the procedure in earlier sections for problems without constraints. We embed an assumed local minimum $y(x)$ in a family of admissible functions. A *one-parameter* family $y(x) + \varepsilon h(x)$ is not, however, a suitable choice, because those curves may not maintain the constancy of W. Therefore we introduce a *two-parameter* family

$$z(x) = y(x) + \varepsilon_1 h_1(x) + \varepsilon_2 h_2(x),$$

where h_1, $h_2 \in C^2[a, b]$,

$$h_1(a) = h_1(b) = h_2(a) = h_2(b) = 0, \tag{6.7}$$

and ε_1 and ε_2 are real parameters ranging over intervals containing the origin. We assume that W does not have an extremum at y. Then for any choice of h_1 and h_2 there will be values of ε_1 and ε_2 in the neighborhood of $(0,0)$ for which $W(z) = k$. Evaluating J and W at z gives

$$\mathcal{J}(\varepsilon_1, \varepsilon_2) = \int_a^b L(x, z, z')\, dx,$$

$$\mathcal{W}(\varepsilon_1, \varepsilon_2) = \int_a^b G(x, z, z')\, dx = k.$$

Because y is the local minimum for (6.4) subject to the constraint (6.5), the point $(\varepsilon_1, \varepsilon_2) = (0,0)$ must be a local minimum for $\mathcal{J}(\varepsilon_1, \varepsilon_2)$ subject to the constraint $\mathcal{W}(\varepsilon_1, \varepsilon_2) = k$. This is just a differential calculus problem and so the Lagrange multiplier rule may be applied. There must exist a constant λ such that

$$\frac{\partial \mathcal{J}^*}{\partial \varepsilon_1} = \frac{\partial \mathcal{J}^*}{\partial \varepsilon_2} = 0 \quad \text{at} \quad (\varepsilon_1, \varepsilon_2) = (0,0), \tag{6.8}$$

where $\mathcal{J}^*$ is defined by

$$\mathcal{J}^* = \mathcal{J} + \lambda \mathcal{W} = \int_a^b L^*(x, z, z')\, dx,$$

with

$$L^* = L + \lambda G. \tag{6.9}$$

We now calculate the derivatives in (6.8), afterward setting $\varepsilon_1 = \varepsilon_2 = 0$. Accordingly,

$$\frac{\partial \mathcal{J}^*}{\partial \varepsilon_i}(0,0) = \int_a^b (L_y^*(x, y, y')h_i + L_{y'}^*(x, y, y')h_i')\, dx, \quad i = 1, 2.$$

Integrating the second term by parts and applying conditions (6.7) give

$$\frac{\partial \mathcal{J}^*}{\partial \varepsilon_i}(0,0) = \int_a^b \left(L_y^*(x, y, y') - \frac{d}{dx} L_{y'}^*(x, y, y') \right) h_i\, dx, \quad i = 1, 2.$$

From (6.8), and because of the arbitrary character of h_1 or h_2, the fundamental lemma implies

$$L_y^*(x, y, y') - \frac{d}{dx} L_{y'}^*(x, y, y') = 0, \tag{6.10}$$

which is a necessary condition for an extremum.

In practice, the solution of the second-order differential equation (6.10) will yield a function $y(x)$ that involves two arbitrary constants and the unknown multiplier λ. These may be evaluated by the two boundary conditions (6.6) and by substituting $y(x)$ into the isoperimetric constraint (6.5).

Example 4.29

(**Shape of a Hanging Rope**) A rope of length l with constant density ρ hangs from two fixed points (a, y_0) and (b, y_1) in the plane. Let $y\,(x)$ be an arbitrary configuration of the rope with the y axis adjusted so that $y(x) > 0$. A small element of length ds at (x, y) has mass ρds and potential energy $\rho g y\,ds$ relative to $y = 0$. Therefore, the total potential energy of the rope hanging in the arbitrary configuration $y = y(x)$ is given by the functional

$$J(y) = \int_0^1 \rho g y \, ds = \int_a^b \rho g y \sqrt{1 + y'^2} \, dx. \tag{6.11}$$

It is known that the actual configuration minimizes the potential energy. Therefore, we are faced with minimizing (6.11) subject to the isoperimetric condition

$$W(y) = \int_a^b \sqrt{1 + y'^2} \, dx = l. \tag{6.12}$$

Hence, we form the auxiliary function

$$L^* = L + \lambda G = \rho g y \sqrt{1 + y'^2} + \lambda \sqrt{1 + y'^2},$$

and write the associated Euler equation (6.10). In this case L^* does not depend explicitly on x, and thus a first integral is

$$L^* - y' L^*_{y'} = C,$$

or

$$(\rho\, g y + \lambda) \left(\sqrt{1 + y'^2} - \frac{y'^2}{\sqrt{1 + y'^2}} \right) = C.$$

Solving for y' and separating variables yields

$$\frac{dy}{\sqrt{(\rho\, g y + \lambda)^2 - C^2}} = \frac{1}{C} \, dx.$$

Letting $u = \rho\, g y + \lambda$ and using the antiderivative formula

$$\int \frac{du}{\sqrt{u^2 - C^2}} = \cosh^{-1} \frac{u}{C}$$

gives

$$\frac{1}{\rho\,g}\cosh^{-1}\frac{u}{C} = \frac{1}{C}x + C_1.$$

Then

$$y = -\frac{\lambda}{\rho g} + \frac{C}{\rho g}\cosh\left(\frac{\rho g x}{C} + C_2\right).$$

Therefore, the shape of a hanging rope is a **catenary**. The constants C, C_2, and λ may be determined from (6.12) and the endpoint conditions $y(a) = y_0$ and $y(b) = y_1$. In practice this calculation is difficult, and there may not be a smooth solution for large values of l. (Why?) □

Example 4.30

Consider the problem of minimizing

$$J(\psi) = \int_{-\infty}^{\infty}\left(\frac{\hbar^2}{2m}(\psi')^2 + V(x)\psi^2\right)dx,$$

subject to the constraint

$$\int_{-\infty}^{\infty}\psi^2(x)\,dx = 1. \tag{6.13}$$

Here $\hbar$ and m are constants, and $V = V(x)$ is a given function. The auxiliary Lagrangian

$$L^* = \frac{\hbar^2}{2m}(\psi')^2 + V(x)\psi^2 - E\psi^2,$$

where $-E$ is a Lagrange multiplier. The Euler equation is

$$L_\psi^* - \frac{d}{dx}L_{\psi'}^* = 0,$$

or

$$-\frac{\hbar^2}{2m}\psi'' + V(x)\psi = E\psi. \tag{6.14}$$

Equation (6.14) is the **Schrödinger equation** *for the wave function ψ in* quantum mechanics for a particle of mass m under the influence of a potential V. Equation (6.13) is a normalization condition for the wave function ψ, whose square is a probability density. In general, solutions ψ of (6.14) will exist only for discrete values of the multiplier E, which are identified with the possible energy levels of the particle and are the eigenvalues. These problems are discussed in Chapter 5. □

EXERCISES

1. Find extremals of the isoperimetric problem

$$J(y) = \int_0^\pi y'^2 \, dx, \quad y(0) = y(\pi) = 0,$$

subject to

$$\int_0^\pi y^2 \, dx = 1.$$

2. Find extremals for the isoperimetric problem

$$J(y) = \int_0^1 (y'^2 + x^2) \, dx, \quad y(0) = 0, \quad y(1) = 0$$

$$\int_0^1 y^2 \, dx = 2.$$

3. Derive a necessary condition for an extremum for the isoperimetric problem: Minimize

$$J(y_1, y_2) = \int_a^b L(x, y_1, y_2, y_1', y_2') \, dx$$

subject to

$$\int_a^b G(x, y_1, y_2, y_1', y_2') \, dx = C$$

and

$$y_1(a) = A_1, \quad y_2(a) = A_2, \quad y_1(b) = B_1, \quad y_2(b) = B_2,$$

where C, A_1, A_2, B_1, and B_2 are constants.

4. Use the result of the preceding exercise to maximize

$$J(x, y) = \int_{t_0}^{t_1} (xy' - yx') \, dt,$$

subject to

$$\int_{t_0}^{t_1} \sqrt{x'^2 + y'^2} \, dt = l.$$

Show that J represents the area enclosed by a curve with parametric equations $x = x(t)$, $y = y(t)$, and the constraint fixes the length of the curve. Thus, this is the problem of determining which curve of a specified perimeter encloses the maximum area. The term *isoperimetric*, meaning *same perimeter*, originated in this context.

Answer: Circles $(x - c_1)^2 + (y - c_2)^2 = \lambda^2$.

5. Determine the equation of the shortest arc in the first quadrant that passes through $(0,0)$ and $(1,0)$ and encloses a prescribed area A with the x axis, where $0 < A \leq \pi/8$.

6. Write down equations that determine the solution of the isoperimetric problem

$$\int_a^b (p(x)y'^2 + q(x)y^2)\, dx \to \text{ min},$$

subject to

$$\int_a^b r(x)y^2\, dx = 1,$$

where p, q, and r are given functions and $y(a) = y(b) = 0$. The result is a *Sturm–Liouville eigenvalue problem* (Chapter 5).

7. Consider the problem of minimizing the functional

$$J(y, z) = \int_a^b L(t, y, z, \dot{y}, \dot{z})\, dt, \qquad (6.15)$$

subject to the algebraic constraint

$$G(t, y, z) = 0, \quad G_z \neq 0 \qquad (6.16)$$

and boundary conditions

$$y(a) = y(b) = z(a) = z(b) = 0,$$

with all of the functions possessing the usual differentiability conditions. If $y(t)$ and $z(t)$ provide a local minimum, show that there exists a function $\lambda(t)$ such that

$$L_y - \frac{d}{dt}L_{\dot{y}} = \lambda(t)G_y, \quad L_z - \frac{d}{dt}L_{\dot{z}} = \lambda(t)G_z.$$

(*Hint:* Solve (6.16) to get $z = g(t, y)$, and substitute into (6.15) to obtain a functional depending only upon y. Find the Euler equation and use the condition $(\partial/\partial y)G(t, y, g(t, y)) = 0$.)

REFERENCES AND NOTES

The calculus of variations is an old subject dating back to the late 17th century. As such, the literature is vast, and there are many excellent treatments of the theory for both necessary and sufficient conditions. The references below represent a few of the recent books that complement this text. A classical reference is Gelfand and Fomin (1963). More recent are Pinch (1993) and van Brunt (2004). A definitive treatise on the canonical formalism is in Rund (1966), and a treatment of invariance of functionals under local Lie groups of transformations, leading to conservation laws and Noether's theorem, is contained in Logan (1977) and Neuenschwander (2010). This material is also covered in detail in the first edition of this text (J. D. Logan, 1987). An important generalization of the calculus of variations is optimal control theory, which is not covered in the present text; see Pinch (1993) for an accessible introduction, and Lenhart and Workman (2007) for applications to biology. Finally, an outstanding, elementary treatment of the basic calculus concepts for functions of a real variable can be found in the classic by Spivak (2008).

Gelfand, I. M. & Fomin, S. V. 1963. *Calculus of Variations*, Prentice-Hall, Englewood Cliffs, NJ (reprinted by Dover Publications, Inc., 2000).

Lenhart, S. & Workman, J. T. 2007. *Optimal Control Applied to Biological Models*, Chapman & Hall/CRC, Boca Raton, FL.

Logan, J. D. 1987. *Applied Mathematics: A Contemporary Approach*, Wiley-Interscience, New York.

Logan, J. D. 1977. *Invariant Variational Principles*, Academic Press, New York.

Neuenschwander, D. E. 2010. *Emmy Noether's Wonderful Theorem*, Johns Hopkins University Press.

Pinch, E. R. 1993. *Optimal Control and the Calculus of Variations*, Oxford University Press, Oxford.

Rund, H. 1966. *The Hamilton-Jacobi Theory in the Calculus of Variations*, Van Nostrand, London.

Spivak, M. 2008. *Calculus*, 4th ed., Publish or Perish Inc., Houston.

van Brunt, B. 2004. *The Calculus of Variations*, Springer-Verlag, New York.

5
Boundary Value Problems and Integral Equations

Topics in this chapter form the traditional core of applied mathematics—boundary value problems, orthogonal expansions, integral equations, Green's functions, and distributions. All of these subjects are interrelated in one of the most aesthetic theories in all of mathematics. The descriptive term 'far-reaching' is an understatement.

We can take a very general overview and summarize our perspective as follows. In an abstract sense, mathematics can be thought of as imposing structures on sets of objects and studying their properties. In linear algebra, for example, the algebraic structure of a vector space is imposed, and addition and scalar multiplication are defined that satisfy certain rules. We can further define a geometric structure on a vector space that imposes concepts like perpendicularity and distance. Finally, linear transformations between vector spaces are introduced and we study how they interrelate with the algebraic and geometric structures. This all leads to the solution of linear systems, the eigenvalue problem, and the decomposition of the vector space and the transformation into fundamental, or canonical, forms. These ideas extend in a straightforward manner to infinite-dimensional function spaces and linear differential and integral operators on those spaces. This theory is not only insightful, but it has far-reaching applications to science, engineering, and mathematics.

To go further, elementary linear algebra deals with finite dimensional vector spaces, such as $\mathbb{R}^n$. The vector space $\mathbb{R}^n$ is the set of all n-tuples of real numbers, and we represent a vector as column vector $\mathbf{x}$. Any set of n independent vectors

Applied Mathematics 4th ed.,
By J. David Logan
Copyright © 2013 John Wiley & Sons, Inc.

forms a basis for $\mathbb{R}^n$, which means every vector in $\mathbb{R}^n$ can be uniquely expanded as a linear combination of those n vectors. Then, a linear transformation A between two vector spaces, $A : \mathbb{R}^n \to \mathbb{R}^m$, can be represented by an $n \times m$ matrix. The important problems can then be formulated using matrices.

1. (**Solvability Problem**) Given a linear transformation $A : \mathbb{R}^n \to \mathbb{R}^m$ and a vector $\mathbf{f} \in \mathbb{R}^m$, find a vector $\mathbf{x} \in \mathbb{R}^n$ for which

$$A\mathbf{x} = \mathbf{f}.$$

   This is the problem of solving m equations in n unknowns and characterizing the solution structure.

2. (**The Eigenvalue Problem**) Given a linear transformation $A : \mathbb{R}^n \to \mathbb{R}^n$ from a vector space into itself, find values λ (eigenvalues) and corresponding nonzero vectors $\mathbf{x}$ (eigenvectors) for which

$$A\mathbf{x} = \lambda\mathbf{x}.$$

Recall that all $n \times n$ matrices have n eigenvalues, but they do not necessarily have a full set of n independent eigenvectors. For special matrices, however, e.g., real symmetric matrices where $A^{\mathrm{T}} = A$, there is a basis of n orthogonal eigenvectors for the entire space $\mathbb{R}^n$. This permits us to decompose the space into orthogonal subspaces (the eigenspaces) and write the matrix itself as a sum of projections onto those eigenspaces.

In this and subsequent chapters we want to extend these ideas from $\mathbb{R}^n$ and $\mathbb{C}^n$ to structures on infinite dimensional function spaces and transformations that lead to ordinary and partial differential equations, and integral equations.

In Chapter 4 we introduced different kinds of vector spaces whose elements are not classical, geometric vectors with magnitude and direction, but rather functions. For example, the set of all continuous functions on an interval $[a, b]$, denoted by $C[a, b]$, is a vector space or, using other language, a linear space. A linear space like $C[a, b]$ is infinite dimensional, not finite dimensional; there is not a finite number of continuous functions on $[a, b]$ for which every continuous function in $C[a, b]$ can be written as a linear combination of those functions. There is no finite basis. We can consider linear transformations on $C[a, b]$ and the solvability problem. For example, define the linear operator $K : C[a, b] \to C[u, b]$ by

$$Ku(x) = \int_a^b k(x, y)u(y)dy, \quad u \in C[a, b],$$

where $k(x, y)$ is a given continuous function for $x, y \in [a, b]$. (In the function space context, we use the term *linear operator* rather than *linear transformation*.) As above, the solvability problem can be represented

$$Ku = f.$$

That is, if $f = f(x)$ is a given continuous function on $C[a, b]$, does there exist a function $u = u(x)$ satisfying this equation? An equation like this, where the unknown u occurs under an integral sign, is called an **integral equation**. Similarly, we can study the eigenvalue problem

$$Ku = \lambda u.$$

Are there values of λ and corresponding continuous functions u for which this equation holds? Integral equations are the subject of Section 5.3.

First, in Section 5.2, we extend these ideas of linear differential operators $L : C^2[a, b] \to C[a, b]$ that associate to every twice continuously differentiable function $u = u(x)$, satisfying conditions $u(a) = u_a$ and $u(b) = u_b$, a continuous function. Then we can consider the solvability problem

$$Lu = f,$$

where $f \in C[a, b]$ is given. This is the problem of solving a nonhomogeneous differential equation with boundary conditions. Similarly, the *differential* eigenvalue problem is to find values of λ and corresponding $C^2[a, b]$ functions u, again satisfying the boundary conditions, for which

$$Lu = \lambda u.$$

A simple example is the differential operator

$$Lu = -u'', \quad u(a) = 0, \ u(b) = 0,$$

where the boundary conditions comprise part of the definition of the operator.

We show that differential and integral operators of special types have an eigenstructure similar to matrices. They have a full set of eigenfuctions that are orthogonal and form a basis for the (infinite dimensional) space. Along with that comes the spectral decomposition, or diagonalization, of the operators, orthogonal projections, and all the machinery of finite dimensional linear algebra. One can envision a large body of literature in mathematics, science, and engineering that addresses all these questions. And these issues only scratch the surface of a vast subject.

5.1 Boundary-Value Problems

Chapters 1 and 2 focused on initial value problems for differential equations, where the auxiliary, initial conditions are given at a single value of time, usually $t = 0$, and the system evolves dynamically from that initial state. Now we bring boundary value problems (BVPs) into focus, where conditions on a differential equation are specified at two different points. We use the independent variable x instead of time t because BVPs are on a spatial domain. We illustrate the idea with a model in heat flow.

Let us consider the following problem in steady-state heat conduction. A cylindrical, uniform, metallic bar of length L and cross-sectional area A is insulated on its lateral side. We assume the left face at $x = 0$ is maintained at T_0 degrees and that the right face at $x = L$ is held at T_L degrees. What is the temperature distribution $u = u(x)$ in the bar after it comes to equilibrium? Here $u(x)$ represents the temperature at each point of the entire cross-section of the bar at position x, where $0 < x < L$. We are assuming that heat flows only in the axial direction along the bar, and we are assuming that any transients caused by initial temperatures in the bar have decayed away. In other words, we have waited long enough for the temperature to reach a steady state. One might conjecture that the temperature distribution is a linear function of x along the bar; that is, $u(x) = T_0 + ((T_L - T_0)/L)x$. This is indeed the case, which we show below. But also we want to consider a more complicated problem where the bar has both a variable conductivity and an internal heat source along its length. An internal heat source, for example, could be resistive heating produced by a current running through the medium.

The physical law that provides the basic model is conservation of energy. If $[x, x + dx]$ is any small section of the bar, then the rate that heat flows in at x, minus the rate that heat flows out at $x + dx$, plus the rate that heat is generated by sources, must equal zero, because the system is in a steady state. See Fig. 5.1.

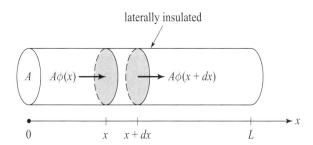

Figure 5.1 Cylindrical bar, laterally insulated, through which heat is flowing in the x-direction. The temperature is uniform in a fixed cross-section.

We denote the heat flux by $\phi(x)$, which is the rate that heat flows to the right at any section x, measured in energy/(area · time), and we let $f(x)$ denote the rate that heat is internally produced at x, measured in energy/(volume ·

time), then

$$A\phi(x) - A\phi(x + dx) + f(x)A\,dx = 0.$$

Cancelling A, dividing by dx, and rearranging gives

$$\frac{\phi(x + dx) - \phi(x)}{dx} = f(x).$$

Taking the limit as $dx \to 0$ yields

$$\phi'(x) = f(x). \tag{1.1}$$

This is an expression of energy conservation in terms of energy flux. But what about temperature? Empirically, the flux $\phi(x)$ at a section x is found to be proportional to the negative temperature gradient $-u'(x)$ (which measures the steepness of the temperature distribution, or profile, at that point), or

$$\phi(x) = -K(x)u'(x). \tag{1.2}$$

This is **Fourier's heat conduction law**. The given proportionality factor $K(x)$ is called the **thermal conductivity**, in units of energy/(length · degrees · time), which is a measure of how well the bar conducts heat at location x. For a uniform bar, K is constant. The minus sign in (1.2) means that heat flows from higher temperatures to lower temperatures. Fourier's law seems intuitively correct and it conforms with the second law of thermodynamics; the larger the temperature gradient, the faster heat flows from high to low temperatures. Combining (1.1) and (1.2) leads to the equation

$$-(K(x)u'(x))' = f(x), \quad 0 < x < L, \tag{1.3}$$

which is the **steady-state heat conduction equation**. When the *boundary conditions*

$$u(0) = T_0, \quad u(L) = T_1, \tag{1.4}$$

are appended to (1.3), we obtain a **boundary value problem** for the temperature $u(x)$. Boundary conditions are conditions imposed on the unknown temperature $u = u(x)$ at two fixed values of the independent variable x, unlike initial conditions that are imposed at a single value. For boundary value problems we usually use x as the independent variable because boundary conditions usually refer to the boundary of a spatial domain.

Note that we could expand the heat conduction equation to

$$-K(x)u''(x) - K'(x)u'(x) = f(x), \tag{1.5}$$

but in general there is little advantage in doing so.

Example 5.1

If there are no sources ($f(x) = 0$) and if the thermal conductivity $K(x) = K$ is constant, then the boundary value problem reduces to

$$u'' = 0, \quad 0 < x < L,$$
$$u(0) = T_0, \quad u(L) = T_L.$$

Thus the bar is homogeneous and can be characterized by a constant conductivity. The general solution of $u'' = 0$ is $u(x) = c_1 x + c_2$; applying the boundary conditions determines the constants c_1 and c_2 and gives the linear temperature distribution $u(x) = T_0 + (T_L - T_0)/L)x$, as we previously conjectured.   □

The thermal conductivity K and heat source f may also depend on the temperature u as well. In this case the steady-state heat conduction equation (1.3) takes the more general form

$$-(K(x, u)u')' = f(x, u),$$

which is a nonlinear second-order equation for the temperature distribution $u = u(x)$.

Boundary conditions at the ends of the bar may also specify the flux rather than the temperature. For example, if heat is injected at $x = 0$ at a rate of N calories per area per time, then the left boundary condition takes the form $\phi(0) = N$, or

$$-K(0)u'(0) = N.$$

Thus, a flux condition at an endpoint imposes a condition on the derivative of the temperature at that endpoint. If the end $x = 0$, say, is insulated, so that no heat passes through that end, then the boundary condition is simply

$$u'(0) = 0,$$

which is called an **insulated boundary condition**. We may also have conditions of the combination

$$u(0) + hu'(0) = c$$

at the boundary, where h and c are given constants. These types of conditions are called **radiation conditions**, and we examine their origin in Chapter 6. As the reader may imagine, there are myriad interesting boundary value problems associated with heat flow. Similar equations arise in diffusion processes in biology and chemistry, for example, in the diffusion of toxic substances where the unknown is the chemical concentration.

Boundary value problems are much different from initial value problems in that they may have no solution, or they may have infinitely many solutions. Consider the following.

Example 5.2

When $K = 1$ and the heat source term is $f(u) = 9u$ and both ends of a bar of length $L = 2$ are held at $u = 0$ degrees, the boundary value problem becomes

$$-u'' = 9u, \quad 0 < x < 2,$$
$$u(0) = 0, \quad u(2) = 0.$$

The general solution to the DE is $u(x) = c_1 \sin 3x + c_2 \cos 3x$, where c_1 and c_2 are arbitrary constants. Applying the boundary condition at $x = 0$ gives $u(0) = c_1 \sin(3 \cdot 0) + c_2 \cos(3 \cdot 0) = c_2 = 0$. So the solution must have the form $u(x) = c_1 \sin 3x$. Next apply the boundary condition at $x = 2$. Then $u(2) = c_1 \sin(6) = 0$, to obtain $c_1 = 0$. We have shown that the only solution is $u(x) = 0$. There is no nontrivial steady state. However, if we make the bar length π, then we obtain the boundary value problem

$$-u'' = 9u, \quad 0 < x < \pi,$$
$$u(0) = u(\pi) = 0.$$

The reader should check that this boundary value problem has infinitely many solutions $u(x) = c_1 \sin 3x$, where c_1 is any number. If we change the right boundary condition, one can check that the boundary value problem

$$-u'' = 9u, \quad 0 < x < \pi.$$
$$u(0) = 0, \quad u(\pi) = 1,$$

has no solution at all. □

The next example illustrates the type of problem that occupies much of the remaining part of the book—an **eigenvalue problem**. In an eigenvalue problem we seek values of a parameter in a boundary value problem so that a solution exists. This is the differential equation analog of the eigenvalue problem for matrices.

Example 5.3

Problem: Find all real values of λ for which the boundary value problem

$$-u'' = \lambda u, \quad 0 < x < \pi,$$
$$u(0) = 0, \quad u'(\pi) = 0,$$

has a nontrivial solution. These values of λ are called the **eigenvalues**, and the corresponding nontrivial solutions are called the **eigenfunctions**. Interpreted in the heat flow context, the left boundary is held at zero degrees and the right end is insulated. The heat source is $f(u) = \lambda u$. We are trying to find which linear heat sources lead to nontrivial steady states. To solve this problem we consider different cases because the form of the solution will be different for $\lambda = 0$, $\lambda < 0$, $\lambda > 0$. If $\lambda = 0$ then the general solution of $u'' = 0$ is $u(x) = ax + b$. Then $u'(x) = a$. The boundary condition $u(0) = 0$ implies $b = 0$ and the boundary condition $u'(\pi) = 0$ implies $a = 0$. Therefore, when $\lambda = 0$, we get only a trivial solution (and thus, $\lambda = 0$ is not an eigenvalue). Next consider the case $\lambda < 0$ so that the general solution has the form

$$u(x) = a \sinh \sqrt{-\lambda} x + b \cosh \sqrt{-\lambda} x.$$

The condition $u(0) = 0$ forces $b = 0$. Then $u'(x) = a\sqrt{-\lambda} \cosh \sqrt{-\lambda} x$. The right boundary condition becomes $u'(\pi) = a\sqrt{-\lambda} \cosh(\sqrt{-\lambda} \cdot 0) = 0$, giving $a = 0$. Recall that $\cosh 0 = 1$. Again there is only the trivial solution, and there are no negative eigenvalues. Finally, assume $\lambda > 0$. Then the general solution takes the form

$$u(x) = a \sin \sqrt{\lambda} x + b \cos \sqrt{\lambda} x.$$

The boundary condition $u(0) = 0$ forces $b = 0$. Then $u(x) = a \sin \sqrt{\lambda} x$ and $u'(x) = a\sqrt{\lambda} \cos \sqrt{\lambda} x$. Applying the right boundary condition gives

$$u'(\pi) = a\sqrt{\lambda} \cos \sqrt{\lambda} \pi = 0.$$

Now we do not have to choose $a = 0$ (which would again give the trivial solution) because we can satisfy this last condition with

$$\cos \sqrt{\lambda} \pi = 0$$

by choosing the yet unknown values of λ. The cosine function is zero at the values $\pi/2 \pm n\pi$, $n = 0, 1, 2, 3, \ldots$. Therefore

$$\sqrt{\lambda} \pi = \pi/2 \pm n\pi, \quad n = 0, 1, 2, 3, \ldots$$

Solving for λ yields

$$\lambda = \lambda_n = \left(\frac{2n+1}{2}\right)^2, \quad n = 0, 1, 2, 3, \ldots.$$

Consequently, the values of λ for which the original boundary value problem has a nontrivial solution are $\frac{1}{4}, \frac{9}{4}, \frac{25}{4}, \ldots$. These are the eigenvalues. The corresponding solutions are

$$u(x) = u_n(x) = a \sin \left(\frac{2n+1}{2}\right) x, \quad n = 0, 1, 2, 3, \ldots.$$

These are the eigenfunctions associated with the eigenvalues. Notice that the eigenfunctions are unique only up to a constant multiple. In terms of heat flow, the eigenfunctions represent possible steady-state temperature profiles in the bar. The eigenvalues are those values λ for which the boundary value problem will have steady-state profiles. $\square$

Boundary value problems are of great interest in applied mathematics, science, and engineering, occurring in many contexts other than heat flow, for example, in wave motion, quantum mechanics, and the partial differential equations.

EXERCISES

1. A homogeneous bar of length 40 cm has its left and right ends held at $30°C$ and $10°C$, respectively. If the temperature in the bar is in steady state, what is the temperature in the cross-section 12 cm from the left end? If the thermal conductivity is K, what is the rate that heat is leaving the bar at its right face?

2. The thermal conductivity of a bar of length $L = 20$ and cross-sectional area $A = 2$ is $K(x) = 1$, and an internal heat source is given by $f(x) = 0.5x(L - x)$. If both ends of the bar are maintained at zero degrees, what is the steady-state temperature distribution in the bar? Sketch a graph of $u(x)$. What is the rate that heat is leaving the bar at $x = 20$?

3. For a metal bar of length L with no heat source and thermal conductivity $K(x)$, show that the steady temperature in the bar has the form

$$u(x) = c_1 \int_0^x \frac{dy}{K(y)} + c_2,$$

where c_1 and c_2 are constants. What is the temperature distribution if $u(0) = 0$ and $-K(L)u'(L) = 1$? Find an analytic formula and plot the temperature distribution in the special case that $K(x) = 1 + x$.

4. Determine the values of λ for which the boundary value problem

$$-u'' = \lambda u, \quad 0 < x < 1,$$
$$u(0) = u(1) = 0,$$

has a nontrivial solution.

5. Consider the nonlinear heat flow problem

$$(uu')' = 0, \quad 0 < x < \pi,$$
$$u(0) = 0, \quad u'(\pi) = 1,$$

where the thermal conductivity depends on temperature and is given by $K(u) = u$. Find the steady-state temperature distribution.

6. Show that if there is a solution $u = u(x)$ to the boundary value problem (1.3)–(1.4), then the following condition must hold.

$$-K(L)u'(L) + K(0)u'(0) = \int_0^L f(x)dx.$$

Interpret this condition physically.

7. Consider the boundary value problem

$$u'' + \omega^2 u = 0, \quad u(0) = a, \quad u(L) = b,$$

where a, b, L, and ω are fixed parameters with $a, b \neq 0$. When does a unique solution exist?

8. Find the eigenvalues λ for the boundary value problem

$$-u'' - 2u' = \lambda u, \quad 0 < x < 1,$$
$$u(0) = 0, \quad u(1) = 0.$$

9. Show that the eigenvalues of the boundary value problem

$$-u'' = \lambda u, \quad 0 < x < 1,$$
$$u'(0) = 0, \quad u(1) + u'(1) = 0,$$

are given by the numbers $\lambda_n = p_n^2$, $n = 1, 2, 3, ...$, where the p_n are roots of the equation $\tan p = 1/p$. Plot graphs of $\tan p$ and $1/p$ and indicate graphically the locations of the values p_n. Numerically calculate the first four eigenvalues.

10. Find the eigenvalues λ of the boundary value problem

$$-x^2 u'' - xu' = \lambda u, \quad 1 < x < e^\pi,$$
$$u(1) = 0, \quad u(e^\pi) = 0.$$

[Hint: this is a Cauchy–Euler equation.]

11. Derive the steady-state heat equation in the case that the cross-sectional area of the bar varies a function of x, or $A = A(x)$.

5.2 Sturm–Liouville Problems

The eigenvalue boundary value problem in heat conduction introduced in the last section is a special case of a far more general problem, called a **Sturm–Liouville problem** (SLP). A Sturm–Liouville differential equation has the form

$$-(p(x)y')' + q(x)y = \lambda y, \quad a < x < b, \tag{2.1}$$

where λ is constant and p and q are real-valued functions defined on $[a, b]$. We take $y \in C^2$, and complex-valued. The equation is usually accompanied by homogeneous boundary conditions on $y(x)$ of the form

$$\alpha_1 y(a) + \alpha_2 y'(a) = 0, \quad \beta_1 y(b) + \beta_2 y'(b) = 0, \tag{2.2}$$

where the constants α_1 and α_2 are not both zero, and the constants β_1 and β_2 are not both zero (i.e., the boundary condition at an endpoint does not collapse). Two special cases of the boundary conditions (2.2) are

$$y(a) = 0, \quad y(b) = 0, \quad \text{(Dirichlet conditions)}$$

and

$$y'(a) = 0, \quad y'(b) = 0. \quad \text{(Neumann conditions)}$$

On a bounded interval $[a, b]$, if p, p', and q are continuous functions on the interval $[a, b]$, and p is never zero in $[a, b]$, then the boundary value problem (2.1)–(2.2) is called a **regular Sturm–Liouville problem** (SLP). Otherwise it is called **singular**. Singular problems arise when the interval $[a, b]$ is infinite or when $p(x_0) = 0$ for some $x_0 \in [a, b]$. SLPs are named after J. C. F. Sturm and J. Liouville, who studied such problems in the mid-1800s.

5.2.1 The Eigenvalue Problem

A regular SLP always has the trivial solution $y(x) \equiv 0$. The question is when will it have a nontrivial solution. A value of λ for which there is a nontrivial solution of (2.1)–(2.2) is called an **eigenvalue**, and the corresponding nontrivial solution is called an associated **eigenfunction**. Notice that any constant multiple of an eigenfunction gives another eigenfunction (but not an independent one).

Remark 5.4

(**Quantum mechanics**) Readers of the the section on the WKB approximation in Chapter 4 will recognize the Sturm–Liouville differential equation is a general version of the one-dimensional, time-independent Schrodinger equation

$$-\frac{\hbar^2}{2m} y'' + V(x)y = Ey,$$

where $V(x)$ is the potential function and E is the energy. Under certain boundary conditions, the problem is to find the values of the energy (eigenvalues) for which there is a wave function $y(x)$. The eigenvalue problem is often singular because of an infinite domain. Also, the potential function may be a piecewise continuous function. $\quad \square$

As we observe in the sequel, the interesting fact about regular SLPs is that they have an infinite number of eigenvalues, and the corresponding eigenfunctions, which are in C^2, form an orthonormal set in the space of all square-integrable functions, which makes orthogonal expansions possible. These facts generalize results from finite-dimensional linear algebra, say in $\mathbb{R}^n$ or $\mathbb{C}^n$ regarding the existence of eigenvalues and eigenvectors of a matrix and the representation of vectors in orthogonal coordinates.

Example 5.5

A simple example of a regular SLP is the problem

$$
\begin{aligned}
-y''(x) &= \lambda y(x), \quad 0 < x < \pi, \\
y(0) &= 0, \ y(\pi) = 0,
\end{aligned}
$$

which is similar to the steady-state heat conduction problem from the last section. Here, $p(x) = q(x) = 1$ and the interval is $[0, \pi]$. This problem has eigenvalues $\lambda = \lambda_n = n^2$ and corresponding eigenfunctions $y = y_n(x) = \sin nx$, $n = 1, 2, \ldots$. Thus, the problem has solution $y = \sin x$ when $\lambda = 1$, $y = \sin 2x$ when $\lambda = 4$, and so on. One way to find the eigenvalues and eigenfunctions is to go through a case-by-case argument separately considering $\lambda = 0$, $\lambda > 0$, and $\lambda < 0$. (We prove later that λ cannot be a complex number.) We examine different cases because the solution of the differential equation has a different form depending on the sign of λ. If $\lambda = 0$, then the equation has the form $y'' = 0$, whose general solution is the linear function $y(x) = Ax + B$. But $y(0) = 0$ implies $B = 0$, and $y(\pi) = 0$ implies $A = 0$. Thus we get only the trivial solution in this case and so $\lambda = 0$ is *not* an eigenvalue. If $\lambda < 0$, say for definiteness $\lambda = -k^2$, then the equation has the form $y'' - k^2 y = 0$ with general solution

$$
y(x) = Ae^{kx} + Be^{-kx}.
$$

The boundary conditions require

$$
\begin{aligned}
y(0) &= A + B = 0, \\
y(\pi) &= Ae^{k\pi} + Be^{-k\pi} = 0,
\end{aligned}
$$

which in turn force $A = B = 0$, and therefore we obtain only the trivial solution in this case. Thus, there are no negative eigenvalues. Finally, consider the case $\lambda > 0$, or $\lambda = k^2$. Then the differential equation takes the form

$$
y'' + k^2 y = 0,
$$

which has general solutions

$$
y(x) = A\cos kx + B\sin kx.
$$

First, $y(0) = 0$ forces $A = 0$. Thus $y(x) = B \sin kx$. The right boundary condition yields

$$y(\pi) = B \sin k\pi = 0.$$

But now we are not required to take $B = 0$; rather, we can select k to make this equation hold. Because the sine function vanishes at multiples of π, we can choose $k = n$, a nonzero integer (we have already considered the case when $k = 0$), or

$$\lambda = \lambda_n = n^2, \quad n = 1, 2, \ldots.$$

The corresponding solutions, or eigenfunctions, are

$$y = y_n(x) = \sin nx, \quad n = 1, 2, \ldots.$$

Here we have arbitrarily selected the constant $B = 1$ for each eigenfunction; we can always multiply eigenfunctions by a constant to get another (not independent) eigenfunction. Alternately, we could write the eigenfunctions as, for example, $y_n(x) = B_n \sin nx$, where a different constant is chosen for each eigenfunction. The constants B_n are chosen to normalize the eigenfunction in some manner. We can easily plot the eigenfunctions; they are just sine curves with higher and higher frequencies as n increases. Note that the eigenfunctions are real, and, generally, we can always take the eigenfunctions of a SLP to be real. □

We can always carry out the preceding argument to determine the eigenvalues and eigenfunctions of a SLP. But these calculations can be tedious, or impossible, and so it is advantageous to have some general results that give the properties of eigenvalues and eigenfuntions. We discuss this matter in the sequel, but first we review the finite-dimensional problem for both guidance and comparison.

Finite dimensional problems. Let $\mathbb{C}^n$ denote the vector space of all n-tuples of complex numbers equipped with the usual definitions of addition and scalar multiplication. We regard the elements as column vectors x, y, and so on, without arrows or boldface type. Geometry in $\mathbb{C}^n$ can be imposed by defining an **inner product** (or *dot* product),

$$(x, y) = \sum_{i=1}^{n} x_i \bar{y}_i = x^{\mathrm{T}} \bar{y}.$$

Then, the **norm**, or size, of a vector is defined by

$$\|x\| = \sqrt{x^{\mathrm{T}} \bar{x}},$$

and the **distance** between two vectors is $\|x - y\|$. We say two vectors are **orthogonal** if $(x, y) = 0$. A set of vectors forms an orthogonal set if they are mutually orthogonal, and the set of vectors forms an orthonormal set if it is an orthogonal set and each element in the set has norm 1. Recall that a vector can always be normalized by dividing by its norm. The inner product satisfies the conditions:

(i) $(x, y) = \overline{(x, y)}$.

(ii) $(x, y + z) = (x, y) + (x + z)$.

(iii) $(\alpha x, y) = \alpha(x, y)$, $\alpha \in \mathbb{C}$.

(iv) $(x, x) \geq 0$ and $(x, x) = 0$ if, and only if, $x = 0$.

The properties of the norm are

(i) $\|x\| \geq 0$ for $x \neq 0$, and $\|x\| \geq 0$ implies $x = 0$.

(ii) $\|\alpha x\| = |\alpha| \|x\|$.

(iii) $\|x + y\| \leq \|x\| + \|y\|$.   (triangle inequality)

(iv) $|(x, y)| \leq \|x\| \|y\|$.   (Cauchy–Schwartz inequality)

Now consider the algebraic eigenvalue problem

$$Ax = \lambda x,$$

where A is an $n \times n$ matrix with complex entries. For certain classes of matrices the eigenvalue problem has extraordinarily useful properties and they give direct insight into the treatment of Sturm–Liouville problems; in fact, one can see that both the algebraic EVP and the SLP are really special cases of the same overarching concept.

We say that a complex matrix A is **Hermitian** if

$$A = \overline{A}^{\mathrm{T}}.$$

That is, the matrix is equal to its complex conjugate transposed. Then, a simple calculation shows

$$(Ax, y) = ((Ax)^{\mathrm{T}} \overline{y}) = x^{\mathrm{T}} A^{\mathrm{T}} \overline{y} = (x, \overline{A}^{\mathrm{T}} y) = (x, Ay).$$

Thus, for Hermitian matrices, the matrix operator can be taken off the first vector and put on the second vector in an inner product; that is, the matrix A is symmetric with respect to the inner product. This leads to a key definition.

Definition 5.6

For any matrix A, the **adjoint** of A is the matrix A^* where $(Ax, y) = (x, A^* y)$ for all $x, y \in \mathbb{C}^n$. A matrix is **self-adjoint** if $A = A^*$.   □

Therefore, Hermitian matrices are self-adjoint. In the same way, if A is a real symmetric matrix, that is, $A = A^{\mathrm{T}}$, then A is self-adjoint because $(Ax, y) = (x, A^{\mathrm{T}}y) = (x, Ay)$.

For self-adjoint matrices on finite dimensional vector spaces we can prove the following important results.

Theorem 5.7

If A is a self-adjoint matrix, then:

(a) (Ax, x) is real.

(b) The eigenvalues are real.

(c) Eigenvectors associated with distinct eigenvalues are orthogonal.

(d) There is an orthonormal basis formed by the eigenvectors.

(e) There is a matrix matrix U, whose columns are the eigenvectors, such that $U^* = U^{-1}$

$$U^*AU = D,$$

where D is a diagonal matrix with the eigenvalues on the diagonal.   □

Proof

The proofs of these facts, except (d), are straightforward. To prove (a) observe

$$(Ax, x) = (x, Ax) = \overline{(Ax, A)}.$$

If $Ax = \lambda x$, then it follows that $(x, Ax) = (x, \lambda x) = \overline{\lambda}(x, x)$, which proves (b). To prove (c) assume that $Ax = \lambda x$ and $Ay = \mu y$. Then

$$(Ax, y) = \lambda(x, y), \quad (x, Ay) = \mu(x, y).$$

Subtracting, we get $(\lambda - \mu)(x, y) = 0$. Because λ and μ are distinct, $(x, y) = 0$. If (d) is true, then take U as indicated, a matrix whose columns are the n eigenvectors of A. For the proof of (d), which is more difficult, we refer the reader to a text on linear algebra.   □

Differential operators. The Sturm–Liouville differential equation (2.1) can be written in the same way and produces similar results as in the finite-dimensional case. We write the equation in operator form as

$$Ly = \lambda y,$$

where L is a differential operator defined by

$$Ly = -(p(x)y')' + q(x)y,$$

where y is a C^2, complex-valued function on $[a, b]$. The boundary conditions (2.2) are considered to be a part of the definition of the operator L; in other words, the domain of the operator L is $D_L = \{y \in C^2 : (2.2) \text{ holds}\}$. These functions form a linear space, or vector space (Chapter 4), over the complex numbers. The geometrical concepts of inner product and norm can be defined in a space of functions; these are generalizations of the inner (or dot) product, and the length of vectors in $\mathbb{C}^n$.

Definition 5.8

If $u, v \in C^2$, then the **inner product** of u and v is a complex number defined by

$$(u, v) = \int_a^b u(x)\overline{v}(x)\, dx. \quad \square \tag{2.3}$$

It is easy to check using properties of integrals that the inner product satisfies the same properties of scalar products of complex vectors. This inner product induces a norm defined by

$$\|u\| = \left(\int_a^b u(x)\overline{u}(x)\, dx \right)^{1/2} = \left(\int_a^b |u(x)|^2\, dx \right)^{1/2}.$$

This norm is called the L^2-norm. It satisfies the same properties as the norm of complex vectors. The same definitions of orthogonal, orthogonal sets, and orthonormal sets hold.

Example 5.9

Consider the set of complex-valued exponential functions

$$y_n(x) = e^{inx}, \quad n = 0, \pm 1, \pm 2, \ldots$$

for $-\pi \le x \le \pi$. Clearly, for $m \ne n$,

$$(y_m, y_n) = (e^{imx}, e^{inx}) = \int_{-\pi}^{\pi} e^{imx} e^{-inx}\, dx = \frac{1}{(m-n)i} e^{i(m-n)x}|_{-\pi}^{\pi} = 0.$$

So, the functions are orthogonal. If $m = n$, then obviously $(e^{imx}, e^{inx}) = 2\pi$, and the norm is $\sqrt{2\pi}$. Hence, the set

$$\phi_n(x) = \frac{1}{\sqrt{2\pi}} e^{inx}, \quad n = 0, \pm 1, \pm 2, \ldots$$

is an orthonormal set on $-\pi \le x \le \pi$. $\quad \square$

Example 5.10

In Example 5.5 we found the eigenvalues $\lambda_n = n^2$ and eigenfunctions $y_n(x) = \sin nx$, for $n = 1, 2, \ldots$ for the SLP $-y'' = \lambda y$, $y(0) = y(\pi) = 0$. It is an exercise[1] to show that the eigenfunctions satisfies the property

$$\int_0^\pi \sin nx \sin mx \, dx = 0, \quad m \neq n, \tag{2.4}$$

so the eigenfunctions form an orthogonal set. Each of these eigenfunctions can be normalized by dividing by its norm. We have

$$\| \sin nx \| = \sqrt{(\sin nx, \sin nx)} = \left(\int_0^\pi \sin^2 nx \, dx \right)^{1/2} = \sqrt{\frac{\pi}{2}}.$$

Therefore the functions

$$\phi_n(x) = \sqrt{\frac{2}{\pi}} \sin nx, \quad n = 1, 2, 3, \ldots$$

form an orthonormal set of eigenfunctions for the SLP. Thus,

$$(\phi_m, \phi_n) = 0, \quad m \neq n, \text{ and } (\phi_n, \phi_n) = 1. \quad \square$$

The self-adjoint property of a real or complex matrix was the the important property that led to the symmetric relation $(Ax, y) = (x, Ay)$ with the inner product. Now we show that the same is true for a Sturm–Liouville operator L. We want to show, therefore, that

$$(Lu, v) = (u, Lv),$$

where u and v satisfy the boundary conditions (2.2). We first prove a fundamental identity.

Lemma 5.11

(**Green's identity**) Let u and v be continuously differentiable, complex-valued functions on the interval $[a, b]$, and let p be real and continuously differentiable. Then

$$\int_a^b (pu')' \overline{v} dx = p[u' \overline{v} - u \overline{v}']_a^b + \int_a^b u(p\overline{v}')' dx.$$

---

[1] The trigonometric identities $\sin A \sin B = \frac{1}{2}(\cos(A - B) - \cos(A + B))$ and $\sin A \cos B = \frac{1}{2}(\sin(A + B) + \sin(A - B))$, along with others, are useful in these calculations.

Proof

We can remove the differential operator from u and put it on $\bar{v}$ using two integrations by parts. We have

$$\int_a^b (pu')'\, \bar{v}dx = pu'\bar{v} - \int_a^b pu'\bar{v}'\, dx$$

$$= pu'\bar{v} \mid_a^b -up\bar{v}' \mid_a^b + \int_a^b u(p\bar{v}')'\, dx$$

$$= p[u'\bar{v} - u\bar{v}']_a^b + \int_a^b u(p\bar{v}')'\, dx.$$

$\square$

Applying this lemma to the Sturm–Liouville operator $Lu = -(pu')' + qu$ gives

$$\int_a^b [-(pu')' + qu]\, \bar{v}dx = \int_a^b u[-(p\bar{v}')' + q\bar{v}]\, dx - p(u'\bar{y} - u\bar{v}') \mid_a^b .$$

In terms of the inner product, this is exactly

$$(Lu, v) = (u, Lv) - [p(u\bar{v}' - u'\bar{v})]_a^b. \tag{2.5}$$

Many may recognize the expression

$$W(x) \equiv u(x)\bar{v}'(x) - u'(x)\bar{v}(x),$$

which is the Wronskian of u and $\bar{v}$. Now we apply the boundary conditions (2.2). For simple Dirichlet or Neumann boundary conditions, it is clear that the Wronskian is zero at both $x = a$ and $x = b$ because both eigenfunctions, and their complex conjugates, satisfy the boundary conditions. For the general, mixed boundary conditions (2.2) the Wronskian is zero as well. For example, without loss of generality we can assume $\alpha_1 \neq 0$. Then, the first boundary condition for u and $\bar{v}$ can be written

$$u = \frac{\alpha_2}{\alpha_1}u', \quad \bar{v} = \frac{\alpha_2}{\alpha_1}\bar{v}'.$$

Then, substituting into the right side of (2.5),

$$W(a) = [u(a)\bar{v}'(a) - u'(a)\bar{v}(a) = -\frac{\alpha_2}{\alpha_1}(u'(a)\bar{v}'(a) - u'(a)\bar{v}'(a) = 0.$$

Similarly, we can show $W(b) = 0$.

In summary, we have proved:

Theorem 5.12

For the regular Sturm–Liouville problem (2.1)–(2.2), the differential operator L is self-adjoint, or

$$(Lu, v) = (u, Lv).$$

Finally we show each eigenvalue has only eigenfunction. Let λ be an eigenvalue and u and v be corresponding, independent real eigenfunctions. [Note that we may always take the eigenfunctions to be real, because the real and imaginary parts of a complex-valued eigenfunction are eigenfunctions.] Then, by definition of independence, the only solution of the system

$$c_1 u(x) + c_2 v(x) = 0, \quad c_1 u'(x) + c_2 v'(x) = 0$$

is $c_1 = c_2 = 0$ for all x. This means the coefficient determinant, which is the Wronskian $W(x)$, must be nonzero for all x. But we have $W(a) = 0$, a contradiction. Thus u and v must be dependent.

Now we can state a fundamental theorem for eigenvalues of a regular Sturm–Liouville problem.

Theorem 5.13

The regular Sturm–Liouville problem (2.1)–(2.2) has infinitely many real eigenvalues λ_n, $n = 1, 2, \ldots$, which can be arranged in an increasing sequence

$$\lambda_1 < \lambda_2 < \lambda_3 < \cdots, \quad \lim_{n \to \infty} |\lambda_n| = +\infty.$$

To each eigenvalue corresponds a single real eigenfunction, up to a constant multiple. Finally, eigenfunctions corresponding to distinct eigenvalues are orthogonal on $[a, b]$. $\square$

Proof

For the proof of existence and ordering of the eigenvalues we refer to the references (e.g., see Birkhoff and Rota, 1989). The remaining parts follow from the self-adjointness of L and the proofs of the results in Theorem 5.7. $\square$

In general, a set of boundary conditions for which

$$(u\bar{v}' - u'\bar{v}_2) \,|_a^b = 0$$

is called **symmetric**. Symmetric boundary conditions guarantee that the operator is self-adjoint and has eigenvalues with independent, orthogonal eigenfunctions.

Example 5.14

(**Sign of eigenvalues**) Another issue concerns the sign of the eigenvalues. The following argument, called an **energy argument**, is useful to show that the eigenvalues are of one sign, without resorting to a case argument. We consider a Schrodinger equation with Dirichlet boundary conditions. Other SLPs can be treated similarly. As an illustration we show that the regular SLP

$$-y'' + q(x)y = \lambda y, \quad a < x < b, \tag{2.6}$$
$$y(a) = y(b) = 0, \tag{2.7}$$

where $q > 0$ on $[a, b]$, has only positive eigenvalues. Multiplying the differential equation (2.6) by y and integrating gives

$$-\int_a^b yy'' dx + \int_a^b qy^2 \, dx = \lambda \int_a^b y^2 dx.$$

The first integral can be integrated by parts to obtain

$$-yy' \big|_a^b + \int_a^b y'^2 dx + \int_a^b qy^2 dx = \lambda \int_a^b y^2 dx.$$

Under Dirichlet boundary conditions (2.7), the boundary terms in the last equation are zero. Hence

$$\int_a^b y'^2 \, dx + \int_a^b qy^2 dx = \lambda \int_a^b y^2 \, dx.$$

Because q is positive, the second integral on the left side is positive. Thus, the eigenvalues λ for (2.6)–(2.7) must be positive. □

Example 5.15

(**Periodic boundary conditions**) Now we consider a SLP differential equation on the interval $[-\pi, \pi]$ with periodic boundary conditions:

$$-y'' = \lambda y, \quad -\pi < x < \pi, \tag{2.8}$$
$$y(-\pi) = y(\pi), \ y'(-\pi) = y(\pi). \tag{2.9}$$

These types of problems are useful in Chapter 6, where we examine partial differential equations with periodic boundary conditions. Moreover, periodic boundary conditions are symmetric. We can either use a case argument or an energy argument to show that the eigenvalues are nonnegative. We leave this as an exercise. The energy argument does not eliminate $\lambda = 0$. But it is easy to see from the problem that $\lambda = 0$ is an eigenvalue with a constant eigenfunction;

so $\lambda_0 = 0$, $y_0(x) = 1$ is an eigenpair. Now assume $\lambda = k^2 > 0$. The differential equation (2.8) has general solution

$$y(x) = a \cos kx + b \sin kx,$$

for arbitrary constants a and b. The first boundary condition in (2.9) forces

$$a \cos k\pi - b \sin k\pi = a \cos k\pi + b \sin k\pi,$$

or

$$b \sin k\pi = 0.$$

This boundary condition can be satisfied if $k = \pm n$, or

$$\lambda = n^2, \quad n = 1, 2, 3, \ldots.$$

With these values the second boundary condition is satisfied automatically. Hence the eigenvalues are $\lambda_n = n^2$, $n = 1, 2, 3, \ldots$, and with each eigenvalue corresponds to two independent eigenfunctions, $\cos nx$ and $\sin nx$. These, along with 1, form an orthonormal set. At this point we can check orthogonality. Using trigometic identities we can show, for example, that

$$\int_{-\pi}^{\pi} \sin nx \sin mx \, dx = 0, \quad \int_{-\pi}^{\pi} \cos nx \sin mx \, dx = 0, \quad m \neq n,$$

$$\int_{-\pi}^{\pi} \sin nx \cos mx \, dx = 0, \quad n, m = 1, 2, 3, \ldots,$$

$$\int_{-\pi}^{\pi} \sin nx \, dx = \int_{-\pi}^{\pi} \cos nx \, dx, \quad n = 1, 2, 3, \ldots.$$

Thus the functions 1, $\cos nx$, $\sin nx$ form an orthogonal set of eigenfunctions for a periodic SLP. Observe that the operator L defined in (2.8)–(2.9) is self-adjoint. Later, we see that these functions form the basis of classical Fourier series. □

5.2.2 Eigenfunction Expansions and Bases

In the last section we proved a key result for regular SLPs regarding the existence of eigenvalues and orthogonal eigenfunctions. Now we take up the basis concept. Having a set of orthonormal eigenfunctions for a regular SLP suggests the possibility of expanding a given function in terms of those functions. We recall the finite dimensional case stated in Theorem 5.7. If A is a self-adjoint matrix, there are n real eigenvalues and a set of n orthonormal eigenvectors, say, $\mathbf{v}_1, \mathbf{v}_2, \ldots \mathbf{v}_n$, and the eigenvectors form a basis for the space; that is, every vector $\mathbf{x}$ can be expanded as a linear combination of those eigenvectors, or

$$\mathbf{x} = c_1 \mathbf{v}_1 + c_2 \mathbf{v}_2 + \cdots + c_n \mathbf{v}_n,$$

and the coefficients, or coordinates, c_n are the projections of the vector $\mathbf{x}$ onto the coordinate directions, or

$$c_n = (\mathbf{x}, \mathbf{v}_n), \quad n = 1, 2, \ldots, n.$$

If the basis $\mathbf{v}_n$ were only an independent set, and not orthogonal, then it would be very difficult to compute the coefficients in the representation of $\mathbf{x}$!

We seek an analogous result for regular SLPs. To illustrate the point we revisit an example.

Remark 5.16

In Example 5.5 we solved the eigenvalue problem

$$-y'' = \lambda y, \quad y(0) = y(\pi) = 0$$

and found orthogonal eigenfunctions $\sin nx$, $n = 1, 2, \ldots$. Which functions f can be expanded in the sine series

$$f(x) = \sum_{n=1}^{\infty} c_n \sin nx = c_1 \sin x + c_2 \sin 2x + c_3 \sin 3x + \cdots,$$

for some set of coefficients c_n, and how can we compute the coefficients c_n? This expansion is not a finite sum, so the answer to this question must come to terms with the notion of convergence of the series. As stated previously, the L^2 norm defines a distance measure, and that will be used to define convergence. □

We now cut to the chase and state the basis theorem for regular SLPs. We give at this point only an informal proof and a few examples, with the more analytic result coming later.

Theorem 5.17

Let $\phi_n(x)$, $n = 1, 2, \ldots$, be the set of orthonormal eigenfunctions for a regular Sturm–Liouville problem. Then every square-integrable function f on $[a, b]$ can be expanded as

$$f(x) = \sum_{n=1}^{\infty} c_n \phi_n(x), \tag{2.10}$$

where the convergence is in L^2 norm. This means, in terms of the partial sums of the series,

$$s_N(x) = \sum_{n=1}^{N} c_n \phi_n(x),$$

that $\|f - s_N\| \to 0$ as $N \to 0$, or

$$\int_a^b |f(x) - s_N(x)|^2 dx \to 0 \text{ as } N \to 0.$$

The coefficients are given by

$$c_n = (f, \phi_n), \quad n = 1, 2, 3 \dots . \quad \square \tag{2.11}$$

The series (2.10) is called the **(generalized) Fourier series** for f, and the coefficients (2.11) are called the **Fourier coefficients**, and we interpret them as projection coefficients of f onto the subspace generated by the eigenfunction ϕ_n. Convergence in L^2 is often called **mean-square convergence**, or **convergence in the mean**.

Proof

Assume the series (2.10) converges in the mean. Multiplying (2.10) by $\phi_m(x)$ and integrating gives

$$
\begin{aligned}
\int_a^b f(x)\phi_m(x)dx &= \int_a^b \left[\sum_{n=1}^{\infty} c_n \phi_n(x)\phi_m(x) \right] dx \\
&= \sum_{n=1}^{\infty} c_n \int_a^b \phi_n(x)\phi_m(x)dx,
\end{aligned}
$$

where we formally pulled the integral under the sum. (This can be shown to be a valid operation). By orthogonality, the sum collapses to a single term, occurring when $n = m$. Thus,

$$\int_a^b f(x)\phi_m(x)dx = c_n \int_a^b \phi_m(x)\phi_m(x)dx = c_m.$$

But this is true for all m. Thus, in terms of the inner product, this gives the Fourier coefficients (2.11). Note that orthogonality was essential in this argument. $\square$

Remark 5.18

If the orthogonal set $y_n(x)$ is used in the expansion instead of the orthonormal set $\phi_n(x)$, that is, if

$$f(x) = \sum_{n=1}^{\infty} c_n y_n(x),$$

then the same argument as in the proof shows that the Fourier coefficients are given by

$$c_n = \frac{1}{\|y_n\|^2}(f, y_n). \quad \Box \tag{2.12}$$

Usually, for theoretical calculations it is easier to work with the orthonormal set; for practical examples we work with the orthogonal set because we can avoid the complicated normalizing constants.

Example 5.19

Expand the function $f(x) = x$, $0 < x < \pi$, in terms of the orthogonal eigenfunctions $\sin nx$ determined in Example 5.5. The Fourier coefficients are, from (2.12),

$$c_n = \frac{1}{\|\sin nx\|^2}(x, \sin nx) = \frac{1}{\|\sin nx\|^2}\int_0^\pi x \sin nx \, dx.$$

Now,

$$\|\sin nx\|^2 = \int_0^\pi \sin^2 nx \, dx = \frac{\pi}{2},$$

and, using integration by parts,

$$c_n = \frac{2}{\pi}\int_0^\pi x \sin nx \, dx = \frac{2}{\pi}\left(-\frac{\pi}{n}\cos n\pi\right) = \frac{2}{n}(-1)^{n+1}.$$

Therefore the generalized Fourier series for $f(x) = x$ is

$$x = \sum_{n=1}^{\infty} \frac{2}{n}(-1)^{n+1}\sin nx = 2\left(\sin x - \frac{1}{2}\sin 2x + \frac{1}{3}\sin 3x + \cdots\right).$$

We are guaranteed that this series converges in the mean-square sense. Observe that the series cannot be differentiated term-by-term; the derived series does not even converge, much less to 1. $\Box$

Convergence. Consider an orthogonal expansion

$$f(x) = \sum_{n=1}^{\infty} c_n f_n(x). \tag{2.13}$$

When we write a series like (2.13), a natural question is, "What is meant by convergence of the infinite series?" There are many answers to this question, depending on how we measure the error of an N-term approximation $s_N(x) \equiv \sum_{n=1}^{N} c_n f_n(x)$ to $f(x)$. On one hand, by the **pointwise error** $E_N(x)$ we mean the function

$$E_N(x) \equiv f(x) - s_N(x) \equiv f(x) - \sum_{n=1}^{N} c_n f_n(x).$$

Therefore, for a fixed number N (giving the number of terms), $E_N(x)$ is a *function* whose value gives the error at each point x. On the other hand, we can define an *integrated* error; the number e_N defined by

$$e_N \equiv \int_a^b |\, f(x) - s_N(x) \,|^2 \, dx \equiv \int_a^b |\, f(x) - \sum_{n=1}^{N} c_n f_n(x) \,|^2 \, dx$$

is called the **mean-square error** in the approximation. Note that e_N is just the square of the pointwise error, integrated over the interval $[a, b]$. Thus, the mean-square error is a type of average over all the pointwise errors. Each error expressions leads to a definition of convergence.

If for each *fixed* x in $[a, b]$ we have $\lim_{N \to \infty} E_N(x) = 0$, then we say that the infinite series (2.13) converges **pointwise** to f on the interval $[a, b]$; in this case, for each fixed x the series is a numerical series that converges to the numerical value $f(x)$, at that x; pointwise convergence means that the error goes to zero at each point. We say that the infinite series (2.13) converges to f in the **mean-square** sense if

$$\lim_{N \to \infty} e_N = 0.$$

Mean-square convergence is *convergence in* $L^2[a, b]$, or $\|s_N - f\| \to 0$ as $N \to \infty$, and it requires that the integrated pointwise error-squared go to zero as more and more terms are taken.

There is a stronger type of convergence, called uniform convergence. The series (2.13) **converges uniformly** to $f(x)$ on $[a, b]$ if for any given tolerance $\varepsilon > 0$, we can find a number N (representing the number of terms) such that $n > N$ implies $|E_n(x)| < \varepsilon$ for all x in $[a, b]$. That is, the error can be made uniformly small over the entire interval by choosing the number of terms large enough. For uniform convergence, N (the number of terms) depends only on the tolerance ε and not on where in the interval the error is taken. Uniform convergence can be expressed practically as

$$\sup_{x \in [a,b]} |s_N - f| \to 0 \text{ as } N \to \infty.$$

To prove uniform convergence of the series on the right side of (2.13) it is sufficient to show[2] there is a numerical sequence M_n such that $\sum M_n$ converges and $|c_n f_n(x)| < M_n$ for all $x \in [a, b]$.

Remark 5.20

One can show that *pointwise convergence does not imply mean-square convergence, and conversely.* However, *uniform convergence implies both mean-square and pointwise convergence.*   □

A novice may not appreciate the subtle differences in convergence and wonder why we have defined three types (actually, there are others). Suffice it to say that the error may not go to zero in a pointwise or uniform manner, but that does not mean the approximation is not useful; a weaker form of convergence, like mean-square convergence, may be all that is required. By the way, the definitions above are valid whether or not the functions $f_n(x)$ are orthogonal. However, mean-square convergence and the orthogonality of the f_n make computation of the coefficients in the series easier.

EXERCISES

1. Show that the SLP

$$-y''(x) = \lambda y(x), \quad 0 < x < l,$$
$$y(0) = 0, \quad y(l) = 0,$$

    has eigenvalues $\lambda_n = n^2\pi^2/l^2$ and corresponding eigenfunctions $y_n(x) = \sin(n\pi x/l)$, $n = 1, 2, \ldots$

2. Show that the SLP

$$-y''(x) = \lambda y(x), \quad 0 < x < l,$$
$$y'(0) = 0, \quad y(l) = 0,$$

    with mixed Dirichlet and Neumann boundary conditions has eigenvalues

$$\lambda_n = \left(\frac{(1+2n)\pi}{2l}\right)^2$$

    and corresponding eigenfunctions

$$y_n(x) = \cos\frac{(1+2n)\pi x}{2l}$$

    for $n = 0, 1, 2, \ldots$.

---

[2] This result is the Weierstrass M-test for uniform convergence; e.g., see Spivak (2008), p 507.

3. Find the eigenvalues and eigenfunctions for the problem with *periodic* boundary conditions:

$$-y''(x) = \lambda y(x), \quad 0 < x < l,$$
$$y(0) = y(l), \quad y'(0) = y'(l).$$

4. Consider the SLP

$$-y'' = \lambda y, \quad 0 < x < 1,$$
$$y(0) + y'(0) - 0, \quad y(1) = 0.$$

Is $\lambda = 0$ an eigenvalue? Are there any negative eigenvalues? Show that there are infinitely many positive eigenvalues by finding an equation whose roots are those eigenvalues, and show graphically that there are infinitely many roots.

5. Show that the SLP

$$-y'' = \lambda y, \quad 0 < x < 2,$$
$$y(0) + 2y'(0) = 0, \quad 3y(2) + 2y'(2) = 0,$$

has exactly one negative eigenvalue. Is zero an eigenvalue? How many positive eigenvalues are there?

6. For the SLP

$$-y'' = \lambda y, \quad 0 < x < l,$$
$$y(0) - ay'(0) = 0, \quad y(l) + by'(l) = 0,$$

show that $\lambda = 0$ is an eigenvalue if and only if $a + b = -l$.

7. Use an energy argument to show that the eigenvalues for the problem

$$-(x^2 y')' = \lambda y, \quad 1 < x < e,$$
$$y(1) = y(e) = 0$$

must be nonnegative. Find the eigenvalues and eigenfunctions.

8. Find eigenvalues and eigenfunctions for the problem

$$-y'' - 2by' = \lambda y, \quad 0 < x < 1, \ b > 0,$$
$$y(0) = y(1) = 0.$$

9. What, if anything, can be deduced about the sign of the eigenvalues for the problem

$$-y'' + xy = \lambda y, \quad 0 < x < 1,$$
$$y(0) = y(1) = 0.$$

What are the eigenvalues and eigenfunctions? (Use a computer algebra system to investigate this problem.)

10. Consider the regular problem

$$-y'' + q(x)y = \lambda y, \quad 0 < x < l,$$
$$y(0) = y(l) = 0,$$

where $q(x) > 0$ on $[0, l]$. Show that if λ and y are an eigenvalue and eigenfunction pair, then

$$\lambda = \frac{\int_0^l (y'^2 + qy^2)\, dx}{||y||^2}.$$

Is $\lambda > 0$? Can $y(x) = $ constant be an eigenfunction?

11. Prove the Cauchy–Schwartz inequality $|\, (f, g)\, | \leq Vertf||\, ||g||$ using the fact that $Q(t) = (f + tg, f + tg) \geq 0$ for any real number t. Expand out to get a polynomial in t.

5.2.3 Best Approximation and Hilbert Spaces

The last section outlined a recipe for expanding an L^2 function f in a basis consisting of the orthonormal eigenfunctions of a regular Sturm–Liouville problem. In the next chapter we examine underpinnings of the origin and applications of SLP in the subject of partial differential equations; in this chapter we examine its utility in solving integral equations.

To dig deeper into the mathematical foundations of orthogonal expansions in function spaces we require some definitions. A vector space on which there is a norm is called a **normed linear space**, and a vector space on which there is defined an inner product is an **inner product space**. An inner product space is a normed linear space under the natural norm associated with the inner product. A normed linear space may not be an inner product space. We have seen examples of both.

In a normed linear space X, with norm $||\cdot||$, a sequence y_n is said to converge to $y \in X$ if $||y_n - y|| \to 0$ as $n \to +\infty$. But we can make the following definition without knowledge of the limit.

Definition 5.21

A sequence y_n in X is a **Cauchy sequence** if for every $\varepsilon > 0$ there in an integer N, independent of ε, such that $\|y_n - y_m\| < \varepsilon$ for all $m, n > N$. In other words, the terms of the sequence get as close as we desire far out in the sequence.   □

Must convergent sequences be Cauchy sequences? Yes. Must Cauchy sequences converge to a limit? Maybe surprisingly, no. For example,

$$x_n = \left(1 + \frac{1}{n}\right)^n$$

is a sequence of rational numbers, and it is a Cauchy sequence. However it does not converge to a rational number. (It in fact converges to the number e, which is not rational.) Thus the rational numbers are deficient in some way. This leads to the following definition.

Definition 5.22

A normed linear space is **complete** if every Cauchy sequence in X has a limit in X. A complete normed linear space is called a **Banach space**. An inner product space that is complete in its natural norm is called a **Hilbert space**. □

Thus, the rational numbers are not complete. They are useful for calculations, by hand or on computers, but are not good if we are interested in convergence. So, what we do is *to complete* the set of rational numbers by adding in all possible limits of Cauchy sequences. We get the real numbers when this is done, and then we can do, for example, calculus. This construction of completion, by the way, is not trivial.

These concepts may be familiar to students in physics, where in the foundations of quantum mechanics the wave functions are assumed, axiomatically, to belong to a Hilbert space.

Example 5.23

Consider the normed linear space of continuous functions with the L^2 norm. It is an inner product space with

$$(f, g) = \int_a^b f(x)\overline{g}(x)dx.$$

But it is not a Banach space or Hilbert space. Let $[a, b] = [0, 1]$ and consider the sequence of functions

$$
f_n(x) = \begin{cases} 0, & 0 \leq x < \frac{1}{2} - \frac{1}{n} \\ 1 + n\left(x - \frac{1}{2}\right), & \frac{1}{2} - \frac{1}{n} < x < \frac{1}{2} \\ 1, & \frac{1}{2} \leq x \leq 1. \end{cases}
$$

We leave it to the reader to sketch plots of f_n and f_m for different values m and n and then compute the L^2 norm

$$
\|f_n - f_m\|^2 = \int_0^1 (f_m(x) - f_n(x)^2 dx
$$

showing that the sequence is Cauchy. Yet the limit is $f(x) = 0$ for $x < \frac{1}{2}$; $f(x) = 1$ for $x > \frac{1}{2}$, which is not continuous.   □

This example invites us to append to this normed linear space the set of all possible limits of Cauchy sequences. But this leads to a serious issue. The limits of such sequences can be very bizarre and the integral of those functions is not defined if we use the classical Riemann integral from calculus. H. Lebesgue in the early 1900s remedied this problem by defining a new, more general, integral, called the **Lebesgue integral**, for which all these limiting functions of Cauchy sequences are defined. With this redefinition of the L^2 norm in $C[a, b]$ the space is complete and therefore both a Banach space and a Hilbert space.

   Using the Lebesgue integral we can rigorously treat the convergence of Fourier series. We approach this task by asking about the best approximation in a Hilbert space of a function by a finite sum of orthonormal functions. Specifically, if $f(x)$ is in L^2 and $\phi_1, \ldots, \phi_N$ is an orthonormal set, what is the best approximation f by a linear combination of the ϕ_n?

Theorem 5.24

If $f \in L^2[a, b]$, $\phi_1, \phi_2, \ldots, \phi_N$ is an orthonormal set in $L^2[a, b]$, and c_n are the Fourier coefficients given by

$$
c_n = (f, \phi_n), \quad n = 1, 2, \ldots, N,
$$

then

$$
\left\| f - \sum_{n=1}^{N} c_n \phi_n \right\|^2 \leq \left\| f - \sum_{n=1}^{N} a_n \phi_n \right\|^2 \tag{2.14}
$$

for any set of coefficients a_n.   □

Proof

The demonstration of (2.14) is straightforward using the definition of norm and properties of the inner product. We have

$$\| f - \sum_{n=1}^{N} a_n \phi_n \|^2 \;=\; (f - \sum_{n=1}^{N} a_n \phi_n, f - \sum_{n=1}^{N} a_n \phi_n)$$

$$=\; (f, f) + \sum_{n=1}^{N} [a_n \bar{a}_n - \bar{a}_n c_n - a_n \bar{c}_n].$$

Adding and subtracting $\sum c_n \bar{c}_n = \sum |c_n|^2$, gives

$$\| f - \sum_{n=1}^{N} a_n \phi_n \|^2 \;=\; \|f\|^2 - \sum_{n=1}^{N} |c_n|^2 + \sum_{n=1}^{N} [a_n \bar{a}_n - \bar{a}_n c_n - a_n \bar{c}_n + c_n \bar{c}_n]$$

$$=\; \|f\|^2 - \sum_{n=1}^{N} |c_n|^2 + \sum_{n=1}^{N} |a_n - c_n|^2$$

$$=\; \| f - \sum_{n=1}^{N} c_n \phi_n \|^2 + \sum_{n=1}^{N} |a_n - c_n|^2.$$

The second term on the right is nonnegative, so (2.14) holds.   □

In the last line of the preceding proof we used the fact that

$$\| f - \sum_{n=1}^{N} c_n \phi_n \|^2 = \|f\|^2 - \sum_{n=1}^{N} |c_n|^2, \tag{2.15}$$

which the reader should verify. This equality is actually Pythagoras' theorem: the square of the length of f is the sum of the squares of the lengths of the projection and the error. It leads to another interesting inequality. Because the left side is nonnegative, we have

$$\sum_{n=1}^{N} c_n^2 \leq \|f\|^2.$$

We record a simple, but important, conclusion that follows from this inequality. Because it holds for each N, we have:

Theorem 5.25

(**Bessel's inequality**) If $f \in L^2$ and $\phi_1, \phi_2, \phi_3, \cdots$ is an orthonormal set in L^2, and if $c_n = (f, \phi_n)$, then

$$\sum_{n=1}^{\infty} |c_n|^2 \leq \|f\|^2. \qquad \text{(Bessel's Inequality)} \quad \square$$

Bessel's inequality shows the series of squared Fourier coefficients converges, and thus $c_n \to 0$ as $n \to \infty$. Therefore the Fourier coefficients get smaller and smaller. But, in addition, using completeness, it shows that the infinite series $\sum c_n \phi_n$ converges in L^2.

Theorem 5.26

Under the assumptions of Theorem 5.25, the Fourier series

$$\sum_{n=1}^{\infty} c_n \phi_n \equiv \sum_{n=1}^{\infty} (f, \phi_n) \phi_n$$

converges to a function in L^2.   □

Proof

Let

$$S_n = \sum_{n=1}^{n} c_k \phi_k$$

be a partial sum of the series. We know $\sum |c_n|^2$ converges and so its partial sums form a Cauchy sequence of real numbers and we have

$$\|S_n - S_m\|^2 = \| \sum_{k=m+1}^{n} c_k \phi_k \| = \| \sum_{k=m+1}^{n} |c_k|^2 \| < \varepsilon.$$

Thus the partial sums S_n form a Cauchy sequence in L^2, and, by completeness, the series $\sum c_n \phi_n$ converges to some function $h \in L^2$.   □

The question is: does the limit h of the Fourier series for f equal f? The next example shows this is a valid question.

Example 5.27

Let $f(x) = \cos x$ on $[0, \pi]$ and $\phi_n = \sqrt{\frac{2}{\pi}} \sin nx$ be an orthonormal set. Clearly

$$c_n = \sqrt{\frac{2}{\pi}} \int_0^{\pi} \cos x \sin nx \, dx = 0$$

for all n. So the Fourier series converges to zero, not $\cos x$.   □

This leads to the important definition:

Definition 5.28

A set ϕ_n, $n = 1, 2, \cdots$, is a **complete orthonormal set** if

$$\sum_{n=1}^{\infty} (f_n, \phi_n)\phi_n = f$$

for all $f \in L^2$. □

This definition should not be confused with the earlier definition of completeness of a normed linear space. In summary, the question of whether a Fourier series converges to the given function comes down to proving that the orthonormal system is complete. For specific orthonormal systems, we leave this exercise to more advanced treatments. Many of the proofs rely on one of the conditions given in the following theorem.

Theorem 5.29

The following are equivalent:

(1) An orthonormal system ϕ_n, $n = 1, 2, \cdots$ is complete.

(2) For every $f \in L^2$,

$$\sum_{n=1}^{\infty} c_n^2 = \|f\|^2. \qquad \text{(Parseval's Equality)} \qquad (2.16)$$

(3) If $(f, \phi_n) = 0$ for all n, then $f = 0$. □

Proof

We prove Parseval's condition only. It follows from

$$\|f - \sum_{n=1}^{N} c_n\phi_n\| = \|f\|^2 - \sum_{n=1}^{N} |c_n|^2.$$

In the limiit as $N \to \infty$ the left side goes to zero by completeness. □

Pointwise convergence results and stronger uniform convergence results are more difficult to obtain, and they usually require continuity and smoothness assumptions on the function f. In the next section we discuss convergence results for the classical Fourier system $\{1, \cos(n\pi x/l), \sin(n\pi x/l)\}$, which, by the way, is a complete orthonormal system.

Finally, we mention the modal interpretation of the generalized Fourier series. Thinking of f as a signal and the orthonormal set f_n as the fundamental modes, the Fourier coefficient c_n determines the contribution of the nth mode, and the generalized Fourier series is the decomposition of the signal into fundamental modes. The sequence of squared coefficients, $|c_1|^2$, $|c_2|^2$, $|c_3|^2, \ldots$, is called the **energy spectrum**, and $|c_n|^2$ is called the energy of the nth mode; by Parseval's equality, the **total energy** in a signal is $\|f\|^2$.

EXERCISES

1. Verify that the set of functions $\cos(n\pi x/l)$, $n = 0, 1, 2 \ldots$, form an orthogonal set on the interval $[0, l]$. If

$$f(x) = \sum_{n=0}^{\infty} c_n \cos\left(\frac{n\pi x}{l}\right)$$

   converges in the mean-square sense on $[0, l]$, what are the formulae for the c_n? This series is called the **Fourier cosine series** on $[0, l]$. Find the Fourier cosine series for $f(x) = 1 - x$ on $[0, 1]$.

2. Let f be defined and integrable on $[0, l]$. The orthogonal expansion

$$f(x) = \sum_{n=1}^{\infty} b_n \sin\frac{n\pi x}{l}, \quad b_n = \frac{2}{l}\int_0^l f(x)\sin\left(\frac{n\pi x}{l}\right) dx,$$

   is called the **Fourier sine series** for f on $[0, l]$. Find the Fourier sine series for $f(x) = \cos x$ on $[0, \pi/2]$. What is the Fourier sine series of $f(x) = \sin x$ on $[0, \pi]$?

3. (**Gram–Schmidt orthogonalization**) A result from linear algebra is that any set of linearly independent vectors may be turned into a set of orthogonal vectors by the **Gram–Schmidt** orthogonalization process. The same process works for functions in $L^2[a, b]$. Let $f_1, f_2, f_3, \ldots$ be an independent set of functions in L^2 and define the set g_n by

$$g_1 = f_1, \quad g_2 = f_2 - \frac{(f_2, g_1)}{\|g_1\|^2}g_1, \quad g_3 = f_3 - \frac{(f_3, g_2)}{\|g_2\|^2}g_2 - \frac{(f_3, g_1)}{\|g_1\|^2}g_1, \ldots.$$

   Show that g_n is an orthogonal set.

4. (**Lengendre polynomials**) The functions 1, x, x^2, x^3 are independent functions on the interval $[-1, 1]$. Use the Gram–Schmidt method to generate a sequence of four orthogonal polynomials $P_0(x), \ldots, P_3(x)$ on $[-1, 1]$. Find an approximation of e^x on $[-1, 1]$ of the form

$$c_0 P_0(x) + c_1 P_1(x) + c_2 P_2(x) + c_3 P_3(x)$$

that is best in the mean-square sense, and plot e^x and the approximation. Compute the pointwise error, the maximum pointwise error over $[\;1,1]$, and the mean-square error.

5. (Haar wavelets) Let ϕ be a function defined by $\phi(x) = 1$ for $x \in [0,1)$, and $\phi(x) = 0$ otherwise. Let $\psi(x) = \phi(2x) - \phi(2x-1)$. Then the Haar wavelets are the functions
$$\psi_{mn}(x) = 2^{m/2}\psi(2^m x - n),$$
for $m, n = 0, \pm1, \pm2, \dots$. Sketch a graph of $\psi(x)$, and then sketch a graph of $\psi_{mn}(x)$ for $m,\ n = 0, \pm1, \pm2$. Generally, what is the graph of $\psi_{mn}(x)$? If f is square integrable and
$$f(x) = \sum_{m=-\infty}^{\infty} \sum_{n=-\infty}^{\infty} c_{mn}\psi_{mn}(x),$$
find coefficients c_{mn}.

6. (Hermite polynomials) The Schrodinger equation on the real line for a simple harmonic oscillator with potential energy $\frac{1}{2}Kx^2$, $K > 0$, can be written in scaled coordinates as
$$-y'' + x^2 y = Ey, \quad x \in \mathbb{R}, \tag{2.17}$$
where E is the energy level. Here,
$$\int_{\mathbb{R}} |y|^2 dx < \infty.$$

a) Let $y(x) = w(x)e^{-x^2/2}$ and show that w satisfies
$$w'' - 2xw' + (E-1)w = 0,$$
which is Hermite's differential equation.

b) Assume a power series solution of the form $w(x) = \sum_0^{\infty} a_k x^k$ and derive the recursion relation
$$(a+2)(k+1)a_{k+2} = (2k+1-E)a_k, \quad k = 0,1,2,\dots.$$

c) If $E = 2n+1$, where n is a nonnegative integer, show that solutions are nth degree polynomials $w(x) = H_n(x)$. (Hermite polynomials) Show that, appropriately normalized, the first few Hermite polynomials are
$$H_0(x) = 1,\ H_1(x) = 2x,\ H_2(x) = 4x^2 - 2,\ H_3(x) = 8x^3 - 12x,\dots.$$

d) Plot the resulting solutions $y_0(x), \dots, y_3(x)$. (For other values of energy E the power series for w is an infinite series and behaves like e^{x^2}; thus the normalization condition cannot hold and those energy levels are not eigenvalues.)

5.3 Classical Fourier Series

We introduced the idea of representing a given function $f(x)$ in terms of an infinite series of orthogonal functions arising from a regular Sturm-Liouville problem. Now we focus on a special set of orthogonal functions given by sines and cosines. One can think of these as the orthogonal set of functions arising from a SLP with periodic boundary conditions (Example 5.15). The resulting orthogonal series is called a classical Fourier series. It is a special case of the orthogonal series discussed in the previous sections, but Fourier series have a long history of applications to signal processing, remote sensing, and other areas of modern engineering and physics. Although first investigated by Euler and Bernoulli in the mid-1700s, it was refined by Joseph Fourier in the early 1800s to solve partial differential equations in heat transfer. Euler, Bernoulli, and Fourier all had a vague notion of convergence (and even the definition of a function!), and few of their contemporaries could hardly imagine representing a discontinuous function by sums of continuous ones. Questions about convergence of Fourier series spawned much of the development and underpinnings of classical analysis and modern analysis in the 1800s and 1900s. For that reason Fourier analysis is one of the monuments of mathematics, and the material in this section could be expanded to volumes; and it has.

Here we do not focus on solving BVPs and SLPs. Rather, we are motivated by the approximation of time-periodic signals for all times t. The question is how to resolve such a signal, e.g., think of a musical signal, into its fundamental frequencies and overtones. As a result, we abandon our use of x as an independent variable and we use t.

We work on an arbitrary symmetric interval $(-l, l)$ about the origin. If f is an integrable function on $(-l, l)$, then its **Fourier series** is the trigonometric series

$$f(t) \sim \frac{a_0}{2} + \sum_{n=1}^{\infty} (a_n \cos \frac{n\pi t}{l} + b_n \sin \frac{n\pi t}{l}), \tag{3.1}$$

where the **Fourier coefficients** a_n and b_n are given by the formulas

$$a_n = \frac{1}{l} \int_{-l}^{l} f(t) \cos \frac{n\pi t}{l} dt, \ n = 0, 1, 2, \ldots$$

$$b_n = \frac{1}{l} \int_{-l}^{l} f(t) \sin \frac{n\pi t}{l} dt, \ n = 1, 2, \ldots.$$

Here, the $\sim$ sign only means that the right side is the Fourier series for f and means nothing regarding convergence of the series to f. The set of functions

$$1, \ \cos \frac{n\pi t}{l}, \ \sin \frac{n\pi t}{l}, \quad n = 1, 2, \ldots, \tag{3.2}$$

is orthogonal on the interval $(-l, l)$. It is shown in more advanced texts that the set of functions (3.2) is complete, and therefore the Fourier series (3.1) converges in the mean-square sense to f when $f \in L^2(-l, l)$. As an aside, it does not matter in our discussion if we include the endpoints of the interval $(-l, l)$; the values of the integrals defining the coefficients give the same value whether or not an endpoint is included. However, we are interested in the convergence of the series at the points.

The next examples give interesting observations regarding the behavior of Fourier series.

Example 5.30

Let $f(x) = t$ on $-l < t < l$. The Fourier coefficients are easily computed to be

$$a_n = \frac{1}{l} \int_{-l}^{l} t \cos \frac{n\pi t}{l} dx = 0, \quad n \geq 0,$$

$$b_n = \frac{1}{l} \int_{-l}^{l} t \sin \frac{n\pi t}{l} dx = \frac{2}{l} \int_{0}^{l} t \sin \frac{n\pi t}{l} dx = \frac{2l}{n\pi}(-1)^{n+1}, \quad n \geq 1.$$

Therefore the Fourier series for f is

$$t \sim \frac{2l}{\pi} \left(\sin \frac{\pi t}{l} - \frac{1}{2} \sin \frac{2\pi t}{l} + \frac{1}{3} \sin \frac{3\pi t}{l} - \cdots \right).$$

We note that series is $2l$ periodic and is the *same* as the Fourier sine series of $f(t) = t$ on the interval $(0, l)$. Therefore, the Fourier sine series on $(0, l)$ converges to the odd extension of $f(t) = t$ to the entire interval $(-l, l)$. At $t = \pm l$ the series converges to zero and thus does not converge to $f(t)$ at these two points. Next, the derived series, obtained by differentiating term-by-term, does not converge at all, much less to the derivative $f'(t) = 1$. So the series cannot be differentiated term-by-term. However, the integrated series does converge. □

Example 5.31

Because the Fourier series for $f(t)$ is $2l$ periodic, it is valid on the entire real line $\mathbb{R}$. Whatever behavior we observe on $(-l, l)$ will repeat itself on every interval $(l, 3l), (3l, 5l), \ldots, (-3l, -l), (-5l, -3l), \ldots$. Therefore, the Fourier series is the Fourier series of the extension of $f(t)$ on $(-l, l)$ to a $2l$ periodic function[3] on all of $\mathbb{R}$. Of course, because of our use of an open interval $(-l, l)$, the periodic extension of f remains undefined at odd multiples of l; however, the Fourier

---

[3] A function f is P-periodic, or periodic of period P, if it repeats itself in every interval of length P; that is, $f(x + P) = f(x)$ for all x.

series may be valid at those missing points. To illustrate, consider the odd
function $f(t) = t$, $-l < t < l$ in the last example. The $2l$ periodic extension of
f to all of $\mathbb{R}$ is shown in Fig. 5.2.   □

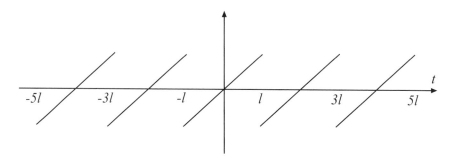

Figure 5.2 The $2l$ periodic extension of $f(t) = t$, $-l < t < l$.

It is clear that the Fourier series simplifies considerably if f is an even,
$f(-t) = f(t)$, or an odd, $f(-t) = -f(t)$, function; for example, the function
$\sin \frac{n\pi x}{l}$ is odd and $\cos \frac{n\pi x}{l}$ is an even function. Moreover, an even function times
an odd function is odd. Therefore, if f is an even function, then the product
$f(x) \sin \frac{n\pi x}{l}$ is an odd function; so all of the coefficients b_n are zero because
an odd function integrated over a symmetric interval about the origin is zero.
Hence, if f is an even function, then its Fourier series reduces to a cosine series
of the form

$$\frac{a_0}{2} + \sum_{n=1}^{\infty} a_n \cos \frac{n\pi x}{l}.$$

Similarly, if f is an odd function, then the coefficients a_n vanish, and the Fourier
series reduces to a sine series

$$\sum_{n=1}^{\infty} b_n \sin \frac{n\pi x}{l}.$$

Reiterating some of the comments in the last example, a Fourier series (3.1),
although computed only on the interval $[-l, l]$, is a $2l$-periodic function because
the sines and cosines are $2l$-periodic. Thus, the series repeats its values in every
interval of length $2l$. Therefore, if f is defined on all of $\mathbb{R}$ and is $2l$-periodic, we
can represent it by its Fourier series on all of $\mathbb{R}$. Relevant to signal processing,
if f is some given signal or data set $2l$-periodic in time (say a signal from an
electrical circuit or an electrocardiogram), then we could digitize the signal
and save it by storing the Fourier coefficients, or some finite subset of them.

Knowledge of the coefficients allows us reproduce the signal via (3.1) and the Fourier coefficients. Of course, the music industry has made great advances in applying these ideas.

In the Fourier series (3.1) the nth **harmonic** is

$$a_n \cos \frac{n\pi t}{l} + b_n \sin \frac{n\pi t}{l}.$$

Each harmonic in the series has a higher frequency, and thus more oscillations, than the preceding harmonic. The **energy spectrum** of the series is the sequence of numbers ω_n^2 defined by

$$\omega_0 = \frac{|a_0|}{\sqrt{2}}, \quad \omega_n = \sqrt{a_n^2 + b_n^2} \quad (n \geq 1).$$

These energies are the squares of the amplitudes of the various harmonics. One can show that Parseval's inequality takes the form

$$\frac{a_0^2}{2} + \sum_{n=1}^{\infty} (a_n^2 + b_n^2) = \frac{1}{l}||f||^2.$$

It is often simpler in calculations to work with the complex form of Fourier series. Using the Euler relation $e^{i\theta} = \cos\theta + i\sin\theta$, we can show that (3.1) can be written equivalently as

$$f(t) \sim \sum_{n=-\infty}^{\infty} c_n e^{in\pi t/l}, \tag{3.3}$$

for coefficients c_n, where f is defined on $(-l, l)$. This is the complex form of the Fourier series for f. Clearly, for $m \neq n$,

$$
\begin{aligned}
(f_m, f_n) &= (e^{im\pi t/l}, e^{in\pi t/l}) = \int_{-l}^{l} e^{im\pi t/l} e^{-in\pi t/l} dx \\
&= \frac{1}{(m-n)i\pi} e^{i(m-n)\pi t/l}\Big|_{-l}^{l} = 0,
\end{aligned}
$$

so the functions $e^{in\pi t/l}$ are orthogonal on (l, l). If $m = n$, then $(e^{in\pi t/l}, e^{-in\pi t/l}) = 2l$, and the norm is $\sqrt{2l}$. Hence, the set

$$\phi_n(t) = \frac{1}{\sqrt{2l}} e^{int}, \quad n = 0, \pm 1, \pm 2, \ldots$$

is an orthonormal set on $-l \leq t \leq l$. The Fourier coefficients are given by

$$c_n = (f, e^{in\pi t/l}) \frac{1}{2l} \int_{-l}^{l} f(t) e^{-in\pi t/l} dt, n = 0, \pm 1, \pm 2. \ldots \tag{3.4}$$

Recall also that the function $e^{i\theta}$ is $2\pi i$ periodic.

Example 5.32

Let $f(t) = e^t$ for $-\pi < t < \pi$. Then the Fourier coefficients are, by direct integration,

$$c_n = (e^t, e^{int}) = \frac{1}{2\pi} \int_{-\pi}^{\pi} e^t e^{-int} dt = \frac{1}{2\pi} \int_{-\pi}^{\pi} e^t e^{(1-in)t} dt = \frac{(-1)^n \sinh \pi}{(1 - in)\pi}.$$

Hence

$$e^t \sim \frac{\sinh \pi}{\pi} \sum_{-\infty}^{\infty} \frac{(-1)^n}{1 - in} e^{int}.$$

We leave it to the reader to obtain the real Fourier series of $f(t) = e^t$.   □

Example 5.33

Fourier series can be used to solve various kinds of equations. In this example we find the solution to the differential–difference equation

$$y' + 2y(t - \pi) = \sin t, \quad t \in \mathbb{R}.$$

Assume the complex form

$$y(t) = \sum c_n e^{int}.$$

Formally differentiating term-by-term and substituting the results into the left side of the given equation yields

$$y' + 2y(t - \pi) = \sum (in + 2(-1)^n) c_n e^{int},$$

where we used $e^{in\pi} = (-1)^n$. Expanding the right side of the equation, $\sin t = \frac{i}{2}(e^{-it} - e^{it})$, and then equating the terms, we get

$$c_{-1} = \frac{i/2}{-2 - i}, \quad c_1 = -\frac{i/2}{2 - i},$$

and otherwise $c_n = 0$, $n \neq -1, 1$. We can solve for c_{-1} and c_1 to get $c_{-1} = -(2i + 1)/10$, $c_1 = (2i - 1)/10$. Then the solution is

$$y(t) - c_{-1} e^{-it} + c_1 e^{it} = -\frac{1}{5}(\cos t + 2 \sin t).$$

This is easily checked for validity.   □

In Section 5.3 we introduced the ideas of pointwise, uniform, and mean-square convergence. We have noted that if f is square integrable on $(-l, l)$, then mean-square convergence is automatic. Pointwise convergence requires additional assumptions about the smoothness of f. It is remarkable, for example, that there exists a continuous function whose Fourier series diverges at infinitely many points in the interval. The case when the graph of f is made up of finitely many continuous, smooth (a continuous derivative) segments will cover most of the interesting functions encountered in science and engineering. A function f is **piecewise continuous** on $[a, b]$ if it is continuous except possibly at a finite number of points in $[a, b]$ where it has simple jump discontinuities; f has a simple jump discontinuity at $x = c$ if both one-sided limits of $f(x)$ exist and are finite at c. The function may or may not be defined at a jump discontinuity. We say f is **piecewise smooth** on $[a, b]$ if both f and f' are piecewise continuous on $[a, b]$, and we say f is piecewise smooth on $(-\infty, \infty)$ if it is piecewise smooth on each bounded subinterval $[a, b]$ of the real line. Then the basic pointwise convergence theorem can be stated as follows.

Theorem 5.34

If f is piecewise smooth on $[-l, l]$ and otherwise $2l$-periodic, then its Fourier series (3.1) converges pointwise for all $x_0 \in \mathbb{R}$ to the value $f(x_0)$ if f is continuous at x_0, and to the mean value of its left and right limits at x_0, namely, to $\frac{1}{2}(f(x_0-) + f(x_0+))$, if f is discontinuous at x_0.

Stronger convergence results, like uniform convergence, require additional smoothness conditions on f. For example, if f and f' are continuous on $[-l, l]$ with $f(-l) = f(l)$ and $f'(-l) = f'(l)$, then the Fourier series converges uniformly to f.

EXERCISES

1. Consider a triangular wave given analytically by: $f(t) = t + 1$ if $-1 < t \leq 0$; $f(t) = 1 - t$ if $0 < x \leq 1$; and otherwise 2-periodic. Compute its Fourier series, and plot f and two-term, three-term, and four-term Fourier approximations.

2. Find the Fourier series for $f(t) = t^2$ on $[-\pi, \pi]$. Sketch a graph of the function defined on $\mathbb{R}$ to which the Fourier series converges pointwise. Use the Fourier series to show

$$\frac{\pi^2}{12} = 1 - \frac{1}{4} + \frac{1}{9} - \frac{1}{16} + \cdots .$$

Graph the frequency spectrum.

3. By considering $f(x) = x^2$ on $(-\pi, \pi)$, show that

$$\frac{1}{\pi} \int_{-\pi}^{\pi} x^4 dx = \frac{1}{2} \left(\frac{2\pi^2}{3} \right)^2 + 16 \sum_{1}^{\infty} \frac{1}{n^4}.$$

4. Find the Fourier series for $f(t) = |\sin t|$ for $t \in (-\pi, \pi)$.

5. On the interval $(0, 4)$ let $f(x) = 0$, $x \in (0, 2)$; $f(x) = 1$, $x \in [2, 4)$.

   a) Sketch the odd extension of f to an 8π periodic function on $\mathbb{R}$.

   b) Sketch the even extension of f to an 8π periodic function on $\mathbb{R}$.

   c) To what value(s) will the Fourier series converge at the points $x = n\pi$, $n = 0, \pm 1, \pm 2, \ldots$, in (a) and in (b)?

   d) Find the Fourier series in (b).

6. Consider the signal defined by $f(t) = t + 1$ for $-2\pi < t \leq 0$ and $f(t) = t$ for $0 < t \leq 2\pi$, and otherwise 4π-periodic. Plot the signal, and find its Fourier series. Plot the frequency spectrum. To what value does the Fourier series converge at $t = 0$, $t = 2\pi$, and $t = \pi$? Plot the sum of the first four harmonics and compare it to f.

7. In Example 5.32 find the Fourier series in terms of sines and cosines.

8. Find the Fourier series for $f(t) = e^{-|t|}$, $|t| < \pi$, with otherwise $f(t + 2\pi) = f(t)$.

9. Show that the Fourier series for $f(x) = \cos at$ on $(-\pi, \pi)$, where a is not an integer, is

$$\cos at = \frac{2a \sin a\pi}{\pi} \left(\frac{1}{2a^2} + \sum_{n=1}^{\infty} (-1)^n \frac{\cos nt}{a^2 - n^2} \right).$$

   Show

$$\csc a\pi = \frac{1}{a\pi} + \frac{2a}{\pi} \sum_{n=1}^{\infty} (-1)^n \frac{1}{n^2 - a^2}.$$

10. What is the sum of the Fourier series for $f(t) = (t + 1) \cos t$, $-\pi < t < \pi$, at $t = 3\pi$?

11. Let $f(t) = -\frac{1}{2}$ on $-\pi < t \leq 0$ and $f(t) = \frac{1}{2}$ on $0 \leq t \leq \pi$. Show that the Fourier series for f is

$$\sum_{n=1}^{\infty} \frac{2}{(2n-1)\pi} \sin(2n-1)t.$$

If $s_N(t)$ denotes the sum of the first N terms, sketch graphs of $s_1(t)$, $s_3(t)$, $s_7(t)$, and $s_{10}(t)$ and compare with $f(t)$. Observe that the approximations overshoot $f(t)$ in a neighborhood of $t = 0$, and the overshoot is not improved regardless of how many terms are taken in the approximation. This overshoot behavior of Fourier series near a discontinuity is called the **Gibbs' phenomenon**.

12. Use Fourier series to solve to find a solution $y = y(t)$ of period 2 of the differential difference equation

$$y'(t) + y(t - 1) = \cos^2 \pi t.$$

13. Find a 4π periodic solution on $\mathbb{R}$ of the differential equation

$$y'' + y' + y = F(t),$$

where $F(t) = 1$, $|t| < \pi$, and $F(t) = 0$, $\pi < |t| < 2\pi$.

5.4 Integral Equations

An integral equation is an equation where the unknown function occurs under an integral sign. Integral equations arise in the analysis of differential equations, and, in fact, many initial and boundary value problems for differential equations may be reformulated as integral equations, and vice-versa. But integral equations naturally arise in modeling as well. The two types of linear equations we discuss are the **Fredholm equation**

$$\int_a^b k(x, y)u(y)\, dy - \lambda u(x) = f(x), \quad a \le x \le b, \tag{4.1}$$

and the **Volterra equation**

$$\int_a^x k(x, y)u(y)\, dy - \lambda u(x) = f(x), \quad a \le x \le b. \tag{4.2}$$

Here u is the unknown function, f is a given continuous function, and λ is a parameter. The given function k is called the **kernel**, and we assume that k is a continuous function on the square $a \le x \le b$, $a \le y \le b$. The interval $[a, b]$ is assumed to be finite. Because Volterra equations have an intimate relationship with initial value problems, we often replace the variables x and y by t and s. The difference between the Fredholm and Volterra equations is the variable upper limit of integration in the Volterra equation. By a solution of (4.1) or (4.2) we mean a continuous function $u = u(x)$ defined on the interval $[a, b]$ that, when substituted into the equation, reduces the equation to an identity on the interval. If $f \equiv 0$, then the equation is called **homogeneous**; otherwise it is **nonhomogeneous**. If $\lambda = 0$, then the unknown appears only under the integral sign and the equation is said to be of the **first kind**; otherwise it is of the **second kind**. If $k(x, y) = k(y, x)$ then the kernel is said to be **symmetric**. Although (4.1) and (4.2) are similar, they require different methods of analysis. Most of the discussion deals with continuous solutions to integral equations; however, a more complete theory can be developed for square integrable functions.

We can streamline the discussion by introducing integral operator notation. For example, if we define the Fredholm operator K by

$$(Ku)(x) = \int_a^b k(x, y)u(y) \, dy, \tag{4.3}$$

then (4.1) can be written simply as

$$Ku - \lambda u = f, \tag{4.4}$$

where we have, as is common practice, suppressed the independent variable x. We think of K as operating on a function u to produce another function Ku. If the kernel k is continuous, then K maps continuous functions to continuous functions. Symbolically, $K : C[a, b] \to C[a, b]$. Ultimately, we want to determine the solution (if any), which could be written symbolically in the form

$$u = (K - \lambda I)^{-1} f.$$

If this were true, then $(K - \lambda I)^{-1}$ would be the inverse of $K - \lambda I$, and we would have succeeded in defining the inverse of an integral operator.

In the same way that we consider eigenvalue problems for matrices and for differential equations (Sturm–Liouville problems), we can consider eigenvalue problems for integral equations. A number λ is called an eigenvalue for the integral operator K if there exists a nontrivial solution u to the integral equation

$$Ku = \lambda u. \tag{4.5}$$

The set of eigenvalues form the **spectrum** of K, and the (geometric) **multiplicity** of an eigenvalue is the dimension of the function space spanned by its corresponding eigenfunctions. Much of the theory of integral equations deals with *solvability* questions for the Fredholm equation (4.4) (or the Volterra equation) and the *solution to the eigenvalue problem* (4.5).

5.4.1 Volterra Equations

Volterra integral equations arise naturally in many settings and we begin with an example.

Example 5.35

(**Inventory control problem**) A shop manager determines that a percentage $k(t)$ of goods remains unsold at time t after she purchased the goods. At what rate should she purchase goods so that the stock remains constant? Typical of inventory problems, as well as population problems, we assume that this process is continuous. Let $u(t)$ be the rate (goods per unit time) at which goods are to be purchased, and let a be the amount of goods initially purchased. In the time interval $[\tau, \tau + \Delta\tau]$ she will buy an amount $u(\tau)\Delta\tau$, and at time t the portion remaining unsold is $k(t - \tau)u(\tau)\Delta\tau$. Therefore the amount of goods in the shop unsold at time t is

$$ak(t) + \int_0^t k(t - \tau)u(\tau)\, d\tau.$$

The first term represents the amount of the initial purchase left unsold at time t. Hence, this is the total amount of goods in the shop left unsold at time t, and so the shop manager should require that

$$ak(t) + \int_0^t k(t - \tau)u(\tau)\, d\tau = a.$$

So the solution to the shop manager's problem satisfies a nonhomogeneous Volterra integral equation. For example, if $k(t) = e^{-bt}$, then we can rearrange the problem as

$$a + \int_0^t e^{b\tau}u(\tau)\, d\tau = ae^{bt}.$$

Differentiating with respect to t gives the solution $u(t) = ab$.   □

Example 5.36

(**Demography**) We set up an integral equation governing the number of newborns in a population where the mortality and fecundity rates are known. Let $u(a, t)$ be the population density of females of age a at time t. Thus, $u(a, t)\, da$ is approximately the number of females between the ages of a and $a + da$ at time t. Further, let $f(a)$ denote the fecundity rate of females at age a, measured in births per female per time. Then $f(a)u(a, t)\, da$ is the rate of births at time

t contributed by females between the ages a and $a + da$. Integrating over all ages gives the total number of births per unit time at time t, or

$$B(t) = \int_0^\infty f(a)u(a,t)\,da.$$

There is no difficulty in using ∞ as the upper limit on the integral because the fecundity will be zero after some age when females cease reproduction. Now we separate ages into the two cases $a < t$ and $a > t$. For $a < t$, females of age a at time t were born at time $t - a$. So $u(a,t)\,da = B(t-a)s(a)\,da$, where $s(a)$ is the survivorship, or the fraction that survive from birth to age a. For $a > t$, females of age a were present at time $t = 0$ having age $a - t$. If $u_0(a)$ is the age density of females at time zero, then $u(a,t)\,da = u_0(a-t)s(t)\,da$ for $a > t$. Consequently, the birthrate must satisfy the Volterra equation

$$B(t) = \int_0^t f(a)s(a)B(t-a)\,da + \int_t^\infty u_0(a-t)f(a)s(t)\,da.$$

The second integral is a known function of time, say $\phi(t)$; thus the birthrate B satisfies the Volterra equation

$$B(t) = \int_0^t f(a)s(a)B(t-a)\,da + \phi(t),$$

which is called the **renewal equation**. The renewal equation occurs often in applied mathematics, especially probability theory.   □

Some integral equations, like the one obtained in the preceding examples, have kernels of a special type that permit a direct solution by Laplace transform methods. The Volterra integral equation

$$u(t) = f(t) + \int_0^t k(t-s)u(s)\,ds \tag{4.6}$$

is said to be of **convolution type** because the integral is the convolution $k * u$ of k and u. We recall that the Laplace transform of a convolution is the product of the Laplace transforms, that is, $\mathcal{L}(k * u) = (\mathcal{L}k)(\mathcal{L}u)$. Therefore, taking the Laplace transform of both sides of (4.6) gives

$$\mathcal{L}u = \mathcal{L}f + (\mathcal{L}k)(\mathcal{L}u).$$

Solving for $\mathcal{L}u$ gives

$$\mathcal{L}u = \frac{\mathcal{L}f}{1 - \mathcal{L}k}.$$

Thus, the solution u is the inverse transform of the right side.

Example 5.37

Consider the integral equation

$$u(t) = t - \int_0^t (t - s)u(s)\, ds.$$

Here $f(t) = t$ and $k(t) = t$, both with transform $1/s^2$. Then, taking the Laplace transform of the equation gives

$$\mathcal{L}u = \frac{1}{s^2} - \frac{1}{s^2}\mathcal{L}u,$$

or

$$\mathcal{L}u = \frac{1}{1+s^2}.$$

Therefore

$$u(t) = \mathcal{L}^{-1}\left(\frac{1}{1+s^2}\right) = \sin t. \quad \square$$

Initial value problems for ordinary differential equations can be reformulated as a Volterra integral equation. Consider

$$u' = f(t, u), \qquad u(t_0) = u_0, \tag{4.7}$$

and assume that f is a continuous function. If $u = u(t)$ is a solution of (4.7), then for all t we have

$$u'(t) = f(t, u(t)).$$

Replacing t by s and integrating from t_0 to t gives

$$\int_{t_0}^t u'(s)\, ds = \int_{t_0}^t f(s, u(s))\, ds,$$

or

$$u(t) = u_0 + \int_{t_0}^t f(s, u(s))\, ds, \tag{4.8}$$

where in the last step the fundamental theorem of calculus was applied. Equation (4.8) is a Volterra integral equation for u, and it is entirely equivalent to the initial value problem (4.7). To recover (4.7) from (4.8), we can differentiate (4.8) with respect to x. The initial condition is automatically contained in the integral form (4.8). One procedure for solving a Volterra equation, which sometimes works, is to recast it into an initial value problem and solve the differential problem.

To reformulate differential equations as integral equations, we can continually integrate the equation until all the derivatives are removed. The following lemma contains a formula for a repeated integral that is useful in carrying out this procedure for second-order equations.

Lemma 5.38

Let f be a continuous function for $x \geq a$. Then

$$\int_a^x \int_a^s f(y)\, dy\, ds = \int_a^x f(y)(x-y)\, dy \quad \square$$

Proof

Let $F(s) = \int_a^s f(y)\, dy$. Then, using integration by parts,

$$\int_a^x \int_a^s f(y)\, dy\, ds = \int_a^x F(s)\, ds$$

$$= \int_a^x \frac{d(s)}{ds} F(s)\, ds$$

$$= xF(x) - aF(a) - \int_a^x sF'(s)\, ds$$

$$= x\int_a^x f(y)\, dy - \int_a^x yf(y)\, dy,$$

which is the result. $\square$

The next example shows how a second-order initial value problem can be transformed into an integral equation.

Example 5.39

Consider the second-order initial value problem:

$$u'' + p(t)u' + q(t)u = f(t), \quad t > a; \quad u(a) = u_0,\ u'(a) = u_1. \tag{4.9}$$

Solving for u'' and integrating from a to t gives

$$u'(t) - u_1 = -\int_a^t p(s)u'(s)\, ds - \int_a^t (q(s)u(s) - f(s))\, ds.$$

If the first integral is integrated by parts the result is

$$u'(t) - u_1 = -p(t)u(t) - \int_a^t [(q(s) - p'(s))u(s) - f(s)]\, ds + p(a)u_0.$$

Integrating again gives

$$u(t) - u_0 = (p(a)u_0 + u_1)(t-a) - \int_a^t p(s)u(s)\, ds$$
$$- \int_a^t \int_a^r [(q(s) - p'(s))u(s) - f(s)]\, ds\, dr.$$

Lemma 5.38 finally implies

$$
\begin{aligned}
u(t) \;=\; & (p(a)u_0 + u_1)(t - a) + u_0 - \int_a^t \{p(s) + (t - s)[q(s) - p'(s)]\} u(s)\, ds \\
& + \int_a^t (t - s) f(s)\, ds,
\end{aligned}
$$

which is a Volterra equation is of the form

$$
u(t) = \int_a^t k(t, s) u(s)\, ds + F(t). \quad \square
$$

Remark 5.40

(**Leibniz rule**) The reverse process, transforming an integral equation into a differential equation, requires the Leibniz rule for differentiating an integral with respect to a variable appearing in the upper limit of the integral and in the integrand:

$$
\frac{d}{dx} \int_{a(x)}^{b(x)} F(x, y)\, dy = \int_{a(x)}^{b(x)} F_x(x, y)\, dy + F(x, b(x)) b'(x) - F(x, a(x)) a'(x)
$$

$$(4.10)$$

(see Exercise 1). $\quad \square$

Iteration. It is common practice in applied science to apply iterative procedures to determine approximate solutions to various kinds of problems for which analytic solutions cannot be found. The integral equation (4.8), and therefore the initial value problem (4.7), can be solved approximately by fixed-point iteration. Beginning with an initial approximation $\phi_0(t)$ we generate a sequence of successive approximations $\phi_1(t)$, $\phi_2(t), \ldots,$ via

$$
\phi_{n+1}(t) = u_0 + \int_a^t f(s, \phi_n(s))\, ds, \quad n = 0, 1, 2, \ldots
$$

This method is called **Picard's method**, and it can be shown under certain assumptions that the sequence $\phi_n(t)$ converges to the unique solution of (4.8). For the linear Volterra equation of the form

$$
u(t) = f(t) + \lambda \int_a^t k(t, s) u(s)\, ds,
$$

a similar type of iteration can be defined. Let us write this equation in operator form as

$$
u = f + \lambda K u, \tag{4.11}
$$

where K is the Volterra integral operator

$$
K u(t) = \int_a^t k(t, s) u(s)\, ds. \tag{4.12}
$$

By making a heuristic calculation we can guess what the solution representation is. First we write the integral equation as

$$(I - \lambda K)u = f.$$

Then, formally inverting the operator,

$$
\begin{aligned}
u &= (I - \lambda K)^{-1} f = \frac{1}{I - \lambda K} f \\
&= (I + \lambda K + \lambda^2 K^2 + \cdots + \lambda^j K^j + \cdots) f.
\end{aligned}
$$

This result is the content of the next theorem. Therefore, letting $u_0 = u_0(t)$ be given, we may define a sequence of successive approximations $u_n(t)$ by

$$u_{n+1} = f + \lambda K u_n, \quad n = 0, 1, 2, \ldots \tag{4.13}$$

We have the following result on convergence of this process.

Theorem 5.41

If u_0, f, k are continuous, then the sequence $u_n(t)$ defined by (4.13) converges uniformly to the unique solution of (4.11) on $[a, b]$ given by

$$u(t) = f(t) + \sum_{j=1}^{\infty} \lambda^j K^j f(t), \tag{4.14}$$

where K^j denotes the composition operator $K(K(\ldots (K) \ldots))$, j times.   □

Remark 5.42

The representation (4.14) of the solution to (4.11) is called the **Neuman series** for u, or the **Born series** in physics. It represents the inverse operator $(I - \lambda K)^{-1}$.   □

Proof

To prove the theorem, note that after repeated iteration we can write

$$u_{n+1} = f + \sum_{j=1}^{n} \lambda^j K^j f + \lambda^{n+1} K^{n+1} u_0. \tag{4.15}$$

To examine the convergence of (4.15) we must look at the remainder term $\lambda^{n+1} K^{n+1} u_0$ as $n \to \infty$. Denoting $M = \max |k|$ and $C = \max |u_0|$, we have

$$|K u_0(t)| \leq \int_a^t |k(t, s) u_0(s)| \, ds \leq (t - a) M C.$$

Next
$$|K^2 u_0(t)| \leq \int_a^t |k(t,s)Ku_0(s)|\,dy \leq \int_a^t M(s-a)MC\,ds = C\frac{(t-a)^2 M^2}{2}.$$

Continuing with an induction argument we obtain
$$|K^{n+1}u_0(t)| \leq C\frac{(t-a)^{n+1}M^{n+1}}{(n+1)!} \leq C\frac{(b-a)^{n+1}M^{n+1}}{(n+1)!}.$$

This bound is independent of t and it approaches zero as $n \to \infty$. Consequently $|\lambda^{n+1}K^{n+1}u_0| \to 0$ uniformly on $[a,b]$, and the unique solution of (4.11) is given by (4.14). $\qquad\square$

Examples are left to the exercises. We point out, however, one important consequence of the Theorem 5.41, namely that the equation $Ku = \mu u$ (set $f=0$ and $\mu = 1/\lambda$) has only the trivial solution. Therefore, *there can be no eigenvalues for a Volterra operator* (4.12).

Corollary 5.43

The Volterra operator (4.12) has no eigenvalues. $\quad\square$

This result is consistent with our comment that Volterra equations come from initial value problems, and linear initial value problems for differential equations have no eigenvalues. The conclusion is in sharp contrast to Fredholm equations, which arise from boundary value problems. We already know that boundary value problems (e.g., Sturm–Liouville problems) have a rich spectral structure, and we expect the same from Fredholm operators.

5.4.2 Fredholm Equations with Degenerate Kernels

Generally, integral equations are problems in infinite-dimensional spaces. However, a special class of Fredholm equations reduces, or degenerates, to a finite-dimensional problem. To this class belongs Fredholm integral equations whose kernel $k(x,y)$ is a finite sum of products of functions of x and y. These are essentially finite-dimensional problems, or problems leading to a linear system of algebraic equations.

Linear Algebra. To understand the mathematical issues for integral equations, we retreat to a familiar problem in linear algebra, the algebraic eigenvalue problem: Given an $n \times n$ matrix A, determine numbers λ for which the matrix equation
$$A\mathbf{u} = \lambda\mathbf{u} \tag{4.16}$$

has a nontrivial solution $\mathbf{u} \in \mathbb{R}^n$. The values of λ for which a solution $\mathbf{u}$ exists are the **eigenvalues**, and the corresponding solutions $\mathbf{u}$ are the associated **eigenvectors**. Eigenvalues may be real or complex numbers.

There are many reasons the algebraic eigenvalue problem is important, but we focus our thinking on a specific notion, namely using the eigenstructure of A to address the nonhomogeneous solvability problem. We know from matrix theory that there are n eigenvalues, not necessarily distinct. Suppose, for our discussion, that the matrix A is a real, symmetric matrix $(A = A^{\mathrm{T}})$. Then A has n real eigenvalues $\lambda_1, \ldots, \lambda_n$, and a full set of n orthonormal eigenvectors $\mathbf{e}_i$, $i = 1, 2, \ldots, n$. Therefore, the eigenvectors from an orthonormal basis for the space $\mathbb{R}^n$ and each vector $\mathbf{u}$ in $\mathbb{R}^n$ can be written uniquely as a linear combination of the $\mathbf{e}_i$, or

$$\mathbf{u} = \sum_{i=1}^{n} c_i \mathbf{e}_i. \tag{4.17}$$

The c_i are the coordinates of $\mathbf{u}$ in this basis and are given by $c_i = \mathbf{u} \cdot \mathbf{e}_i = (\mathbf{u}, \mathbf{e}_i)$, which are the projections of $\mathbf{u}$ onto the ith eigenvector.

The point we make is that knowledge of the eigenvalues and eigenvectors can be used to solve other algebraic problems. Take, for example, the nonhomogeneous linear system

$$A\mathbf{u} = \mu\mathbf{u} + \mathbf{f}, \tag{4.18}$$

where $\mathbf{f}$ is a given vector, and μ is a given constant. If a solution $\mathbf{u}$ exists, it must have the form (4.17) for some choice of the constants c_i, which are to be determined. Further, because $\mathbf{f}$ is a given vector, it has a known expansion in terms of the eigenvectors,

$$\mathbf{f} = \sum_{i=1}^{n} f_i \mathbf{e}_i,$$

where the coefficients $f_i = (\mathbf{f}, \mathbf{e}_i)$ are known. Substituting these two expressions for $\mathbf{u}$ and $\mathbf{f}$ into (4.18) gives

$$A\left(\sum_{i=1}^{n} c_i \mathbf{e}_i\right) = \mu \sum_{i=1}^{n} c_i \mathbf{e}_i + \sum_{i=1}^{n} f_i \mathbf{e}_i.$$

Combining terms,

$$\sum_{i=1}^{n} c_i A\mathbf{e}_i = \sum_{i=1}^{n} (\mu c_i + f_i)\mathbf{e}_i.$$

But $A\mathbf{e}_i = \lambda_i \mathbf{e}_i$, because λ_i and $\mathbf{e}_i$ form an eigen-pair. Consequently,

$$\sum_{i=1}^{n} c_i \lambda_i \mathbf{e}_i = \sum_{i=1}^{n} (\mu c_i + f_i)\mathbf{e}_i.$$

Equating coefficients of the independent vectors $\mathbf{e}_i$ then gives

$$c_i\lambda_i = \mu c_i + f_i, \quad i = 1, \ldots, n.$$

Solving for the unknown coefficients c_i gives

$$c_i = \frac{f_i}{\lambda_i - \mu}, \quad i = 1, \ldots, n.$$

Therefore, if μ is *not* an eigenvalue, then there is a unique solution to (4.18) given by

$$\mathbf{u} = \sum_{i=1}^{n} \frac{f_i}{\lambda_i - \mu}\mathbf{e}_i, \quad \mu \neq \lambda_i.$$

The right side represents $(A - \mu I)^{-1}\mathbf{f}$, and so we have inverted the finite-dimensional operator $A - \mu I$.

If μ is an eigenvalue, say $\mu = \lambda_J$, then

$$c_i = \frac{f_i}{\lambda_i - \mu}, \quad i \neq J,$$

and $c_J\lambda_J = \mu c_J + f_J$. If $f_J = (\mathbf{f}, \mathbf{e}_J) \neq 0$, that is, $\mathbf{f}$ is not orthogonal to the Jth eigenspace, then (4.18) cannot have a solution. On the other hand, if $f_J = 0$, then c_J can be any arbitrary constant and (4.18) has infinitely many solutions of the form

$$\mathbf{u} = c_J\mathbf{e}_J + \sum_{i=1,i\neq J}^{n} \frac{f_i}{\lambda_i - \mu}\mathbf{e}_i.$$

We have proved the following important result about the solution structure of (4.18).

Theorem 5.44

(**Fredholm Alterative theorem**) Consider the system (4.18), where A is a real symmetric matrix.

(a) If μ is not an eigenvalue of A, then (4.18) has a unique solution.

(b) If $\mu = \lambda_J$ is an eigenvalue of A, then either there is no solution, or there are infinitely many solutions. In other words, in this case there are solutions if, and only if, f is orthogonal to the the eigenspace corresponding to λ_J.

$\square$

In summary, knowledge of the eigen-structure of a matrix A makes the solution of other problems straightforward. This same idea, based on expansions in terms of eigenfunctions, remains applicable to other operators, such as integral operators and differential operators. Knowledge of the eigen-structure makes it possible to invert such operators.

Separable kernels. Let us now consider the integral equation of the form

$$Ku - \lambda u = f, \tag{4.19}$$

where λ is a parameter, f is a given continuous function, and K is the operator given by (4.3) with kernel

$$k(x, y) = \sum_{j=1}^{n} \alpha_j(x)\beta_j(y), \tag{4.20}$$

where the α_j and β_j are real-valued continuous functions. A kernel of this form is called **separable**, or **degenerate**.

Example 5.45

The Fredholm integral operator

$$Ku(x) = \int_0^1 (1 - 3xy)u(y)\, dy$$

has a separable kernel with

$$\alpha_1(x) = 1, \alpha_2(x) = -3x, \beta_1(y) = 1, \beta_2(y) = y. \quad \square$$

Without loss of generality, we can always assume that the α_j are independent, and the β_j are independent. When we write out equation (4.19), after substituting the kernel, we obtain

$$\sum_{j=1}^{n} \alpha_j(x) \int_a^b u(y)\beta_j(y)\, dy - \lambda u(x) = f(x).$$

The inner product notation is convenient for the remaining discussion. We notice that this last equation is just

$$\sum_{j=1}^{n} \alpha_j(x)c_j - \lambda u(x) = f(x), \tag{4.21}$$

where the unknown constants c_j are defined by the inner products

$$c_j = (u, \beta_j), \qquad j = 1, 2, \ldots. \tag{4.22}$$

If we multiply (4.21) by $\beta_i(x)$ and integrate (this is the same as taking the inner product), we obtain

$$\sum_{j=1}^{n} (\beta_i, \alpha_j)c_j - \lambda c_i = (f, \beta_i), \qquad i = 1, 2, \ldots.$$

These equations represent n linear, algebraic equations in the unknowns $c_1, c_2, \ldots, c_n$. The system can be written concisely in matrix format by introducing the unknown column vector $\mathbf{c} = (c_1, \ldots, c_n)^T$, the column vector $\mathbf{F} = ((f, \beta_1), \ldots, (f, \beta_n))^T$, and the coefficient matrix A with entries (β_i, α_j). Thus we have reduced the solution of the integral equation (4.19) to the solution of the linear system

$$(A - \lambda I)\mathbf{c} = \mathbf{F}. \tag{4.23}$$

Determining $\mathbf{c}$ gives the c_j, and therefore (4.21) can be used (provided $\lambda \neq 0$) to determine the solution $u(x)$ of the separable integral equation (4.19).

Notice that there are several cases to be considered, depending upon whether λ is an eigenvalue or not, and whether $\lambda = 0$. The problem $Ku - \lambda u = f$ contains both the eigenvalue problem $Ku = \lambda u$ and the solvability problem $Ku = f$.

(1) The case $\lambda \neq 0$. First consider equation (4.19) for $\lambda \neq 0$. In this case

$$u(x) = \frac{1}{\lambda} \left(-f(x) + \sum_{j=1}^{n} \alpha_j(x)c_j \right), \tag{4.24}$$

where the c_j are obtained from (4.23). If λ is not an eigenvalue of the matrix A, then (4.23) has a unique solution $\mathbf{c}$; then the solution of $Ku - \lambda u = f$ is given by (4.24). If is an eigenvalue of A, then (4.23) either has no solution or infinitely many solutions, depending on whether $\mathbf{F}$ is orthogonal to the eigenspace generated by λ.

If $f = 0$, then (4.19) reduces to the eigenvalue problem $Ku = \lambda u$ for K; so the nonzero eigenvalues of the matrix A coincide with the eigenvalues of K. There are finitely many such eigenvalues, and the corresponding eigenfunctions $u(x)$ are defined by (4.21) with $f = 0$, where the c_j are the components of the corresponding eigenvectors of A. We can summarize our conclusions in the following alternative theorem.

Theorem 5.46

Consider the Fredholm integral equation (4.19), where the kernel is given by (4.20) and $\lambda \neq 0$. Let A be the matrix $A \equiv ((\beta_i, \alpha_j))$. If λ is not an eigenvalue of A, then (4.19) has a unique solution given by (4.24); if λ is an eigenvalue of A, then (4.19) either has no solution or infinitely many solutions. In particular, the eigenvalue problem $Ku = \lambda u$ has finitely many eigenvalues and eigenfunctions. □

(2) The case $\lambda = 0$. Next consider the case $\lambda = 0$. First, if $f = 0$, then (4.21) becomes

$$\sum_{j=1}^{n} c_j \alpha_j(x) = 0,$$

which implies $c_j = (u, \beta_j) = 0$ for all j, or u is orthogonal to all the β_j. There are infinitely many independent functions u that satisfy this condition. Suffice it to say that the n functions β_i span a finite-dimensional space, whereas u lies generally in an infinite dimensional space, the square integrable functions. The space of functions orthogonal to the span of $\beta_1, \ldots, \beta_n$ is infinite dimensional. Thus, there are infinitely many nontrivial solutions to the equation $Ku = 0$. This means, of course, that zero is an eigenvalue of K with *infinite multiplicity*; that is, the dimension of the eigenspace corresponding to $\lambda = 0$ is infinite.

If $\lambda = 0$ and $f \neq 0$, then (4.21) becomes

$$\sum_{j=1}^{n} c_j \alpha_j(x) = f(x).$$

Now it is clear that no solution can exist if f is not a linear combination of the α_j. If f is a linear combination of the α_j, say $f(x) = \sum_{j=1}^{n} f_j \alpha_j(x)$, then we must have $c_j = (u, \beta_j) = f_j$ for all j. Again, there are infinitely many solutions. We summarize the discussion in the following alternative theorem.

Theorem 5.47

Consider the equation $Ku = f$ with separable kernel (4.20). If f is not a linear combination of the α_j then there is no solution; if f is a linear combination of the α_j, then there are infinitely many solutions. In particular, there are infinitely many eigenfunctions corresponding to a zero eigenvalue. □

In summary, the preceding discussion implies the following theorem regarding the spectrum of a Fredholm operator with a separable kernel.

Theorem 5.48

The integral operator K with separable kernel (4.20) has finitely many nonzero eigenvalues of finite multiplicity, and zero is an eigenvalue of infinite multiplicity. □

Example 5.49

Find the nonzero eigenvalues of the operator

$$Ku(x) = \int_0^1 (1 - 3xy)u(y)\,dy.$$

Here the kernel is degenerate with

$$\alpha_1(x) = 1, \alpha_2(x) = -3x, \beta_1(y) = 1, \beta_2(y) = y.$$

The matrix A is given by

$$A = \begin{pmatrix} 1 & -\frac{3}{2} \\ \frac{1}{2} & -1 \end{pmatrix},$$

and $\det(A - \lambda I) = \lambda^2 - \frac{1}{4} = 0$ gives the eigenvalues $\lambda = \pm\frac{1}{2}$. An eigenvector corresponding to $\lambda = \frac{1}{2}$ is $\mathbf{c} = (3, 1)^{\mathrm{T}}$ and an eigenvector corresponding to $\lambda = -\frac{1}{2}$ is $(1, 1)^{\mathrm{T}}$. By formula (4.21), with $f = 0$, the eigenfunctions are

$$u(x) = 2(3\alpha_1(x) + \alpha_2(x)) = 6(1 - x), \quad u(x) = -2(\alpha_1(x) + \alpha_2(x)) = -2(1 - 3x).$$

But eigenfunctions are unique only up to a constant multiple, and therefore the eigenfunctions corresponding to $\lambda = \frac{1}{2}$ and $\lambda = -\frac{1}{2}$ are $u(x) = a(1 - x)$ and $u(x) = b(1 - 3x)$, respectively, where a and b are arbitrary constants. Next consider the nonhomogeneous equation

$$\int_0^1 (1 - 3xy)u(y)\,dy - \lambda u(x) = f(x). \qquad (4.25)$$

From the alternative theorem, if $\lambda \neq \pm\frac{1}{2}$, then (4.25) has a unique solution. □

5.4.3 Symmetric Kernels

Many problems in the physical sciences lead naturally to Fredholm equations with continuous, symmetric kernels. First we consider the eigenvalue problem

$$Ku = \lambda u, \qquad (4.26)$$

where K is the integral operator given by (4.3) and the kernel is real, symmetric, and continuous on the square $a \leq x, y \leq b$. That is, $k(x, y) = k(y, x)$. We need the following lemma.

Lemma 5.50

If k is a real, continuous, symmetric kernel, then the integral operator K defined by (4.3) has the symmetry property

$$(Ku, v) = (u, Kv)$$

for all continuous functions u, v. □

Proof

The proof is a straightforward calculation using the symmetry of k definition of and interchanging order of integration. We have

$$(Ku, v) = \int_a^b \left(\int_a^b k(x, y)u(y)\,dy \right) v(x)\,dx$$

$$= \int_a^b u(x) \left(\int_a^b k(x, y)v(y)\,dy \right) dx = (u, Kv).$$

□

Integral operators that have the properties assumed in the lemma are called **symmetric operators**. Such operators, as we have noted in the case of symmetric matrices and Sturm–Liouville operators, have an extremely nice spectral structure. First we prove the following lemma.

Lemma 5.51

Let K be the integral operator defined by (4.3) with a real, symmetric, continuous kernel. Then the eigenvalues, if they exist, are real, and eigenfunctions corresponding to distinct eigenvalues are orthogonal. □

Proof

First observe that (Ku, u) is real for all u. This follows from

$$(Ku, u) = (u, Ku) - \overline{(Ku, u)}.$$

If a number equals its complex conjugate, then it must be real. Now let λ, u be an eigenpair; that is, $Ku = \lambda u$. Then $(Ku, u) = (\lambda u, u) = \lambda(u, u)$. Since (Ku, u) and (u, u) are real, λ must be real. Finally, let $Ku = \lambda u$ and $Kv = \nu v$, where $\lambda \neq \nu$. Then $(u, v) = 0$ follows from

$$\lambda(u, v) = (Ku, v) = (u, Kv) = (u, \nu v) = \nu(u, v),$$

which completes the proof. □

Just as for Sturm–Liouville problems, the question of existence of eigenvalues is more difficult and beyond the scope of this text; we refer the reader to other sources for a proof (e.g., see Stakgold, 1998).

We already know the spectral structure of Fredholm operators with degenerate kernels; there are finitely many nonzero eigenvalues of finite multiplicity, and zero is an eigenvalue of infinite multiplicity. Now we formulate a theorem for nonseparable kernels; the result is the **Hilbert–Schmidt theorem.**[4]

Theorem 5.52

Consider the integral operator K defined by

$$(Ku)(x) = \int_a^b k(x,y)u(y)dy$$

where the kernel k is real, continuous, symmetric, and not degenerate. Then K has infinitely many eigenvalues $\lambda_1, \lambda_2, \ldots$, each with finite multiplicity, and they can be ordered as

$$0 < \cdots \leq |\lambda_n| \leq \cdots \leq |\lambda_2| \leq |\lambda_1|,$$

with $\lim \lambda_n = 0$. Moreover, any square integrable function f can be expanded in terms of the set of orthonormal eigenfunctions $\phi_k(x)$ associated with the eigenvalues as

$$f(x) = \sum_{k=1}^{\infty} f_k \phi_k(x),$$

where the series converges to f in $L^2[a,b]$, and the coefficients are given by

$$f_k = (f, \phi_k), \qquad k = 1, 2, \ldots. \quad \square \tag{4.27}$$

Having eigenfunctions opens the possibility of solving integral equations using eigenfunction expansions. Consider the integral equation

$$Ku - \mu u = f, \tag{4.28}$$

---

[4] David Hilbert (1862–1943) was one of the great mathematicians in the early 20th century. His formalist approach to problems became a key idea in studying abstract concepts in functional analysis. Hilbert spaces are now core structures in mathematics and science, especially quantum theory.

where the kernel of K is real, continuous, and symmetric, and f is a given continuous function. First we expand f and u in terms of the eigenfunctions of K as

$$f(x) = \sum_{k=1}^{\infty} f_k \phi_k(x), \qquad u(x) = \sum_{k=1}^{\infty} u_k \phi_k(x),$$

where the f_k are known and given by (4.27), and where the u_k are unknown and to be determined. Substituting these expansions into (4.28) gives

$$K\left(\sum_{k=1}^{\infty} u_k \phi_k\right) - \mu \sum_{k=1}^{\infty} u_k \phi_k = \sum_{k=1}^{\infty} f_k \phi_k.$$

But

$$K\left(\sum_{k=1}^{\infty} u_k \phi_k\right) = \sum_{k=1}^{\infty} u_k K \phi_k = \sum_{k=1}^{\infty} u_k \lambda_k \phi_k.$$

Therefore, equating the coefficients of ϕ_k yields

$$(\lambda_k - \mu) u_k = f_k, \quad k = 1, 2, 3, \ldots. \tag{4.29}$$

Therefore, if μ is not an eigenvalue, we have the unique solution representation of (4.28) given by

$$u(x) = \sum_{k=1}^{\infty} \frac{f_k}{\lambda_k - \mu} \phi_k(x).$$

From the definition of the inner product, we can write this formula as

$$u(x) = \int_a^b \left(\sum_{k=1}^{\infty} \frac{\phi_k(x)\phi_k(y)}{\lambda_k - \mu}\right) f(y)\, dy.$$

The right side is a representation of $(K - \mu I)^{-1}$, the inverse operator of $K - \mu I$, acting upon f. (In this discussion we have ignored questions of convergence, but the calculations can be made rigorous.)

If $\mu = \lambda_m$ for some fixed m, then in order to have a solution we must have $f_m = (f, \phi_m) = 0$, or f orthogonal to ϕ_m. In this case, then, u_m is arbitrary and we obtain infinitely many solutions of (4.28) given by

$$u(x) = c\phi_m(x) + \sum_{k=1, k \neq m}^{\infty} \frac{f_k}{\lambda_k - \mu} \phi_k(x),$$

where c is an arbitrary constant.

Numerical solution. When there is a unique solution, a Fredholm integral equation of the form (4.28) can be resolved numerically by discretizing the domain and approximating the integral by a quadrature formula (e.g., the trapezoid rule). To this end, let $x_i = a + (i-1)h$ for $i = 1, 2, \ldots, N$, where

$h = (b-a)/(N-1)$ is the step size, and there are $N-1$ subintervals; note that $a = x_1$ and $b = x_N$. Then, evaluating the integral equation at the x_i gives

$$\int_a^b k(x_i, y)u(y)\,dy - \mu u(x_i) = f(x_i), \quad i = 1, 2, \ldots, N.$$

Now recall the trapezoid rule:

$$\int_a^b g(y)\,dy \approx \frac{b-a}{2(N-1)}\left(g(a) + 2\sum_{j=2}^{N-1} g(x_j) + g(b) \right).$$

Replacing the integral in the integral equation by its trapezoidal approximation, and introducing the notation u_i for the approximation to $u(x_i)$, we obtain

$$\frac{b-a}{2(N-1)}\left(k(x_i, a)u_1 + 2\sum_{j=2}^{N-1} k(x_i, x_j)u_j + k(x_i, b)u_N \right) - \mu u_i = f(x_i),$$

$$(4.30)$$

for $i = 1, 2, \ldots, N$. This is a system of N linear equations for the N unknowns $u_1, u_2, \ldots, u_N$, and it can be solved easily by matrix methods using a computer algebra system.

EXERCISES

1. Prove Leibniz rule (4.10) Hint: Call the integral $I(x, a, b)$, where a and b depend on x, and differentiate, using the chain rule.

2. Let

$$A = \begin{pmatrix} 1 & 2 & 0 \\ 2 & 4 & 0 \\ 0 & 0 & 5 \end{pmatrix}.$$

   a) Find the eigenvalues and an orthonormal set of eigenvectors, and then determine the solvability of the following systems.

   b) $A\mathbf{u} = 5\mathbf{u} + (1, -\frac{1}{2}, 0)^T$.

   c) $A\mathbf{u} = \mathbf{u} + (\frac{1}{2}, -1, 0)^T$.

   d) $A\mathbf{u} = 5\mathbf{u} + (5, 2, 0)^T$.

3. Let

$$A = \begin{pmatrix} 2 & 2 \\ 1 & 3 \end{pmatrix},$$

   and consider the system $(A - 4I)\mathbf{u} = \mathbf{b}$. Find conditions on the vector $\mathbf{b}$ for which this system has a solution.

4. Discuss the solvability of the following integral equations, and solve if possible.

a) $u(x) = f(x) + \lambda \int_0^{1/2} u(y)\, dy$.

b) $u(x) = b + \lambda \int_0^1 u(y)^2\, dy$.

c) $u(x) = f(x) + \int_0^1 xyu(y)\, dy$.

5. Find the eigenvalues and orthonormal eigenfunctions associated with the following integral operators:

a) $Ku = \int_{-\pi}^{\pi} k(x+y)u(y)\, dy$ where $k(x) = \frac{k_0}{2} + \sum_1^\infty k_n \cos nx$, where k_n is a strictly decreasing sequence of positive numbers. Note that the set $\{1, \cos nx, \sin nx\}$ forms an orthogonal set on the interval.

b) $Ku = \int_0^1 \min(x,y)u(y)\, dy$.

c) $Ku = \int_0^\pi k(x,y)u(y)\, dy$ where $k(x,y) = y(\pi-x)$ if $x > y$, and $k(x,y) = x(\pi - y)$ if $x < y$.

d) $Ku = \int_0^x \sin x \cos 2y\, u(y)\, dy$.

6. Find the expansion of the function

$$f(x) = \begin{cases} x, & 0 \le x < \pi/2, \\ \pi - x & \pi/2 \le x \le \pi, \end{cases}$$

in terms of the eigenfunctions of the operator K in Exercise 5(c).

7. Formulate the integral equation

$$u(t) = 1 + \int_0^t s \ln\left(\frac{s}{t}\right) u(s)\, ds$$

as an initial value problem.

8. Let K be the operator in Exercise 4(c). Solve the following problems, or state why there is no solution.

a) $Ku - 2u = 0$.

b) $Ku - \frac{\pi}{9}u = x(\pi - x)$.

c) $Ku - 2u = x(\pi - x)$.

d) $Ku - \frac{\pi}{9}u = \cos 3x$.

9. Find the first three nonzero terms of the Neumann series for the solution to the integral equation

$$u(t) = t + \mu \int_0^t (t - s)u(s)\, ds.$$

10. Solve the integral equation $u(t) = \int_0^t e^{t-s} u(s)\, ds$.

11. Reformulate the initial value problem

$$u'' - \lambda u = f(x), \quad x > 0; \quad u(0) = 1,\ u'(0) = 0,$$

as a Volterra integral equation.

12. Solve the integral equation

$$\int_0^t y u(y)\, dy - \lambda u(t) = f(t), \quad 0 \le t \le 1$$

, where f is continuous and $\lambda \ne 0$. Hint: as an artifice, assume f' exists and then write the solution only in terms of f.

13. Consider the operator $Ku(x) = \int_{-1}^1 k(x + y)u(u)\, dy$, where $k(x)$ is a continuous, even, periodic function with Fourier expansion

$$k(x) = \frac{k_0}{2} + \sum_1^\infty k_n \cos nx, \quad k_{n+1} < k_n.$$

Find the eigenvalues and eigenfunctions of K.

14. Determine if the integral operator

$$Ku(x) = \int_0^\pi \sin x \sin 2y\, u(y)\, dy$$

has eigenvalues.

15. Solve the integral equation

$$\int_0^1 k(x, y)u(y)\, dy - \lambda u(x) = x$$

using eigenfunction expansions, where $k(x, y) = x(1 - y)$ if $x < y$, and $k(x, y) = y(1 - x)$ if $x > y$. Hint: Transform the problem into a Sturm–Liouville problem.

16. Replace $\sin xy$ by the first two terms in its power series expansion to obtain an approximate solution to

$$u(x) = x^2 + \int_0^1 \sin(xy)u(y)\, dy.$$

17. Investigate the solvability of

$$u(x) = \sin x + 3 \int_0^\pi (x + y)u(y)\, dy.$$

18. Solve the integral equation

$$\int_0^1 e^{x+y} u(y)\, dy - \lambda u(x) = f(x),$$

considering all cases.

19. Solve the nonlinear integral equation $u(x) = \lambda \int_0^1 y u(y)^2\, dy$.

20. Solve the integral equation

$$u(x) = f(x) + \int_0^1 (x\xi^2 + x^2\xi) u(\xi)\, d\xi,$$

considering all cases.

21. For the following integral equation, find a condition on μ for which the equation has a solution, and find the solution.

$$y(x) = \sin x + \mu \int_0^{\pi/2} \sin x \cos \xi\, y(\xi) d\xi.$$

22. For which value(s) of λ is there a unique solution to the integral equation

$$\int_0^\pi (\sin \xi + \cos x) y(\xi)\, d\xi - \lambda y(x) = 1?$$

23. Reformulate the boundary value problem

$$u'' + \lambda u = 0, \quad 0 < x < l; \quad u(0) = u(l) = 0,$$

as a Fredholm integral equation.

24. Reformulate the initial value problem

$$u'' + u = t^2, \quad u(0) = 1, \quad u'(0) = 0,$$

as a Volterra integral equation of the form $u + Ku = f$, and find f and the kernel of K.

25. Find all solutions to the integral equation

$$y(x) = \mu \int_0^1 (x^4 + \xi^4) y(\xi) d\xi$$

that correspond to negative values of μ.

26. Find the eigenvalues and eigenfunctions of the integral operator

$$Ku(x) = \int_{-1}^1 (1 - |x - y|) u(y)\, dy.$$

27. Solve the Hammerstein integral equation

$$u(x) = \frac{\lambda}{2} \int_0^1 \beta(x)\beta(y)(1 + u(y)^2)dy,$$

where λ is a real parameter and β is a continuous function.

28. Consider the differential-integral equation

$$y'' + \left(\lambda - \int_0^1 y'^2 ds\right) y(x) = 0, \quad y(0) = y(1) = 0.$$

Treating λ as a bifurcation parameter, determine how the amplitude of the solution depends on λ and sketch a bifurcation diagram of the amplitude vs. λ.

29. (This exercise requires perturbation methods from Chapter 3.) Consider the differential-integral equation

$$\varepsilon u' = u - u^2 - u \int_0^t u(s)ds, \quad u(0) = a < 1,$$

where ε is a small, positive parameter, which models a population $u = u(t)$ undergoing logistics growth and the cumulative effect of a toxin on the population. Find a uniformly valid approximation for $t > 0$. [See K. G. TeBeest, 1997. *SIAM Review*, **39**(3), 484-493.]

30. Consider the differential–integral operator

$$Ku = -u'' + 4\pi^2 \int_0^1 u(s)\,ds, \quad u(0) = u(1) = 0.$$

Prove that eigenvalues of K, provided they exist, are positive. Find the eigenfunctions corresponding to the eigenvalue $\lambda = 4\pi^2$.

31. Use (4.30) to numerically solve the integral equation

$$\int_0^1 (1 - 3xy)u(y)\,dy - u(x) = x^3.$$

5.5 Green's Functions

What is a Green's function?[5] Mathematically, it is the kernel of an integral operator that represents the inverse of a differential operator; physically, it is the response of a system when a unit point source is applied to the system.

---
[5] George Green, 1793–1841.

For example, it is the electric field induced by a single point charge. In this section we show how these two apparently different interpretations are actually the same.

5.5.1 Inverses of Differential Operators

To fix the notion we consider a regular, nonhomogeneous Sturm–Liouville problem (SLP), which we write in the form

$$Lu \equiv -(pu')' + qu = f, \quad a < x < b, \tag{5.1}$$

$$B_1 u(a) \equiv \alpha_1 u(a) + \alpha_2 u'(a) = 0, \tag{5.2}$$

$$B_2 u(b) \equiv \beta_1 u(b) + \beta_2 u'(b) = 0, \tag{5.3}$$

where p, p', q, and f are continuous on $[a, b]$, and $p > 0$. So that the boundary conditions do not disappear, we assume that not both α_1 and α_2 are zero, and similarly for β_1 and β_2. For conciseness we represent this problem in operator notation

$$Lu = f, \tag{5.4}$$

where we consider the differential operator L as acting functions in $C^2[a, b]$ that satisfy the boundary conditions (5.2)–(5.3). Thus, L contains the boundary conditions in its definition.

Recall the procedure in matrix theory when we have a *matrix* equation $L\mathbf{u} = \mathbf{f}$, where $\mathbf{u}$ and $\mathbf{f}$ are vectors and L is a square, invertible matrix. We immediately have the solution $\mathbf{u} = L^{-1}\mathbf{f}$, where L^{-1} is the inverse matrix. The inverse matrix exists if $\lambda = 0$ is not an eigenvalue of L, or when $\det L \neq 0$. We want to perform a similar calculation with the differential equation (5.4) and write the solution

$$u = L^{-1}f,$$

where L^{-1} is the inverse operator of L. Because L is a differential operator, we expect that the inverse operator to be an integral operator of the form

$$(L^{-1}f)(x) = \int_a^b g(x, \xi) f(\xi) \, d\xi, \tag{5.5}$$

with kernel g. Again drawing inferences from matrix theory, we expect the inverse to exist when $\lambda = 0$ is not an eigenvalue of L, that is, when there are no nontrivial solutions to the differential equation equation $Lu = 0$. If there are nontrivial solutions to $Lu = 0$, then L is not a one-to-one transformation ($u = 0$ is always a solution) and so L^{-1} does not exist.

If the inverse L^{-1} of the differential operator L exists, then the kernel function $g(x, \xi)$ in (5.5) is called the **Green's function** associated with L

(recall that L contains in its definition the boundary conditions). This is the mathematical characterization of a Green's function. Physically, as we observe subsequently, the Green's function $g(x, \xi)$ is the solution to (5.4) when f is a unit point source acting at the point ξ; thus, it is the response of the system at x to a unit, point source at ξ.

The approach here is to first present the methodology and a heuristic physical motivation. Then the mathematical result unfolds with a precise representation of the Green's function.

Example 5.53

(**Inversion**) In this example we compute the inverse operator directly using integration. Consider the boundary value problem

$$-u'' = f(x), \quad 0 < x < 1; \quad u(0) = u(1) = 0. \tag{5.6}$$

Here $Lu = -u''$ is the differential operator, and it includes the boundary conditions as part of its definition. We can invert L by direct integration. Integrating from 0 to x gives

$$-u'(x) = \int_0^x f(y)dy + C.$$

Integrating again,

$$-u(x) = \int_0^x (x - \xi)f(\xi)d\xi + Cx,$$

where we used the left boundary condition and Lemma 5.38. Applying the right boundary condition gives

$$C = -\int_0^1 (1 - \xi)f(\xi)d\xi.$$

Therefore,

$$u(x) = \int_0^x (\xi - x)f(\xi)d\xi + \int_0^1 x(1 - \xi)f(\xi)d\xi.$$

Now we break up the last integral from 0 to x and x to 1, and then combine the two integrals from 0 to x to get

$$u(x) = \int_0^x \xi(1 - x)f(\xi)d\xi + \int_x^1 x(1 - \xi)f(\xi)d\xi.$$

We can combine the two integrals by defining the function $g(x, \xi)$ by

$$g(x, \xi) = \begin{cases} x(1 - \xi) & x < \xi, \\ \xi(1 - x) & x > \xi. \end{cases}$$

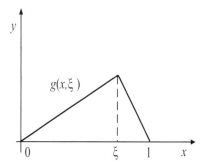

Figure 5.3 Green's function for (5.6).

Then,

$$u(x) = \int_0^1 g(x,\xi) f(\xi) d\xi \equiv L^{-1} f.$$

The function $g(x,\xi)$ is the kernel of the inverse differential operator L and is the **Green's function** for L. Using the Heaviside function, the Green's function can be written

$$g(x,\xi) = \xi(1-x) H(x-\xi) + x(1-\xi) H(\xi-x).$$

The reader should verify that $u(x)$ is indeed the solution to the differential. [Break up the integral into two integrals over $[0,x]$ and $[x,1]$, and then differentiate using Leibniz rule.] The Green's function is shown in Fig. 5.3. In $g(x,\xi)$ we are regarding x as the variable and ξ as a parameter.   □

It is important to note the properties of the Green's function in the last example: (a) It satisfies the differential equation for all $x \neq \xi$. (b) It satisfies both boundary conditions. (c) It is continuous for all x, and in particular, at $x = \xi$. (d) The derivative of $g(x,\xi)$ has a simple jump discontinuity at $x = \xi$. We shall see later that these properties are shared by all Green's functions for Sturm–Liouville operators.

5.5.2 Physical Interpretation

Next we investigate the physical interpretation of the Green's function. Our discussion is heuristic and intuitive, but it is made precise in the next section on distributions. To fix the context we consider the steady-state heat flow problem that we derived in Section 5.1. Refer to Fig. 5.1. We briefly repeat the derivation because it is relevant here. We have a cylindrical bar of length L and cross-sectional area A, and $u = u(x)$ denote the temperature at cross section x. Further, $\phi = \phi(x)$ denotes the energy flux across the face at x, measured in energy/(area $\cdot$ time), and $f(x)$ is a given, distributed heat source over the length of the bar, measured in energy/(volume $\cdot$ time). If x and $x + dx$ denote two arbitrary locations in the bar, then conservation of energy yields

$$A\phi(x) - A\phi(x + dx) + f(x)A\,dx = 0,$$

or, rearranging,

$$\frac{\phi(x + dx) - \phi(x)}{dx} = f(x).$$

Taking the limit as $dx \to 0$ gives $\phi'(x) = f(x)$. Fourier's heat conduction law states that

$$\phi(x) = -Ku'(x),$$

where K is the thermal conductivity of the bar, a physical constant. Then

$$-Ku''(x) = f(x),$$

which is the steady-state heat equation. For simplicity, we assume that the ends of a bar of unit length are held at zero degrees, and $K = 1$. Then the steady temperature distribution $u = u(x)$ along the length of the bar satisfies the boundary value problem

$$-u'' = f(x), \quad 0 < x < 1; \quad u(0) = u(1) = 0, \tag{5.7}$$

which is the same as (5.6).

Next imagine that $f = \delta(x, \xi)$ is an idealized heat source of unit strength that acts only at a single point $x = \xi$ in $(0, 1)$[6]. Thus, it is assumed to have the properties

$$\delta(x, \xi) = 0, \quad x \neq \xi$$

and

$$\int_{\xi-\varepsilon}^{\xi+\varepsilon} \delta(x, \xi)\,dx = 1 \text{ for all } \varepsilon > 0.$$

Moreover, we assume that

$$\int_{\xi-\varepsilon}^{\xi+\varepsilon} \theta(x)\delta(x, \xi)\,dx = \theta(\xi)$$

for all $\varepsilon > 0$ and all nice functions $\theta(\xi)$.

A little reflection reveals that there is no ordinary function δ with these two properties; a function that is zero everywhere but one point must have zero integral. Further, there is no integrable function $\delta(x, \xi)$ that 'sifts out'

---

[6] The unit source function $\delta(x, \xi)$ is also denoted by $\delta_\xi(x)$ or $\delta(x - \xi)$.

the value of a function $\theta(x)$ when integrated. Nevertheless, this δ symbol has been used since the inception of quantum mechanics in the late 1920s (it was introduced by the mathematician-physicist P. Dirac[7]; in spite of not having a rigorous definition of the symbol until the early 1950s, when the mathematician L. Schwartz[8] gave a precise characterization of point sources, both physicists and engineers used the "δ function" with great success. In the next section we give a careful definition of the delta function, but for the present we continue with an intuitive discussion.

The differential equation for steady heat flow becomes, symbolically,

$$-u'' = \delta(x, \xi). \tag{5.8}$$

Thus, for $x \neq \xi$ we have $-u'' = 0$, which has solutions of the form $u = Ax + B$. To satisfy the boundary conditions we take

$$u = Ax, \quad x < \xi; \qquad u = B(1 - x), \quad x > \xi.$$

We summarize the calculation we made in the last example. To determine the two constants A and B we use the fact that the temperature should be continuous at $x = \xi$, giving the relation $A\xi = B(1 - \xi)$. To obtain another condition on A and B we integrate the symbolic differential equation (5.8) over an interval $[\xi - \varepsilon, \xi + \varepsilon]$ containing ξ to get

$$-\int_{\xi-\varepsilon}^{\xi+\varepsilon} u''(x)\, dx = \int_{\xi-\varepsilon}^{\xi+\varepsilon} \delta(x, \xi)\, dx.$$

The right side is unity because of the unit heat source assumption; the fundamental theorem of calculus is then applied to the left side to obtain

$$-u'(\xi + \varepsilon) + u'(\xi - \varepsilon) = 1.$$

Taking the limit as $\varepsilon \to 0$ and multiplying by -1 yields

$$u'(\xi^+) - u'(\xi^-) = -1.$$

This condition is a **jump condition** on the derivative at the point $x = \xi$; because the flux is $\phi(x) = -Ku'(x)$, it requires a jump in the flux. This last condition forces $-B - A = -1$. Therefore, solving the two equations for A and B simultaneously gives

$$A = (1 - \xi), \quad B = \xi.$$

Therefore the steady-state temperature in the bar, caused by a point source at $x = \xi$ is

$$u = (1 - \xi)x, \quad x < \xi; \quad u = \xi(1 - x), \quad x > \xi,$$

---

[7] Paul Dirac (1902-1984) won the Nobel Prize in physics in 1933 for his groundbreaking work in quantum mechanics.
[8] Laurent Schwartz (1915-2002).

or

$$u(x,\xi) = \xi(1-x)H(x-\xi) + x(1-\xi)H(\xi-x).$$

This is precisely the Green's function for the operator $L = -d^2/dx^2$ with boundary conditions $u(0) = u(1) = 0$ that we calculated in Example 5.53. Consequently, this gives a physical interpretation for the Green's function; it is the response of a system (in this case the steady temperature response) caused by a point source (in this case a unit heat source).

The method above is characteristic of calculations of Green's function. We solve the homogeneous differential equation (with no source) for $x < \xi$ and for $x > \xi$. We apply the two appropriate boundary conditions at the endpoints and then require continuity at $x = \xi$. Finally we obtain the jump condition on the derivative by integrating the point source differential equation over an interval $[\xi - \varepsilon, \xi + \varepsilon]$ containing the singular point ξ, using the properties of the delta function. These four conditions determine the Green's function.

This idea is fundamental in applied mathematics. The Green's function $g(x,\xi)$ is the solution to the *symbolic* boundary value problem

$$Lg(x,\xi) \equiv (-p(x)g')' + q(x)g = \delta(x,\xi), \tag{5.9}$$

$$B_1g(a,\xi) = 0, \quad B_2g(b,\xi) = 0, \tag{5.10}$$

where δ represents a point source at $x = \xi$ of unit strength. Furthermore, the solution $Lu = f$ to the boundary value problem (5.7) can therefore be regarded as a superposition of point sources of magnitude $f(\xi)$ over the entire interval $a < \xi < b$.

Example 5.54

Consider the differential operator $L = -d^2/dx^2$ on $0 < x < 1$ with the boundary conditions $u'(0) = u'(1) = 0$ being part of the definition of L. In this case the Green's function does not exist because the equation $Lu = 0$ has nontrivial solutions (any constant function will satisfy the differential equation and the boundary conditions); stated differently, $\lambda = 0$ is an eigenvalue. Physically we can also see why the Green's function does not exist. This problem can be interpreted in the context of steady-state heat flow. The zero-flux boundary conditions imply that both ends of the bar are insulated, and so heat cannot escape from the bar. Thus it is impossible to impose a point source, which would inject heat energy at a constant, unit rate, and have the system respond with a time-independent temperature distribution; energy would build up in the bar, precluding a steady state. □

In this heuristic discussion we noted that δ is not an ordinary function. Furthermore, the Green's function g is not differentiable at $x = \xi$, so it remains to determine the meaning of applying a differential operator to g as in formulas (5.9)–(5.10). Our goal in the next section is to put these notions on firm ground.

Example 5.55

(**Causal Green's function**) The **causal Green's function** is the Green's function for an initial value problem. Consider the problem

$$Lu \equiv -(pu')' + qu = f(t), \quad t > 0; \quad u(0) = u'(0) = 0. \tag{5.11}$$

We assume p, p', and q are continuous for $t \geq 0$ and $p > 0$. The causal Green's function, also called the *impulse response function*, is the solution to (5.11) when f is a unit impulse applied at time τ, or, in terms of the delta function notation, when $f = \delta(t, \tau)$. Thus, symbolically, $Lg(t, \tau) = \delta(t, \tau)$, where g denotes the causal Green's function. We give a physical argument to determine g. Because the initial data is zero, the response of the system is zero up until time τ; therefore, $g(t, \tau) = 0$ for $t < \tau$. For $t > \tau$ we require $Lg(t, \tau) = 0$, and we demand that g be continuous at $t = \tau$, or

$$g(\tau^+, \tau) = 0.$$

At $t = \tau$, the time when the impulse is given, we require that g have a jump in its derivative of magnitude

$$g'(\tau^+, \tau) = -1/p(\tau).$$

This jump condition is derived in the same way that the jump condition was obtained earlier, namely, by integrating $Lg = \delta(t, \tau)$ over an interval $(\tau - \varepsilon, \tau + \varepsilon)$. The continuity condition and the jump condition, along with the fact that g satisfies the homogeneous differential equation, are enough to determine $g(x, \tau)$ for $t > \tau$. $\quad \Box$

Finally, we present the general result.

Theorem 5.56

Consider the SLP (5.1)–(5.3), and assume that $\lambda = 0$ is *not* an eigenvalue of L. Then L^{-1} exists and is given by (5.5) with

$$g(x, \xi) = \begin{cases} -\dfrac{u_1(x)u_2(\xi)}{p(\xi)W(\xi)}, & x < \xi \\[2ex] -\dfrac{u_1(\xi)u_2(x)}{p(\xi)W(\xi)}, & x > \xi \end{cases} \tag{5.12}$$

Here, $u_1 = u_1(x)$ and $u_2 = u_2(x)$ are two linearly independent solutions of the homogeneous differential equation $Lu = -(pu')' + qu = 0$ with $B_1 u_1(a) = 0$ and $B_2 u_2(b) = 0$, and $W(x) = u_1 u_2' - u_1' u_2$ is the Wronskian of u_1 and u_2. Finally, the solution to the nonhomogeneous problem $Lu = f$ is given by

$$u(x) = L^{-1} f(x) = \int_a^b g(x, \xi) f(\xi) d\xi. \quad \square \qquad (5.13)$$

Clearly, the Green's function can be expressed as a single equation in terms of the Heaviside unit step function $H(x)$ by

$$g(x, \xi) = -\frac{1}{p(\xi) W(\xi)} (H(x - \xi) u_1(\xi) u_2(x) + H(\xi - x) u_1(x) u_2(\xi)). \qquad (5.14)$$

In general, the basic properties of the Green's function are straightforward calculations from the expression (5.12).

(a) $g(x, \xi)$ satisfies the differential equation $Lg(x, \xi) = 0$ for $x \neq \xi$.

(b) $g(x, \xi)$ satisfies the boundary conditions (5.2) and (5.3).

(c) $g(x, \xi)$ is a continuous function of x on $[a, b]$, in particular, at $x = \xi$.

(d) $g(x, \xi)$ is not differentiable at $x = \xi$; rather, there is a jump in the derivative at $x = \xi$ given by

$$g'(\xi^+, \xi) - g'(\xi^-, \xi) = -\frac{1}{p(\xi)}.$$

Here, prime denotes differentiation with respect to x. Thus g is a continuous curve with a corner at $x = \xi$.

Now we return to the SLP problem (5.1)–(5.3) and ask what can be said if $\lambda = 0$ is an eigenvalue, that is, if the homogeneous problem has a nontrivial solution. In this case there may not be a solution, and if there is a solution, it is not unique. The following theorem gives a partial result.

Theorem 5.57

Consider the Sturm–Liouville problem

$$Lu \equiv -(pu')' + qu = f, \qquad a < x < b,$$
$$B_1 u(a) \equiv \alpha_1 u(a) + \alpha_2 u'(a) = 0,$$
$$B_2 u(b) \equiv \beta_1 u(b) + \beta_2 u'(b) = 0,$$

and assume there exists a nontrivial solution ϕ of the homogeneous problem $L\phi = 0$. Then, if the SLP has a solution, then f is orthogonal to ϕ, or

$$(\phi, f) \equiv \int_a^b \phi f \, dx = 0. \quad \square$$

Proof

Assume a solution u exists. Then

$$(\phi, f) = (\phi, Lu) = -\int_a^b \phi(pu')'\, dx + \int_a^b \phi qu\, dx.$$

The first integral may be integrated by parts twice to remove the derivatives from u and put them on ϕ. Performing this calculation gives, after collecting terms,

$$\begin{aligned}
(\phi, f) &= [p(u\phi' - \phi u')]_a^b + \int_a^b u(-(p\phi')' + q\phi)\, dx \\
&= [p(u\phi' - \phi u')]_a^b + \int_a^b uL\phi\, dx \\
&= [p(u\phi' - \phi u')]_a^b,
\end{aligned}$$

because $L\phi = 0$. Both ϕ and u satisfy the boundary conditions, so one easily shows that $[p(u\phi' - \phi u')]_a^b = 0$. □

5.5.3 Green's Function via Eigenfunctions

In this section we show that the eigenvalues and eigenvalues of a regular Sturm–Liouville operator L determine the Green's function, or the solution to the nonhomogeneous problem $Lu = f$. This is not surprising because we have already seen that solutions to a nonhomogeneous problem in matrix theory and in integral equation can be determined from the eigenstructure of the operator.

As usual, we study the more general problem

$$Lu - \mu u = f, \tag{5.15}$$

where L is the regular Sturm–Liouville operator, which includes the boundary conditions (5.3), and f is continuous. We know from Section 5.2 that the SLP problem

$$Lu = \lambda u, \quad a < x < b, \tag{5.16}$$

has infinitely many eigenvalues and corresponding orthonormal eigenfunctions λ_n and $\phi_n(x), n = 1, 2, \ldots$, respectively. Moreover, the eigenfunctions form a basis for the square-integrable functions on (a, b). Therefore we assume that the solution u of (5.15) is given in terms of the eigenfunctions as

$$u(x) = \sum_{n=1}^{\infty} c_n \phi_n(x), \tag{5.17}$$

where the coefficients c_n are to be determined. Further, we write the given function f in terms of the eigenfunctions as

$$f(x) = \sum_{n=1}^{\infty} f_n \phi_n(x), \quad f_n = \int_a^b f(\xi)\phi_n(\xi)\, d\xi.$$

Next we multiply (5.15) by ϕ_n and integrate to obtain, using inner product notation,

$$(Lu, \phi_n) - \mu(u, \phi_n) = f_n.$$

Because L is a symmetric operator,

$$(Lu, \phi_n) = (u, L\phi_n) = (u, \lambda_n \phi_n) = \lambda_n(u, \phi_n),$$

and therefore

$$\lambda_n c_n - \mu c_n = f_n.$$

Hence

$$(\lambda_n - \mu)c_n = f_n, \quad n = 1, 2, 3, \ldots.$$

There are two cases. If μ is not an eigenvalue, then the c_n are determined uniquely by

$$c_n = \frac{f_n}{\lambda_n - \mu}, \quad n = 1, 2, 3, \ldots \tag{5.18}$$

Therefore we have a unique solution to (5.15). If $\mu = 0$, and if zero is not an eigenvalue of L, then the solution collapses into the inverse operator and Green's function as follows:

$$
\begin{aligned}
u(x) &= \sum_{n=1}^{\infty} c_n \phi_n(x) \\
&= \sum_{n=1}^{\infty} \frac{f_n}{\lambda_n} \phi_n(x) \\
&= \sum_{n=1}^{\infty} \frac{\int_a^b f(\xi)\phi_n(\xi)\, d\xi}{\lambda_n} \phi_n(x) \\
&= \int_a^b \left[\sum_{n=1}^{\infty} \frac{\phi_n(\xi)\phi_n(x)}{\lambda_n} \right] f(\xi)\, d\xi.
\end{aligned}
$$

Consequently, we have inverted the operator L and so we must have

$$g(x, \xi) = \sum_{n=1}^{\infty} \frac{\phi_n(x)\phi_n(\xi)}{\lambda_n},$$

which gives the Green's function in terms of the eigenvalues and eigenfunctions of the operator L. This expansion for g is called the **bilinear expansion.**

Now case two: μ is an eigenvalue; that is, $\mu = \lambda_J$ for some index J. Then, using (5.18), we observe that $c_J \cdot 0 = f_J$. Thus, if $f_J \neq 0$, there is no solution to (5.15). On the other hand, if $f_J = 0$, then c_J is arbitrary, and there are infinitely many solutions to (5.15) of the form

$$u(x) = c_J \phi_J(x) + \sum_{n \neq J}^{\infty} \frac{f_n}{\lambda_n} \phi_n(x).$$

The key result is the differential equation version of the Fredholm alternative: If μ is not an eigenvalue of the operator L in (5.15), then there is a unique solution (and a Green's function if $\lambda \neq 0$). If $\mu = \lambda_J$ is an eigenvalue of L, then there exists a solution if, and only if, f is othogonal to the eigenspace spanned by ϕ_J.

Example 5.58

Consider the operator $L = -d^2/dx^2$ with Dirichlet boundary conditions $u(0) = u(\pi) = 0$. The eigenvalues and eigenfucntions are

$$\lambda_n = n^2, \quad \phi_n(x) = \sqrt{\frac{2}{\pi}} \sin nx, \ n = 1, 2, \ldots.$$

Therefore the Green's function is

$$g(x, \xi) = \frac{2}{\pi} \sum_{n=1}^{\infty} \frac{\sin(nx)\sin(n\xi)}{n^2}. \quad \square$$

EXERCISES

1. A function $y = f(x)$ has a second derivative $f''(x)$ that is continuous at all points except $x = \xi$, where it has a simple jump discontinuity (finite left and right derivatives at ξ). Explain carefully why $f'(x)$ is continuous at ξ.

2. Consider the steady-state heat conduction equation with a piecewise continuous conductivity $K(x)$:

$$-(K(x)u')' = f(x), \quad u(0) = 0, \ u(1) = 1,$$

where $K(x) = K_1(x)$ for $1 \leq \xi$, $K(x) = K_2(x)$, for $\xi < x \leq 1$, where ξ is a fixed value in $(0, 1)$. What are the conditions on the temperature u and flux ϕ at ξ? Hint: Integrate over a small interval about ξ.

3. Using direct integration, show that the boundary value problem

$$-u'' = f(x), \quad 0 < x < 1, \quad u'(0) = u'(1) = 0$$

has a solution only if f is orthogonal to the constant function 1. Find $u(x)$ in this case.

4. Discuss the solvability of the boundary value problem

$$u'' + \pi^2 u = f(x), \quad 0 < x < 1; \quad u(0) = u(1) = 0.$$

5. Determine if there is a Green's function associated with the operator $Lu = u'' + 4u$, $0 < x < \pi$, with $u(0) = u(\pi) = 0$. Find the solution to the boundary value problem

$$u'' + 4u = f(x), \quad 0 < x < \pi; \quad u(0) = u(\pi) = 0.$$

6. Consider the boundary value problem

$$u'' - 2xu' = f(x), \quad 0 < x < 1; \quad u(0) = u'(1) = 0.$$

Find Green's function or explain why there isn't one.

7. Consider the boundary value problem

$$u'' + u' - 2u = f(x), \quad 0 < x < 1; \quad u(0) = u'(1) = 0.$$

Find Green's function or explain why there isn't one.

8. Use the method of Green's function to solve the problem

$$-(K(x)u')' = f(x), \quad 0 < x < 1; \quad u(0) = u(1) = 0; \quad K(x) > 0.$$

(Note: This differential equation is a steady-state heat equation in a bar with variable thermal conductivity $K(x)$.)

9. Consider a spring–mass system governed by the initial value problem

$$mu'' + ku = f(t), \quad t > 0; \quad u(0) = u'(0) = 0,$$

where $u = u(t)$ is the displacement from equilibrium, f is an applied force, and m and k are the mass and spring constants, respectively.

a) Show that the causal Green's function is

$$g(t, \tau) = \frac{1}{\sqrt{km}} \sin \sqrt{\frac{k}{m}} (t - \tau), \quad t > \tau.$$

b) Find the solution to the initial value problem and write it in the form

$$u(t) = \frac{1}{\sqrt{km}} \int_0^t \sin \sqrt{\frac{k}{m}} (t - \tau) f(\tau) \, d\tau.$$

10. By finding Green's function in two different ways, evaluate the sum

$$\sum_{n=1}^{\infty} \frac{\sin\ nx \sin\ n\xi}{n^2}, \quad 0 < x, \xi < \pi.$$

11. Find the inverse of the differential operator $Lu = -(x^2 u')'$ on $1 < x < e$ subject to $u(1) = u(e) = 0$.

12. Consider a symbolic eigenvalue problem with a 'delta function' coefficient $a\delta(x,0)$, $a > 0$, at the origin:

$$-u'' - a\delta(x,0)u = \lambda u, \quad -\infty < x < \infty,$$

subject to the boundary conditions $u(x) \to 0$ as $|x| \to \infty$. Find the negative eigenvalues and eigenfunctions. Hint: To obtain a jump condition, integrate the equation about a small interval surrounding $x = 0$ and then use the properties of the delta function.

13. Consider the eigenvalue problem for the Schrodinger equation

$$-y'' + V(x)y = \lambda y, \quad x > 0; \ y(0) = 0, \ \lim_{x \to \infty} y(x) = 0,$$

where the potential V is given by $V(x) = 0$, $x < 1$, and $V(x) = V_0 > 0$, $x \geq 1$. Find the eigenvalues in the range $0 < \lambda < V_0$ and the associated wave functions. Sketch a generic wave function and associated probability density. Hints: At discontinuities require continuity of y and y' (Why?); determine the eigenvalues graphically.

5.6 Distributions

In the last section we showed, in an intuitive manner, that the Green's function satisfies a differential equation with a unit point source. Because the Green's function is not a smooth function, it leads us to the question of what it means to differentiate such a function. Also, we have not pinned down the properties of a point source (a delta function) in a precise way. If a point source is not a function, then what is it? The goal of this section is to answer these questions. New notation, terminology, and concepts are required.

5.6.1 Test Functions

Let K be a set of real numbers. A real number c is said to be a **limit point** of set K if every open interval containing c, no matter how small, contains at

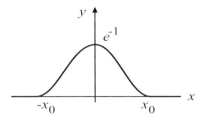

Figure 5.4 Plot of the test function (6.1).

least one point of K. If K is a set, then the **closure** of K, denoted by $\overline{K}$, is the set K along with all its limit points. A set is called **closed** if it contains all of its limit points. For example, the closure of the half-open interval $(a, b]$ is the closed interval $[a, b]$. A useful set in characterizing properties of a function ϕ is the set of points where it takes on nonzero values. The closure of this set is called the **support** of the function; precisely, we define supp $\phi \equiv \overline{\{x | \phi(x) \neq 0\}}$. So, the support of a function is the set of points where it is nonzero, along with the limit points of that set.

Now we introduce an important class of functions that plays a crucial role in our discussion. In the following, (a, b) denotes an open interval, which may be infinite. By the set $C_0^\infty(a, b)$ we mean the set of all continuous functions on (a, b) whose derivatives of all order exist and are continuous on (a, b), and which have their support contained in a closed, bounded subset of (a, b). These functions are called **test functions**. An example of a test function on $\mathbb{R}$ is the bell-shaped curve (see Fig. 5.4).

$$\phi(x) = \exp\left(-\frac{x_0^2}{x_0^2 - x^2}\right), |x| < x_0; \quad \phi(x) = 0, \quad |x| \geq x_0. \tag{6.1}$$

Therefore, a test function in a very smooth function on (a, b) with no corners or kinks in its graph or in any of its derivatives; its support must be entirely contained in (a, b) and, in fact, must lie in a closed, bounded interval strictly contained in (a, b); if a and b are finite, then a test function must be zero at a and b.

Finally, a function f is said to be **locally integrable** on (a, b) if

$$\int_c^d |f(x)|\, dx < \infty$$

for all subintervals $[c, d]$ of (a, b). Recall from calculus that an integrable function need not be differentiable, or even continuous.

As we know, a function with a corner (such as Green's function or the function $|x|$), does not have a derivative (tangent line) at that corner. However, it is possible to generalize the notion of derivative so that it makes sense to discuss the concept in a broader context. First we introduce the idea of a weak derivative. Let $u = u(x)$ be a continuously differentiable function on (a, b) and let $f = f(x)$ be its derivative. That is,

$$u' = f.$$

Multiplying both sides by a test function $\phi \in C_0^\infty(a, b)$ and integrating from a to b gives

$$\int_a^b u' \phi \, dx = \int_a^b f\phi \, dx.$$

Integrating the left side by parts and using the fact that $\phi(a) = \phi(b) = 0$ gives

$$-\int_a^b u\phi' \, dx = \int_a^b f\phi \, dx. \qquad (6.2)$$

Now observe that equation (6.2) does not require u to be differentiable, just integrable. The integration by parts removed the derivative from u and put it on ϕ, which has derivatives of all order. We can take (6.2) as the basis of a definition for a derivative of an integrable function. If u and f are locally integrable functions, we say that f is the **weak derivative** of u on (a, b) if (6.2) holds for all $\phi \in C_0^\infty(a, b)$. Clearly, if u is continuously differentiable, then it has a weak derivative (the ordinary one); but a function can have a weak derivative without having an ordinary derivative. In fact, this idea can be extended so as not to require even local integrability of u or f.

Example 5.59

On the interval $(a, b) = (-1, 1)$ let $u(x) = |x|$ and $f(x) = H(x) - H(-x)$, where H is the Heaviside function. The function u is not differentiable at $x = 0$ in the usual sense. Yet $u' = f$ in the weak sense on $(-1, 1)$ because, from (6.2),

$$-\int_{-1}^1 |x| \phi'(x) \, dx = \int_{-1}^1 (H(x) - H(-x))\phi(x) \, dx, \quad \forall \phi \in C_0^\infty(-1, 1).$$

This is easily checked. The left side is

$$\int_{-1}^0 x\phi'(x) \, dx - \int_0^1 x\phi'(x) \, dx = -\int_{-1}^0 \phi(x) \, dx + \int_0^1 \phi(x) \, dx,$$

where we have integrated by parts and used the fact that ϕ vanishes at $x = -1$, 1. The right side is

$$\int_{-1}^0 -H(-x)\phi(x) \, dx + \int_0^1 H(x)\phi(x) \, dx = -\int_{-1}^0 \phi(x) \, dx + \int_0^1 \phi(x) \, dx,$$

and therefore (6.2) holds and $(d/dx)|x| = H(x) - H(-x)$ in a weak sense. □

Example 5.60

Again take $(a, b) = (-1, 1)$ and let $u(x) = H(x)$. What is u' in the weak sense? Suppose $u' = f$ weakly. Then, by definition (6.2), we should have

$$-\int_{-1}^{1} H(x)\phi'(x)\,dx = \int_{-1}^{1} f(x)\phi(x)\,dx, \quad \forall \phi \in C_0^\infty(-1, 1).$$

The left side simplifies to

$$-\int_{0}^{1} H(x)\phi'(x)\,dx = -\int_{0}^{1} \phi'(x)\,dx = \phi(0).$$

So we require that

$$\int_{-1}^{1} f(x)\phi(x)\,dx = \phi(0), \quad \forall \phi \in C_0^\infty(-1, 1). \tag{6.3}$$

But, as the following calculation shows, there is no locally integrable function f for which (6.3) holds for all ϕ in $C_0^\infty(-1, 1)$. Take the test function $\phi_a(x) = \exp(-a^2/(a^2 - x^2))$ for $|x| < a < 1$, and $\phi_a(x) = 0$ otherwise. Then

$$e^{-1} = \phi_a(0) = \int_{-a}^{a} f(x)\exp(-a^2/(a^2 - x^2))\,dx \leq e^{-1}\int_{-a}^{a} |f(x)|\,dx.$$

The right side goes to zero as $a \to 0$ (because f is locally integrable), and so we obtain the contradiction $e^{-1} \leq 0$. Therefore the weak derivative of the Heaviside function cannot be a locally integrable function. Intuitively, the derivative should be zero everywhere except at $x = 0$, where there is a unit jump in H; so the derivative should be localized at the point $x = 0$, similar to a point source. In fact, as we shall show, the derivative of the Heaviside function is the "delta function," in a sense to be defined. □

5.6.2 Distributions

The last example showed that the concept of a weak derivative needs to be generalized further if we are to define the derivative of any locally integrable function. To accomplish this task we need to expand our idea of function and not think of the values of a function at points in its domain, but rather think of a function in terms of its action on other objects. For example, suppose $u = u(x)$ represents the distributed temperature at locations x in a bar. This is an idealized concept, because when we measure temperature at a point x, for example, we use a thermometer that has finite extent; it is not just localized at x. So, in fact, we are really measuring the average temperature $\bar{u}(x)$ of the points near x, or more precisely, a *weighted average*

$$\bar{u}(x) = \int u(\xi)\theta(x,\xi)d\xi$$

over some interval containing x. We cannot measure the exact, idealized $u(x)$, just the average. The weight function $\theta(x,\xi)$ is characterized by properties of the thermometer placed at x. So sometimes it is better to think of a function in terms of its integrated values rather than its point values. In fact, often we have to give up our idea of requiring our functions to have values or local properties. To extend the idea further, a 'delta function' $\delta(x,\xi)$ is just an idealized thermometer that measures the temperature exactly; that is,

$$\bar{u}(x) = \int u(\xi)\delta(x,\xi)d\xi = u(\xi).$$

This idea leads us to define the notion of a *distribution,* or *generalized function.* A **distribution** is a mapping (transformation) that associates with each test function ϕ in $C_0^\infty(a,b)$ a real number. (Think of the ϕ as a temperature profile.) We met the idea of mapping a function space to $\mathbb{R}$ in Chapter 4 on calculus of variations, where we considered functionals. A distribution is just a continuous linear functional on the set of test functions. If we denote the set of distributions by D', then $f \in D'$ implies $f \colon C_0^\infty(a,b) \to \mathbb{R}$. We denote the image of $\phi \in C_0^\infty(a,b)$ under f by (f,ϕ) rather than the usual function notation $f(\phi)$. Linearity implies $(f,\alpha\phi) = \alpha(f,\phi)$ and $(f,\phi_1 + \phi_2) = (f,\phi_1) + (f,\phi_2)$.[9] The requirement that f is continuous is a slightly more technical concept, namely that of convergence in $C_0^\infty(a,b)$. We say that a sequence ϕ_n of test functions in $C_0^\infty(a,b)$ converges to zero in $C_0^\infty(a,b)$ (and we write $\phi_n \to 0$ in $C_0^\infty(a,b)$), if there is a single, closed bounded interval K in (a,b) containing the supports of all the ϕ_n, and on that interval K the sequence ϕ_n and the sequence of the derivatives $\phi_n^{(m)}$ converge uniformly to zero on K as $n \to \infty$. Then we say that f is **continuous** if $(f,\phi_n) \to 0$ for all sequences $\phi_n \to 0$ in $C_0^\infty(a,b)$.

Example 5.61

With each locally integrable function u on (a,b) there is a natural distribution u defined by

$$(u,\phi) \equiv \int_a^b u(x)\phi(x)\,dx, \quad \phi \in C_0^\infty(a,b). \tag{6.4}$$

---

[9] In physics, the notation for the action of a distribution f on a test function ϕ is $< f \mid \phi >$; this is the bracket notation; $< f \mid$ denotes the 'bra', or the functional, and $\mid \phi >$ denotes the 'ket', or the function (or, vector).

Note that we speak of the distribution u and use the same symbol as for the function. One can verify that u defined by (6.4) satisfies the linearity and continuity properties for a distribution. Therefore, every locally integrable function is a distribution via (6.4). □

Example 5.62

The distribution δ_ξ, $\xi \in (a, b)$, defined by

$$(\delta_\xi, \phi) = \phi(\xi), \quad \phi \in C_0^\infty(a, b), \tag{6.5}$$

is called the **Dirac** or **delta distribution** with pole at ξ. It is easily checked that δ_ξ is linear and continuous, and thus defines a distribution. The Dirac distribution δ_ξ acts by sifting out the value of ϕ at ξ. Note that we cannot represent (6.5) by an integral

$$(\delta_\xi, \phi) = \int_a^b \delta_\xi(x)\phi(x)\,dx = \phi(\xi),$$

because of the argument following equation (6.3); there we showed that there is no locally integrable function with this property. A distribution, like the Dirac distribution, that cannot be represented as (6.4) is called a **singular distribution**. □

  The customary notation used in distribution theory is confusing. Even when f is a singular distribution and no locally integrable function f exists for which $(f, \phi) = \int_a^b f(x)\phi(x)\,dx$, we still write the integral, even though it makes no sense. The integral expression here is only symbolic notation for the mathematically correct expression (f, ϕ). Further, even though it is impermissible to speak of the values of a distribution f at points x, we still on occasion write $f(x)$ to denote the distribution. For example, the Dirac distribution δ_0 with pole at zero is often written $\delta(x)$, and also δ_ξ is written $\delta_\xi(x)$ or $\delta(x - \xi)$.

Example 5.63

Consider $u(x) = H(x)$, the Heaviside function. We found earlier, in order for $u' = f$ to exist in a weak sense, it is necessary that

$$-\int_{-1}^{1} H(x)\phi(x)\,dx = \int_{-1}^{1} f(x)\phi(x)\,dx = \phi(0) \tag{6.6}$$

for all test functions ϕ. There is no integrable function f that satisfies this equation. However, if f is the Dirac distribution $f = \delta_0$ with pole at zero, then according to (6.5), equation (6.6) is satisfied if we interpret the integral symbolically. Thus, in a *distributional* sense, $H' = \delta_0$, or $H'(x) = \delta(x)$. □

Now let us take a more formal approach and underpin the issues we raised. For conciseness, we use $D \equiv C_0^\infty(a, b)$ to denote the set of test functions on (a, b), and D' to denote the set of distributions (continuous linear functionals) on D. The interval (a, b) can be $(-\infty, \infty)$ or $(0, \infty)$. Equality of two distributions is defined by requiring the two distributions have the same action on all the test functions; we say that two distributions f_1 and f_2 are **equal** if $(f_1, \phi) = (f_2, \phi)$ for all $\phi \in D$, and we write $f_1 = f_2$ in D'.

It is possible to do algebra and calculus in the set D' of distributions just as we do with ordinary functions. It is easy to see that D' is a linear space under the definition $(f_1 + f_2, \phi) = (f_1, \phi) + (f_2, \phi)$ of addition, and $(cf, \phi) = c(f, \phi), c \in \mathbb{R}$, of scalar multiplication. To define rules for distributions we use the guiding principle that any general definition should be consistent with what is true for locally integrable functions, since locally integrable functions are distributions. For example, let $\alpha = \alpha(x)$ be an infinitely differentiable function (i.e., $\alpha \in C^\infty(a, b)$), and suppose we wish to define the multiplication αf, where $f \in D'$. If f is a locally integrable function, then

$$(\alpha f, \phi) = \int_a^b (\alpha f) \phi \, dx = \int_a^b f(\alpha \phi) \, dx = (f, \alpha \phi). \tag{6.7}$$

Observe that $\alpha \phi \in D$. Therefore we make the definition

$$(\alpha f, \phi) = (f, \alpha \phi)$$

for *any* $f \in D'$; we have then defined the distribution αf by its action on test functions through the action of f.

We can also define the derivative of a distribution f. Again, assume f and f' are locally integrable. Using the same argument that led to (6.2), namely integration by parts and the fact that test functions have support in the interval (a, b), we are led to

$$(f', \phi) = \int_a^b f' \phi \, dx = -\int_a^b f \phi' \, dx = -(f, \phi').$$

Therefore, if $f \in D'$ we define the derivative $f' \in D'$ by

$$(f', \phi) = -(f, \phi'), \quad \phi \in D,$$

f' is called the **distributional derivative** of f. In the same way (integrating by parts n times), we define the nth distributional derivative $f^{(n)}$ of a distribution f by

$$(f^{(n)}, \phi) = (-1)^n (f, \phi^{(n)}), \quad \phi \in D.$$

It is routine to verify that these definitions do indeed satisfy the requirements of a distribution (linearity and continuity). Thus, we have the interesting fact that a distribution has derivatives of all orders.

Example 5.64

Calculate the derivative of the Dirac distribution δ_ξ. By definition,

$$(\delta_\xi', \phi) = -(\delta_\xi, \phi') = -\phi'(\xi). \quad \square$$

Example 5.65

A function like $f(x) = 1/x$ is not locally integrable. If we try to define a regular distribution by

$$\left(\frac{1}{x}, \phi \right) = \int_{-\infty}^{\infty} \frac{1}{x} \phi(x) dx,$$

then it is invalid because the integral does not exist in any interval containing the origin. We can avoid this difficulty if we define the distribution using the principal value of the integral. That is, define

$$
\begin{aligned}
\int_{-\infty}^{\infty} \frac{1}{x} \phi(x) dx &= \lim_{\varepsilon \to 0} \left(\int_{-\infty}^{-\varepsilon} \frac{1}{x} \phi(x) dx + \int_{\varepsilon}^{+\infty} \frac{1}{x} \phi(x) dx \right) \\
&= \lim_{\varepsilon \to 0} \left(\int_{\varepsilon}^{\infty} \frac{\phi(x) - \phi(-x)}{x} dx \right) \\
&= 2\phi'(0).
\end{aligned}
$$

One can check that these integrals exist. So we define

$$\left(PV \frac{1}{x}, \phi \right) = 2\phi'(0),$$

which is a distribution. Note that it is linear and continuous because $\left(PV \frac{1}{x}, \phi_n \right) = 2\phi_n'(0) \to 0$ as $\phi_n \to 0$. $\quad \square$

Other properties of distributions are useful as well. For example, if $c \in \mathbb{R}$ and $f(x) \in D'(\mathbb{R})$, then we can define the translated distribution $f(x - c) \in D'(\mathbb{R})$ by

$$(f(x - c), \phi(x)) = (f(x), \phi(x + c)), \quad \phi \in D.$$

(See Exercise 5.) For example, if $f(x) = \delta(x)$, the Dirac distribution with pole at zero, then $\delta(x - c)$ is defined by $(\delta(x - c), \phi(x)) = (\delta(x), \phi(x + c)) = \phi(c)$. But also, $(\delta_c(x), \phi(x)) = \phi(c)$. Therefore $\delta(x - c)$ has the same action on test functions as $\delta_c(x)$ and so

$$\delta_c(x) = \delta(x - c).$$

Thus we have yet another way to write the delta distribution with pole at c. The sifting property of the delta "function" is often written, as we have already done, symbolically as

$$\int_a^b \delta(x - \xi) \phi(x) \, dx = \phi(\xi).$$

In summary, a distribution is a generalization of the idea of a function. Rather than defined at points with local properties as functions are, distributions are global objects defined in terms of their integrated values, or their action on test functions.

In Chapter 6 we show how Laplace transforms of distributions can be defined.

5.6.3 Distribution Solutions to Differential Equations

We are now poised to define what is meant by a weak solution and a distribution solution to a differential equation. Consider the differential operator L defined by

$$Lu \equiv \alpha u'' + \beta u' + \gamma u, \tag{6.8}$$

where $\alpha, \beta,$ and γ are C^∞ functions on (a, b). By a classical solution, or **genuine solution,** to the differential equation

$$Lu = f, \tag{6.9}$$

where f is continuous, we mean a twice, continuously differential function $u = u(x)$ that satisfies the differential equation identically for all x. In other words, a genuine solution has enough derivatives so that it makes sense to substitute it into the differential equation to check, pointwise, if it is indeed a solution. We can also interpret (6.9) in a distributional sense. That is, if u and f are distributions, then Lu is a distribution; if $Lu = f$, as distributions, then we say u is a **distribution solution** of (6.9). Recall, for $Lu = f$ as distributions, that we mean $(Lu, \phi) = (f, \phi)$ for all test functions $\phi \in D$. Clearly, if u is a classical solution, then it is a distribution solution, but the converse is not true.

There is still more terminology. A **fundamental solution** associated with L is a distribution solution to

$$Lu = \delta(x - \xi),$$

where the right side of (6.9) is a Dirac distribution. A fundamental solution must satisfy only a differential equation in a distributional sense, and it need not be unique.

To elaborate, let u be a distribution. Then Lu is a distribution and it is defined by the action

$$
\begin{aligned}
(Lu, \phi) &= (\alpha u'', \phi) + (\beta u', \phi) + (\gamma u, \phi) \\
&= (u'', \alpha\phi) + (u', \beta\phi) + (u, \gamma\phi) \\
&= (u, (\alpha\phi)'') - (u, (\beta\phi)') + (u, \gamma\phi) = (u, L^*\phi),
\end{aligned}
$$

where L^* is the **formal adjoint** operator defined by

$$L^*\phi \equiv (\alpha\phi)'' - (\beta\phi)' + \gamma\phi.$$

It follows that if u is a distribution solution to $Lu = f$, then

$$(u, L^*\phi) = (f, \phi), \quad \forall \phi \in D. \tag{6.10}$$

If both u and f are locally integrable, then (6.10) can be written

$$\int_a^b u(x)L^*\phi(x)\,dx = \int_a^b f(x)\phi(x)\,dx, \quad \forall \phi \in C_0^\infty(a,b), \tag{6.11}$$

and we say u is a **weak solution** of (6.9) if (6.11) holds. Note that (6.11) does not require u to have derivatives in the ordinary sense. A weak solution is a special case of a distribution solution; a weak solution is a function solution whose (weak) derivatives are functions, and (6.11) holds. Obviously, a classical solution is a weak solution, but the converse is not true.

In passing, we call an operator L **formally self-adjoint** if $L^* = L$. Specifically, a Sturm–Liouville operator $Lu \equiv -(pu')' + qu$ (regardless of boundary conditions) is formally self-adjoint, as integration by parts shows (see Exercise 9). In other words,

$$\begin{aligned}
\int_a^b (Lu)v\,dx &= \int_a^b u(L^*v)\,dx + B(u,v) \\
&= \int_a^b u(Lv)\,dx + B(u,v),
\end{aligned}$$

where $B(u,v)$ denotes the boundary conditions. Given boundary conditions on u, the boundary conditions on v that force $B(u,v) = 0$ are called the **adjoint boundary conditions**. If $L = L^*$ and the adjoint boundary coincide with those of L, then L is said to be **self-adjoint**.

What we accomplished in the preceding discussion is significant. We replaced the usual question "Does this problem have any function solutions?" with the question "Does this problem have any distribution solutions or weak solutions?" Thus, we are not discarding physically meaningful solutions because of minor technicalities, for example, not having a derivative at some point. Now the symbolic differential equation for the Green's function,

$$Lg(x, \xi) = \delta(x - \xi),$$

as discussed in Section 5.4, is meaningful. It is to be interpreted as an equation for distributions and actually means

$$(Lg(x, \xi), \phi) = (\delta(x - \xi), \phi), \quad \forall \phi \in D,$$

or, equivalently,

$$(g(x, \xi), L^*\phi) = \phi(\xi). \quad \square$$

Example 5.66

We verify that the function

$$u = g(x, \xi) = \xi(1 - x)H(x - \xi) + x(1 - \xi)H(\xi - x)$$

is the Green's function associated with the operator $L = -d^2/dx^2$ on $(0, 1)$ subject to the boundary conditions $u(0) = u(1) = 0$. That is, we show g satisfies the problem

$$-g''(x, \xi) = \delta(x - \xi), \tag{6.12}$$
$$g(0, \xi) = g(1, \xi) = 0, \tag{6.13}$$

in a distributional sense. We must show

$$(g(x, \xi), -\phi'') = \phi(\xi) \tag{6.14}$$

for all $\phi \in D(0, 1)$. To this end,

$$(g(x, \xi), -\phi'') = -\int_0^1 g(x, \xi)\phi''(x)\, dx$$

$$= -(1 - \xi)\int_0^\xi x\phi''(x)\, dx - \xi\int_\xi^1 (1 - x)\phi''(x)\, dx.$$

Now we integrate the two integrals by parts to get

$$\int_0^\xi x\phi''\, dx = \xi\phi'(\xi) - \phi(\xi),$$

and

$$\int_\xi^1 (1 - x)\phi''\, dx = (\xi - 1)\phi'(\xi) - \phi(\xi).$$

Upon substitution we obtain (6.14). So, interpreted properly, the Green's function is the solution to the boundary value problem (6.12)–(6.13) with a point source. □

EXERCISES

1. Does $(u, \phi) = \phi(0)^2$ define a distribution in $D'(\mathbb{R})$?

2. Is the function $\phi(x) = x(1 - x)$ on $(0, 1)$ a test function? Why or why not?

3. Is $f(x) = 1/x$ locally integrable on $(0, 1)$?

4. Let $\phi \in D(\mathbb{R})$ be a test function. For which of the following does $\psi_n \to 0$ in $D(\mathbb{R})$?

   (a) $\psi_n(x) = \frac{1}{n}\phi(x)$.

(b) $\psi_n(x) = \frac{1}{n}\phi\left(\frac{x}{n}\right)$.

(c) $\psi_n(x) = \frac{1}{n}\phi(nx)$.

5. Prove the following statements:

(a) $x\delta'(x) = -\delta(x)$.

(b) $\alpha(x)\delta'(x) = -\alpha'(0)\delta(x) + \alpha(0)\delta'(x)$, where $\alpha \in C^\infty(\mathbb{R})$.

6. Show that $u(x,\xi) = \frac{1}{2}|x-\xi|$ is a fundamental solution for the operator $L = d^2/dx^2$ on $\mathbb{R}$.

7. Let $c \neq 0$ be a constant and $f(x)$ a distribution in $D'(\mathbb{R})$. Show that it is appropriate to define the distributions $f(x-c)$ and $f(cx)$ by

$$(f(x-c),\phi(x)) = (f(x),\phi(x+c)),$$

and

$$(f(cx),\phi(x)) = \frac{1}{|c|}\left(f(x),\phi\left(\frac{x}{c}\right)\right).$$

8. Find $\frac{d}{dx}\,|\,x^2 - 1\,|$.

9. Compute the distributional derivative of $H(x)\cos x$, where H is the Heaviside function. Does the derivative exist in a weak sense?

10. Show that the Sturm–Liouville operator $Lu \equiv -(pu')' + qu$ is formally self-adjoint.

11. Find a fundamental solution associated with the operator L defined by $Lu = -x^2u'' - xu' + u$, $0 < x < 1$, such that $u(x,\xi) = x$ for $0 < x < \xi$.

12. In $D'(\mathbb{R})$ compute $\left(\frac{d}{dx} - \lambda\right)(H(x)e^{\lambda x})$.

13. Find the adjoint of the derivative operator $L = -d/dx$ on $x > 0$. Is L formally self-adjoint?

14. Consider the operator $L = d^2/dx^2 - x^2$ on $0 < x < 1$. (a) Find the formal adjoint of L? Is L formally self-adjoint? (b) Suppose the boundary conditions accompanying L are $u(0) = u'(1) = 0$. Find the adjoint boundary conditions. Is L self-adjoint?

REFERENCES AND NOTES

This chapter presented the core of classical applied mathematics—boundary value problems, Green's functions, integral equations, and distributions. There are many texts that specialize in one or more of these subjects. Comprehensive treatments, all more advanced than the present text, are given in Keener (2000) and Stakgold (1998). For elementary linear algebra Kwak and Hong (2004) is a highly readable text, and Kelley and Peterson (2010) is an upper-level differential equations text that discusses boundary value problems. Spivak's (2008) classic calculus text provides an unsurpassed treatment of the basic results and techniques in analysis. Strauss (1992) and Vretblat (2006) both have excellent discussions on Fourier analysis and their applications. Quantum mechanics is a rich source of examples of all the ideas in this chapter; Griffiths (2005) is an excellent introduction to this material.

Birkhoff, G. & Rota, G.-C. 1978. *Ordinary Differential Equations*, 2nd ed., John Wiley & Sons, New York.

Griffiths, D. J. 2005. *Introduction to Quantum Mechanics*, 2nd ed., Pearson Prentice-Hall, Upper Saddle River, NJ.

Keener, J. P. 2000. *Principles of Applied Mathematics*, revised edition, Perseus Press, Cambridge, MA.

Kelley, W. G. & Peterson, A. 2010. *The Theory of Differential Equations*, 2nd ed., Springer-Verlag, New York.

Kwak, J. H. & Hong, S. 2004. *Linear Algebra*, 2nd ed., Birkhauser, Boston.

Spivak, M. 2008. *Calculus*, 4th ed., Publish or Perish, Inc., Houston.

Stakgold, I. 1998. *Green's Functions and Boundary Value Problems*, 2nd, ed., Wiley–Interscience, New York.

Strauss, W. A. 1992. *Partial Differential Equations*, John Wiley & Sons, New York.

Vretblad, A. 2006. *Fourier Analysis and Its Applications*, Springer-Verlag, NY.

6

Partial Differential Equations

Partial differential equations is one of the most fundamental areas in applied analysis, and it is hard to imagine any area of science and engineering where the impact of this subject is not felt. Partial differential equation models, for example, form the basis of electrodynamics (Maxwell's equations), quantum mechanics (the Schrödinger equation), continuum mechanics (equations governing fluid, solid, and gas dynamics; acoustics), and reaction–diffusion systems in biological and chemical sciences, only to mention a few. The goal in this and the next chapter are to introduce some important models and key techniques that expose the solution structure of such equations.

6.1 Basic Concepts

Ordinary differential equations describe how systems evolve in time. Yet many systems, perhaps most, evolve in both space and time, giving rise to two or more independent variables. For example, the temperature u in a slender, laterally insulated, metal bar of length l, subject to prescribed temperatures at its ends, depends upon the location x in the bar and upon the time t. That is, $u = u(x, t)$. Later we show, using conservation of energy, that the temperature u must satisfy a partial differential equation

$$u_t(x, t) - ku_{xx}(x, t) = 0, \qquad \text{(heat equation)}$$

Applied Mathematics 4th ed.,
By J. David Logan

where $k > 0$ is a numerical constant characterizing the thermal properies of
the material composing the bar; the equation holds for $t > 0$ and $0 < x < l$. It
is called the **heat equation**, and it is the basic equation of heat transfer.

Consider another example from population dynamics. Imagine, for example,
toxic bacteria distributed throughout a canal of length L. If we ignore spatial
variation then we may model the total population $u = u(t)$ of bacteria by the
usual logistic equation

$$\frac{du(t)}{dt} = ru(t)\left(1 - \frac{u(t)}{K}\right),$$

where r is the growth rate and K is the carrying capacity. Here, the population
depends only on time t and we have an ordinary differential equation. However,
in a nonhomogeneous environment the parameters r and K may depend upon
geographical location x. Even if the environment is homogeneous, the bacteria
may be distributed nonuniformly in the canal, and they may move from high
population densities to low densities (this is called diffusion), giving changes
in spatial dependence. If we wish to consider how the spatial distribution of
bacteria changes in time, then we must include a spatial independent variable
and write $u = u(x,t)$, where now u is interpreted as a population density,
or population per unit length; that is, $u(x,t)\Delta x$ represents approximately the
number of bacteria between x and $x + \Delta x$. If we include diffusion of the bacteria
then the model modifies to

$$u_t(x,t) = Du_{xx}(x,t) + ru(x,t)\left(1 - \frac{u(x,t)}{K}\right),$$

where D is the diffusion constant. This is **Fisher's equation**,[1] one of the
fundamental equations in structured population dynamics.

It is common, as in ordinary differential equations, to drop the independent
variables in the notation and just write, for example, the heat equation as
$u_t - ku_{xx} = 0$, the dependence of u on x and t being understood. In the next
section we derive these model equations.

In general, a **second-order partial differential equation** in two independent variables is an equation of the form

$$G(x, t, u, u_x, u_t, u_{xx}, u_{xt}, u_{tt}) = 0, \tag{1.1}$$

where (x, t) lies in some domain D in $\mathbb{R}^2$. By a *solution* we mean a twice
continuously differentiable function $u = u(x,t)$ on D, which when substituted
into (1.1), reduces it to an identity for (x,t) in D. We assume that u is twice

---

[1] R. A. Fisher (1890–1962), who made significant contributions to statistics and to
evolutionary biology and genetics, formulated the equation in 1937 as a model for
the spread of an advantageous gene throughout a population.

continuously differentiable so that it makes sense to calculate the second-order derivatives and substitute them into (1.1). A solution of (1.1) may be represented graphically as a smooth surface in three-dimensional xtu space lying above the domain D. We regard x as a position or spatial coordinate and t as time. The domain D in $\mathbb{R}^2$ where the problem is defined is referred to as a space-time domain, and problems that include time as an independent variable are called **evolution problems**. Such problems model how a system evolves in time, particularly those that govern diffusion processes and wave propagation. When two spatial coordinates, say x and y, are the independent variables we refer to the problem as an **equilibrium** or **steady-state** problem; time is not involved. A large portion of the literature on theoretical partial differential equations deals with questions of regularity of solutions. That is, given an equation and auxiliary conditions, how smooth are the solutions? We do not address many of these issues here, but we refer to an advanced text (see References and Notes).

A partial differential equation of type (1.1) has infinitely many solutions. Similar to the general solution of an ordinary differential equation, which depends on arbitrary constants, the general solution of a partial differential equation depends on arbitrary functions.

Example 6.1

Consider the simple partial differential equation

$$u_{tx} = lx.$$

Integrating with respect to x gives

$$u_t = \tfrac{1}{2}tx^2 + f(t),$$

where f is an arbitrary function. Integrating with respect to t gives the general solution

$$u = \tfrac{1}{4}t^2x^2 + g(t) + h(x),$$

where h is an arbitrary function and $g(t) = \int f(t)dt$ is also an arbitrary function. Thus the general solution depends on two arbitrary functions; any choice of g and h yields a solution provided they are suitably differentiable. □

A general **first-order partial differential equation** has the form

$$H(x, t, u, u_x, u_t) = 0, \quad (x, t) \in D,$$

where H is a given function. These equations are examined in detail in Chapter 7. Here we give only an example.

Example 6.2

Notice that the second-order equation

$$u_{tx} + 2u_x = 0$$

reduces immediately to the first-order equation

$$v_t + 2v = 0$$

by making the substitution $v = u_x$. The v equation looks much like an ordinary differential equation because there is only a t derivative. Let's multiply by the integrating factor e^{2t} as we would do for an ODE. The left side becomes

$$\frac{\partial}{\partial t}\left(ve^{2t}\right) = 0.$$

When we integrate both sides with respect to t, we obtain

$$ve^{2t} = C(x),$$

where the 'constant' of integration is a function $C = C(x)$ of the other variable x, as indicated. Then $v(x,t) = C(x)e^{-2t}$. Therefore, $u_x = C(x)e^{-2t}$, and we can integrate with respect to x to obtain the general solution of the original equation,

$$u(x,t) = \left(\int C(x)\partial x\right)e^{-2t} + B(t) = \tilde{C}(x)e^{-2t} + B(t),$$

where $\tilde{C}$ and B are arbitrary functions (an integral of an arbitrary function is an arbitrary function). Any PDE that has only partial derivatives with respect to only one of the variables can be treated like an ODE in that variable with the other variable as a parameter.   □

For ordinary differential equations, initial or boundary conditions fix values for the arbitrary constants of integration and thus often pick out a unique solution. Similarly, partial differential equations have auxiliary conditions. They are usually accompanied by initial or boundary conditions that select out one of its many solutions. A condition given at $t = 0$ along some segment of the x axis is called an **initial condition**. A condition given along any other curve in the xt plane is called a **boundary condition**. Initial or boundary conditions involve specifying values of u, its derivatives, or combinations of both along the given curves in the xt plane.

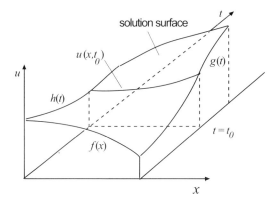

Figure 6.1   A solution surface $u = u(x,t)$ representing the temperature at (x,t) and a time snapshot $u = u(x,t_0)$ at time $t = t_0$. The functions h and g define the boundary conditions, and f specifies the initial condition. We think of the heat equation as propagating the initial temperature profile into the region at future times.

Example 6.3

Heat flow in a bar of length l is governed by the heat equation

$$u_t - ku_{xx} = 0, \quad t > 0, \quad 0 < x < l,$$

where k is a physical constant and $u = u(x,t)$ is the temperature in the bar at location x at time t. An auxiliary condition of the form

$$u(x,0) = f(x), \quad 0 < x < l,$$

is an **initial condition** because it is given at $t = 0$. We regard $f(x)$ as the initial temperature distribution in the bar. Conditions of the form

$$u(0,t) = h(t), \quad u(l,t) = g(t), \quad t > 0,$$

are **boundary conditions**, and $h(t)$ and $g(t)$ represent specified temperatures imposed at the boundaries $x = 0$ and $x = l$ for $t > 0$, respectively. These functions are depicted in Fig. 6.1. Graphically, the surface $u = u(x,t)$ representing the solution has f, g, and h as its boundaries.We can visualize the solution by sketching the solution surface or by plotting several time snapshots, or spatial profiles frozen in time. Figure 6.1 shows a graph of one profile $u(x,t_0)$ for some fixed t_0. It is a cross section of the solution surface. Often a sequence of profiles $u(x,t_1), u(x,t_2), \ldots$, for $t_1 < t_2 < \ldots$, are plotted on the same set of xu axes to indicate how the temperature profiles change in time.   □

The general solution of a partial differential equation is usually difficult or impossible to find. Therefore, for partial differential equations, we seldom solve a boundary value problem by determining the general solution and then finding the arbitrary functions from the initial and boundary data. This is in sharp contrast to ordinary differential equations where the general solution is found and the arbitrary constants are evaluated from the initial or boundary conditions. Generalizations of the partial differential equation (1.1) can be made in various directions—higher-order derivatives, several independent variables, and several unknown functions (governed by several equations). Later we observe that there are fundamentally three types of partial differential equations, those that govern diffusion processes, those that govern wave propagation, and those that model equilibrium phenomena. These types are termed **parabolic**, **hyperbolic**, and **elliptic** equations, respectively. Mixed types also occur.

6.1.1 Linearity and Superposition

The separation of partial differential equations into the classes of *linear* and *nonlinear* equations is a significant one. Linear equations have a linear algebraic structure to their solution set; that is, the sum of two solutions of a linear homogeneous equation is again a solution, as is a constant multiple of a solution. These facts, termed the **superposition principle**, often aid in constructing solutions. Superposition for linear equations often allows one to construct solutions that can meet diverse boundary or initial requirements. This observation is the basis of the Fourier method, or method of eigenfunction expansions, for linear equations. Linear equations are also amenable to transform methods for finding solutions, for example, Laplace transforms and Fourier transforms. On the other hand, none of these principles or methods are valid for nonlinear equations. In summary, there is a profound difference between these two classes of problems. (See Logan (2008), for example, for an introductory treatment of nonlinear partial differential equations.)

To formulate these concepts more precisely, we can regard the partial differential equation (1.1) as defining a differential operator L acting on the unknown function $u(x, t)$ and we write (1.1) as

$$Lu(x, t) = f(x, t), \quad (x, t) \in D.$$

Or, suppressing the independent variables,

$$Lu = f, \quad (x, t) \in D. \tag{1.2}$$

In (1.2) all terms involving u are put on the left in the term Lu, and f is a given function of x and t, often called the source term. If $f = 0$ on D, then (1.2) is

homogeneous; if f is not identically zero, then (1.2) is **nonhomogeneous**. The heat equation $u_t - ku_{xx} = 0$ can be written $Lu = 0$, where L is the partial differential operator $\partial/\partial t - k\partial^2/\partial x^2$, and it is clearly homogeneous. The partial differential equation $uu_t + 2txu - \sin tx = 0$ can be written $Lu = \sin tx$, where L is the differential operator defined by $Lu = uu_t + 2txu$. This equation is nonhomogeneous. The definition of linearity depends on the operator L in (1.2). We say that (1.2) is a **linear** equation if L has the properties

(i) $L(u + w) = Lu + Lw$,

(ii) $L(cu) = cLu$,

where u and w are functions and c is a constant. If (1.2) is not linear, then it is **nonlinear** .

Example 6.4

The heat equation is linear because

$$L(u + w) = (u + w)_t - k(u + w)_{xx}$$
$$= u_t + w_t - ku_{xx} - kw_{xx}$$
$$= Lu + Lw,$$

and

$$L(cu) = (cu)_t - k(cu)_{xx}$$
$$= cu_t - cku_{xx}$$
$$= cLu. \qquad \square$$

Example 6.5

The differential equation $uu_t + 2txu - \sin tx = 0$ is nonlinear because

$$L(u + w) = (u + w)(u + w)_t + 2tx(u + w)$$
$$= uu_t + wu_t + ww_t + uw_t + 2txu + 2txw,$$

yet
$$Lu + Lw = uu_t + 2txu + ww_t + 2txw.$$
Note that the nonhomogeneous term $\sin tx$ does not affect linearity or nonlinearity. $\square$

It is clear that $Lu = f$ is linear if Lu is first-degree in u and its derivatives; that is, no products involving u and its derivatives occur. Hence the most general linear equation of second order is of the form

$$au_{tt} + bu_{xt} + cu_{xx} + du_t + eu_x + gu = f(x,t), \quad (x,y) \in D, \qquad (1.3)$$

where the functions a, b, c, d, e, g, and f are given continuous functions of (x, t) on D. If any of the coefficients $a, \ldots, g$, or f depend on u, we say that the equation is **quasi-linear**. Referring to an earlier remark, (1.3) is **hyperbolic** (wave-like), **parabolic** (diffusion-like), or **elliptic** (equilibrium type) on a domain D if the discriminant $b(x, t)^2 - 4a(x, t)c(x, t)$ is positive, zero, or negative, respectively, on that domain. For example, the heat equation $u_t - ku_{xx} = 0$ has $b^2 - 4ac = 0$ and is parabolic on all of $\mathbb{R}^2$. Two other well-known equations of mathematical physics, the wave equation $u_{tt} - c^2 u_{xx} = 0$ and Laplace's equation $u_{xx} + u_{yy} = 0$, are hyperbolic and elliptic, respectively.

If $Lu = 0$ is a linear homogeneous equation and u_1 and u_2 are two solutions, then it obviously follows that $u_1 + u_2$ is a solution because $L(u_1 + u_2) = Lu_1 + Lu_2 = 0 + 0 = 0$. Also cu_1 is a solution because $L(cu_1) = cLu_1 = c \cdot 0 = 0$. A simple induction argument shows that if $u_1, \ldots, u_n$ are solutions of $Lu = 0$ and $c_1, \ldots, c_n$ are constants, then the finite sum, or linear combination, $c_1 u_1 + \cdots + c_n u_n$ is also a solution; this is the **superposition principle** for linear equations. If certain convergence properties hold, the superposition principle can be extended to infinite sums $c_1 u_1 + c_2 u_2 + \cdots$.

Another form of a superposition principle is a continuous version of the one just cited. In this case let $u(x, t; \xi)$ be a family of solutions on D, where ξ is a real parameter ranging over some interval I. That is, suppose $u(x, t; \xi)$ is a solution of $Lu = 0$ for each $\xi \in I$. Then we formally superimpose these solutions by forming

$$u(x, t) = \int_I c(\xi) u(x, t; \xi) \, d\xi,$$

where $c(\xi)$ is a function representing a continuum of coefficients, the analog of $c_1, c_2, \ldots, c_n$. If we can write

$$Lu = L \int_I c(\xi) u(x, t; \xi) \, d\xi$$

$$= \int_I c(\xi) Lu(x, t; \xi) \, d\xi$$

$$= \int_I c(\xi) \cdot 0 \, d\xi = 0,$$

then $u(x, t)$ is also a solution. This sequence of steps, and hence the superposition principle, depends on the validity of pulling the differential operator L inside the integral sign. For such a principle to hold, these formal steps would need careful analytical verification.

Example 6.6

Consider the heat equation

$$u_t - ku_{xx} = 0, \quad t > 0, \quad x \in \mathbb{R}. \tag{1.4}$$

It is straightforward to verify that

$$u(x,t;\xi) = \frac{1}{\sqrt{4\pi kt}} e^{-(x-\xi)^2/4kt}, \quad t > 0, \quad x \in \mathbb{R},$$

is a solution of (1.4) for any $\xi \in I = \mathbb{R}$. This solution is called the **fundamental solution** of the heat equation. We can formally superimpose these solutions to obtain

$$u(x,t) = \int_{-\infty}^{\infty} c(\xi) \frac{1}{\sqrt{4\pi kt}} e^{-(x-\xi)^2/4kt} \, d\xi,$$

where $c(\xi)$ is some function representing the coefficients. It can be shown that if $c(\xi)$ is continuous and bounded, then differentiation under the integral sign can be justified and $u(x,t)$ is therefore a solution. The fundamental solution is recognized as a normal probability density function (bell-shaped curve) with mean ξ and variance $2kt$. Therefore, as time increases the spatial profiles, or temperature distributions, spread. We should not be surprised that a probability density is related to solutions to the heat equation because heat transfer involves the random collisions of the particles making up the medium. We discuss the connection between diffusion and randomness in a later section. □

Example 6.7

This example shows how we can use superposition to construct solutions that satisfy auxiliary conditions. The pure initial value problem for the heat equation,

$$\begin{aligned} u_t - k u_{xx} &= 0, \quad t > 0, \quad x \in \mathbb{R}, \\ u(x,0) &= f(x), \quad x \in \mathbb{R}, \end{aligned}$$

is called the **Cauchy problem**. If f is bounded and continuous, then the solution is given by the integral representation

$$u(x,t) = \int_{-\infty}^{\infty} f(\xi) \frac{1}{\sqrt{4\pi kt}} e^{-(x-\xi)^2/4kt} \, d\xi.$$

Later we derive this formula by the Fourier transform method. But we can observe formally that the solution is valid by the superposition principle and the fact that $u(x,0) = f(x)$. This latter fact comes from making the change of variables $r = (x - \xi)/\sqrt{4kt}$ and rewriting the solution as

$$u(x,t) = \frac{1}{\sqrt{\pi}} \int_{-\infty}^{\infty} f(x - r\sqrt{4kt}) e^{-r^2} \, dr.$$

Taking the limit as $t \to 0^+$ and formally pulling the limit under the integral gives $u(x,0) = \frac{1}{\sqrt{\pi}} \int_{-\infty}^{\infty} f(x) e^{-r^2} \, dr = f(x) \frac{1}{\sqrt{\pi}} \int_{-\infty}^{\infty} e^{-r^2} \, dr = f(x)$. □

EXERCISES

1. Use software to sketch the solution surface and temperature time profiles of the fundamental solution of the heat equation (1.4)

$$u(x,t) = \frac{1}{\sqrt{4\pi kt}} e^{-x^2/4kt}$$

for different times t. What is the behavior of the profiles as t approaches zero? Comment on the differences one would observe in the profiles for large values of k and for small values of k.

2. Find the general solution of the following partial differential equations in terms of arbitrary functions.

   a) $u_{xx} + u = 6y$, where $u = u(x,y)$.

   b) $tu_{xx} - 4u_x = 0$, where $u = u(x,t)$. (Hint: Let $v = u_x$.)

   c) $u_{xt} + (1/x)u_t = t/x^2$, where $u = u(x,t)$.

   d) $u_{tx} + u_x = 1$, where $u = u(x,t)$.

   e) $uu_t = x - t$, where $u = u(x,t)$. (Hint: $(u^2)_t = 2uu_t$.)

3. Find a formula for the solution to the

$$u_{xt} = f(x,t), \quad x,\ t > 0$$

   that satisfies the auxiliary conditions $u(x,0) = g(x)$, $x > 0$ and $u(0,t) = h(t)$, $t > 0$, where f, g, and h are given, well-behaved functions with $g(0) = h(0)$, $g'(0) = h'(0)$.

4. Find the general solution of the first-order linear equation

$$u_t + cu_x = 0$$

   by changing variables to the new spatial coordinate $z = x - ct$, where c is a constant. (Hint: take $\tau = t$, $z = x - ct$.)

5. Show that the general solution $u = u(x,y)$ of the equation

$$yu_x - xu_y - 0$$

   is $u = \psi(x^2 + y^2)$, where ψ is an arbitrary function.

6. Determine regions in the plane where the equation $yu_{xx} - 2u_{xy} + xu_{yy} = 0$ is hyperbolic, elliptic, or parabolic.

7. Show that a parabolic equation $\alpha u_{tt} + \beta u_{tx} + \gamma u_{xx} = F(x, t, u, u_x, u_t)$, where α, β, γ are constant, can be reduced to an equation of the form $U_{\eta\eta} = G(\xi, \eta, U, U_\xi, U_\eta)$ for $U = U(\xi, \eta)$ by a linear transformation of the variables, $\xi = ax + bt$, $\eta = cx + dt$.

8. Find all solutions of the heat equation $u_t = ku_{xx}$ of the form $u(x, t) = U(z)$, where $z = x/\sqrt{kt}$.

6.2 Conservation Laws

Many fundamental equations in the natural and physical sciences are obtained from **conservation laws**. Conservation laws are balance laws, or equations that express the fact that some quantity is balanced throughout a process. In thermodynamics, for example, the first law states that the change in internal energy in a given system is equal to, or is balanced by, the total heat added to the system plus the work done on the system. Thus the first law of thermodynamics is an energy balance law, or conservation law. As another example, consider a fluid flowing in some region of space that consists of chemical species undergoing chemical reaction. For a given chemical species, the time rate of change of the total amount of that chemical in the region must equal the rate at which the chemical flows into the region, minus the rate at which it flows out, plus the rate at which the species is created, or consumed, by the chemical reactions. This is a verbal statement of a conservation law for the amount of the given chemical species. Similar balance or conservation laws occur in all branches of science. In population ecology, for example, the rate of change of a given animal population in a certain region must equal the birthrate, minus the deathrate, plus the migration rate into or out of the region.

6.2.1 One Dimension

Mathematically, conservation laws translate into differential equations, which are then regarded as the *governing equations* or *equations of motion* of the process. These equations dictate how the process evolves in time. First we formulate the basic one-dimensional conservation law, out of which will come some of the basic models and concepts.

Let us consider some quantity (mass, animals, energy, momentum, or whatever) distributed along a tube of cross-sectional area A (see Fig. 6.2), and let $u = u(x, t)$ denote its density, or concentration, measured in the amount of the quantity per unit volume. An implicit assumption is that there is enough of the

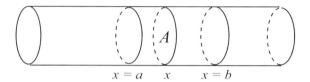

Figure 6.2 Cylindrical tube of cross-sectional area A showing a cross section at x and a finite section I: $a \le x \le b$. $u(x,t)$ is the concentration at x at time t.

quantity distributed in space that it makes sense to talk about its density. (If there were only a few items, then a probability model would be more appropriate.) By definition, u varies in only one spatial direction, the axial direction x. Therefore, by assumption, u is constant in any cross section of the tube, which is the one-dimensional assumption. Next consider an *arbitrary* segment of the tube denoted by the interval $I = [a, b]$. The total amount of the quantity u inside I at time t is

$$\text{Total amount of the quantity in } I = \int_a^b u(x,t)A\, dx.$$

Assume there is motion of the quantity in the tube in the axial direction. We define the **flux** of u at x at time t to be the scalar function $J(x,t)$; that is, $J(x,t)$ is the amount of the quantity u flowing through the cross section at x at time t, per unit area, per unit time. Thus the dimensions of J are $[J] =$ amount/(area·time). Notice that flux is density times velocity. By convention we take J to be positive if the flow at x is in the positive x direction, and J is negative at x if the flow is in the negative x direction. Therefore, at time t the *net* rate that the quantity is flowing into the interval I is the rate it is flowing in at $x = a$ minus the rate it is flowing out at $x = b$. That is,

$$\text{Net rate of quantity flowing into } I = AJ(a,t) - AJ(b,t).$$

Finally, the quantity u may be created or destroyed inside I by some external or internal source (e.g., by a chemical reaction if u were a species concentration, or by birth or death if u were a population density, or by a heat source or sink if u were energy density—a heat source could be resistive heating if the tube were a wire through which a current is flowing). We denote this *source function*, which is a local function acting at each x, by $f(x,t,u)$ with dimensions $[f] =$ amount /(vol·time). Consequently, f is the rate that u is created (or destroyed) at x at time t, per unit volume. Note that the source function f may depend on u itself, as well as space and time. If f is positive, we say that it is a *source*, and if f is negative, we say that it is a *sink*. Given f, we may calculate the total rate that u is created in I by integration. We have

Rate that quantity is produced in I by sources $= \displaystyle\int_a^b f(x,t,u(x,t))A\,dx.$

The fundamental conservation law may now be formulated for the quantity u. For any interval I, we have

Rate of change of total amount of the quantity in I

$=$ Net rate of quantity flowing into I

$+$ Rate that quantity is produced by sources in I.

In terms of the expressions we introduced above, after canceling the constant cross-sectional area A, we have

$$\frac{d}{dt}\int_a^b u(x,t)\,dx = J(a,t) - J(b,t) + \int_a^b f(x,t,u)\,dx. \qquad (2.1)$$

In summary, (2.1) states that the rate that u changes in I must equal the net rate at which u flows into I plus the rate that u is produced in I by sources. Equation (2.1) is called a **conservation law in integral form**, and it holds even if u, J, or f are not smooth (continuously differentiable) functions. The latter remark is important when we consider in subsequent chapters physical processes giving rise to shock waves, or discontinuous solutions.

If some restrictions are place on the triad u, J, and f, (2.1) may be transformed into a single partial differential equation. Two results from elementary integration theory are required to make this transformation: (i) the fundamental theorem of calculus, and (ii) the result on differentiating an integral with respect to a parameter in the integrand. Precisely,

(i) $\int_a^b J_x(x,t)\,dx = J(b,t) - J(a,t).$

(ii) $\frac{d}{dt}\int_a^b u(x,t)\,dx = \int_a^b u_t(x,t)\,dx.$

These results are valid if J and u are continuously differentiable functions on $\mathbb{R}^2$. Of course, (i) and (ii) remain correct under less stringent conditions, but the assumption of smoothness is all that is required in the subsequent discussion. Therefore, assuming smoothness of u and J, as well as continuity of f, equations (i) and (ii) imply that the conservation law (2.1) may be written

$$\int_a^b [u_t(x,t) + J_x(x,t) - f(x,t,u)]\,dx = 0, \quad \text{for all intervals } I = [a,b]. \quad (2.2)$$

Because the integrand is a continuous function of x, and because (2.2) holds for all intervals of integration I, it follows that the integrand must vanish identically; that is,

$$u_t + J_x = f(x, t, u), \quad x \in \mathbb{R}, \quad t > 0. \tag{2.3}$$

Equation (2.3) is a partial differential equation relating the density $u = u(x, t)$ and the flux $J = J(x, t)$. Both are regarded as unknowns, whereas the form of the source f is given. Equation (2.3) is called a **conservation law in differential form**, in contrast to the integral form (2.2). The J_x term is called the *flux term* because it arises from the movement, or transport, of u through the cross section at x. The source term f is often called a *reaction term* (especially in chemical contexts) or a *growth* or *interaction* term (in biological contexts). Finally, we have defined the flux J as a function of x and t, but this dependence on space and time may occur through dependence on u or its derivatives. For example, a physical assumption may require us to posit $J(x, t) = J(x, t, u(x, t))$, where the flux is dependent upon u itself.

If there are several quantities with densities u_1, u_2,..., then we can write down a conservation law for each. The fluxes and source terms may involve several of the densities, leading to a coupled system of conservation laws.

Remark 6.8

(**Small box method**) We derived the differential form of the conservation law (2.3) by balancing mass in an interval and then appealing to the arbitrariness of the interval. This is sometimes referred to as the **large box** method. Another common approach is to use a **small box**, or small interval $[x, x+dx]$, where dx is a small increment. Then we could write a conservation law on this interval as

$$\frac{\partial}{\partial t}(u(\tilde{x}_1, t)A\,dx) = AJ(x, t) - AJ(x + dx, t) + f(\tilde{x}_2, t)A\,dx,$$

where $x < \tilde{x}_1$, $\tilde{x}_2 < x + dx$. Dividing by the volume of the box $A\,dx$ gives

$$\frac{\partial}{\partial t}u(\tilde{x}_1, t) = \frac{J(x, t) - J(x + dx, t)}{dx} + f(\tilde{x}_2, t).$$

Now, taking the limit as $dx \to 0$ gives the conservation law

$$u_t = -J_x + f. \qquad \square$$

6.2.2 Several Dimensions

It is straightforward to formulate conservation laws in several spatial dimensions. For review, and to set the notation, we introduce some basic definitions.

In $\mathbb{R}^n$, points are denoted by $\mathbf{x} = (x_1, ..., x_n)$, and $d\mathbf{x} = dx_1 \cdots dx_n$ represents a volume element. In $\mathbb{R}^2$ we often use (x, y), and in $\mathbb{R}^3$ we use (x, y, z). If $u = u(\mathbf{x}, t)$ is a scalar function (a density), then its integral over a suitable region (volume) Ω is

$$\int_\Omega u(\mathbf{x}, t) \, d\mathbf{x}.$$

In $\mathbb{R}^2$ we have $d\mathbf{x} = dx dy$, and in $\mathbb{R}^3$ we have $d\mathbf{x} = dx dy dz$. The boundary of Ω is denoted by $\partial\Omega$. We always assume that the domain Ω has a smooth boundary $\partial\Omega$, or piecewise smooth boundary composed of finitely many smooth components. We say Ω is **open** if it includes none of its boundary, and we denote $\overline{\Omega} = \Omega \cup \partial\Omega$ as the **closure** of Ω. This assumption will cover the cases of spheres, rectangles, parallelepipeds, cylinders, and so on, which are the domains important in most science and engineering problems. We refer to these type of domains as *well behaved*. The set of continuous functions on a domain Ω is denoted by $C(\Omega)$. Surface integrals are denoted by

$$\int_{\partial\Omega} u(\mathbf{x}, t) \, dA.$$

Nearly always, surface integrals are flux integrals (discussed below). Partial derivatives are represented by subscripts, or $\frac{\partial u}{\partial x_k} = u_{x_k}$. The **gradient** of u is the vector

$$\nabla u = (u_{x_n}, ..., u_{x_n}).$$

The gradient points in the direction of the maximum increase of the function. The **directional derivative** of u in the direction of a unit vector $\mathbf{n} = (n_1, ..., n_n)$ is denoted by

$$\frac{du}{dn} = \mathbf{n} \cdot \nabla u.$$

It measures the change in u in the direction $\mathbf{n}$. If $\mathbf{J}(\mathbf{x}, t) = (\mathbf{x}, t) = (J_1(\mathbf{x}, t), ..., J_n(\mathbf{x}, t))$ is a vector field, then its **divergence** is the scalar function

$$\nabla \cdot \mathbf{J} = \frac{\partial J_1}{\partial x_1} + \cdots + \frac{\partial J_n}{\partial x_n}.$$

Two common, alternate notations for the divergence and gradient are div $\mathbf{J}$ and grad u, respectively.

The **flux** of a vector field $\mathbf{J}(\mathbf{x}, t)$ through an oriented surface element $\mathbf{n} dA$, where $\mathbf{n}$ is an outward oriented unit normal vector (see Fig. 6.3 for a diagram in $\mathbb{R}^3$), is defined to be

$$\text{flux} = \mathbf{J}(\mathbf{x}, t) \cdot \mathbf{n} \, dA.$$

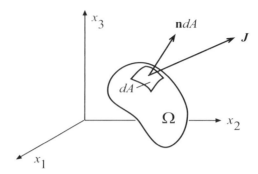

Figure 6.3   A surface with surface element $\mathbf{n}dA$, oriented by its outward unit normal vector $\mathbf{n}$, along with the flux vector $\mathbf{J}$.

There are two useful differential identities that are straightforward to prove. The first is a formula for the divergence of a gradient, and the second is a product rule for the divergence of a scalar function times a vector field:

$$\nabla \cdot \nabla u \;=\; \Delta u \equiv \frac{\partial^2 u}{\partial x_1^2} + \cdots + \frac{\partial^2 u}{\partial x_n^2}$$
$$\nabla \cdot (u\mathbf{J}) \;=\; u\nabla \cdot \mathbf{J} + \nabla u \cdot \mathbf{J}.$$

The first expression defines the **Laplacian** Δu as the sum of the second partial derivatives. Another common notation for the Laplacian of u is

$$\Delta u = \nabla^2 u.$$

Next we formally catalog two useful integral theorems in $\mathbb{R}^n$ that are analogs of the one-dimensional results from the preceding section.

Theorem 6.9

(a) Let f be a continuous function on a well-behaved domain Ω with

$$\int_D f(\mathbf{x})\,d\mathbf{x} = 0$$

for every subdomain $D \subset \Omega$. Then $f(\mathbf{x}) = 0$ for all $x \in \Omega$.

(b) Let $u = u(\mathbf{x}, t)$ and $u_t(\mathbf{x}, t)$ be continuous on a well-behaved domain $x \in \Omega$, and for all $t \in I$, where I is an open interval. Then the time derivative can be pulled under the integral sign, or

$$\frac{d}{dt} \int_\Omega u(\mathbf{x}, t)\,d\mathbf{x} = \int_\Omega u_t(\mathbf{x}, t)\,d\mathbf{x}. \quad \square$$

The most important integral identity in multivariable calculus in the **divergence theorem**, which is an n-dimensional version of the fundamental theorem of calculus. It states that the integral of a partial derivative over a volume is a boundary, or surface, integral. It has been proved for very general domains Ω, but we only require it for well-behaved domains. A natural mathematical question concerns conditions on the functions for which the divergence theorem holds. Generally, we assume that all functions are continuous on the closed domain consisting of Ω and its boundary $\partial \Omega$, and have as many continuous partial derivatives in Ω that occur in the formula. We do not formulate the most general conditions. Let Ω be an open, well-behaved domain. By $u \in C^1(\overline{\Omega})$ we mean that u is continuous on Ω, and its first partial derivatives are continuous on Ω and can be extended continuously to the boundary of Ω. By $u \in C^2(\Omega)$ to mean that u has continuous second partial derivatives in Ω. Many of our theorems require $u \in C^1(\overline{\Omega}) \cap C^2(\Omega)$, that is, u is twice continuously differentiable in the open domain Ω and its first derivatives can be extended to the closure. With all this said, we in fact do not require detailed mathematical analysis in our work; so these technical issues do not enter the discussion.

Theorem 6.10

(**Divergence theorem**) Let Ω be a well-behaved, open, bounded region in $\mathbb{R}^n$ and let $u \in C^1(\overline{\Omega}) \cap C^2(\Omega)$. Then

$$\int_{\Omega} u_{x_k} d\mathbf{x} = \int_{\partial \Omega} u n_k dA,$$

where n_k is the kth component of the outward unit normal vector $\mathbf{n}$ on the bounding surface. □

If $\mathbf{J}$ is a vector field, we can apply this result to each component and add to obtain the standard vector version of the theorem:

$$\int_{\Omega} \nabla \cdot \mathbf{J} dx = \int_{\partial \Omega} \mathbf{J} \cdot \mathbf{n} dA. \qquad \text{(Divergence theorem)}$$

The divergence theorem leads immediately to fundamental integral identities. We state these important results.

Corollary 6.11

Under the assumptions of the Theorem 6.10:

(a) **Integration by parts formula**

$$\int_\Omega w u_{x_k}\, d\mathbf{x} = -\int_\Omega u w_{x_k}\, d\mathbf{x} + \int_{\partial\Omega} u w n_k\, dA.$$

(a) **Green's first identity**

$$\int_\Omega (u\Delta w + \nabla u \cdot \nabla w)d\mathbf{x} = \int_{\partial\Omega} u \frac{dw}{dn}\, dA.$$

(c) **Green's second identity**

$$\int_\Omega u\Delta w\, d\mathbf{x} = \int_\Omega w\Delta u\, d\mathbf{x} + \int_{\partial\Omega}\left(u\frac{dw}{dn} - w\frac{du}{dn}\right) dA.$$

It is impossible to overestimate the utility of integration by parts formulae in partial differential equations. Green's second identity is just an integration by parts formula for the Laplacian operator. The proofs of these relations are requested in the Exercises.

With this preparation we are ready to write down the conservation law in several variables. Let Ω be a region in $\mathbb{R}^n$ and let $u = u(\mathbf{x}, t)$ be the density, or concentration, of some quantity distributed throughout Ω. Let $J(\mathbf{x}, t)$ be the flux vector for that quantity, where $\mathbf{J} \cdot \mathbf{n} dA$ measures the rate that the quantity crosses an oriented surface element in the medium. In this domain Ω let V be an arbitrary, well-behaved bounded region (for example, a sphere). Then the total amount of the quantity in V is given by the volume integral

$$\text{Total amount in } V = \int_V u(\mathbf{x}, t)\, d\mathbf{x}.$$

The time rate of change of the total amount of the quantity in V must be balanced by the rate that the quantity is produced in V by sources, plus the net rate that the quantity flows through the boundary of V. We let $f(\mathbf{x}, t, u)$ denote the source term, so that the rate that the quantity is produced in V is given by

$$\int_V f(\mathbf{x}, t, u)\, d\mathbf{x}.$$

The net *outward* flux of the quantity u through the boundary ∂V is given by the surface integral

$$\int_{\partial V} \mathbf{J} \cdot \mathbf{n}\, dA,$$

where $\mathbf{n} = \mathbf{n}(\mathbf{x})$ is the outward unit normal. Therefore, the **conservation law**, or balance law, is given by

$$\frac{d}{dt} \int_V u \, d\mathbf{x} = - \int_{\partial V} \mathbf{J} \cdot \mathbf{n} \, dA + \int_V f(\mathbf{x}, t, u) \, d\mathbf{x}. \qquad (2.4)$$

The minus sign on the flux term occurs because outward flux decreases the rate that u changes in V. This integral law is the higher-dimensional analog of (2.2).

The integral form of the conservation law (2.4) can be reformulated as a local condition, that is, a partial differential equation, provided that u and $\mathbf{J}$ are sufficiently smooth functions. In this case the surface integral can be rewritten as a volume integral over V using the divergence theorem, and the time derivative can be brought under the integral sign by Theorem 6.9(b). Hence,

$$\int_V u_t \, d\mathbf{x} = \int_V \nabla \cdot \mathbf{J} \, d\mathbf{x} + \int_V f(\mathbf{x}, t, u) \, d\mathbf{x}.$$

From Theorem 6.9(a), the arbitrariness of V implies the **local form of the conservation law** throughout the domain Ω:

$$u_t + \nabla \cdot \mathbf{J} = f(\mathbf{x}, t, u), \quad x \in \Omega, \quad t > 0. \qquad (2.5)$$

Equation (2.5) is the higher-dimensional version of equation (2.3), the local conservation law in one dimension.

6.2.3 Constitutive Relations

Because the one-dimensional conservation law (2.3) [or (2.5)] is a single partial differential equation for two unknown quantities (the density u and the flux J), our intuition indicates that another equation is required to have a well-determined system. This additional equation is often an equation that is based on an assumption about the physical properties of the medium, which, in turn, is based on empirical reasoning or experimental observation. Equations expressing these assumptions are called **constitutive relations** or **equations of state**. Thus constitutive equations are on a different level from the basic conservation law; the latter is a fundamental law of nature connecting the density u to the flux J, whereas a constitutive relation is often an approximate equation whose origin lies in empirics. In the next few paragraphs we examine some possibilities for modeling the flux, or motion of the quantity under examination.

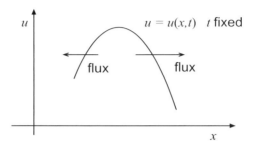

Figure 6.4 Time snapshot of the concentration $u(x,t)$ at a fixed time t. The arrows indicate the direction of the flow, from higher concentrations to lower concentrations. The flow is said to be *down the gradient*. The minus occurs because positive flux requires a negative derivative.

Example 6.12

(**Diffusion**) In many physical problems it is observed that the amount of the substance that moves through a cross section at x at time t is proportional to its density gradient u_x, or $J(x,t) \propto u_x(x,t)$. If $u_x > 0$, then $J < 0$ (the substance moves to the left), and if $u_x < 0$, then $J > 0$ (the substance moves to the right). Figure 6.4 illustrates the situation. The substance moves down the density gradient, from regions of high concentration to regions of low concentrations. We say the motion is "down the gradient." By the second law of thermodynamics, heat behaves in this manner; heat flows from hotter regions to colder regions, and the steeper the temperature distribution curve, the more rapid the flow of heat. This movement is closely connected to the random motion of molecules caused by collisions of the particles. If u represents a concentration of organisms, one may observe that they move from high densities to low densities with a rate proportional to the density gradient. Therefore, we assume the basic constitutive law

$$J(x,t) = -Du_x(x,t), \tag{2.6}$$

which is known as **Fick's law**. It models diffusion. The positive proportionality constant D is called the **diffusion constant** and has dimensions $[D] = \text{length}^2/\text{time}$. Fick's law accurately describes the behavior of many physical and biological systems. This assumption reduces the conservation law (2.3) to a single second-order linear partial differential equation for the unknown density $u = u(x,t)$ given by

$$u_t = Du_{xx} + f(x,t,u). \tag{2.7}$$

In several dimensions Fick's law is

$$\mathbf{J}(\mathbf{x},t) = -D\nabla u, \tag{2.8}$$

giving

$$u_t = D\Delta u + f(\mathbf{x}, t, u). \tag{2.9}$$

Equations (2.7) and (2.9) are called **reaction–diffusion equations**, and they govern processes where the flux, called a diffusive flux, is specified by Fick's law. If the source term vanishes, they are called **diffusion equations**. The diffusion constant D defines a characteristic time, or time scale, τ for a diffusion process. If L is a length scale (e.g., the length of the container), the quantity

$$\tau = \frac{L^2}{D}$$

is the only constant in the process with dimensions of time, and τ gives a measure of the time required for discernible changes in concentration to occur. The diffusion constant may depend upon $\mathbf{x}$ in a nonhomogeneous medium, as well as the density u itself. These cases are mentioned below.   □

Example 6.13

(**Classical heat equation**) In the case where u is an energy density (i.e., a quantity with dimensions of energy per unit volume), the diffusion equation describes the transport of energy in the medium. In a homogeneous medium with density ρ and specific heat C, the energy density is

$$u(\mathbf{x}, t) = C\rho T(\mathbf{x}, t),$$

where T is the temperature. The dimensions are $[C]$ =energy/(mass·degree) and $[\rho]$ =mass/volume. For example, the units of energy could be joules or calories. The conservation law may be written

$$C\rho T_t + \nabla \cdot \mathbf{J} = f,$$

where f is a heat source. In heat conduction Fick's law has the form

$$\mathbf{J} = -K\nabla T, \tag{2.10}$$

where K is the thermal conductivity, measured in energy/(mass·time·degree). In the context of heat conduction, Fick's law is called **Fourier's law** of heat conduction. It follows that the temperature T satisfies the equation

$$C\rho T_t - K\nabla \cdot \nabla T = f,$$

or

$$T_t = k\Delta T + \frac{1}{C\rho}f, \quad k = \frac{K}{C\rho}. \tag{2.11}$$

Equation (2.11) is called the **heat equation** with heat source f, and the constant k, which plays the role of the diffusion constant, is called the **diffusivity**, measured as length-squared per unit time. Thus the diffusivity k in heat flow problems is the analog of the diffusion constant, and the heat equation is just the diffusion equation. In this discussion we have assumed that the physical parameters C, K, and ρ of the medium are constant; however, these quantities could depend on the temperature T, which is an origin of nonlinearity in the problem. If these parameters depend on location $\mathbf{x}$, then the medium is nonhomogeneous and the conservation law in this case becomes

$$T_t = \frac{1}{C(\mathbf{x})\rho(\mathbf{x})}\nabla\cdot(K(\mathbf{x})\nabla T) + \frac{f}{C(\mathbf{x})\rho(\mathbf{x})}.$$

The variable conductivity cannot be pulled outside the divergence. Notice that if the system is in a steady-state where the temperature is not changing in time and the source is independent of t, then $T_t = 0$ and the temperature depends on $\mathbf{x}$ alone, $T = T(\mathbf{x})$. Then the last equation becomes

$$-\nabla\cdot(K(\mathbf{x})\nabla T) = f(\mathbf{x},T).$$

This is the **steady-state heat equation**, and it is addressed in detail in later sections. □

Example 6.14

(**Advection**) Advection, or convection or drift, occurs when the bulk motion of the medium carries the substance under investigation along with it at a given velocity. For example, wind can carry propagules, water can carry chemicals or animals, air motion can transport heat energy, and so on. The advective flux, which measures the amount of the substance crossing a section, is given by the velocity times the concentration, or

$$\mathbf{J}(\mathbf{x},t) = \mathbf{c}u(\mathbf{x},t),$$

where $\mathbf{c} = \mathbf{c}(\mathbf{x},t)$ is the velocity of the medium. Then, combined with the conservation law, we obtain the **reaction-advection equation**

$$u_t + \nabla\cdot(\mathbf{c}u) = f.$$

If the advection speed $\mathbf{c}$ is a constant vector, this equation reduces to a first-order PDE

$$u_t + \mathbf{c}\cdot\nabla u = f.$$

These types of equations give rise to wave propagation, rather than diffusion, and are examined in Chapter 7. □

Example 6.15

(**Advection–diffusion**) When both advection and diffusion occur, the flux is composed of two parts and the model becomes

$$u_t = \nabla \cdot (D(x)\nabla u) - \nabla \cdot (\mathbf{c}(\mathbf{x})u) + f.$$

Here we have assumed D and $\mathbf{c}$ are spatially dependent. In one dimension, with constant diffusion,

$$u_t = Du_{xx} - cu_x + f.$$

This is the **reaction–advection–diffusion** equation. $\square$

Example 6.16

(**Growth**) If $u = u(x,t)$ is a population density that grows logistically and Fick's law models its diffusion, then the population law in one dimension is

$$u_t = Du_{xx} + ru\left(1 - \frac{u}{K}\right),$$

which is Fisher's equation (see Section 6.1). This important equation is a prototype of nonlinear reaction-diffusion equations. Other types of growth rates are also of great interest in population ecology. For example, the reaction term $f(u) = ru$ models growth or decay, and the Allee growth rate $f(u) = ru(a - u)(u - b)$, $0 < a < b$, models growth where the growth rate is negative for small populations. $\square$

6.2.4 Probability and Diffusion

There is a strong connection between diffusion and randomness. For example, we think of heat flow, governed by the diffusion equation, being caused by random molecular collisions which cause the energy to disperse from high-energy regions to low-energy regions. This idea is based upon the kinetic theory of gases, and the quantities involved are statistical averages of the behavior of large numbers of particles. For animal populations, diffusion is a different type of mechanism, but the constitutive law is the same as for heat flow (Fick's law). If there are only a few particles or a few animals, then it is not sensible to define a density and we cannot apply the direct arguments above to obtain a partial differential equation. In this case we must rely on probability theory and make deductions about the spread of the quantity of interest. We fix the idea by examining animal populations. As we shall observe, information about how an organism moves on a short time scale can be used to determine where it is likely to be at any time. We translate the short time behavior into a probability density function (pdf) $u(x,t)$ for the random variable $X = X(t)$ giving the location of the organism at time t. The analysis leads to a partial differential equation for $u(x,t)$.

Suppose an organism is located at the origin $x = 0$ at time $t = 0$. The movement rule is a simple random walk in one dimension. At each time step the organism jumps a distance h either to the left or right with equal probability. As time advances in equal discrete steps, the organism moves on the lattice of points $0, \pm h, \pm 2h, \pm 3h, \ldots$ on the x axis. We let $u(x,t)h$ be the probability that the organism is located in the interval between x and $x+h$ at time t, given that it was at $x = 0$ at $t = 0$. Rather than calculating the probability density function directly, we show that it must satisfy a special partial differential equation, the Fokker–Planck equation. This result follows from the **master equation**

$$u(x, t + \tau) = \frac{1}{2}u(x - h, \tau) + \frac{1}{2}u(x + h, \tau),$$

which states that the probability that the organism is in the interval $[x, x+h]$ at time $t + \tau$ equals the probability that it was either in $[x - h, x]$ at time t and jumped to the right, or it was in the interval $[x + h, x + 2h]$ at time t and it jumped to the left (we have canceled a h factor in the last equation). Expanding each term in the master equation in Taylor series, we get

$$u(x,t) + u_t(x,t)\tau + \frac{1}{2}u_{tt}(x,t)\tau^2 + \cdots$$
$$= \frac{1}{2}\left(u(x,t) - u_x(x,t)h + \frac{1}{2}u_{xx}(x,t)h^2 + \cdots\right)$$
$$+ \frac{1}{2}\left(u(x,t) + u_x(x,t)h + \frac{1}{2}u_{xx}(x,t)h^2 + \cdots\right),$$

where the "dots" denote higher-order terms in the increments h and τ. Simplifying, we obtain

$$u_t + \frac{\tau}{2}u_{tt}(x,t) = \frac{h^2}{2\tau}u_{xx} + \cdots.$$

Now we pass from a discrete random walk to a continuous random walk. Taking the limit as $h, \tau \to 0$ in such a way that the ratio of h^2 to 2τ remains constant, or $h^2/2\tau = D$, we get

$$u_t = Du_{xx},$$

which is the one-dimensional diffusion equation. If the probabilities of moving to the left or right in the random walk are not equal, then a *drift* term arises in the preceding argument and the probability density function satisfies the advection-diffusion equation (see the Exercises). These equations for the probability densities in the random walk models are special cases of the

Fokker–Planck equation. In passing from a discrete random walk to a continuous one, we have masked a lot of detailed analysis; suffice it to say that this process can be put on a firm mathematical foundation using the theory of stochastic processes.

We have derived the diffusion equation in two ways—from a conservation law and from a probability model. Previously, we defined the **fundamental solution** to the one-dimensional diffusion equation as

$$u(x,t) = \frac{1}{\sqrt{4\pi Dt}} e^{-x^2/4Dt}, \quad t > 0.$$

It will be derived later using Fourier transforms. This solution is the probability density function for the random variable $X = X(t)$, which is the location of an individual performing a continuous random walk beginning at the origin. That is,

$$\Pr(a < X(t) \leq b) = \int_a^b u(x,t)\,dx.$$

It is straightforward to show that $\int_{-\infty}^{\infty} u(x,t)dx = 1$. We recognize the distribution u as a normal density with variance $2Dt$. Therefore, the probability density profiles spread as time increases, and the location of the individual becomes less certain. As $t \to 0^+$ we have $u(x,t) \to 0$ if $x \neq 0$, and $u(0,t) \to +\infty$. In fact, we have $u(x,0) = \delta_0(x)$, in a distributional sense (fundamental solution). The **fundamental solution** to the diffusion equation $u_t = D\Delta u$ in $\mathbb{R}^n$ is

$$u(\mathbf{x},t) = \frac{1}{(4\pi Dt)^{n/2}} e^{-|\mathbf{x}|^2/4Dt}, \quad \mathbf{x} \in \mathbb{R}^n, \; t > 0.$$

This solution, which one can verify, can be derived using Fourier transforms (Section 6.5). The function $u(\mathbf{x} - \xi, t)$ describes the evolution of organisms or heat, for example, from a point release of unit magnitude at a location ξ at time $t = 0$. Thus, the initial condition is a delta distribution $\delta_\xi(\mathbf{x})$. So it is the solution to a problem with a unit point source. An example from population ecology illustrates the utility of this solution.

Example 6.17

(**Invasion of species**) In two spatial dimensions, consider a diffusing animal population whose growth rate is λu. The model is the growth–diffusion equation

$$u_t = D\Delta u + \lambda u, \quad \mathbf{x} \in \mathbb{R}^2, \; t > 0.$$

A change of dependent variables to $v(\mathbf{x},t) = u(\mathbf{x},t)e^{-\lambda t}$ transforms the growth-diffusion equation into the diffusion equation for v,

$$v_t = D\Delta v, \quad \mathbf{x} \in \mathbb{R}^2, \; t > 0.$$

The fundamental solution is, using $R = |\mathbf{x}| = \sqrt{x^2 + y^2}$,

$$v(R, t) = \frac{1}{4\pi Dt} e^{-R^2/4Dt}, \quad R > 0, \ t > 0.$$

Therefore, a point release of u_0 animals at the origin will result in the radial population density

$$u(R, t) = \frac{u_0}{4\pi Dt} e^{\lambda t - R^2/4Dt}, \quad R > 0, \ t > 0.$$

We ask how fast the animals invade the region. Imagine that the population is detectable at a given radius when it reaches a critical threshold density u_c; that is,

$$\frac{u_0}{4\pi Dt} e^{\lambda t - R^2/4Dt} = u_c.$$

This leads to

$$\frac{R}{t} = \sqrt{4\lambda D - \frac{4D}{t} \ln\left(\frac{4\pi u_c}{u_0} t\right)}.$$

For large times t the limit is

$$\frac{R}{t} = \sqrt{4\lambda D}.$$

Therefore the invasion speed approaches a constant determined by the growth rate λ and the diffusion constant D. This result has been confimed by various observations, one of the most famous being the muskrat invasion in Europe and Asia beginning in 1905 by an accidental release in Bohemia.   □

6.2.5 Boundary Conditions

In the context of heat conduction, we now discuss the types of auxiliary conditions that lead to a well-defined mathematical problem that is physically meaningful. The heat conduction equation is the canonical example of a parabolic partial differential equation. Basically, evolution problems require an initial condition, and if spatial boundaries are present, then boundary conditions are needed. We want to formulate problems with conditions that make the mathematical problem **well posed**. This means that a solution must exist, the solution must be unique, and the solution should be stable in the sense that it should depend continuously on initial and boundary data.

Consider a bar of length l and constant cross-sectional area A. The temperature $u = u(x, t)$ (we use u as the dependent variable, rather than T) must satisfy the one-dimensional heat equation (see (2.11))

$$u_t - ku_{xx} = 0, \quad 0 < x < l, \quad t > 0, \tag{2.12}$$

where we have assumed no heat sources.

An auxiliary condition on the temperature u at time $t = 0$ is called an **initial condition** and is of the form

$$u(x, 0) = f(x), \quad 0 < x < l, \tag{2.13}$$

where $f(x)$ is the given initial temperature distribution. Conditions prescribed at $x = 0$ and $x = l$ are **boundary conditions**. If the temperature u at the ends of the bar are prescribed, then the boundary conditions take the form

$$u(0, t) = g(t), \quad u(l, t) = h(t), \quad t > 0, \tag{2.14}$$

where g and h are prescribed functions. Boundary conditions of this type are called **Dirichlet conditions**. Other boundary conditions are possible. It is easy to imagine physically that one end of the bar, say at $x = 0$, is insulated so that no heat can pass through. By Fourier's heat law this means that the flux at $x = 0$ is zero or

$$u_x(0, t) = 0, \quad t > 0. \tag{2.15}$$

Condition (2.15) is called an **insulated boundary condition**. Or, one may prescribe the flux at an end as a given function of t, for example,

$$-K u_x(0, t) = g(t), \quad t > 0, \tag{2.16}$$

which is called a **Neumann condition**. If the conductivity K depends on u, then condition (2.16) is a nonlinear boundary condition. Finally, a boundary condition may be specified by Newton's law of cooling,

$$-K u_x(0, t) = \alpha(u(0, t) - \psi(t)), \quad t > 0,$$

which requires the flux be proportional to the difference between the temperature at the end and the temperature ψ of the outside environment. This is a **Robin condition**, or **radiation condition**.

Remark 6.18

For the advection–diffusion equation

$$u_t = D u_{xx} - c u_x,$$

the flux is $J(x, t) = -D u_x + cu$; so specifying the flux at a boundary amounts to specifying J, and not just the diffusive flux $-D u_x$. □

If an initial condition is given and a boundary condition is prescribed at one end (say at $x = 0$) of a very long rod, then the problem can be considered on a semi-infinite domain $x \geq 0$ on which the heat conduction equation holds. For example, the initial boundary value problem

$$u_t - ku_{xx} = 0, \quad t > 0, \ 0 < x < \infty,$$
$$u(x, 0) = 0, \quad 0 < x < \infty,$$
$$u(0, t) = g(t), \quad t > 0,$$

models heat flow in an infinite medium $x \geq 0$ of constant diffusivity k, initially at zero degrees, subject to the maintenance of the end at $x = 0$ at $g(t)$ degrees. Here we expect the problem to model heat conduction in a long bar for short enough time so that any condition at the far end would not affect the temperature distribution in the portion of the rod that we are interested in studying. As a practical example, heat flow in an infinite medium arises from the study of the underground temperature variations, given the temperature changes at ground level on the earth ($x = 0$).

It seems clear that problems on the infinite interval $-\infty < x < \infty$ are also relevant in special physical situations where the medium is very long in both directions. In this case only an initial condition can be prescribed. To restrict the class of solutions to those that are physically meaningful, a condition is sometimes prescribed at infinity; for example, u is bounded at infinity or $\lim_{x \to +\infty} u(x, t) = 0$, $t > 0$.

In summary, by an **initial boundary value problem** for the heat conduction equation we mean the problem of solving the heat equation (2.12) subject to the initial initial condition (2.13) and some type of boundary conditions imposed at the endpoints. It may be clear from the physical context which auxiliary conditions should be prescribed to obtain a unique solution; in some cases it is not clear. A large body of mathematical literature is devoted to proving existence and uniqueness theorems for various kinds of partial differential equations subject to sundry auxiliary data. An example of a uniqueness theorem is the following:

Theorem 6.19

The initial boundary value problem

$$u_t - ku_{xx} = 0, \quad 0 < x < l, \ 0 < t < T,$$
$$u(x, 0) = f(x), \quad 0 < x < l,$$
$$u(0, t) = g(t), \quad u(l, t) = h(t), \ 0 < t < T,$$

where $f \in C[0, l]$ and $g, h \in C[0, T]$, has a solution $u(x, t)$ on the rectangle $R \colon 0 \leq x \leq l, \ 0 \leq t \leq T$, then the solution is unique.   □

Proof

We prove this theorem by an **energy argument**. By way of contradiction, assume solutions are not unique and there are two distinct solutions $u_1(x,t)$ and $u_2(x,t)$. Then their difference $w(x,t) \equiv u_1(x,t) - u_2(x,t)$ must satisfy the boundary value problem

$$w_t - kw_{xx} = 0, \quad 0 < x < l, \ 0 < t < T,$$
$$w(x,0) = 0, \quad 0 < x < l,$$
$$w(0,t) = w(l,t) = 0, \quad 0 < t < T.$$

If we show $w(x,t) \equiv 0$ on R, then $u_1(x,t) = u_2(x,t)$ on R, which is a contradiction. To this end define the **energy integral**

$$E(t) = \int_0^l w^2(x,t)\,dx.$$

Taking the time derivative,

$$E'(t) = \int_0^l 2ww_t\,dx = 2k\int_0^l ww_{xx}\,dx = 2kww_x\big|_0^l - 2k\int_0^l w_x^2\,dx,$$

where the last equality was obtained using integration by parts. Because the boundary term vanishes,

$$E'(t) = -2k\int_0^l w_x^2(x,t)dx \leq 0.$$

Thus $E(t)$ is nonincreasing, which along with the facts that $E(t) \geq 0$ and $E(0) = 0$, implies $E(t) = 0$. Therefore $w(x,t) = 0$ identically in $0 \leq x \leq l$, $0 \leq t \leq T$, since w is continuous in both its arguments. Existence of a solution is established later by actually exhibiting the solution. $\square$

   The energy method is an important technique in proving uniqueness and obtaining estimates of solutions in partial differential equations in one and in several dimensions. In addition to existence and uniqueness questions the notion of continuous dependence of the solution on the initial or boundary data is important. From a physical viewpoint it is reasonable that small changes in the initial or boundary temperatures should not lead to large changes in the overall temperature distribution. A mathematical model should reflect this stability in that small changes in the auxiliary data should lead to only small changes in the solution. Stated differently, the solution should be stable under small perturbations of the initial or boundary data. (We remark, parenthetically, that although stability seems desirable, many important physical processes are unstable, and so **ill-posed** problems are often studied as well.)

Example 6.20

(**Hadamard's**[2] **example**) Consider the partial differential equation

$$u_{tt} + u_{xx} = 0, \quad t > 0, \quad x \in \mathbb{R} \tag{2.17}$$

subject to the initial conditions

$$u(x,0) = 0, \quad u_t(x,0) = 0, \quad x \in \mathbb{R}. \tag{2.18}$$

Note that (2.17) is elliptic and not parabolic like the heat equation. The solution of this pure initial value problem is clearly the zero solution $u(x,t) \equiv 0$ for $t \geq 0$, $x \in \mathbb{R}$. Now let us change (2.18) to

$$u(x,0) = 0, \quad u_t(x,0) = 10^{-n} \sin 10^n x, \tag{2.19}$$

which represents a very small change in the initial data, provided n is large. As one can check, the solution of (2.17) subject to this new boundary condition is

$$u(x,t) = 10^{-2n} \sin(10^n x) \sinh(10^n t). \tag{2.20}$$

For large values of t, the function $\sinh(10^n t)$ behaves like $\exp(10^n t)$. Therefore the solution grows exponentially with t. Therefore, for the initial value problem (2.17)–(2.18), an arbitrarily small change in the initial data leads to an arbitrarily large solution, and the problem is not well posed. Observe that the PDE here is Laplace's equation which models steady state phenomena; an initial value problem for Laplace's equation is not well posed.   □

Initial value problems for elliptic equations are not well posed. In numerical calculations stability is essential. For example, suppose a numerical scheme is devised to propagate the initial and boundary conditions. Those conditions can never be represented exactly in the computer, as small errors will exist because of roundoff or truncation of the data. If the problem itself is unstable, then these small errors in the data may be propagated in the numerical scheme in such a way that the calculation becomes meaningless.

EXERCISES

1. Solve the nonlinear Cauchy problem

$$\begin{aligned} u_t &= k u_{xx} + u_x^2, \quad t > 0, \quad x \subset \mathbb{R}, \\ u(x,0) &= f(x), \quad x \in \mathbb{R}, \end{aligned}$$

   by making the transformation $w = e^u$.

---

[2] Jacques Hadamard (1865–1963) was a French mathematician who contributed to analysis, number theory, and partial differential equations.

2. Use the energy method to prove that a solution to the initial boundary value problem

$$u_t - ku_{xx} = 0, \quad 0 < x < l, \; 0 < t < T,$$
$$u(x,0) = f(x), \quad 0 < x < l,$$
$$u_x(0,t) = 0, \quad u(l,t) = h(t), \; 0 < t < T,$$

must be unique.

3. Use the energy method to prove uniqueness for the problem

$$
\begin{aligned}
u_t &= \Delta u, & \mathbf{x} \in \Omega, \; t > 0, \\
u(\mathbf{x},0) &= f(\mathbf{x}), & \mathbf{x} \in \Omega, \\
u(\mathbf{x},t) &= g(\mathbf{x}), & \mathbf{x} \in \partial\Omega, \; t > 0.
\end{aligned}
$$

4. A homogeneous (constant ρ, C, and K) metal rod has cross-sectional area $A(x)$, $0 < x < l$, and there is only a small variation of $A(x)$ with x, so that the assumption of constant temperature in any cross section remains valid. There are no sources and the flux is given by $-Ku_x(x,t)$. From a conservation law obtain a partial differential equation for the temperature $u(x,t)$ that reflects the area variation of the bar.

5. In the absence of sources, derive the diffusion equation for radial motion in the plane,

$$u_t = \frac{D}{r}(ru_r)_r,$$

from first principles. That is, take an arbitrary domain between circles $r = a$ and $r = b$ and apply a conservation law for the density $u = u(r,t)$, assuming the flux is $J(r,t) = -Du_r$. Assume no sources.

6. A fluid, having density ρ, specific heat C, and conductivity K, flows at a constant velocity V in a cylindrical tube of length L and radius R. The temperature at position x is $T = T(x,t)$, and diffusion of heat is ignored. As it flows, heat is lost through the lateral side at a rate jointly proportional to the area and to the difference between the temperature T_e of the external environment and the temperature $T(x,t)$ of the fluid (Newton's law of cooling). Derive a partial differential differential equation model for the temperature $T(x,t)$. Find the general solution of the equation by transforming to a moving coordinate system $z = x - Vt$, $\tau = t$.

7. Show that the growth–diffusion equation $u_t = D\Delta u + ru$ can be transformed into a pure diffusion equation via the transformation $v = ue^{-rt}$.

8. Repeat the random walk argument in Section 6.2.4 in the case that the probabilities of moving to the right and to the left are p and $1 - p$, respectively, and show in appropriate limits that one obtains the advection-diffusion equation $u_t = -cu_x + Du_{xx}$.

9. Show that the nonhomogeneous problem

$$u_t - Du_{xx} = f(x), \ \ 0 < x < l, \ t > 0,$$
$$u(0, t) = g(t), \ \ u(l, t) = h(t), \ t > 0,$$
$$u(x, 0) = u_0(x), \ \ 0 < x < l,$$

can be transformed into a problem with homogeneous boundary conditions by subtracting from u a linear function of x that satisfies the boundary conditions for all time.

10. Consider the following initial boundary value problem for the diffusion equation.

$$u_t - Du_{xx} = f(x), \ \ 0 < x < l, \ t > 0,$$
$$u_x(0, t) = A, \ \ u_x(l, t) = B, \ t > 0,$$
$$u(x, 0) = u_0(x), \ \ 0 < x < l.$$

Show that if the solution is independent of time, i.e., $u = u(x)$, then $DA - DB = \int_0^l f(x) \, dx$. Give a physically meaningful interpretation of this condition.

11. Solve the initial, boundary value problem

$$u_t - ku_{xx} = 0, \ x > 0, \ t > 0$$
$$u(0, t) = 1, \ u(\infty, t) = 0, \ t > 0,$$
$$u(x, 0) = 0, \ x > 0,$$

for the heat equation by assuming a solution of the form $u(x, t) = U(z)$ where $z = x/\sqrt{kt}$. Write the solution in terms of the error function erf.

12. Green's theorem in the plane is usually stated in calculus texts as

$$\int\int_R (Q_x(x, y) - P_y(x, y)) \, dx dy = \int_C P(x, y) \, dx + Q(x, y) \, dy,$$

where P and Q are smooth functions in a well-behaved domain R in the xy plane with boundary C. The line integral is taken counterclockwise. Verify that Green's theorem is really the divergence theorem in two dimensions.

13. Show that the nonlinear initial boundary value problem

$$u_t = u_{xx} - u^3, \quad 0 < x < l, \ t > 0,$$
$$u(0, t) = 0, \quad u(l, t) = 0, \ t > 0,$$
$$u(x, 0) = 0, \quad 0 < x < l,$$

has only the trivial solution. (Hint: multiply by u and integrate.)

14. Let $u = u(\mathbf{x}, t)$ be a positive solution to the equation

$$u_t = \Delta u + f(\mathbf{x}), \quad \mathbf{x} \in \Omega, \ t > 0,$$

where Ω is a well-behaved, bounded domain, and let $s = \ln u$ be the local **entropy** of u. Let $S = \int_\Omega s d\mathbf{x}$ denote the total entropy in Ω, let $J = -\int_{\partial\Omega} \nabla s \cdot \mathbf{n} dA$ be the net entropy flux through the boundary, and let $Q = \int_\Omega f u^{-1} d\mathbf{x}$ be the entropy change due to the source f in Ω. Prove that

$$\frac{dS}{dt} + J \geq Q.$$

15. Two homogeneous rods of length L_1 and L_2 have the same density and specific heats, but their thermal conductivities are K_1 and K_2, which differ. The two rods are welded together with a weld joint that puts them in perfect contact. The composite rod is laterally insulated and the temperature of the left end is held at 0 degrees, while the right end is held at τ degrees. Assume the system is allowed to come to a steady state where the temperature distribution in the system depends only upon x.

  a) Discuss why the temperature and the flux should be continuous at the joint.

  b) Determine the steady-state temperature distribution in the composite rod.

6.3 Equilibrium Equations

6.3.1 Laplace's Equation

Laplace's equation, the prototype of elliptic equations, is one of the basic equations of mathematics. Elliptic equations model equilibrium, or time-independent, phenomena. One way to motivate equilibrium states is to imagine that these states arise from a long-time, or asymptotic, limit of transient states, which are solutions to evolution problems. For example, suppose that conditions on the boundary $\partial\Omega$ are fixed and do not depend on time, and sources, if present, are time-independent. Then, after a long time we expect that the effects of the initial condition in the region Ω will decay away, giving an equilibrium state $u = u(\mathbf{x})$ that satisfies a steady-state type equation.

Consider the initial boundary value problem for the diffusion equation,

$$
\begin{aligned}
u_t - D\Delta u &= f(\mathbf{x}), & \mathbf{x} \in \Omega,\ t > 0, \\
u(\mathbf{x}, t) &= g(\mathbf{x}), & \mathbf{x} \in \partial\Omega,\ t > 0, \\
u(\mathbf{x}, 0) &= u_0(\mathbf{x}), & \mathbf{x} \in \Omega,\ t > 0.
\end{aligned}
$$

Over the long run we may expect the effects of the initial data to be small and an equilibrium solution $u = u(\mathbf{x})$ to emerge that satisfies the equilibrium problem

$$
\begin{aligned}
-D\Delta u &= f(\mathbf{x}), & \mathbf{x} \in \Omega, & \qquad (3.1) \\
u(\mathbf{x}) &= g(\mathbf{x}), & \mathbf{x} \in \partial\Omega. & \qquad (3.2)
\end{aligned}
$$

Equation (3.1) is **Poisson's**[3] **equation**.

If there are no sources, then the equation (3.1) reduces to

$$
\Delta u = 0, \quad \mathbf{x} \in \Omega, \qquad\qquad (3.3)
$$

which is **Laplace's equation**. This partial differential equation is possibly the most analyzed equation in analysis. It also arises in many other natural settings. Readers who have studied complex analysis may recall that the two-dimensional Laplace equation is satisfied by both the real and imaginary parts of a holomorphic (analytic) function $F(z) = u(x, y) + iv(x, y)$ on a domain Ω in the complex plane; that is, $\Delta u = 0$ (or $u_{xx} + u_{yy} = 0$) and $\Delta v = 0$ in Ω. Thus Laplace's equation plays an important role in complex analysis; conversely, complex analysis can be the basis of an approach to study Laplace's equation in two dimensions. These equations also arise naturally in potential theory. For example, if $\mathbf{E}$ is a static electric field in Ω induced by charges lying outside Ω, then $\nabla \times \mathbf{E} = 0$. If $V(\mathbf{x})$ is a potential function, which means $\mathbf{E} = -\nabla V$, then it follows that the potential function V satisfies Laplace's equation. Laplace's equation is also important in fluid mechanics and many other areas of application. Solutions to Laplace's equation are called **harmonic** functions. In one dimension Laplace's equation is just $-u''(x) = 0$, which has linear solutions $u(x) = ax + b$. So, steady-state temperature profiles in a rod are linear. The constants a and b are determined from boundary conditions.

---

[3] S. Poisson (1781–1840), a French mathematician and scientist, made great contributions to physics, mathematics, and probability. He ranks as one of the most prolific and gifted scientists of all time.

What kinds of auxiliary conditions are appropriate for Laplace's equation, or equation (3.1)? From the preceding remarks we expect to impose only boundary conditions along $\partial\Omega$, and not initial conditions (in Hadamard's example we observed that the initial value problem for Laplace's equation in two dimensions was not well posed). Therefore, we impose only boundary conditions, such as the Dirichlet condition (3.2), the Neuman condition,

$$\frac{du}{dn} = g(\mathbf{x}), \quad \mathbf{x} \in \partial\Omega, \tag{3.4}$$

which specifies the flux across the boundary, or a Robin condition,

$$\frac{du}{dn} + a(\mathbf{x})u = g(\mathbf{x}), \quad \mathbf{x} \in \partial\Omega. \tag{3.5}$$

In some applications in three dimensions, it is convenient to work in either cylindrical or spherical coordinates. These coordinates are defined by the equations

$$x = r\cos\theta$$
$$y = r\sin\theta \qquad \text{(cylindrical coordinates } r, \theta, z)$$
$$z = z$$

and

$$x = r\sin\phi\cos\theta$$
$$y = r\sin\phi\sin\theta \qquad \text{(spherical coordinates } r, \phi, \theta)$$
$$z = r\cos\phi$$

where θ is the polar angle and ϕ is the azimuthal angle. An application of the chain rule permits us to write the Laplacian in cylindrical and spherical coordinates as

$$\Delta u = u_{rr} + \frac{1}{r}u_r + \frac{1}{r^2}u_{\theta\theta} + u_{zz} \qquad \text{(cylindrical)}$$

and

$$\Delta u = \frac{1}{r^2}\frac{\partial}{\partial r}(r^2 u_r) + \frac{1}{r^2\sin\phi}\frac{\partial}{\partial\phi}(\sin\phi u_\phi) + \frac{1}{r^2\sin^2\phi}u_{\theta\theta} \qquad \text{(spherical)}$$

The details of these calculations are left as an exercise that should be done once in everyone's life; but not twice. For two-dimensional problems with circular symmetry, polar coordinates are appropriate.

Example 6.21

Find all radial solutions to Laplace's equation in two dimensions. We write Laplace's equation in polar coordinates as

$$u_{rr} + \frac{1}{r}u_r = 0, \quad u = u(r).$$

Notice that the partial derivatives are actually ordinary derivatives. We can immediately write

$$v_r + \frac{1}{r}v = 0, \quad v = u_r(r).$$

This equation is separable and can be integrated to get

$$v(r) = \frac{a}{r},$$

where a is a constant. Then

$$u(r) = a \ln r + b,$$

where b is another arbitrary constant. For example, equilibrium temperatures between two concentric circles, with constant values on the boundaries, vary logarithmically. This is in contrast to the linear distribution in a bar, or linear geometry. □

Example 6.22

(**Gradient in polar coordinates**) It is straightforward, yet tedious, to calculate the standard differential operators (gradient, divergence, curl) in curvilinear coordinates. For example, the gradient operator in polar coordinates is

$$\text{grad} = \frac{\partial}{\partial r}\mathbf{e}_r + \frac{1}{r}\frac{\partial}{\partial \theta}\mathbf{e}_\theta, \tag{3.6}$$

where $\mathbf{e}_r$ and $\mathbf{e}_\theta$ are the unit polar vectors. This follows from the chain rule for derivatives and the fact that

$$\mathbf{i} = \cos\theta\,\mathbf{e}_r - \sin\theta\,\mathbf{e}_\theta, \quad \mathbf{j} = \sin\theta\,\mathbf{e}_r + \cos\theta\,\mathbf{e}_\theta.$$

Verification is requested in the exercises. Expressions for the gradient, divergence, and curl in all coordinate systems can easily be found in handbooks or on the web. □

6.3.2 Basic Properties

The identities in the previous subsection permit us to obtain in an easy manner some basic properties of equilibrium problems. For example, we can prove that solutions to the Dirichlet problem are unique.

Theorem 6.23

Let h be a continuous function on $\partial\Omega$ and f be continuous on $\overline{\Omega}$. If the Dirichlet problem

$$\Delta u = f, \quad \mathbf{x} \subset \Omega; \quad u - h, \quad \mathbf{x} \in \partial\Omega$$

has a solution $u \in C^1(\overline{\Omega}) \cap C^2(\Omega)$, then it is unique. $\quad\square$

Proof

Assume that there are two solutions, u_1 and u_2. Then

$$\Delta u_1 = 0, \quad \mathbf{x} \in \Omega, \ u_1 = h, \ \mathbf{x} \in \partial\Omega; \quad \Delta u_2 = 0, \quad \mathbf{x} \in \Omega, \ u_2 = h, \ \mathbf{x} \in \partial\Omega.$$

Therefore the difference $w \equiv u_1 - u_2$ must satisfy the homogeneous problem

$$\Delta w = 0, \quad \mathbf{x} \in \Omega; \ w = 0, \ \mathbf{x} \in \partial\Omega.$$

Now, in Green's identity, take $u = w$ to obtain

$$\int_\Omega \nabla w \cdot \nabla w \, d\mathbf{x} = 0.$$

Hence $\nabla w = 0$ and so $w = const.$ in Ω; since w is continuous in $\overline{\Omega}$ and zero on the boundary we must have $u_1 - u_2 = w = 0$ in $\overline{\Omega}$, or $u_1 = u_2$. $\quad\square$

In the same way we can prove the following theorem.

Theorem 6.24

Let g be continuous on $\partial\Omega$ and f be continuous on $\overline{\Omega}$. If the Neumann problem

$$\Delta u = f, \quad \mathbf{x} \in \Omega; \quad \frac{du}{dn} = g, \quad \mathbf{x} \in \partial\Omega,$$

has a solution $u \in C^1(\overline{\Omega}) \cap C^2(\Omega)$, then necessarily

$$\int_\Omega f \, dx = \int_{\partial\Omega} g \, dS. \quad\square \tag{3.7}$$

The proof follows immediately from setting $w = 1$ in Green's identity. Physically, (3.7) means that, in a steady state, the net flux (of heat, for example) through the boundary should be balanced by the total amount of heat being created in the region by the sources.

Additional applications of the Green's identities are found in the Exercises.

Another important property of (nonconstant) solutions to Laplace's equation is that they attain their maximum and minimum values on the boundary of the domain, and not the interior. This result is called the **maximum principle**, and we refer to the references for a proof.

Theorem 6.25

(**Maximum principle**) Let Ω be an open, bounded, well-behaved domain in $\mathbb{R}^n$. If $\Delta u = 0$ in Ω and u is continuous in $\overline{\Omega}$, then the maximum and minimum values of u are attained on $\partial\Omega$. $\quad\square$

EXERCISES

1. Start with the product rule for derivatives and derive the integration by parts formula in Corollary 6.11.

2. Use the divergence theorem to derive Green's identities stated in Corollary 6.11.

3. Consider the initial boundary value problem for the n-dimensional heat equation given by

$$u_t - k\Delta u = \rho(\mathbf{x}, t), \quad \mathbf{x} \in \Omega,\ t > 0,$$
$$u(\mathbf{x}, t) = g(\mathbf{x}, t), \quad \mathbf{x} \in \partial\Omega,\ t > 0,$$
$$u(\mathbf{x}, 0) = f(\mathbf{x}), \quad \mathbf{x} \in \Omega,$$

where ρ, g, and f are continuous. Use an energy argument to prove that solutions are unique under reasonable smoothness conditions on u.

4. Let $q > 0$ and continuous in $\overline{\Omega} \subset \mathbb{R}^n$, f be continuous in $\overline{\Omega}$, and g be continuous on $\partial\Omega$. Prove that the Neumann problem

$$-\Delta u + q(\mathbf{x})u = f(\mathbf{x}), \quad \mathbf{x} \in \Omega,$$
$$\frac{du}{dn} = g(\mathbf{x}), \quad \mathbf{x} \in \partial\Omega,$$

can have at most one solution in $C^1(\overline{\Omega}) \cap C^2(\Omega)$.

5. A radially symmetric solution to Laplace's equation $\Delta u = 0$ in $\mathbb{R}^3$ is a solution of the form

$$u = u(r), \quad r = \sqrt{x_1^2 + x_2^2 + x_3^2},$$

where u depends only on the distance from the origin. Find all radially symmetric solutions in $\mathbb{R}^3$.

6. Prove that if the Dirichlet problem

$$-\Delta u = \lambda u, \quad \mathbf{x} \in \Omega,$$
$$u = 0, \quad \mathbf{x} \in \partial\Omega,$$

has a nontrivial solution, then the constant λ must be positive.

7. Let $\mathbf{f}$ be a smooth vector field on $\mathbb{R}^3$ with the property that

$$\|\mathbf{f}(\mathbf{x})\| \leq \frac{1}{|\mathbf{x}|^3 + 1}, \quad \mathbf{x} \in \mathbb{R}^3.$$

Prove that $\int_{\mathbb{R}^n} \nabla \cdot \mathbf{f}\, d\mathbf{x} = 0$.

8. (Calculus of variations) Problems in the calculus of variations (Chapter 4) can be generalized to multiple integral problems. Consider the problem of minimizing the functional

$$J(u) = \int_\Omega L(\mathbf{x}, u, \nabla u)\, d\mathbf{x},$$

over all $u \in C^2(\Omega)$ with $u(\mathbf{x}) = f(\mathbf{x})$, $\mathbf{x} \in \partial\Omega$, where f is a given function on the boundary; Ω is a bounded well-behaved region in $\mathbb{R}^n$.

a) Show that the first variation is

$$\delta J(u, h) = \int_\Omega [L_u h + L_{\nabla u} \cdot \nabla h]\, d\mathbf{x}$$
$$= \int_\Omega (L_u - \nabla \cdot L_{\nabla u}) h\, d\mathbf{x} - \int_{\partial\Omega} h L_{\nabla u} \cdot \mathbf{n}\, dA,$$

where $L_{\nabla u} = (L_{u_{x_1}}, ..., L_{u_{x_n}})$, and where $h \in C^2(\Omega)$ with $h(\mathbf{x}) = 0$, $\mathbf{x} \in \partial\Omega$.

b) Show that a necessary condition for u to minimize J is that u must satisfy the Euler equation

$$L_u - \nabla \cdot L_{\nabla u} = 0, \quad \mathbf{x} \in \Omega.$$

c) If u is not fixed on the boundary $\partial\Omega$, find the natural boundary condition.

9. Find the Euler equation for the following functionals:

   a) $J(u) = \int_\Omega (\nabla u \cdot \nabla u)\, d\mathbf{x}$.

   b) $J(u) = \int_\Omega (x^2 u_x^2 + y^2 u_y^2)d\mathbf{x}$, where $u = u(x, y)$, $d\mathbf{x} = dxdy$.

   c) $J(u) = \frac{1}{2}\int_0^l \int_\Omega (u_t^2 - c^2 \nabla u \cdot \nabla u - mu^2)d\mathbf{x}dt$ where $u = u(\mathbf{x}, t)$, and c and m are constants. (Hint: Treat time t as an additional independent variable and work in $\mathbb{R}^{n+1}$.)

10. The biomass density u (mass per unit volume) of zooplankton in a very deep lake varies as a function of depth x and time t. Zooplankton diffuse vertically with diffusion constant D, and buoyancy effects cause them to migrate toward the surface at a constant speed of ag, where g is the acceleration due to gravity and a is a positive constant. At any time, in any vertical tube of unit cross-sectional area, the total biomass of zooplankton is a constant U.

    a) From first principles derive a partial differential equation model for the biomass density of zooplankton. (Take $x = 0$ at the surface.)

    b) Find the *steady-state* biomass density as a function of depth. (Formulate any reasonable boundary conditions, or other auxiliary conditions, that are needed to solve this problem.)

11. Consider the initial boundary value problem

$$
\begin{aligned}
u_t &= u_{xx} - u^3, \quad x \in (0, 1),\ t > 0, \\
u(x, 0) &= 0, \quad x \in (0, 1), \\
u(0, t) &= u(1, t) = 0, \quad t > 0.
\end{aligned}
$$

Use an energy method to show that the solution must be identically zero.

12. Verify the formula (3.6) for the gradient in polar coordinates. Derive the formula for the divergence in polar coordinates.

6.4 Eigenfunction Expansions

In Chapter 5 we noted the importance of the eigenvalue problem for transformations on $\mathbb{R}^n$ (matrices), for Sturm–Liouville operators, and for integral operators. The same theme carries over to partial differential operators. Basically, these problems have the form

$$
\begin{aligned}
Lu &= \lambda u, \quad \mathbf{x} \in \Omega, & (4.1) \\
B(u) &= 0, \quad \mathbf{x} \in \partial\Omega, & (4.2)
\end{aligned}
$$

where L is a linear partial differential operator and B is a linear operator on the boundary. Specification of the operator includes both L and B.

If there is a value of λ for which (4.1)–(4.2) has a nontrivial solution $u(\mathbf{x})$, then λ is called an eigenvalue and $u(\mathbf{x})$ is a corresponding eigenfunction. Any constant multiple of an eigenfunction is also an eigenfunction with respect to the same eigenvalue. The set of all eigenfunctions form an natural orthogonal basis for the function spaces involved in solving boundary value problems.

6.4.1 Spectrum of the Laplacian

Consider

$$-\Delta u = \lambda u, \quad \mathbf{x} \in \Omega, \tag{4.3}$$

$$u = 0, \quad \mathbf{x} \in \partial\Omega. \tag{4.4}$$

To find the **spectrum** (eigenvalues) we use the method of **separation of variables**. It is a widely used method for solving linear boundary value problems on bounded domains with boundaries along the coordinate directions. Here the method is illustrated on the Laplacian operator, and in the next section we show how it equally applies to evolution problems. The negative sign on the Laplacian will give positive eigenvalues. As an aside, the spectrum of an operator can actually contain values other than eigenvalues. There is a part of the spectrum, particularly for operators on infinite spatial domains, called the continuous spectrum. Therefore, in a broader context, we should refer to the spectrum for the Laplacian operator above as the *discrete* spectrum.

Example 6.26

Take $\Omega = [0, \pi] \times [0, \pi] \subset \mathbb{R}^2$. Then the partial differential equation is $-(u_{xx} + u_{yy}) = \lambda u$ for $u = u(x, y)$. The Dirichlet boundary conditions become $u(0, y) = u(\pi, y) = 0$ for $0 \le y \le \pi$, and $u(x, 0) = u(x, \pi) = 0$ for $0 \le x \le \pi$. We assume a separable solution of the form $u(x, y) = X(x)Y(y)$ for some functions X and Y to be determined. Substituting we get

$$-X''(x)Y(y) - X(x)Y''(y) = \lambda X(x)Y(y),$$

which can be rewritten as

$$-\frac{X''(x)}{X(x)} = \frac{Y''(y) + \lambda Y(y)}{Y(y)},$$

with a function of x on one side and a function of y on the other. Now comes a crucial observation. A function of x can equal a function of y for all x and y

only if both are equal to a constant. That is,

$$-\frac{X''(x)}{X(x)} = \frac{Y''(y) + \lambda Y(y)}{Y(y)} = \mu$$

for some constant μ, called the *separation constant*. Therefore the partial differential equation separates into two ordinary differential equations for the two spatial variables:

$$-X''(x) = \mu X(x), \quad -Y''(y) = (\lambda - \mu)Y(y).$$

Next we substitute the assumed form of u into the boundary conditions to obtain $X(0) = X(\pi) = 0$, and $Y(0) = Y(\pi) = 0$. Consequently we have obtained two boundary value problems of Sturm–Liouville type,

$$\begin{aligned} -X''(x) &= \mu X(x), \quad X(0) = X(\pi) = 0, \\ -Y''(y) &= (\lambda - \mu)Y(y), \quad Y(0) = Y(\pi) = 0. \end{aligned}$$

The first problem was solved in Chapter 5 and we have eigenpairs

$$X_n(x) = \sin nx, \quad \mu_n = n^2, \quad n = 1, 2, 3, ...$$

Then the Y problem becomes

$$-Y''(y) = (\lambda - n^2)Y(y), \quad Y(0) = Y(\pi) = 0.$$

This problem is again of Sturm-Liouville type and will have nontrivial solutions $Y_k(y) = \sin ky$ when $\lambda - n^2 = k^2$, for $k = 1, 2, 3, ...$ Therefore, double indexing λ, we get eigenvalues and eigenfunctions for the negative Laplacian with Dirichlet boundary conditions as

$$u_{n,k}(x, y) = \sin nx \sin ky, \quad \lambda_{n,k} = n^2 + k^2, \quad n, k = 1, 2, 3, ...$$

Observe that there are infinitely many positive, real eigenvalues, whose limit is infinity, and the corresponding eigenfunctions are orthogonal, that is,

$$\int_\Omega (\sin nx \sin ky)(\sin mx \sin ly)dA = 0 \quad \text{if} \quad n \neq m \text{ or } k \neq l. \quad \square$$

This example illustrates what to expect from the Dirichlet problem (4.3)–(4.4). In fact, we have, in general, the following properties.

1. The eigenvalues are real.

2. There are infinitely many eigenvalues that can be ordered as $0 < \lambda_1 \leq \lambda_2 \leq \lambda_3 \leq \cdots$ with $\lambda_n \to +\infty$ as $n \to \infty$.

3. Eigenfunctions corresponding to distinct eigenvalues are orthogonal in the inner product $(u, v) = \int_\Omega u(\mathbf{x})v(\mathbf{x})d\mathbf{x}$.

4. The set the eigenfunctions $u_n(\mathbf{x})$ is complete in the sense that any square-integrable function $f(\mathbf{x})$ on Ω can be uniquely represented in its generalized Fourier series

$$f(\mathbf{x}) = \sum_{n=1}^{\infty} c_n u_n(\mathbf{x}), \quad c_n = \frac{(f, u_n)}{||u_n||^2},$$

where the c_n are the Fourier coefficients, and the norm is $||u_n|| = \sqrt{(u_n, u_n)}$. Convergence of the series is in the $L^2(\Omega)$ sense, meaning

$$\int_{\Omega} \left(f(\mathbf{x}) - \sum_{n-1}^{N} c_n u_n(\mathbf{x}) \right)^2 d\mathbf{x} \to 0 \quad \text{as} \quad N \to \infty.$$

The results are exactly the same when the Dirichlet boundary condition (4.4) is replaced by a Neumann boundary condition $\frac{du}{dn} = 0$, with the exception that $\lambda = 0$ is also an eigenvalue, with eigenfunction $u_0(x) = $ constant. When a Robin boundary condition $\frac{du}{dn} + a(\mathbf{x})u = 0$ is imposed, again there is a zero eigenvalue with constant eigenfunction provided $a(x) \geq 0$. If $a(x)$ fails to be nonnegative, then there may also be negative eigenvalues.

Remark 6.27

The previous results extend to the eigenvalue problem

$$-\nabla \cdot (p\nabla u) + qu = \lambda w u, \quad \mathbf{x} \in \Omega.$$

Here, $w = w(\mathbf{x}) > 0$, $p = p(\mathbf{x}) > 0$, $q = q(\mathbf{x})$, p, q and w are continuous on $\overline{\Omega}$, and p has continuous first partial derivatives on Ω. The boundary conditions may be Dirichlet ($u = 0$ on $\partial\Omega$) Neumann ($\frac{du}{dn} = 0$ on $\partial\Omega$), or Robin ($\frac{du}{dn} + a(\mathbf{x})u = 0$ on $\partial\Omega$, with $a(x) \geq 0$). In this case the eigenfuctions are orthogonal $u_n(x)$ with respect to the inner product

$$(u, v)_w = \int_{\Omega} u(\mathbf{x})v(\mathbf{x})w(\mathbf{x}) d\mathbf{x},$$

and $||u||_w = \sqrt{(u, u)_w}$. The Fourier series takes the form

$$f(\mathbf{x}) = \sum_{n}^{\infty} c_n u_n(\mathbf{x}), \quad c_n = \frac{(f, u_n)_w}{||u_n||_w^2},$$

where convergence is

$$\int_{\Omega} \left(f(\mathbf{x}) - \sum_{n=1}^{N} c_n u_n(\mathbf{x}) \right)^2 w(\mathbf{x}) d\mathbf{x} \to 0 \quad \text{as} \quad N \to \infty. \quad \square$$

Just as for one-dimensional Sturm–Liouville problems and for integral equations, the completeness of the set of eigenfunctions allows us to solve the non-homogeneous problem.

Theorem 6.28

(**Fredholm alternative**) Consider the boundary value problem

$$-\nabla \cdot (p\nabla u) + qu = \mu u + f(\mathbf{x}), \quad \mathbf{x} \in \Omega,$$

with homogeneous Dirichlet, Neumann, or Robin boundary conditions, where the coefficient functions satisfy the conditions in Remark 6.22, and f is a given function. Then,

(a) If μ is not an eigenvalue of the corresponding homogeneous problem ($f = 0$), then there is a unique solution for all functions f with $\int_\Omega f(\mathbf{x})d\mathbf{x} < \infty$.

(b) If μ is an eigenvalue of the homogenous problem, then there is no solution or infinitely many solutions, depending upon the function f.

The proof of this theorem is straightforward and follows exactly the format of the proofs in Chapter 5, namely, to expand u and f in eigenfunction expansions and solve for the coefficients of u.

6.4.2 Evolution Problems

The method of separation of variables is, in general, a way of constructing solutions to partial differential equations in the form of an eigenfuction expansion.

Let $u = u(\mathbf{x}, t)$ satisfy the initial boundary value problem

$$u_t = k\Delta u, \quad \mathbf{x} \in \Omega, \ t > 0,$$
$$u(\mathbf{x}, t) = 0, \quad \mathbf{x} \in \partial\Omega, \ t > 0,$$
$$u(\mathbf{x}, 0) = f(\mathbf{x}) \quad \mathbf{x} \in \Omega,$$

on a well-behaved bounded domain Ω. The strategy is simple. We seek functions of the form $u(\mathbf{x}, t) = F(\mathbf{x})T(t)$ which satisfy the partial differential equation and the Dirichlet boundary condition, but not necessarily the initial condition. We find that there are many such solutions, and because the problem is linear, any linear combination of them will satisfy the equation and boundary condition. We then form the linear combination and select the constants so that the sum also satisfies the initial condition.

To this end substitute $u(\mathbf{x}, t) = F(\mathbf{x})T(t)$ into the partial differential equation to obtain

$$F(\mathbf{x})T'(t) = kT(t)\Delta F(\mathbf{x}),$$

which we write as

$$\frac{T'(t)}{kT(t)} = \frac{\Delta F(\mathbf{x})}{F(\mathbf{x})} = -\lambda,$$

for some λ. The crucial part of the argument is to note that λ is a constant because it cannot be both a function of only t and only $\mathbf{x}$. When this occurs, we say the differential equation is separable, and λ is called the separation constant. Consequently, we have

$$-\Delta F = \lambda F, \quad T' = -\lambda kT. \tag{4.5}$$

The partial differential equation is separated into a spatial part and a temporal part. The sign we put on λ does not matter because it all comes out correctly in the end. But, note that the choice of a negative sign gives the standard negative Laplacian examined previously and makes the temporal part decay, which is expected from a diffusion process. Next, the boundary condition implies $u(\mathbf{x}, t) = F(\mathbf{x})T(t) = 0$ for $x \in \partial\Omega$, or

$$F(\mathbf{x}) = 0, \quad \mathbf{x} \in \partial\Omega. \tag{4.6}$$

(A choice of $T(t) = 0$ would give the uninteresting trivial solution.)

The first equation in (4.5) along with the Dirichlet boundary condition (4.6) is the eigenvalue problem for the negative Laplacian that we studied in the previous section. The eigen-pairs are λ_n, $F_n(\mathbf{x})$, $n = 1.2, 3, ...$. The eigenvalues are positive and the eigenfunctions are orthogonal. In the context of an evolution problem, the spatial eigenfunctions are called the *normal modes*. With these eigenvalues, the temporal equation $T' = -\lambda_n kT$ has solutions $T_n(t) = e^{-\lambda_n kt}$. Therefore we have constructed solutions $u_n(\mathbf{x}, t) = F_n(\mathbf{x})e^{-\lambda_n kt}$, $n = 1, 2, 3, ...$, that satisfy the partial differential equation and the Dirichlet boundary condition. Next we form the linear combination

$$u(\mathbf{x}, t) = \sum_{n=1}^{\infty} a_n F_n(\mathbf{x})e^{-\lambda_n kt}.$$

The initial condition now forces

$$f(\mathbf{x}) = \sum_{n=1}^{\infty} a_n F_n(\mathbf{x}),$$

which is the **generalized Fourier series** for f. The coefficients must be the **Fourier coefficients**

$$a_n = \frac{1}{(F_n, F_n)} \int_\Omega f(\mathbf{x})F_n(\mathbf{x})\,d\mathbf{x}, \quad n = 1, 2, 3, ...$$

Recall that $(F_n, F_n) = ||F_n||^2 = \int_\Omega F_n(\mathbf{x})^2 d\mathbf{x}$.

Therefore we have derived a formal solution to the initial boundary value problem in the form of an infinite series. The validity of such a series representation must be proved by checking that it does indeed satisfy the problem. This requires substitution into the partial differential equation and boundary conditions, so conditions that guarantee term-by-term differentiation must be verified.

Example 6.29

In one spatial dimension, consider the initial boundary value problem for the heat equation:

$$u_t = k u_{xx}, \quad 0 < x < l, \ t > 0, \tag{4.7}$$

$$u(0, t) = 0, \quad u(l, t) = 0, \ t > 0, \tag{4.8}$$

$$u(x, 0) = f(x), \quad 0 < x < l. \tag{4.9}$$

This models heat flow in a bar of length l with the ends held at constant zero temperature. The function f defines the initial temperature distribution. Substituting $u(x, t) = F(x)T(t)$ into the partial differential equation and boundary conditions leads to the equation $T'(t) = -\lambda k T(t)$, and the eigenvalue problem

$$-F'' = \lambda F, \quad 0 < x < 1, \quad F(0) = F(l) = 0, \tag{4.10}$$

which is the Sturm–Liouville problem discussed in Chapter 5. The eigenvalues and eigenfunctions are

$$\lambda_n = \frac{n^2 \pi^2}{l^2}, \quad F_n(x) = \sin \frac{n \pi x}{l}, \quad n = 1, 2, \ldots$$

The solution to the time equation is $T_n(t) = e^{-n^2 \pi^2 k t / l^2}$. We form the series

$$u(x, t) = \sum_{n=1}^{\infty} a_n e^{-n^2 \pi^2 k t / l^2} \sin \frac{n \pi x}{l},$$

where the a_n are to be determined. But the initial condition implies

$$f(x) = \sum_{n=1}^{\infty} a_n \sin \frac{n \pi x}{l},$$

which is the generalized Fourier series for f. The a_n are the Fourier coefficients

$$a_n = \frac{1}{(F_n, F_n)} \int_0^l f(\xi) \sin \frac{n \pi \xi}{l} \, d\xi, \quad n = 1, 2, 3, \ldots$$

We have $(F_n, F_n) = \int_0^l \sin^2 \frac{n\pi x}{l}\, dx = \frac{1}{2}$. Therefore the solution to (4.7)–(4.9) is given by the infinite series

$$u(x,t) = \frac{2}{l} \sum_{n=1}^{\infty} \left(\int_0^l f(\xi) \sin \frac{n\pi\xi}{l}\, d\xi \right) e^{-n^2\pi^2 t/l^2} \sin \frac{n\pi x}{l}.$$

By rearranging the terms and formally switching the order of summation and integration, we can write the solution in a particularly nice form as

$$u(x,t) = \int_0^l G(x,\xi,t) f(\xi)\, d\xi,$$

where

$$G(x,\xi,t) = \frac{2}{l} \sum_{n=1}^{\infty} e^{-n^2\pi^2 t/l^2} \sin \frac{n\pi\xi}{l} \sin \frac{n\pi x}{l}. \quad \square$$

This last example shows how to proceed when there is one spatial variable. In higher dimensions the eigenvalue problem will be higher-dimensional, and it can be solved by separating the spatial variables.

Remark 6.30

The eigenfunction expansion method performs equally well on any evolution equation of the form

$$u_t = \nabla \cdot (p\nabla u) - qu, \quad \mathbf{x} \in \Omega,\ t > 0,$$

with initial condition $u(\mathbf{x}, 0) = f(x)$, $\mathbf{x} \in \Omega$. Here, $p = p(\mathbf{x}) > 0$, $q = q(\mathbf{x})$, p, q are continuous on $\overline{\Omega}$, and p has continuous first partial derivatives on Ω. All of these equations are separable. The boundary conditions may be Dirichlet ($u = 0$ on $\partial\Omega$), Neumann ($\frac{du}{dn} = 0$ on $\partial\Omega$), or Robin ($\frac{du}{dn} + a(x)u = 0$ on $\partial\Omega$, with $a \geq 0$). The domain Ω must be bounded, and the boundary conditions have to be given on $x -$ constant planes. In polar, cylindrical, and spherical coordinates the boundary conditions must be given on constant coordinate curves and surfaces. The associated eigenvalue problem will have the form

$$-\nabla \cdot (p\nabla F) + qF = \lambda F, \quad \mathbf{x} \in \Omega,$$

with either Dirichlet, Neumann, or Robin boundary conditions. $\square$

Remark 6.31

The eigenfunction method is also applicable to equations of the form

$$u_{tt} = \nabla \cdot (p\nabla u) - qu, \quad \mathbf{x} \in \Omega,\ t > 0,$$

with the same caveats and limitations as in the last Remark. These are wave equations and have second-order time derivatives. When variables are separated, the temporal equation will be second-order, and in this case two initial conditions, $u(\mathbf{x},0) = f(\mathbf{x})$ and $u_t(\mathbf{x},0) = g(\mathbf{x})$, $\mathbf{x} \in \Omega$, are required. Wave equations are discussed in the next chapter. □

Nonhomogenous problems. Generally, a partial differential equation with a source term,

$$u_t = \nabla \cdot (p\nabla u) - qu + f(\mathbf{x}), \quad \mathbf{x} \in \Omega, \ t > 0,$$

can be solved by assuming a solution of the form

$$u(\mathbf{x},t) = \sum_{n=1}^{\infty} g_n(t)F_n(\mathbf{x}),$$

where the $F_n(x)$ are the eigenfunctions of the nonhomogeneous spatial problem. Upon substitution into the nonhomogeneous equation, the unknown coefficients $g_n(t)$ can be determined by solving the resulting ordinary differential equations for g_n, subject to the initial condition.

When problems have both nonhomogeneous boundary conditions and a source term, the strategy is to first homogenize the boundary conditions, and then homogenize the PDE.

EXERCISES

1. Use the eigenfunction expansion method to solve the following problems:

    a)

    $$u_t = ku_{xx}, \quad 0 < x < l, \ t > 0,$$
    $$u_x(0,t) = u_x(l,t) = 0, \ t > 0,$$
    $$u(x,0) = f(x), \ 0 < x < l.$$

    b)

    $$u_{tt} = c^2 u_{xx} - a^2 u, \quad 0 < x < l, \ t > 0,$$
    $$u(0,t) = u(l,t) = 0, \quad t > 0,$$
    $$u(x,0) = f(x), \ u_t(x,0) = 0, \ 0 < x < l.$$

    c)

    $$u_{xx} + u_{yy} = 0, \quad 0 < x < \pi, \ y > 0,$$
    $$u(0,y) = u(\pi,y) = 0, \ y > 0,$$
    $$u(x,0) = x(\pi - x), \quad 0 < x < \pi; \ u \text{ bounded}$$

2. Transform the problem

$$u_t = u_{xx}, \quad 0 < x < \pi, \ t > 0,$$
$$u(0,t) = 3, \ u(\pi,t) = 1,$$
$$u(x,0) = f(x),$$

into one with homogeneous boundary conditions. (Hint: Let $u(x,y) = v(x,y) + A(x)$, where A is chosen to satisfy the boundary conditions.)

3. Transform the problem with a source term,

$$u_t = u_{xx} + \sin\left(\frac{x}{2}\right), \quad 0 < x < \pi, \ t > 0,$$
$$u(0,t) = 0, \ u(\pi,t) = 0,$$
$$u(x,0) = f(x),$$

into a nonhomogeneous problem. Solve the problem when $f(x) = 1$.

4. Solve the problem

$$t u_t = u_{xx}, \ x \in (0,\pi), \ t > 1,$$
$$u(0,t) = u(\pi,t) = 0,$$
$$u(x,1) = \sin x + \sin 3x.$$

5. Solve the problem

$$u_{xx} + u_{xx} = 0, \ x,y \in (0,\pi),$$
$$u(0,y) = u(\pi,y) = 0,$$
$$u(x,0) = \sin 5x - 2\sin 3x, \ u(x,\pi) = 0.$$

6. Organisms of density $u(x,t)$ are distributed in a patch of length l. They diffuse with diffusion constant D, and their growth rate is ru. At the ends of the patch a zero density is maintained, and the initial distribution is $u(x,0) = f(x)$. Because there is competition between growth in the interior of the patch and escape from the boundaries, it is interesting to know whether the population increases or collapses. Show that the condition on the patch size, $l < \pi\sqrt{D/r}$, ensures death of the population.

7. Consider Laplace's equation on the unit circle with given boundary condition:

$$r(ru_r)_r + u_{\theta\theta} = 0, \quad 0 < r < 1, \ 0 < \theta \le 2\pi,$$
$$u(1,\theta) = f(\theta), \ 0 < \theta \le 2\pi.$$

a) Assuming $u = R(r)Y(\theta)$, along with the implicit periodic boundary conditions $u(r,0) = u(r,2\pi)$, $u_\theta(r,0) = u_\theta(r,2\pi)$, use separation of variables to find an infinite series representation of the solution.

b) Show that the solution can be written in integral for as

$$u(r,\theta) = \frac{1}{2}\int_0^{2\pi} \frac{(1-r^2)f(\phi)}{1+r^2-2r\cos(\theta-\phi)}\,d\phi.$$

This representation is **Poisson's integral formula**.

8. Consider Laplace's equation on a unit sphere.

a) Assume a solution of the form $u(r,\theta,\phi) = S_n(\phi)r^n$, $n = 0,1,2,\ldots$, and show that $S_n(\phi)$, the *spherical harmonics*, satisfies the equation

$$S_n'' + \frac{\cos\phi}{\sin\phi}S_n' + n(n+1)S_n = 0.$$

b) Change the independent variable to $x = \cos\phi$ with $P_n(x) = S_n(\phi)$, and show

$$(1-x^2)P_n'' - 2xP_n' + n(n+1)P_n = 0.$$

This is **Legendre's differential equation**.

c) Assume a power series solution of Legendre's equation of the form

$$P_n(x) = \sum_{k=0}^{\infty} a_k x^k,$$

and show that for each fixed n,

$$a_{k+2} = \frac{k(k+1)-n(n+1)}{(k+2)(k+1)}a_k, \quad k = 0,1,2,\ldots.$$

d) Show that for each fixed n, there is a polynomial solution. These are called the Legendre polynomials. Up to a constant multiple, find the first four Legendre polynomials $P_0(x),\ldots P_3(x)$. [Note: It can be shown that the Legendre polynomials are orthogonal on $-1 < x < 1$.]

9. Consider the partial differential operator

$$Lu = -\nabla\cdot(p\nabla u) - qu, \quad \mathbf{x}\in\Omega,$$

where $p = p(\mathbf{x}) > 0$, $q = q(\mathbf{x})$, p and q are continuous on $\overline{\Omega}$, and p has continuous first partial derivatives on $\overline{\Omega}$.

a) Prove the integration by parts formula

$$\int_\Omega vLu\,dx = \int_\Omega uLv\,dx + \int_{\partial\Omega} p\left(u\frac{dv}{dn} - v\frac{du}{dn}\right)dA.$$

b) Consider the eigenvalue problem $Lu = \lambda u$, $\mathbf{x} \in \Omega$ with a Dirichlet boundary condition $u = 0$, $\mathbf{x} \in \partial\Omega$. Prove that the eigenvalues are positive and that distinct eigenvalues have corresponding orthogonal eigenfunctions.

10. Give a statement and proof of the Fredholm alternative theorem for the problem

$$-\Delta u + qu = \mu u + f(\mathbf{x}), \quad \mathbf{x} \in \Omega,$$

with a homogeneous Dirichlet boundary condition.

6.5 Integral Transforms

The eigenfunction expansion method is, for the most part, applicable to problems on bounded spatial domains. Transform methods, on the other hand, are usually applied to problems on infinite or semi-infinite spatial domains. In this section we introduce two fundamental integral transforms, the Laplace transform and the Fourier transform. Basically, the idea is to transform a partial differential equation into an ordinary differential equation in a transformed domain, solve the ordinary differential equation, and then return to the original domain by an inverse transform.

6.5.1 Laplace Transforms

Laplace transforms are introduced in elementary differential equations courses as a technique for solving linear ordinary differential equations with constant coefficients. They are particularly useful for nonhomogeneous problems where the source term is piecewise continuous or is an impulse function (a delta function). We assume the reader is familiar with these elementary methods (see any sophomore-level ODE book), and we show that they easily extend to partial differential equations.

First, we review the basic notions for functions of a single variable. If $u = u(t)$ is a piecewise continuous function on $t \geq 0$ that does not grow too fast,[4],

---

[4] For example, take u to be of *exponential order* i.e., $|u(t)| \leq c\exp(at)$ for t sufficiently large, where $a, c > 0$.

then the **Laplace transform** of u is defined by

$$(\mathcal{L}u)(s) \equiv U(s) = \int_0^\infty u(t)e^{-st}\, dt \tag{5.1}$$

In shorthand, we write $U = \mathcal{L}u$. The Laplace transform is an example of an integral transform; it takes a given function $u(t)$ in the time domain and maps it to a new function $U(s)$ in the so-called transform domain. U and s are called the transform variables. The Laplace transform is linear in that $\mathcal{L}(c_1 u + c_2 v) = c_1\mathcal{L}u + c_2\mathcal{L}v$, where c_1 and c_2 are constants. If the transform $U(s)$ is known, then $u(t)$ is called the **inverse transform** of $U(s)$ and we write $(\mathcal{L}^{-1}U)(t) = u(t)$, or just $u = \mathcal{L}^{-1}U$. Pairs of Laplace transforms and their inverses have been tabulated in extensive tables and in computer algebra programs; Table 6.1 gives a brief list.

The importance of the Laplace transform, like other transforms, is that it changes derivative operations in the time domain to multiplication operations in the transform domain. In fact, we have the important operational formulas

$$(\mathcal{L}u')(s) = sU(s) - u(0), \tag{5.2}$$
$$(\mathcal{L}u'')(s) = s^2U(s) - su(0) - u'(0). \tag{5.3}$$

Formulae (5.2) and (5.3) are readily proved using the definition (5.1) and integration by parts.

The strategy for solving differential equations is simple. Taking the Laplace transform of an ordinary differential equation for $u(t)$ results in an algebraic equation for $U(s)$ in the transformed domain. Solve the algebraic equation for $U(s)$ and then recover $u(t)$ by inversion.

Determining $u(t)$ from knowledge of its transform $U(s)$ would take us into the realm of complex contour integration (which is not a prerequisite for this book). However, we can indicate the general formula for the inverse transform. The **inversion formula** is

$$u(t) = (\mathcal{L}^{-1}U)(t) = \frac{1}{2\pi i}\int_{a-i\infty}^{a+i\infty} U(s)e^{st}\, ds.$$

The integral is a complex contour integral taken over the infinite vertical line, called a Bromwich path, in the complex plane from $a - i\infty$ to $a + i\infty$. The number a is any real number for which the resulting Bromwich path lies to the right of any singularities (poles, essential singular points, or branch points and cuts) of the function $U(s)$. Calculating inverse transforms using the Bromwich integral usually involves a difficult contour integration in the complex plane and using the residue theorem. In this text we only use a table or a computer algebra system.

Table 6.1 Laplace Transforms

$u(t)$	$U(s)$
1	$s^{-1}, \quad s > 0$
e^{at}	$\frac{1}{s-a}, \quad s > a$
t^n, n a positive integer	$\frac{n!}{s^{n+1}}, \quad s > 0$
$\sin at$ and $\cos at$	$\frac{a}{s^2+a^2}$ and $\frac{s}{s^2+a^2}, \quad s > 0$
$\sinh at$ and $\cosh at$	$\frac{a}{s^2-a^2}$ and $\frac{s}{s^2-a^2}, \quad s > |a|$
$e^{at}\sin bt$	$\frac{b}{(s-a)^2+b^2}, \quad s > a$
$e^{at}\cos bt$	$\frac{s-a}{(s-a)^2+b^2}, \quad s > a$
$t^n\exp(at)$	$\frac{n!}{(s-a)^{n+1}}, \quad s > a$
$H(t-a)$	$s^{-1}\exp(-as), \quad s > 0$
$\delta(t-a)$	$\exp(-as)$
$H(t-a)f(t-a)$	$F(s)\exp(-as)$
$f(t)e^{-at}$	$F(s+a)$
$H(t-a)f(t)$	$e^{-as}\mathcal{L}(f(t+a))$
$\operatorname{erf}\sqrt{t}$	$s^{-1}(1+s)^{-1/2}, \quad s > 0$
$\frac{1}{\sqrt{t}}\exp\left(\frac{-a^2}{4t}\right)$	$\sqrt{\pi/s}\exp(-a\sqrt{s}), \quad (s > 0)$
$1 - \operatorname{erf}\left(\frac{a}{2\sqrt{t}}\right)$	$s^{-1}\exp(-a\sqrt{s}), \quad s > 0$
$\frac{a}{2t^{3/2}}\exp\left(\frac{-a^2}{4t}\right)$	$\sqrt{\pi}\exp(-a\sqrt{s}), \quad s > 0$
$u^{(n)}(t)$	$s^n U(s) - s^{n-1}u(0) - s^{n-2}u'(0) - \cdots - u^{(n-1)}(0)$
$\int_0^t u(\tau)v(t-\tau)\,d\tau$	$U(s)\,V(s)$

One of the most useful tools is the **convolution theorem**. Often, solving a differential equation results in having to invert the product of two transforms in the transform domain. The convolution theorem tells what the inverse transformation is.

Theorem 6.32

(**Convolution theorem**) Let u and v be piecewise continuous on $t \geq 0$ and of exponential order. Then

$$\mathcal{L}(u * v)(s) = U(s)V(s),$$

where

$$(u * v)(t) \equiv \int_0^t u(t - y)v(y)\, dy$$

is the **convolution** of u and v, and $U = \mathcal{L}u, V = \mathcal{L}v$. Furthermore,

$$\mathcal{L}^{-1}(U(s)V(s)) = (u * v)(t). \quad \square$$

Whereas the Laplace transform is additive, it is not multiplicative. The convolution theorem tells what to take the transform of in order to get a product of Laplace transforms, namely the convolution. The convolution theorem is easy to prove using the definition of the transform and interchanging the order of integration; it is proved in most elementary texts (e.g., Logan 2010).

Remark 6.33

We leave it to an exercise to show that convolution is commutative, that is,

$$u * v = v * u. \quad \square$$

Example 6.34

Solve the initial value problem

$$u'' + k^2 u = f(t), \quad t > 0; \ u(0) = u'(0) = 0.$$

Taking the transform of both sides we get

$$s^2 U - su(0) - u'(0) + k^2 U = F(s),$$

where $F = \mathcal{L}(f)$. Applying the initial conditions and then solving for U gives

$$U(s) = \frac{1}{s^2 + k^2} F(s).$$

To invert, we note the right side is the product of two transforms and we can apply the convolution theorem. From the table,

$$\mathcal{L}^{-1}\left(\frac{1}{s^2 + k^2}\right) = \frac{1}{k}\sin kt,$$

and $f = \mathcal{L}^{-1}F$. Therefore

$$u(t) = \mathcal{L}^{-1}U(s) = \frac{1}{k}\sin kt * f(t) = \frac{1}{k}\int_0^t \sin k(t - \tau)t f(\tau)\, d\tau. \quad \square$$

The same transform strategy applies to partial differential equations where the unknown is a function of two variables, for example, $u = u(x,t)$. Now we transform on t, as before, with the variable x being a parameter unaffected by the transform. In particular, we define the Laplace transform of $u(x,t)$ by

$$(\mathcal{L}u)(x,s) \equiv U(x,s) = \int_0^\infty u(x,t)e^{-st}\, dt.$$

Then time derivatives transform as in (5.2) and (5.3); for example,

$$(\mathcal{L}u_t)(x,s) \quad - \quad sU(x,s) - u(x,0),$$
$$(\mathcal{L}u_{tt})(x,s) \quad = \quad s^2U(x,s) - su(x,0) \quad u_t(x,0).$$

On the other hand, spatial derivatives are left unaffected, for example,

$$(\mathcal{L}u_x)(x,s) = \int_0^\infty \frac{\partial}{\partial x} u(x,t)e^{-st}\, dt = \frac{\partial}{\partial x} \int_0^\infty u(x,t)e^{-st}\, dt = U_x(x,s),$$

and $(\mathcal{L}u_{xx})(x,s) = U_{xx}(x,s)$. Therefore, taking Laplace transform of a partial differential equation in x and t reduces it to an ordinary differential equation in x with s as a parameter; all the t derivatives are turned into multiplication in the transform domain.

Example 6.35

Let $u = u(x,t)$ denote the concentration of a chemical contaminant, and let $x > 0$ be a semi-infinite region that initially contains no contaminant. For times $t > 0$ we impose a constant, unit concentration on the boundary $x = 0$, and we ask how the contaminant diffuses into the region. Assuming a unit diffusion constant (the problem can be rescaled to make the diffusion constant unity), the mathematical model is

$$u_t - u_{xx} = 0, \ x > 0, \ t > 0,$$
$$u(x,0) = 0, \ x > 0,$$
$$u(0,t) = 1, \ t > 0, \ u(x,t) \text{ bounded.}$$

Taking Laplace transforms of both sides of the equation yields

$$sU(x,s) - U_{xx}(x,s) = 0.$$

This is an ordinary differential equation with x as the independent variable, and the solution is

$$U(x,s) = a(s)e^{-\sqrt{s}\,x} + b(s)e^{\sqrt{s}\,x}.$$

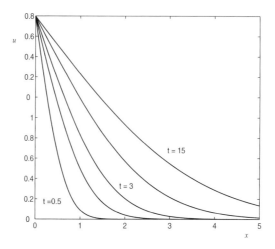

Figure 6.5 Concentration profiles $u = u(x, t)$ for different times t.

Because solutions are bounded we set $b(s) = 0$. Then

$$U(x, s) = a(s)e^{-\sqrt{s}\,x}.$$

Next we take Laplace transforms of the boundary condition to get $U(0, s) = 1/s$, where we used $\mathcal{L}(1) = 1/s$. Therefore $a(s) = 1/s$ and the solution in the transform domain is

$$U(x, s) = \frac{1}{s}e^{-\sqrt{s}\,x}.$$

Consulting Table 6.1, we find that the solution is

$$u(x, t) = 1 - \operatorname{erf}\left(\frac{x}{\sqrt{4t}}\right) = 1 - \frac{2}{\sqrt{\pi}}\int_0^{x/\sqrt{4t}} e^{-r^2}\,dr,$$

where erf is the error function. Figure 6.5 shows several time snapshots of the concentration of the contaminant as it diffuses into the medium.   □

The Laplace transform of a distribution can be defined as well, and we may formally transform delta functions and thus solve equations with point source functions. See Table 6.1 for the relevant entries. The theoretical basis for these calculations is contained in Section 6.7.2.

Example 6.36

(**Transform of the delta function**) Here we give an intuitive calculation based on approximating a delta distribution $\delta_a(t)$ by the limit of a sequence of unit pulses

$$f_\varepsilon(t) = \frac{1}{2\varepsilon}(H(a-\varepsilon) - H(a+\varepsilon))$$

as $\varepsilon \to 0$. (H is the Heaviside function.) On the left, for ease of writing, we suppress the dependence on a. This is a sequence of rectangular, unit pulses of width 2ε and height $1/2\varepsilon$ that get narrower and taller as $\varepsilon \to 0$, yet all have area 1. In the limit it approaches a unit source at $t = a$, or a delta 'function'. That is, intuitively,

$$\lim_{\varepsilon \to 0} f_\varepsilon(t) = \delta_a(t), \quad \lim_{\varepsilon \to 0} \int_0^\infty f_\varepsilon(t)dt = 1.$$

We calculate the Laplace transform of $f_\varepsilon(t)$ and take the limit.

$$\begin{aligned} \mathcal{L}(f_\varepsilon(t)) &= \frac{1}{2\varepsilon}\left(\frac{e^{(a-\varepsilon)s}}{s} - \frac{e^{(a+\varepsilon)s}}{s}\right) \\ &= \frac{1}{2\varepsilon}e^{-as}\left(\frac{e^{\varepsilon s} - e^{-\varepsilon s}}{s}\right) \\ &= \frac{1}{2}e^{-as}\frac{2\sin(\varepsilon s)}{\varepsilon s}. \end{aligned}$$

In the limit as $\varepsilon \to 0$ we get

$$\lim_{\varepsilon \to 0} \mathcal{L}(f_\varepsilon(t)) = e^{-as},$$

leading us to conclude $\mathcal{L}(\delta_a(x)) = e^{-as}$. $\square$

EXERCISES

1. Find the Laplace transform of $u(t) = 1/\sqrt{t}$. (Hint: make the substitution $st = r^2$ in the integral.)

2. The gamma function is defined by

$$\Gamma(t) \int_0^\infty e^{-x}x^{t-1}dx, \ t > -1.$$

   a) Show $\Gamma\left(\frac{1}{2}\right) = \sqrt{\pi}$.

   b) Show $\mathcal{L}(t^a) = \frac{\Gamma(a+1)}{s^{a+1}}$, $s > 0$.

3. Show that $u * v = v * u$.

4. Solve the following initial value problems.

   a) $u' + au = f(t)$, $u(0) = u_0$, where a is constant.

   b) $u'' - k^2 u = f(t)$, $u(0) = 0$, $u'(0) = 1$.

c) $u' + 3u = \delta_1(t) + H_4(t)$, $u(0) = 1$.

5. If $f(t)$ is periodic of period p for $t \geq 0$, i.e., $f(t + P) = f(t)$ for all t, show that

$$F(s) = \frac{1}{1 - e^{-ps}} \int_0^p f(\tau) e^{-s\tau} d\tau.$$

6. Solve the integral equation

$$u(t) = \frac{1}{\sqrt{\pi}} \int_0^t \frac{u(\tau)}{\sqrt{t - \tau}} d\tau.$$

7. Show that the transform of $t^n f(t)$ is $(-1)^n U^{(n)}(s)$, where n is a positive integer.

8. Use a transform method to solve the diffusion problem

$$\begin{aligned}
u_t &= Du_{xx}, & 0 < x < l, \, t > 0, \\
u(0, t) &= u(l, t) = 1, & t > 0, \\
u(x, 0) &= 1 + \sin\left(\frac{\pi x}{l}\right), & 0 < x < l.
\end{aligned}$$

9. A toxic chemical diffuses into a semi-infinite domain $x \geq 0$ from its boundary $x = 0$, where the concentration is maintained at $g(t)$. The model is

$$\begin{aligned}
u_t &= Du_{xx}, & x > 0, \, t > 0, \\
u(x, 0) &= 0, & x > 0, \\
u(0, t) &= g(t), & t > 0.
\end{aligned}$$

Determine the concentration $u = u(x, t)$ and write the solution in the form of $u(x, t) = \int_0^t K(x, t - \tau) g(\tau) d\tau$, identifying the kernel K.

10. The small, transverse deflections from equilibrium of an elastic string under the influence of gravity satisfy the forced wave equation

$$u_{tt} = c^2 u_{xx} - g,$$

where c^2 is a constant and g is the acceleration due to gravity; $u(x, t)$ is the actual deflection at location x at time t. (Wave equations are derived in Chapter 7.) At $t = 0$ a semi-infinite string $(x \geq 0)$, supported underneath, lies motionless along the x axis. Suddenly the support is removed and the string falls under the force of gravity with the deflection at $x = 0$ held fixed at the origin; the initial velocity $u_t(x, 0)$ of the string is zero. Determine the displacement of the string and sketch several time snapshots of the solution.

11. Solve $u_{tt} = c^2 u_{xx}$ on $t > 0$, $x > 0$, subject to $u(x,0) = u_t(x,0) = 0$ and boundary condition $u(0,t) = g(t)$.

12. Solve $u_{tt} = c^2 u_{xx}$ on $t > 0$, $x \in (0,1)$, subject to boundary conditions $u(0,t) = u(1,t) = 0$ and initial conditions $u(x,0) = \sin \pi x$, $u_t(x,0) = -\sin \pi x$.

6.5.2 Fourier Transforms

It is impossible to overestimate the important role that Fourier analysis played in the evolution of mathematical analysis over the last two centuries. Much of this development was motivated by theoretical issues associated with problems in engineering and the sciences. Now, Fourier analysis is a monument of applied mathematics, forming a base of signal processing, imaging, probability theory, and differential equations, only to mention a few such areas.

There are many ways to introduce the Fourier transform. Many texts show that it is the limit of a complex Fourier series as the period of the function approaches infinity. Here we examine its origin from a PDE viewpoint. In Section 7.2 we show that the Cauchy problem for a linear, homogeneous, constant coefficient PDE may admit a solution $u(x,t)$ that is the superposition of plane waves; that is,

$$u(x,t) = \int_{-\infty}^{\infty} c(\xi) e^{i(\xi x - \omega(\xi)t)} \, d\xi,$$

where $\omega - \omega(\xi)$ is the dispersion relation, and the coefficient function, the amplitude $c(\xi)$, is to be determined. [Note that we used ξ in lieu of k as the wave number.] If $u(x,0) = f(x)$, we find

$$f(x) = \int_{-\infty}^{\infty} c(\xi) e^{i\xi x} \, d\xi. \tag{5.4}$$

This looks very much like a continuous version of the complex Fourier series

$$f(x) = \sum_{\xi=-\infty}^{\infty} c_\xi e^{i\xi x}.$$

We know that the c_ξ are the Fourier coefficients of f given by

$$c_\xi = \frac{1}{2\pi}(f, e^{i\xi x}).$$

Therefore, we suspect in the continuous case that

$$c(\xi) = \frac{1}{2\pi} \int_{-\infty}^{\infty} f(x) e^{-i\xi x} \, dx. \tag{5.5}$$

Compare the last four formulas and observe the strong similarity! The formulas (5.4) and (5.5) are basically the Fourier transform and the inverse Fourier transform, respectively, with a little rearrangement of variables.

From a PDE viewpoint, it turns out that the Fourier transform is an integral operator with properties similar to the Laplace transform in that derivatives are turned into multiplication operations in the transform domain. Thus the Fourier transform, like the Laplace transform, is useful as a computational tool in solving differential equations. The **Fourier transform** of a function in one dimension $u = u(x)$, $x \in \mathbb{R}$, is defined by the equation

$$(\mathcal{F}u)(\xi) \equiv \hat{u}(\xi) = \int_{-\infty}^{\infty} u(x)e^{i\xi x}\, dx.$$

The variable ξ is called the frequency (or wave number) variable. There is one important remark about notation. There is no standard convention on how to define the Fourier transform; some put a factor of $1/2\pi$ or $1/\sqrt{2\pi}$ in front of the integral, and some have a negative power in the exponential or even a factor of 2π. Physicists and engineers often use k, ω, or p (meaning wave number, frequency, or momentum) in place of ξ. Therefore one needs to be aware of these differences when consulting other sources because the formulas change depending upon the convention.

Another important, even crucial, issue is to decide what set of functions on which to define the transform. Certainly, if u is integrable on $\mathbb{R}$, that is, $\int_{-\infty}^{\infty} |u|\, dx < \infty$, then $|\hat{u}(\xi)| = \left|\int_{-\infty}^{\infty} u(x)e^{i\xi x}\, dx\right| \leq \int_{-\infty}^{\infty} |u|\, dx < \infty$, and $\hat{u}$ exists. But, as the example below shows, if u is integrable then $\hat{u}$ is not necessarily integrable; this causes problems, therefore, with inversion of the transform. Consequently, in the theory of Fourier transforms one often takes the domain to be a much smaller class of functions, for example, the square-integrable functions $L^2(\mathbb{R})$, or the **Schwartz class** of functions $\mathcal{S}$. The Schwartz class consists of very smooth functions that decay, along with all their derivatives, very rapidly at infinity. Specifically, $u \in S$ if $u \in C^\infty(\mathbb{R})$ and u and all its derivatives $u^{(k)}$ have the property $\lim_{|x|\to\infty} x^p u^{(k)}(x) = 0$, for any positive integer p. Thus, Schwartz class functions decay faster at infinity than any algebraic polynomial. This assumption makes the theory go through in a highly aesthetic way (see, for example, Strichartz 1994).

Example 6.37

Compute the Fourier transform of the characteristic function

$$u(x) = \chi_{[-a,a]}(x) = \begin{cases} 1, & |x| \leq a, \\ 0, & |x| > a. \end{cases}$$

We have

$$\hat{u}(\xi) = \int_{-\infty}^{\infty} \chi_{[-a,a]}(x)e^{i\xi x}\, dx. = \int_{-a}^{a} e^{i\xi x}\, dx = \frac{2\sin a\xi}{\xi}.$$

Notice that u has its support, or nonzero values, in a bounded interval, and it is integrable. Its Fourier transform is in $C^{\infty}(\mathbb{R})$, yet it is not absolutely integrable.
□

Remark 6.38

In general, one can show that if u is absolutely integrable, then (i) $\hat{u}$ is continuous and bounded, (ii) $\lim_{|\xi|\to\infty} \hat{u}(\xi) = 0$, and (iii) if $u \in S$, then $\hat{u} \in S$.
□

Our goal in this text is not a thorough analysis of transforms; rather, we want to study the mechanics and applications of Fourier transforms. Our strategy is to perform the calculations formally, making what smoothness assumptions that are required to do the calculations. Once a result, say, a solution formula, is obtained, we could then make assumptions that are needed to validate it. For example, we may require a high degree of smoothness and rapid decay at infinity to apply the transform method and invert it, but at the end we may find that the solution is valid under much less strenuous conditions.

A basic property of the one-dimensional Fourier transform is

$$(\mathcal{F}u^{(k)})(\xi) = (-i\xi)^{k}\hat{u}(\xi), \tag{5.6}$$

confirming our comment that derivatives are transformed to multiplication by a factor of $(-i\xi)^{k}$. This operational formula is easily proved using integration by parts (like for the Laplace transform), assuming u and its derivatives are continuous and integrable.

For functions of two variables, say $u = u(x,t)$, the variable t acts as a parameter and we define the **Fourier transform** by

$$(\mathcal{F}u)(\xi,t) \equiv \hat{u}(\xi,t) = \int_{-\infty}^{\infty} u(x,t)e^{i\xi x}\, dx.$$

Then, under Fourier transformation, x derivatives turn into multiplication and t derivatives remain unaffected; for example,

$$(\mathcal{F}u_{x})(\xi,t) = (-i\xi)\hat{u}(\xi,t),$$
$$(\mathcal{F}u_{xx})(\xi,t) = (-i\xi)^{2}\hat{u}(\xi,t),$$
$$(\mathcal{F}u_{t})(\xi,t) = \hat{u}_{t}(\xi,t).$$

Solving a differential equation for u first involves transforming the problem into the transform domain and then solving for $\hat{u}$. Then one is faced with the inversion problem, or the problem of determining the u for which $\mathcal{F}u = \hat{u}$. One of the nice properties of the Fourier transform is the simple form of the **inversion formula**, or **inverse transform**:

$$(\mathcal{F}^{-1}\hat{u})(x) \equiv u(x) = \frac{1}{2\pi} \int_{-\infty}^{\infty} \hat{u}(\xi)e^{-i\xi x} \, d\xi. \tag{5.7}$$

This result is the content of the **Fourier integral theorem**. For example, if $u \in S$, then $\hat{u} \in S$, and conversely. Under the weaker condition that both u and $\hat{u}$ are integrable, then (5.7) holds at all points where u is continuous.

Some Fourier transforms can be calculated directly, as in the preceding example; many others require complex contour integration. A table of Fourier transforms appears at the end of this section (Table 6.2). In the next example we calculate the transform of Gaussian function $u(x) = e^{-ax^2}, a > 0$, using a differential equation technique.

Example 6.39

(**The Gaussian**) We want to calculate $\hat{u}$ where

$$\hat{u}(\xi) = \int_{-\infty}^{\infty} e^{-ax^2} \, e^{i\xi x} \, dx.$$

Differentiating with respect to ξ and then integrating by parts gives

$$\hat{u}'(\xi) = i \int_{-\infty}^{\infty} xe^{-ax^2} e^{i\xi x} \, dx$$

$$= \frac{-i\xi}{2a} \int_{-\infty}^{\infty} e^{-ax^2} \, e^{i\xi x} \, dx$$

$$= \frac{-\xi}{2a}\hat{u}(\xi).$$

Therefore we have a differential equation $\frac{d\hat{u}}{d\xi} = (-\xi)/(2a)\hat{u}$ for $\hat{u}$. Separating variables and integrating gives the general solution

$$\hat{u}(\xi) = Ce^{-\xi^2/4a},$$

and the constant C can be determined by noticing

$$\hat{u}(0) = \int_{-\infty}^{\infty} e^{-ax^2} \, dx = \sqrt{\frac{\pi}{a}}.$$

Consequently,

$$\mathcal{F}\left(e^{-ax^2}\right) = \sqrt{\frac{\pi}{a}} \, e^{-\xi^2/4a}. \tag{5.8}$$

So, the Fourier transform of a Gaussian (a normal probability distribution) function is a Gaussian; conversely, the inverse transform of a Gaussian is a Gaussian. Observe that a Gaussian with large spread (variance) gets transformed to a Gaussian with a narrow spread, and conversely.   □

Similar to Laplace transforms, a convolution relation holds for Fourier transforms. If u and v are absolutely integrable, then we define the **convolution** of u and v by

$$(u * v)(x) = \int_{-\infty}^{\infty} u(x - y)v(y)\, dy.$$

Then we have the following theorem.

Theorem 6.40

(**Convolution theorem**) If u and v are in $\mathcal{S}$, then $u * v \in \mathcal{S}$ and

$$\mathcal{F}\,(u * v)(\xi) = \hat{u}(\xi)\hat{v}(\xi).$$

Then

$$(u * v)(x) = \mathcal{F}^{-1}(\hat{u}(\xi)\hat{v}(\xi)).   \square$$

This formula, which can also be stated under less strenuous conditions, asserts that the inverse transform of a product is a convolution. As with Laplace transforms, this is a useful relationship in solving differential equations because we frequently end up with a product of transforms in the frequency domain.

Proof

The convolution theorem follows from formally interchanging the order of integration in the following string of equalities:

$$\mathcal{F}(u * v)(\xi) = \int_{-\infty}^{\infty} \left(\int_{-\infty}^{\infty} u(x - y)v(y)\, dy \right)\, e^{i\xi x}\, dx$$

$$= \int_{-\infty}^{\infty} \int_{-\infty}^{\infty} u(x - y)v(y)e^{i\xi x}\, dx\, dy$$

$$= \int_{-\infty}^{\infty} \int_{-\infty}^{\infty} u(r)v(y)e^{i\xi r}\ e^{i\xi y}\, dr\, dy$$

$$= \int_{-\infty}^{\infty} u(r)e^{i\xi r}\, dr \int_{-\infty}^{\infty} v(y)e^{i\xi y}\, dy$$

$$= \hat{u}(\xi)\hat{v}(\xi).$$

□

Example 6.41

Consider the ordinary differential equation

$$u'' - u = f(x), \quad x \in \mathbb{R}.$$

Assume that u and its derivatives, and f, are sufficiently smooth and decay rapidly at infinity. Proceeding formally, we take the transform of both sides to obtain

$$(-i\xi)^2 \hat{u} - \hat{u} = \hat{f},$$

Hence

$$\hat{u}(\xi) = -\frac{1}{1 + \xi^2} \hat{f}(\xi).$$

In the transform domain the solution is a product of transforms, and so we apply the convolution theorem. From Table 6.2

$$\mathcal{F}\left(\tfrac{1}{2} e^{-|x|} \right) = \frac{1}{1 + \xi^2}.$$

Therefore

$$u(x) = -\tfrac{1}{2} e^{-|x|} * f(x) = -\frac{1}{2} \int_{-\infty}^{\infty} e^{-|x-y|} f(y)\, dy.$$

We can show, in fact, that this is a solution if f is continuous and integrable. □

Example 6.42

Use Fourier transforms to solve the pure initial value problem for the diffusion equation:

$$u_t - k u_{xx} = 0, \quad x \in \mathbb{R},\, t > 0; \quad u(x,0) = f(x),\, x \in \mathbb{R}. \tag{5.9}$$

We assume the functions u and f are in $\mathcal{S}$. Taking Fourier transforms of the equation gives

$$\hat{u}_t = -\xi^2 k \hat{u},$$

which is an ordinary differential equation in t for $\hat{u}(\xi, t)$, with ξ as a parameter. Its solution is

$$\hat{u}(\xi, t) = C(\xi) e^{-\xi^2 kt}.$$

The initial condition gives $\hat{u}(\xi, 0) = \hat{f}(\xi)$, and so $C(\xi) = \hat{f}(\xi)$. Therefore

$$\hat{u}(\xi, t) = e^{-\xi^2 kt} \hat{f}(\xi).$$

Replacing a by $1/4kt$ in formula (5.8) gives

$$\mathcal{F}\left(\frac{1}{\sqrt{4\pi kt}} e^{-x^2/4kt} \right) = e^{-\xi^2 kt}.$$

By the convolution theorem we have

$$u(x,t) = \int_{-\infty}^{\infty} \frac{1}{\sqrt{4\pi kt}} e^{-(x-y)^2/4kt} f(y)\, dy. \quad \square \qquad (5.10)$$

This solution was derived under the assumption that u, $f \in S$. But, now that we have it, we can show that it is a solution under milder restrictions on f. For example, one can prove that (5.10) is a solution to (5.9) if f is a continuous, bounded function on $\mathbb{R}$. If f is only piecewise continuous, then u remains a solution for $t > 0$, but at point of discontinuity of f, say x_0, we have $\lim_{t \to 0^+} u(x_0, t) = \frac{1}{2}(f(x_0^-) + f(x_0^+))$.

Example 6.43

Next we consider the problem of finding an electrical potential $u(x, y)$ in the upper-half plane caused by a given potential on the lower edge $y = 0$. Therefore, consider the boundary value problem

$$\begin{aligned} u_{xx} + u_{yy} &= 0, \quad x \in \mathbb{R},\ y > 0, \\ u(x,0) &= f(x), \quad x \in \mathbb{R}. \end{aligned}$$

We append the condition that the solution u stay bounded as $y \to \infty$. This example is similar to the last example. Taking the transform (on x, with y as a parameter) of the equation we obtain

$$\hat{u}_{yy} - \xi^2 \hat{u} = 0,$$

which has general solution

$$\hat{u}(\xi, y) = a(\xi)e^{-\xi y} + b(\xi)e^{\xi y}.$$

The first term will grow exponentially if $\xi < 0$, and the second term will grow if $\xi > 0$. To maintain boundedness for all ξ we can take

$$\hat{u}(\xi, y) = c(\xi)e^{-|\xi| y}.$$

Applying the Fourier transform to the boundary condition yields $c(\xi) = \hat{f}(\xi)$. Therefore the solution in the transform domain is

$$\hat{u}(\xi, y) = e^{-|\xi| y} \hat{f}(\xi).$$

By the convolution theorem,

$$u(x, y) = \frac{y}{\pi} \frac{1}{x^2 + y^2} * f = \frac{y}{\pi} \int_{-\infty}^{\infty} \frac{f(\eta)\, d\eta}{(x - \eta)^2 + y^2}. \quad \square$$

Table 6.2   Fourier Transforms

$u(x)$	$\widehat{u}(\xi)$
$u^{(n)}(x)$	$(-i\xi)^n\widehat{u}(\xi)$
$u(ax)$	$\frac{1}{a}\widehat{u}\left(\frac{\xi}{a}\right)$
$u(x-a)$	$e^{ia\xi}\widehat{u}(\xi)$
$\int_0^x u(y)dy$	$-\frac{1}{i\xi}\widehat{u}(\xi)$
$u*v$	$\widehat{u}(\xi)\,\widehat{v}(\xi)$
$e^{-a|x|}$	$\frac{2a}{a^2+\xi^2}$
$H(a-|x|)$	$\frac{2\sin(a\xi)}{\xi}$
e^{-ax^2}	$\sqrt{\frac{\pi}{a}}e^{-\xi^2/4a}$
$\delta_a(x)$	$e^{ia\xi}$
$H(x)$	$\pi\delta(\xi)-\frac{i}{\xi}$
$H(x)e^{-x}$	$\frac{1}{1-i\xi}$
1	$2\pi\delta(\xi)$

As in the case of Laplace transforms, the Fourier transform can be extended to distributions; a few examples are given in Table 6.2. In Section 6.7 we define the Fourier transform of a distribution.

We end this section with some brief comments on eigenvalue problems on infinite domains. For example, consider the eigenvalue problem

$$-\Delta u = \lambda u, \quad \mathbf{x} \in \mathbb{R}^n.$$

On bounded domains there are accompanying boundary conditions (Dirichlet, Neumann, Robin) on u. Now we must decide what kind of boundary conditions to impose at infinity. For example, do we assume u is square-integrable, bounded, or goes to zero as $|\mathbf{x}| \to \infty$? A one-dimensional problem reveals the issues. Consider

$$-u'' = \lambda u, \quad x \in \mathbb{R}.$$

The general solution is: $u(x) = ax + b$ if $\lambda = 0$, $u(x) = ae^{-\sqrt{-\lambda}x} + be^{\sqrt{-\lambda}x}$ if $\lambda < 0$, and $u(x) = ae^{-i\sqrt{\lambda}x} + be^{i\sqrt{\lambda}x}$, if $\lambda > 0$. None of these solutions is square-integrable on $\mathbb{R}$ or goes to zero at $x = \pm\infty$; so in these cases there would be no eigenvalues. However, if we impose a boundedness condition on all of $\mathbb{R}$, then any $\lambda \geq 0$ gives a nontrivial solution $u(x, \lambda)$. In this case we say λ belongs to the **continuous spectrum** of the operator $-\frac{d^2}{dx^2}$, and the associated eigenfunctions are $e^{-i\sqrt{\lambda}x}$, $e^{i\sqrt{\lambda}x}$, $\lambda \geq 0$. These are sometimes given the unfortunate misnomers *improper eigenvalues* and *improper eigenfunctions*. Later, in a quantum mechanical context, we see that these functions can be interpreted as traveling wave packets.

We can write this set of eigenfunctions as e^{-ikx}, where $-\infty < k < \infty$. Then, if f is a given function, we can think of representing it as an as eigenfunction expansion, but this time not as a sum (as for the case of discrete eigenvalues), but as an integral

$$f(x) = \frac{1}{2\pi} \int_{-\infty}^{\infty} c(k) e^{-ikx} dk.$$

But this means that the continuum $c(k)$ of Fourier coefficients is given by the Fourier transform of f. In a distributional sense, the eigenfunctions e^{-ikx} are orthogonal. Therefore, the entire theory of eigenfunction expansions seems to remain valid for operators with continuous spectrum. Indeed, this is true and there is a close connection to eigenfunction expansions and transform theory.

These ideas have their origin in quantum theory. The eigenvalue problem for the Schrödinger operator

$$-\Delta \psi + V(\mathbf{x})\psi = \lambda\psi, \quad \mathbf{x} \in \mathbb{R}^n,$$

where V is the potential function, can have both eigenvalues (with square-integrable eigenfunctions, called bound states) and continuous spectrum (with bounded eigenfunctions, called unbound states). The eigenfunction expansion then involves both Fourier series (sums) and Fourier integrals. These topics are beyond our scope and we refer the reader to Keener (2000) or Stakgold (1998).

EXERCISES

1. Show that the inverse Fourier transform of $e^{-a|\xi|}$, $a > 0$, is

$$\frac{a}{\pi} \frac{1}{x^2 + a^2}.$$

2. Verify the following properties of the Fourier transform:

   a) $(\mathcal{F}u)(\xi) = 2\pi(\mathcal{F}^{-1}u)(-\xi)$.

   b) $\mathcal{F}(e^{iax}u)(\xi) = \hat{u}(\xi + a)$.

   c) $\mathcal{F}(u(x+a)) = e^{-ia\xi}\hat{u}(\xi)$.

3. Find the Fourier transform of the following functions:

   a) $u(x) = H(x)e^{-ax}$, where H is the heaviside function.

   b) $u(x) = xe^{-ax^2}$.

   c) $u(x) = f(x)\cos ax$, where f is given.

4. Give an argument similar to that in Example 6.36 to find the Fourier transform of the delta distribution $\delta_a(x)$.

5. Show that the solution of the Cauchy problem

$$u_t = Du_{xx}, \quad x \in \mathbb{R}, \ t > 0,$$
$$u(x,0) = \delta(x), \quad x \in \mathbb{R},$$

where $\delta(x)$ is a point source at the origin is the fundamental solution to the diffusion equation,

$$u(x,t) = \frac{1}{\sqrt{4\pi Dt}} e^{-(x-a)^2/4Dt}.$$

6. What is the solution to the diffusion equation when, at $t = 0$, there two point sources, one at $x = -4$ and the other at $x = 5$, of magnitudes 3 units and 1 unit, respectively? Choose a value of the diffusion constant D and plot several solution profiles showing how the concentration spreads.

7. If $\mathcal{F}\left(xe^{-|x|}\right) = \frac{4i\xi}{(1+\xi^2)^2}$, find $\mathcal{F}\left(\frac{x}{(1+x^2)^2}\right)$.

8. Compute $u * u$ where $u(x) = e^{-|x|}$, and then find the Fourier transform of the function $f(x)$ such that $\hat{f}(\xi) = \frac{1}{(1+\xi^2)^2}$.

9. If $\hat{f}(\xi) = e^{-\xi^2}$, find $\mathcal{F}\left(f(2x+1)\right)$.

10. Use Fourier transforms to find the solution to the initial value problem for the advection–diffusion equation

$$u_t - cu_x - u_{xx} = 0, \quad x \in \mathbb{R}, \quad t > 0;$$
$$u(x,0) = f(x), \quad x \in \mathbb{R}.$$

11. Solve the Cauchy problem for the nonhomogeneous heat equation:

$$u_t = u_{xx} + F(x,t), \quad x \in \mathbb{R}, \ t > 0; \quad u(x,0) = 0, \ x \in \mathbb{R}.$$

12. The Fourier transform and its inverse in $\mathbb{R}^n$ are given by

$$\hat{u}(\xi) = \int_{\mathbb{R}^n} u(\mathbf{x}) e^{i\xi \cdot \mathbf{x}} \, d\mathbf{x}, \quad \mathbf{x}, \xi \in \mathbb{R}^n,$$
$$u(\mathbf{x}) = \frac{1}{(2\pi)^n} \int_{\mathbb{R}^n} \hat{u}(\mathbf{x}) e^{-i\xi \cdot \mathbf{x}} \, d\mathbf{x}.$$

a) Show that the Fourier transform of the Laplacian is $\mathcal{F}(\Delta u) = -|\xi|^2 \hat{u}(\xi)$.

b) Find the Fourier transform of the n-dimensional Gaussian $u(\mathbf{x}) = e^{-a|\mathbf{x}|^2}$.

c) Solve the Cauchy problem for the diffusion equation in $\mathbb{R}^n$.
$$u_t = D\Delta u, \quad \mathbf{x} \in \mathbb{R}^n, \quad u(\mathbf{x}, 0) = f(\mathbf{x}), \quad \mathbf{x} \in \mathbb{R}^n.$$

13. If u and v are absolutely integrable, show that
$$\int_{\mathbb{R}} u(x)\widehat{v}(x)dx = \int_{\mathbb{R}} \widehat{u}(x)v(x)dx.$$
Hint: the given conditions allow interchange of the order of integration.

14. Prove the **Plancherel relation**
$$\int_{-\infty}^{\infty} |u(x)|^2 dx - \frac{1}{2\pi}\int_{\mathbb{R}} |\widehat{u}(\xi)|^2 d\xi,$$
or, $\|u\|^2 = \frac{1}{2\pi}\|\widehat{u}\|^2$. Hint: set $\overline{u} = \widehat{v}$ in the previous problem. (Recall that $\|u\|$ is the L^2 norm.)

15. Use the Plancherel relation to evaluate the integral
$$\int_{-\infty}^{\infty} \frac{dx}{(1+x^2)^2}.$$

16. Find a formula for the solution to the free Schrödinger equation
$$\psi_t = i\psi_{xx}, \quad x \in \mathbb{R}, \ t > 0; \quad \psi(x, 0) = e^{-x^2}, \ x \in \mathbb{R}.$$

17. (Uncertainty principle) In quantum mechanics, the square of the wave function, $|\psi(x)|^2$ is probability density for the location of a particle; that is, $\Pr(X \in I) = \int_I |\psi(x)|^2 dx$, where X is a random variable for the position of the particle, and I is an interval. The quantity $\|x\psi\|^2$ therefore measures the spread of the density. The spread of the momentum of a particle is $\|x\widehat{\psi}\|^2$. Heisenberg's uncertainty principle states that
$$\|x\psi\|^2\|x\widehat{\psi}\|^2 \geq \frac{1}{2}, \tag{5.11}$$
which is to say that both the position and momentum of a particle cannot be measured simultaneously as close as desired. Prove (5.11) using the Cauchy–Schwartz inequality and Plancherel's formula and the following steps:
$$\frac{1}{2} = |\,(x\psi, \psi')\,| \leq \|x\psi\|^2\|\psi'\|^2$$
$$= \|x\psi\|^2\|\widehat{\psi'}\|^2$$
$$= \|x\psi\|^2\|ix\widehat{\psi}\|^2.$$
The first equality follows from integration by parts and the fact that $\|\psi\|^2 = 1$.

18. Solve the initial boundary value problem:

$$u_t = Du_{xx}, \quad x > 0, \ t > 0,$$
$$u(x,0) = f(x), \quad x > 0,$$
$$u(0,t) = 0, \quad t > 0,$$

by extending f to the entire real line as an odd function and then using Fourier transforms.

19. Solve the following initial value problem for the heat equation.

$$u_t = ku_{xx}, \quad x \in \mathbb{R}, \ t > 0,$$
$$u(x,0) = u_0, \quad |x| \le a; \quad u(x,0) = 0, \quad |x| > a,$$

where u_0 is a constant. Write your solution in terms of the erf function. For large times t, show that u decays like $\frac{1}{\sqrt{t}}$, or more specifically, $u \sim u_0 a / \sqrt{\pi k t}$.

20. Show that the solution to Laplace's equation in the upper half-plane with a Neumann boundary condition,

$$u_{xx} + u_{yy} = 0, \quad x \in \mathbb{R}, \ y > 0,$$
$$u_y(x,0) = f(x), \quad x \in \mathbb{R},$$

is given by

$$u(x,y) = \frac{1}{2\pi} \int_{-\infty}^{\infty} \ln[(x-\eta)^2 + y^2] f(\eta) d\eta.$$

Hint: let $v(x,y) = u_y(x,y)$.

21. Use Plancherel's formula to show

$$\int_{\mathbb{R}} \frac{\sin^2 x}{x^2} dx = \pi.$$

22. Solve the boundary value problem

$$u_{xx} + u_{yy} = 0, \ x \in \mathbb{R} \ y > 0,$$

with each of the two boundary conditions: (a) $u(x,0) = H(x)$, $x \in \mathbb{R}$, and (b) $u(x,0) = \frac{1}{1+x^2}$, $x \in \mathbb{R}$.

23. Solve the integral equation

$$\int_{\mathbb{R}} (x-y) e^{-y^2/4} dy = e^{-x^2/4}.$$

6.6 Stability of Solutions

6.6.1 Reaction–Diffusion Equations

We already noted the broad occurrence of diffusion problems in biological systems. In this section we investigate another aspect of such problems, namely the persistence of equilibrium states in systems governed by **reaction–diffusion** systems. At issue is the stability of those states: if a system is in an equilibrium state and it is slightly perturbed, does the system return to that state, or does it evolve to a completely different state?

Underpinned by the seminal work of Alan Turing[5] in 1952 on the chemical basis of morphogenesis, it has been shown in extensive research that diffusion-induced instabilities can give rise to spatial patterns in all sorts of biological systems. Reaction-diffusion models have been used to explain the evolution of form and structure in developmental biology (morphogenesis), tumor growth, ecological spatial patterns, aggregation in slime molds, patterns on animal coats, and many other observed processes in molecular and cellular biology.

In this section we introduce the basic idea of stability in reaction–diffusion models and we observe that such systems can have instabilities that lead to density variations and patterns.

A similar theory of local stability of equilibrium solutions can be developed for partial differential equations. In the interval $0 < x < L$ consider the reaction–diffusion equation

$$u_t = Du_{xx} + f(u), \tag{6.1}$$

with no-flux boundary conditions $u_x(0,t) = u_x(\pi, t) = 0$. Let $u(x,t) = u_e$ be a constant equilibrium solution (so that $f(u_e) = 0$). To fix the idea let $a = f'(u_e) > 0$. Note that the equilibrium solution satisfies the partial differential equation and the boundary conditions. Next, let $U(x,t)$ be a small perturbation from the equilibrium solution, or $u(x,t) = u_e + U(x,t)$. The *perturbation equation* is determined by substituting this expression into (6.1) and the boundary conditions:

$$U_t = DU_{xx} + f(u_e + U), \quad U_x(0,t) = U_x(L,t) = 0.$$

To linearize this equation we expand the right side in a Taylor series and discard the nonlinear terms:

$$f(u_e + U) = f(u_e) + f'(u_e)U + \cdots = f'(u_e)U + \cdots.$$

---

[5] A. Turing was a British mathematician who was instrumental in breaking the German codes during World War II.

Then we obtain the *linearized perturbation equation*

$$U_t = DU_{xx} + aU, \quad a = f'(u_e) > 0, \tag{6.2}$$

subject to boundary conditions

$$U_x(0,t) = U_x(L,t) = 0. \tag{6.3}$$

This problem, which is on a bounded interval, can be solved by separation of variables. We assume $U = g(x)h(t)$ and then substitute into (6.2) and boundary conditions (6.3) to obtain

$$gh' = Dg''h + agh,$$

or

$$\frac{h'}{h} = \frac{Dg''}{g} + a = \lambda.$$

Then $h = Ce^{\lambda t}$ and

$$g'' + \left(\frac{a - \lambda}{D}\right)g = 0, \quad g'(0) = g'(L) = 0.$$

This equation cannot have nontrivial exponential solutions that satisfy the boundary conditions; therefore $a - \lambda \geq 0$. In the case $\lambda = a$ we obtain a constant solution, and in the case $a - \lambda > 0$ we obtain

$$g(x) = A \sin\sqrt{\frac{a - \lambda}{D}}x + B \cos\sqrt{\frac{a - \lambda}{D}}x.$$

Applying $g'(0) = 0$ forces $A = 0$; then $g'(L) = 0$ implies

$$\sin\sqrt{\frac{a - \lambda}{D}}L = 0,$$

or

$$\sqrt{\frac{a - \lambda}{D}}L = n\pi, \quad n = 1, 2, \ldots.$$

Therefore, incorporating the case $\lambda = a$, we obtain eigenvalues

$$\lambda = \lambda_n = a - \frac{Dn^2\pi^2}{L^2}, \quad n = 0, 1, 2, \ldots. \tag{6.4}$$

The modal solutions are therefore

$$U_n(x,t) = e^{\lambda_n t}\cos\frac{n\pi}{L}x, \quad n = 0, 1, 2, \ldots. \tag{6.5}$$

What conclusions can we draw from this calculation? Because the general solution $U(x,t)$ to the boundary value problem (6.2)–(6.3) is a linear combination of the modal solutions (6.5),

$$U(x,t) = c_0 e^{at} + \sum_{n=1}^{\infty} c_n e^{\lambda_n t} \cos \frac{n\pi}{L} x,$$

it will decay if all of the modes decay, and it will grow if one of the modes grows. The constants c_n are determined by the initial perturbation. The spatial part of the modal solutions, or the cosine, remains bounded. The amplitude factor $e^{\lambda_n t}$, and therefore the eigenvalues λ_n, will determine growth or decay. Let us examine the modes. In the case $n = 0$ the eigenvalue is $\lambda_n = a$ and the modal solution is $c_0 e^{at}$, which grows exponentially. In fact, any mode satisfying $a \geq \frac{Dn^2\pi^2}{L^2}$ is unstable. Therefore, instabilities are likely to occur for modes of low frequency (n small), or in systems of large size L or low diffusion properties D. Oppositely, small systems with large diffusion constants are stabilizing, as are the high–frequency modes and geometrically small regions. In the most general case the initial perturbation will nearly always contain all modes and the steady state will be locally unstable.

Above we examined the stability of a constant solution, but the idea applies to nonconstant equilibrium solutions as well.

6.6.2 Pattern Formation

The mechanism causing instabilities in the preceding problem causes patterns to form. We apply this idea to a problem in cell aggregation. A slime mold population is a collection of unicellular amoeboid cells that feed on bacteria in the soil. When the food supply is plentiful, the bacteria are generally distributed uniformly throughout the soil. But, as the food supply becomes depleted and starvation begins, the amoeba start to secrete a chemical (cyclic AMP) that acts as an attractant to the other amoeba and aggregation sites form. The rest of the story is even more interesting as the aggregation sites evolve into slugs that ultimately develop into sporangiophores consisting of a stalk and head containing new spores. The spores are released and the process begins anew. We are interested here in only the first part of this complicated problem, the onset of aggregation. We work in one spatial dimension and imagine the amoeba are distributed throughout a tube of length L.

Let $a = a(x,t)$ and $c = c(x,t)$ denote the density and concentration of the cellular amoeba and cAMP, respectively. The fundamental conservation laws are (see Section 6.2)

$$a_t = -J_x^{(a)}, \quad c_t = -J_x^{(c)} + F,$$

where $J^{(a)}$ and $J^{(c)}$ are the fluxes of the amoeba and the chemical, respectively. There are no source terms in the amoeba equation because we will not consider birth and death processes on the time scale of the analysis. The source term in the chemical equation consists of two parts: production of the chemical by the amoeba and degradation of the chemical in the soil. We assume the chemical is produced by the amoeba at a rate proportional to the density of the amoeba, and the chemical degrades at a rate proportional to its concentration; that is,

$$F = fa - kc.$$

The motion of the chemical is caused by diffusion only, and we assume Fick's law:

$$J^{(c)} = -\delta c_x,$$

where δ is the diffusion constant. The amoeba are also assumed to randomly diffuse via Fick's law, but there is another flux source for the amoeba, namely attraction to the chemical. We assume this attraction is *up the chemical gradient*, toward high concentrations of c. Additionally, the rate of diffusion should depend on the amoeba population because that will increase the magnitude of the chemical concentration released. This type of flow, induced by chemical gradients, is called **chemotaxis**. Therefore we assume

$$
\begin{aligned}
J^{(a)} &= \quad \text{random flux} + \text{chemotatic flux} \\
&= \quad -\mu a_x + \nu a c_x,
\end{aligned}
$$

where μ is the amoeba motility and ν is the strength of the chemotaxis, both assumed to be positive constants. Note that the random flux, having the form of Fick's law, has a negative sign because flow is "down the gradient" (from high to low densities), and the chemotatic flux has a positive sign because that term induces flow "up the chemical gradient" (from low to high concentrations). Putting all these equations all together gives a system of reaction–diffusion equations,

$$a_t = \mu a_{xx} - \nu(a c_x)_x, \quad c_t = \delta c_{xx} + fa - kc. \tag{6.6}$$

Both a and c satisfy no-flux boundary conditions (i.e., $a_x = c_x = 0$ at $x = 0$, L,), which means there is no escape from the medium.

Notice that there will be an equilibrium solution $a = \bar{a}$, $c = \bar{c}$ to (6.6) provided

$$f\bar{a} = k\bar{c}.$$

That is, the production of the chemical equals its degradation. This equilibrium state represents the spatially uniform state in the soil before aggregation begins.

To determine the local stability of this uniform state we let

$$a = \bar{a} + u(x, t), \quad c = \bar{c} + v(x, t),$$

where u and v are small perturbations. Substituting these quantities into (6.6) gives, after simplification, the perturbation equations

$$u_t = \mu u_{xx} - \nu((\bar{a} + u)v_x)_x, \quad v_t = \delta v_{xx} + fu - kv.$$

These equations are nonlinear because of the uv_x term in the amoeba equation. If we discard the nonlinear terms on the assumption that the product of small terms is even smaller, then we obtain the linearized perturbation equations

$$u_t = \mu u_{xx} - \nu \bar{a} v_{xx}, \quad v_t = \delta v_{xx} + fu - kv. \tag{6.7}$$

Easily one can see that the perturbations satisfy Neumann, no-flux boundary conditions.

Motivated by our knowledge of the form of solutions to linear equations, we assume there are modal solutions of the form

$$u(x,t) = c_1 e^{\sigma t} \cos rx, \quad v(x,t) = c_2 e^{\sigma t} \cos rx, \tag{6.8}$$

where r and σ are to be determined, and c_1 and c_2 are some constants. Notice the form of these assumed solutions. The spatial part is bounded and periodic with wave number, or frequency, r, and have wavelength $2\pi/r$. The temporal part is exponential with growth factor σ, which may be a real or complex number. If σ is negative or has negative real part, then the perturbation will decay and the equilibrium state will return (the uniform state is stable); if σ is positive or has positive real part, then the perturbations will grow and the equilibrium will be unstable. Let us ask why solutions of (6.7) should be of this form (6.8) without going through the entire separation of variables argument. Equations (6.7) are linear with constant coefficients, and both equations contain both unknowns u and v and their derivatives. If we are to have a solution, then all the terms must match up in one way or another in order to cancel. So both u and v must essentially have the same form. Because there is a first derivative in t, the time factor must be exponential so it will cancel, and the spatial part must be a sine or a cosine because of the appearance of a second spatial derivative. We anticipate the cosine function because of the no-flux boundary conditions, as in Chapter 5. If we substitute (6.8) into (6.7) we obtain the two algebraic equations

$$(\sigma + \mu r^2)c_1 - \nu \bar{a} r^2 c_2 = 0, \quad -fc_1 + (\sigma + k + \delta r^2)c_2 = 0,$$

which relate all the parameters. We may regard these as two linear, homogeneous equations for the constants c_1 and c_2. If we want a nontrivial solution for c_1 and c_2, then matrix theory dictates that the determinant of the coefficient matrix must be zero. That is,

$$(\sigma + \mu r^2)(\sigma + k + \delta r^2) - f\nu \bar{a} r^2 = 0,$$

which is an equation relating the temporal growth factor σ, the wave number r, and the other constants in the problem. Expanded out, this equation is a quadratic in σ,

$$\sigma^2 + \gamma_1 \sigma + \gamma_2 = 0,$$

where

$$\gamma_1 = r^2(\mu + \gamma) + k > 0, \quad \gamma_2 = r^2[\mu(\delta r^2 + k) - f\nu\bar{a}].$$

The roots of the quadratic are

$$\sigma = \frac{1}{2}\left(-\gamma_1 \pm \sqrt{\gamma_1^2 - 4\gamma_2}\right).$$

Clearly one of the roots (taking the minus sign) is always negative or has negative real part. The other root can have positive or negative real part, depending upon the value of the discriminant $\gamma_1^2 - 4\gamma_2$. We are interested in determining if there are parameter choices that lead to an instability; so we ask when σ positive. Hence, γ_2 must be negative, or

$$\mu(\delta r^2 + k) < f\nu\bar{a}.$$

If this inequality holds, there is an unstable mode and perturbations will grow. We analyze this further.

The number r is the wave number of the perturbations. Applying the no-flux boundary conditions forces

$$r = \frac{n\pi}{L}, \quad n = 0, 1, 2, 3, \ldots.$$

For each value of n we obtain a wave number $r = r_n$ and a corresponding growth factor σ_n. The nth mode will therefore grow and lead to local instability when

$$\mu\left(\delta\frac{n^2\pi^2}{L^2} + k\right) < f\nu\bar{a}. \tag{6.9}$$

We can now ask what factors destabilize the uniform, equilibrium state in the amoeba–cAMP system and therefore promote aggregation. That is, when is (6.9) likely to hold? We can list the factors that may make the left side of the inequality (6.9) smaller than the right side: low motility μ of the bacteria; low degradation rate k or large production rate f of cAMP; large chemotactic strength ν; large dimensions L of the medium; a small value of n (thus, low wave number, or long wavelength, perturbations are less stabilizing than short wavelength, or high wave number, perturbations); decreasing the diffusion constant of the cAMP. Figure 6.6 shows time snapshots of the amoeba density for the mode $n = 2$ when it is unstable. The regions where the amplitude is high correspond to higher concentrations of amoeba, or regions of aggregation.

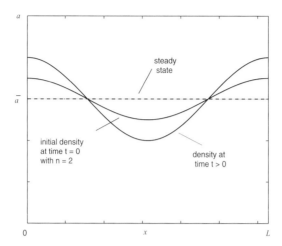

Figure 6.6 Plot showing the growing amoeba density at time t when the uniform state is unstable to local perturbations in the mode $n = 2$. This instability gives rise to the aggregation sites in the regions where the solution has greater amplitude.

EXERCISES

1. Consider the problem for Fisher's equation with Dirichlet boundary conditions:

$$u_t = u_{xx} + u(1 - u), \quad -\frac{\pi}{2} < x < \frac{\pi}{2},$$

$$u = 3 \quad \text{at} \quad x = \pm\frac{\pi}{2}.$$

a) Show that $u_e(x) = \frac{3}{1+\cos x}$ is an equilibrium solution.

b) Define perturbations $U(x,t)$ by the equation $u = u_e(x) + U(x,t)$ and find the linearized perturbation equation and boundary conditions for $U(x,t)$.

c) Assume a solution to the linearized equation of the form $U = e^{\sigma t}g(x)$ and show that g must satisfy

$$g'' + \frac{\cos x - 5}{1 + \cos x}g = \sigma g, \quad g = 0 \quad \text{at} \quad x = \pm\frac{\pi}{2}. \tag{6.10}$$

d) Show that if (6.10) has a nontrivial solution, then $\sigma < 0$, thereby showing local stability of the steady solution. (Hint: Consider two cases, when g is positive and when g is negative on the interval and then examine the signs of g'' and the other terms in (6.10) at a maximum or minimum point.)

2. (**Turing system**). Consider the system of reaction-diffusion equations on the spatial domain $0 < x < L$ given by

$$u_t = \alpha u_{xx} + f(u,v), \quad v_t = \beta v_{xx} + g(u,v)$$

with no-flux boundary conditions $u_x = v_x = 0$ at $x = 0, L$. Let $u = \bar{u}$, $v = \bar{v}$ be an equilibrium solution and define small perturbations U and V from equilibrium given by

$$u = \bar{u} + U(x,t), v = \bar{v} + V(x,t).$$

a) Show that U and V satisfy no-flux boundary conditions and the linearized perturbation equations

$$
\begin{aligned}
U_t &= \alpha U_{xx} + f_u(\bar{u},\bar{v})U + f_v(\bar{u},\bar{v})V, & (6.11) \\
V_t &= \beta V_{xx} + g_u(\bar{u},\bar{v})U + g_v(\bar{u},\bar{v})V. & (6.12)
\end{aligned}
$$

b) Introduce matrix notation

$$\mathbf{w} = \begin{pmatrix} U \\ V \end{pmatrix}, \quad D = \begin{pmatrix} \alpha & 0 \\ 0 & \beta \end{pmatrix}, \quad J = \begin{pmatrix} f_u(\bar{u},\bar{v}) & f_v(\bar{u},\bar{v}) \\ g_u(\bar{u},\bar{v}) & g_v(\bar{u},\bar{v}) \end{pmatrix},$$

and show that (6.11)–(6.12) can be written as

$$\mathbf{w}_t = D\mathbf{w}_{xx} + J\mathbf{w}. \qquad (6.13)$$

c) Assume modal solutions to (6.13) of the form

$$\mathbf{w} = \mathbf{c}e^{\sigma_n t}\cos\frac{n\pi x}{L}, \quad \mathbf{c} = \begin{pmatrix} c_{1n} \\ c_{2n} \end{pmatrix}, \quad n = 0, 1, 2, ...,$$

and show that for a nontrivial solution we must have

$$\det\left(\sigma_n I + \frac{n^2\pi^2}{L^2}D - J\right) = 0. \qquad (6.14)$$

(When expanded, this equation is a quadratic equation for the growth factor σ_n of the nth mode. The roots σ_n depend upon the diffusion constants α, β, the equilibrium solution $\bar{u}, \bar{v}$, the size of the medium L, and the wavelength $\pi L/n$ of the perturbation. If one can find values of the parameters that make one of the roots positive or have positive real part, then there is an unstable mode.)

3. Apply the method of Exercise 2 to examine the stability of the steady state of the Turing system

$$u_t = Du_{xx} + 1 - u + u^2 v, \quad v_t = v_{xx} + 2 - u^2 v, \quad 0 < x < \pi,$$

under no-flux boundary conditions. Specifically, find the condition (6.14) and determine values of D for which various modes (n) are unstable.

6.7 Distributions

6.7.1 Elliptic Problems

The notions of Green's functions and distributions introduced in Chapter 5 carry over in a straightforward manner to several variables and to partial differential equations. We briefly state some of the main ideas. Let Ω be an open set in $\mathbb{R}^n$ (for definiteness the reader may take $n = 2$ or $n = 3$). We denote by $D(\Omega) \equiv C_0^\infty(\Omega)$ the set of functions defined on Ω that have continuous derivatives of all orders and that vanish outside a closed, bounded subset of Ω. The set $D(\Omega)$ is called the set of **test functions** on Ω. Then, a **distribution** u on $D(\Omega)$ is a continuous linear functional on $D(\Omega)$, and we denote the value, or action, of the distribution u by (u, ϕ), where $\phi \in D(\Omega)$. The set of all distributions defined on $D(\Omega)$ is denoted by $D'(\Omega)$. Every function u that is **locally integrable** on Ω, namely $\int_K |u(x)| \, dx < \infty$ for all closed bounded subsets K of Ω, generates a distribution via the formula

$$(u, \phi) = \int_\Omega u(x)\phi(x) \, dx, \quad \phi \in D(\Omega). \tag{7.1}$$

The **Dirac (delta) distribution** with pole at $\xi \in \mathbb{R}^n$ is defined by

$$(\delta_\xi, \phi) = \phi(\xi), \quad \phi \in D(\Omega). \tag{7.2}$$

As in the one-dimensional case we often denote a locally integrable function and its associated distribution by the same symbol. Even if a distribution u is not generated by a locally integrable function as in (7.1), we often write it as $u(x)$ even though it is not a pointwise function of x; and we write its action on ϕ symbolically in the integral form (7.1), even though the integral is not defined. Thus, for example, we write $\delta_\xi(x)$ or $\delta(x-\xi)$ for the delta distribution, and we define its action (7.2) symbolically as

$$\int_\Omega \delta(x - \xi)\phi(x) \, dx = \phi(\xi).$$

The multiplication of a distribution $u \in D'(\Omega)$ by a $C^\infty(\Omega)$ function a is defined by $(au, \phi) = (u, a\phi), \phi \in D(\Omega)$; moreover, the partial derivative of a distribution u is defined by

$$\left(\frac{\partial}{\partial x_k} u, \phi \right) = - \left(u, \frac{\partial}{\partial x_k} \phi \right), \quad \phi \in D(\Omega).$$

Second-order derivatives are defined by

$$\left(\frac{\partial^2}{\partial x_j \partial x_k} u, \phi \right) = \left(u, \frac{\partial^2}{\partial x_j x_k} \phi \right), \quad \phi \in D(\Omega).$$

All of these definitions for general distributions are motivated, as in the one-dimensional case, by the action of locally integrable functions. Higher-order distributional derivatives can be defined as well. Thus it makes sense to regard a linear, second-order partial differential operator L with C^∞-coefficients as a distribution in $D'(\Omega)$. If f is a distribution, then the differential equation

$$Lu = f \tag{7.3}$$

can be regarded as an equation in distributions, and any distribution u that satisfies (7.3) is called a **distribution solution**. Thus, u is a distribution solution to (7.3) if, and only if,

$$(Lu, \phi) = (f, \phi), \quad \forall \phi \in D(\Omega). \tag{7.4}$$

In particular, if $f \equiv \delta(x - \xi)$ is the delta distribution, then u is called a **fundamental solution** associated with the operator L. The formal adjoint operator associated with L is denoted by L^* and is defined by the relation

$$(Lu, \phi) = (u, L^*\phi), \quad \phi \in D(\Omega). \tag{7.5}$$

Therefore, u is a distribution solution of (7.3) if, and only if,

$$(u, L^*\phi) = (f, \phi), \quad \forall \phi \in D(\Omega), \tag{7.6}$$

and u is a fundamental solution (with pole at ξ) associated with the operator L if, and only if,

$$(u, L^*\phi) = \phi(\xi), \quad \forall \phi \in D(\Omega).$$

If u and f are locally integrable functions on Ω that satisfy (7.5), then we say u is a **weak solution** to (7.3). Thus, u is a weak solution to (7.3) if, and only if,

$$\int_\Omega u(x) L^*\phi(x) \, dx = \int_\Omega f(x)\phi(x) \, dx, \quad \forall \phi \in D(\Omega).$$

A solution to (7.3) that is twice continuously differentiable (we are restricting the discussion to second-order PDEs) on Ω is called a classical, or **genuine**, solution.

So, there are three levels of solutions: classical, weak, and distribution. Classical solutions are functions that have sufficiently many derivatives and satisfy the PDE (7.3) pointwise on the domain of interest. Weak solutions of (7.3), where f is locally integrable, are locally integrable functions that satisfy (7.5). Distribution solutions of (7.3) may not be functions or even generated by functions. Classical solutions are weak solutions, and weak solutions are distribution solutions, but the converses are not true.

Example 6.44

Show that the function

$$g(x, y) = \frac{1}{4\pi} \ln(x^2 + y^2) \tag{7.7}$$

is a distributional solution of the equation

$$\Delta u = \delta(x, y),$$

where $\Delta \equiv \partial^2/\partial x^2 + \partial^2/\partial y^2$ is the two-dimensional Laplacian, and $\delta(x, y)$ is the Dirac distribution with pole at $(0,0)$. Thus we have to show that $(\Delta g, \phi) \equiv (g, \Delta\phi) = \phi(0,0)$ for all $\phi \in D(\mathbb{R}^2)$. To this end, using polar coordinates, we have

$$(4\pi g, \Delta\phi) = \int_{R^2} \ln(x^2 + y^2)(\phi_{xx} + \phi_{yy}) \, dx \, dy$$

$$= \int_0^{2\pi} \int_0^{\infty} \ln r^2 \left(\phi_{rr} + \frac{1}{r}\phi_r + \frac{1}{r^2}\phi_{\theta\theta} \right) r \, dr \, d\theta$$

$$= \lim_{\epsilon \to 0} \int_\epsilon^{\infty} \int_0^{2\pi} \left(r \ln r^2 \phi_{rr} + \ln r^2 \phi_r + \frac{1}{r} \ln r^2 \phi_{\theta\theta} \right) d\theta \, dr.$$

The integral of the last term is zero by the θ periodicity of ϕ and its θ derivatives. So we need to calculate the other two terms. Integrating by parts gives

$$\int_\epsilon^{\infty} \ln r^2 \phi_r \, dr = \phi \ln r^2 |_\epsilon^{\infty} - \int_\epsilon^{\infty} \frac{2}{r}\phi \, dr$$

$$= -\phi(\epsilon, \theta) \ln \epsilon^2 - \int_\epsilon^{\infty} \frac{2}{r}\phi \, dr$$

and integrating by parts twice gives

$$\int_\epsilon^{\infty} r \ln r^2 \phi_{rr} \, dr = r\phi_r \ln r^2 - \int_\epsilon^{\infty} (2 + \ln r^2)\phi_r \, dr$$

$$= -\phi_r(\epsilon, \theta)\epsilon \ln \epsilon^2 + (2 + \ln \epsilon^2)\phi(\epsilon, \theta) + \int_\epsilon^{\infty} \frac{2}{r}\phi \, dr.$$

Putting these results together gives

$$
\begin{aligned}
(4\pi g, \Delta\phi) &= \lim_{\epsilon \to 0} \int_0^{2\pi} \left(-\phi(\epsilon, \theta)\ln\epsilon^2 - \int_\epsilon^\infty \frac{2}{r}\phi\,dr \right) d\theta \\
&+ \lim_{\epsilon \to 0} \int_0^{2\pi} \left((2 + \ln\epsilon^2)\phi(\epsilon, \theta) + \int_\epsilon^\infty \frac{2}{r}\phi\,dr \right) d\theta \\
&= 2\lim_{\epsilon \to 0} \int_0^{2\pi} \phi(\epsilon, \theta)\,d\theta = 4\pi\phi(0,0).
\end{aligned}
$$

Therefore, the function defined by (7.7), which is called the **logarithmic potential**, is a fundamental solution associated with the two-dimensional Laplacian. Note that it is not correct to say (7.7) is a genuine solution or a weak solution to Laplace's equation $\Delta u = 0$ in $\mathbb{R}^2$. Observe that the Laplacian operator is invariant under a translation of coordinates, and so a fundamental solution associated with the Laplacian with pole at (ξ, η) is

$$
g(x, y; \xi, \eta) = \frac{1}{4\pi} \ln((x - \xi)^2 + (y - \eta)^2). \quad \square \tag{7.8}
$$

Example 6.45

A fundamental solution associated with the Laplacian in $\mathbb{R}^3$ is the function $g(x, y, z) \equiv 1/4\pi r$, where $r = \sqrt{x^2 + y^2 + z^2}$. This means $\Delta g = \delta(x, y, z)$ in the sense of distributions. Again, the distribution g can be translated to any point (ξ, η, ζ) to obtain a fundamental solution with pole (ξ, η, ζ). In three dimensions g is called the **Newtonian potential**. We remark that the fundamental solutions associated with the Laplacian are radial solutions in that they depend only on the distance from the pole. $\quad \square$

The Green's function associated with a partial differential operator, such as the Laplacian, is a fundamental solution that also satisfies the homogeneous boundary conditions. As such, the Green's function is the equilibrium response of the physical system under investigation caused by a unit point source. Mathematically, the Green's function is the kernel of the integral operator that represents the inverse of the partial differential operator. There is no single, universal method to construct Green's function. On finite domains one may use separation of variables, and on infinite domains one may try transform methods. Often geometric or physical arguments are helpful.

Example 6.46

Find the Green's function associated with the Laplacian operator on the two-dimensional domain $-\infty < x < \infty, y > 0$ with a Dirichlet boundary condition on $y = 0$. This means that we seek a distributional solution $G = G(x, y; \xi, \eta)$ to the problem

$$G_{xx} + G_{yy} = \delta(x, y; \xi, \eta), \quad x, \xi \in \mathbb{R}; \ y, \eta > 0, \tag{7.9}$$

$$G(x, 0; \xi, \eta) = 0, \quad x, \xi \in \mathbb{R}; \ \eta > 0. \tag{7.10}$$

Physically, G could represent the potential in the half-plane $y > 0$ of an static electric field generated by a unit point positive charge at (ξ, η), with the condition that the potential vanish along $y = 0$. We shall construct the Green's function G by the so-called **method of images**, or reflection principle. We know that the fundamental solution g from equation (7.8) in Example 6.44 satisfies the equation $\Delta g = \delta(x, y; \xi, \eta)$ for a positive charge at (ξ, η), but it does not satisfy the boundary condition. To compensate we locate an "image" charge of opposite sign at $(\xi, -\eta)$, the reflection of the point (ξ, η) through the y axis. The potential due to that charge would be $\tilde{g} \equiv -g(x, y; \xi, -\eta)$. Now take $G \equiv g + \tilde{g}$, or

$$G(x, y, \xi, \eta) = \frac{1}{4\pi} \ln \frac{(x - \xi)^2 + (y - \eta)^2}{(x - \xi)^2 + (y + \eta)^2}. \tag{7.11}$$

Clearly G satisfies the homogeneous boundary condition (7.10). Moreover, $\Delta G = \Delta g + \Delta \tilde{g} = \delta(x, y, \xi, \eta) + 0 = \delta(x, y, \xi, \eta)$ in the upper half-plane $y > 0$. Here we used the fact that $\Delta \tilde{g} = 0$ in the *upper* half-plane; its pole is in the lower half-plane, and everywhere but at its pole it satisfies Laplace's equation. $\square$

Knowing Green's function allows us to solve the nonhomogeneous problem

$$u_{xx} + u_{yy} = \rho(x, y), \quad x \in \mathbb{R}, \ y > 0, \tag{7.12}$$

$$u(x, 0) = 0, \quad x \in \mathbb{R}. \tag{7.13}$$

The Green's function G given by (7.11) is the response of the system for a unit point source (say a unit charge) at (ξ, η). Therefore, to find the response to a problem with a continuum of sources (a charge density) of magnitude ρ we can superimpose all the point source responses over all ξ and η to obtain

$$u(x, y) = \int_0^\infty \int_{-\infty}^\infty G(x, y; \xi, \eta) \rho(\xi, \eta) \, dx \, dy. \tag{7.14}$$

One can show that (7.14) is the solution to (7.12)–(7.13) under reasonable conditions on ρ (see, e.g., McOwen (2003)).

6.7.2 Fourier Transforms of Distributions

In problems involving point sources we are faced with a Dirac distribution in the partial differential equation. If we want transform methods to apply, then we must make sense of the transform of a distribution. A formal calculation gives the correct answer. For example, consider taking the Fourier transform of the delta distribution $\delta(x-a)$ with pole at $x=a$; we have

$$(\mathcal{F}(\delta(x-a))(\xi) = \int_{-\infty}^{\infty} \delta(x-a)e^{i\xi x}\,dx = (\delta(x-a), e^{i\xi x}) = e^{ia\xi}.$$

Thus

$$\widehat{\delta(x-a)}(\xi) = e^{ia\xi}.$$

Many readers will be satisfied with this formal calculation. However, to take some of the mystery out of it we give a cursory explanation of its underpinnings. Suppose u is a distribution in $D'(\mathbb{R})$ that is generated by a locally integrable function u, and assume that the Fourier transform $\hat{u}$ is also locally integrable. (Here we are using the convention of denoting a function and its associated distribution by the same symbol.) Then, for any $\phi \in D(\mathbb{R})$ we have

$$(\hat{u}, \phi) = \int_{-\infty}^{\infty} \left(\int_{-\infty}^{\infty} u(x)e^{i\xi x}\,dx \right) \phi(\xi)\,d\xi$$

$$= \int_{-\infty}^{\infty} \left(\int_{-\infty}^{\infty} \phi(\xi)e^{i\xi x}\,d\xi \right) u(x)\,dx = (u, \hat{\phi}).$$

This seems correct, but it would require that both ϕ and its Fourier transform $\hat{\phi}$ to be test functions in $D(\mathbb{R})$. If both are test functions, then one can prove that ϕ must be identically zero. Therefore we cannot define $\hat{u}$ on the set of test functions $D(\mathbb{R})$. The way out of this problem is to consider distributions on a larger class of test functions, namely the Schwartz class $\mathcal{S}$; Schwartz class functions contain the C^∞ functions with compact support. A **tempered distribution** is a continuous linear functional on $\mathcal{S}$. With this approach, it is possible to define the **Fourier transform** of a tempered distribution u to be the tempered distribution, $\hat{u}$ whose action on $\mathcal{S}$ is given by

$$(\hat{u}, \phi) = (u, \hat{\phi}), \quad \phi \in \mathcal{S}.$$

Example 6.47

(**Delta distribution**) Thus, for example, the Fourier transform of the Dirac distribution is defined by

$$(\widehat{\delta(x-a)}, \phi) = (\delta(x-a), \hat{\phi}), \quad \phi \in \mathcal{S}.$$

Therefore,

$$(\widehat{\delta(x-a)}, \phi) = \hat{\phi}(a) = \int_{-\infty}^{\infty} \phi(\xi) e^{ia\xi}\, d\xi = (e^{ia\xi}, \phi)$$

for all $\phi \in \mathcal{S}$. We conclude that, in the sense of tempered distributions,

$$\widehat{\delta(x-a)} = e^{ia\xi}.$$

In particular, if $a = 0$, then we have

$$\widehat{\delta(x)} = 1. \qquad \square$$

Example 6.48

In this example we use the definition of the transform and the result on transforms of derivatives, $\mathcal{F}\phi^{(n)}(x) = (-ix)^n \hat{\phi}(x)$, to show

$$\mathcal{F}(x^n) = 2\pi i^n \delta_0^{(n)}, \quad n = 1, 2, \ldots.$$

We let the notation work for us.

$$
\begin{aligned}
(\widehat{x^n}, \phi) = (x^n, \hat{\phi}) &= (1, x^n \hat{\phi}) \\
&= (\hat{1}, (-i)^n \phi^{(n)}) \\
&= (2\pi\delta_0, (-i)^n \phi^{(n)}) \\
&= 2\pi(-i)^n ((-1)^n \delta_0^{(n)}, \phi) = (2\pi i^n \delta_0^{(n)}, \phi). \qquad \square
\end{aligned}
$$

6.7.3 Diffusion Problems

Now let us consider problems involving time. We examine the heat operator in one spatial dimension and time defined by

$$Lu \equiv u_t - ku_{xx}.$$

Based on our discussion of elliptic problems, we say that u is a fundamental solution associated with L if u is a distributional solution of

$$Lu = \delta(x, t; \xi, \tau), \tag{7.15}$$

where the right side is interpreted as a unit heat source applied at $x = \xi$ and at the instant of time $t = \tau$. Here, the domain Ω is a space–time domain consisting of coordinates $(x, t) \in \mathbb{R}^2$. To find the fundamental solution we take the source to be at $\xi = 0$ and $\tau = 0$. We can then examine the initial value problem

$$u_t - ku_{xx} = 0, \quad t > 0, \tag{7.16}$$

$$u(x, 0) = \delta(x). \tag{7.17}$$

(If we want to think about this carefully, we can regard (7.16) as a one-parameter family of equations for $u(\cdot, t)$ in $D'(\mathbb{R})$ for each value of the parameter $t > 0$, and (7.17) as an equation in $D'(\mathbb{R})$). Now let us solve (7.16)–(7.17) using Fourier transforms, assuming that the distributions are tempered. Taking the transform of (7.16) gives, using ζ for the transform variable,

$$\hat{u}_t(\zeta, t) + k\zeta^2\hat{u}(\zeta, t) = 0.$$

Solving this ordinary differential equation for $\hat{u}$ yields

$$\hat{u}(\zeta, t) = c(\zeta)e^{-k\zeta^2 t},$$

where $c(\zeta)$ is arbitrary. The Fourier transform of the initial condition (7.17) gives $\hat{u}(\zeta, 0) = 1$, and therefore $c(\zeta) = 1$. Consequently

$$\hat{u}(\zeta, t) = e^{-k\zeta^2 t}.$$

So the Fourier transform of the solution is a Gaussian function. We know that the inverse transform is also a Gaussian and is in fact given by

$$u(x, t) = \frac{1}{\sqrt{4\pi kt}}e^{-x^2/4kt}.$$

This is the solution to the initial value problem (7.16)–(7.17). Finally, translating the source to $x = \xi, t = \tau$ and multiplying by a Heaviside function to turn on the solution at $t = \tau$, we obtain

$$K(x - \xi, t - \tau) \equiv \frac{H(t - \tau)}{\sqrt{4\pi k(t - \tau)}}e^{-(x-\xi)^2/4k(t-\tau)}, \tag{7.18}$$

which is the fundamental solution associated with the heat operator. One can verify that (7.18) is a distributional solution to (7.15).

To understand the fundamental solution of the heat equation, let us fix $\tau = 0$ and consider

$$K(x - \xi, t) \equiv \frac{1}{\sqrt{4\pi kt}}e^{-(x-\xi)^2/4kt}, \quad t > 0. \tag{7.19}$$

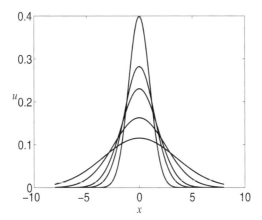

Figure 6.7 Temperature profiles for different times t caused by a unit energy source at $x = \xi = 0$ and $t = 0$.

We observe that

$$\lim_{t \to 0^+} K(x - \xi, t) = 0, \quad x \neq \xi,$$

$$\lim_{t \to \infty} K(x - \xi, t) = 0, \quad x \in \mathbb{R},$$

$$\lim_{t \to 0^+} K(x - \xi, t) = +\infty, \quad x = \xi.$$

Therefore, for large times the fundamental solution goes to zero for all x. As $t \to 0^+$, however, the solution develops an infinite spike at $x = \xi$ while tending to zero for $x \neq \xi$. So, physically, we interpret the fundamental solution $K(x - \xi, t)$ as the temperature distribution in an infinite bar initially at zero degrees with an instantaneous unit heat source applied at $x = \xi$ and $t = 0$. It is a unit heat source because, as one can check,

$$\int_{-\infty}^{\infty} K(x - \xi, t) \, dx = 1, \quad t > 0.$$

That is, the area under the temperature profile curve, which is proportional to the energy, is unity. Stated differently, the function $K(x - \xi, t)$ is the temperature effect at (x, t) caused by a unit heat source applied initially at $x = \xi$. Figure 6.7 shows time snapshots of the function K for different values of t. The fundamental solution (7.19) associated with the heat operator is useful in constructing solutions to a variety of initial boundary value problems involving the heat equation.

Example 6.49

Consider the pure initial value problem for the heat equation

$$u_t - k u_{xx} = 0, \quad x \in \mathbb{R}, \quad t > 0; \quad u(x,0) = f(x), \quad x \in \mathbb{R}. \quad (7.20)$$

By our previous discussion $K(x - \xi, t)$ is the heat response of the system when a unit point source is applied at $x = \xi$ and $\tau = 0$. Thus, to solve (7.20) we superimpose these solutions for point sources of magnitude $f(\xi)$ for all $\xi \in \mathbb{R}$. Therefore the solution to (7.20) is

$$u(x,t) = \int_{-\infty}^{\infty} K(x - \xi, t) f(\xi) \, d\xi$$

$$= \int_{-\infty}^{\infty} \frac{1}{\sqrt{4\pi kt}} e^{-(x-\xi)^2/4kt} f(\xi) \, d\xi. \qquad \square$$

Example 6.50

Consider the nonhomogeneous problem

$$u_t - k u_{xx} = F(x,t), \quad x \in \mathbb{R}, \ t > 0; \quad u(x,0) = 0, \quad x \in \mathbb{R}. \quad (7.21)$$

The solution for a point source at (ξ, τ) is $K(x - \xi, t - \tau)$. To find the solution for distributed sources $F(x,t)$ we superimpose to obtain

$$u(x,t) = \int_{-\infty}^{\infty} \int_{-\infty}^{\infty} \frac{H(t - \tau)}{\sqrt{4\pi k(t - \tau)}} e^{-(x-\xi)^2/4k(t-\tau)} F(\xi, \tau) \, d\xi \, d\tau$$

$$= \int_{0}^{t} \int_{-\infty}^{\infty} \frac{1}{\sqrt{4\pi k(t - \tau)}} e^{-(x-\xi)^2/4k(t-\tau)} F(\xi, \tau) \, d\xi \, d\tau,$$

which is the solution to (7.21). $\square$

EXERCISES

1. Is the function e^{x^2} locally integrable on $\mathbb{R}$? Does it generate a distribution in $D'(\mathbb{R})$? Does it generate a tempered distribution?

2. Show that for any locally integrable function f on $\mathbb{R}$ the function $u(x,y) = f(x - y)$ is a weak solution to the equation $u_x + u_y = 0$ on $\mathbb{R}^2$.

3. In the quarter plane in $\mathbb{R}^2$ find the Green's function associated with the boundary value problem

$$\Delta u = \delta(x, y; \xi, \eta), \quad x, y, \xi, \eta > 0,$$
$$u(x, 0) = 0, \ x > 0; \quad u(0, y) = 0, \ y > 0.$$

(*Hint:* Put image charges in the other quadrants.)

4. In the upper half plane in $\mathbb{R}^2$, use an image charge find Green's function for the Neumann problem

$$\Delta u = \delta(x, y; \xi, \eta), \quad x, \xi \in \mathbb{R}, \ y, \eta > 0,$$
$$u_y(x, 0) = 0, \quad x \in \mathbb{R}.$$

5. At $t = 0$ and at a point in a homogeneous medium an amount of heat energy E is released and allowed to diffuse outward. Deduce that the temperature $u(r, t)$ at time t and distance r from the release point satisfies the initial boundary value problem

$$u_t - \frac{k}{r^2}(r^2 u_r)_r = 0, \quad r > 0, \quad t > 0,$$
$$u(r, 0) = 0, \quad r > 0,$$
$$\lim_{r \to \infty} u(r, t) = 0, \quad t > 0,$$
$$4\pi c \int_0^\infty r^2(r, t)\, dr = E, \quad t > 0,$$

where k is the diffusivity of the medium and c is its heat capacity. Use dimensional analysis and the Pi theorem to show that

$$u = \frac{E}{c}(kt)^{-3/2} U\left(\frac{r}{\sqrt{kt}}\right)$$

for some function U. Determine U to conclude that

$$u(r, t) = \frac{E/c}{(4\pi k t)^{3/2}} e^{-r^2/4kt}.$$

6. Using the definition of Fourier transform, verify that $\mathcal{F}(1) = 2\pi\delta_0$.

7. In $\mathbb{R}^2$ consider the operator

$$L = \frac{\partial^2}{\partial x^2} - \frac{\partial^2}{\partial x \partial y} - \frac{\partial}{\partial y}.$$

Find the adjoint operator L^*.

REFERENCES AND NOTES

There is a large number of excellent elementary textbooks introducing partial differential equations, too numerous to mention. The list below is a small selection that this author often finds useful. Strauss (1991), Farlow (1993), and Logan (2004) give entry-level introductions. Strauss' text, as well as Zauderer (1989), have thorough treatments of eigenvalue problems. At the next level, Evans (1998) and McOwen (2003) give excellent introductions to the theory. Texts that contain biological applications are Britton (2003), Edelstein-Keshet (1988), Grindrod (1991), and Kot (2001). Nonlinear equations are discussed in Logan (2008), which is an elementary treatment, and in the classic treatise by Whitham (1974), which emphasizes hyperbolic problems. Excellent, elementary treatments of Fourier transforms and distributions can be found in Strichartz (1994). Both Keener (2000) and Stakgold (1998) are advanced texts on applied mathematics that contain theoretical treatments of Green's functions and eigenvalue problems.

Britton, N. 2003. *Essential Mathematical Biology*, Springer-Verlag, New York.

Edelstein-Keshet, L. 1988. *Mathematical Models in Biology*, reprinted by Soc. Industr. Appl. Math., Philadelphia.

Evans, L. 1998. *Partial Differential Equations*, Amer. Math. Soc., Providence.

Farlow, S. 1993. *Partial Differential Equations for Scientists and Engineers*, Dover Publications, New York.

Grindrod, P. 1991. *Patterns and Waves*, Oxford University Press, Oxford.

Keener, J. 2000. *Principles of Applied Mathematics* (revised edition), Westview Press (Perseus Book Group), Cambridge, MA.

Kot, M. 2001. *Elements of Mathematical Ecology*, Cambridge University Press, Cambridge.

Logan, J. D. 2008. *An Introduction to Nonlinear Partial Differential Equations*, Wiley-Interscience, New York.

Logan, J. D. 2004. *Applied Partial Differential Equations*, 2nd ed., Springer-Verlag, New York.

McOwen, R. C. 2003. *Partial Differential Equations*, 2nd ed., Pearson Education, Upper Saddle River, NJ.

Stakgold, I. 1998. *Green's Functions and Boundary Value Problems*, 2nd ed., Wiley-Interscience, New York.

Strauss, W. 1991. *Partial Differential Equations*, John Wiley and Sons, New York.

Strichartz, R. 1994. *A Guide to Distribution Theory and Fourier Transforms*, CRC Press, Boca Raton, FL.

Whitham, G. B. 1974. *Linear and Nonlinear Waves*, Wiley-Interscience, New York.

Zauderer, E. 1989. *Partial Differential Equations of Applied Mathematics*, 2nd ed., Wiley-Interscience, New York.

7

Wave Phenomena

Two of the fundamental processes in nature are diffusion and wave propagation. In the last chapter we studied the equation of heat conduction, a parabolic partial differential equation that is the prototype of equations governing linear diffusion processes. In the present chapter we investigate wave phenomena and obtain equations that govern the propagation of waves first in simple model settings. The evolution equations governing such phenomena are hyperbolic and are fundamentally different from their parabolic counterparts in diffusion and from elliptic equations that govern equilibrium states.

7.1 Waves

A **wave** is defined as an identifiable signal or disturbance in a medium that is propagated in time, carrying energy with it. A few familiar examples are electromagnetic waves, waves on the surface of water, sound waves, and stress waves in solids, as occur in earthquakes. Material or matter is not necessarily convected with the wave; it is the disturbance, which carries energy, that is propagated. In this section we investigate several model equations that occur in wave phenomena and point out basic features that are encountered in the study of the propagation of waves.

One simple mathematical model of a wave is the function

$$u(x,t) = f(x - ct), \tag{1.1}$$

Applied Mathematics 4th ed.,
By J. David Logan

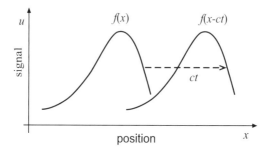

Figure 7.1 Right–traveling wave moving at speed c.

which represents an undistorted **right-traveling wave** moving at constant velocity c. The coordinate x represents position, t time, and u the strength of the disturbance. At $t = 0$ the wave profile is $u = f(x)$ and at $t > 0$ units of time the disturbance has moved to the right ct units of length (see Fig. 7.1). Of particular importance is that the wave profile described by (1.1) moves without distortion. Not all waves have this property; it is characteristic of linear waves, or wave profiles that are solutions to linear partial differential equations. On the other hand, waves that distort, and possibly break, are characteristic of nonlinear processes. To find a model that has (1.1) as the solution we compute u_t and u_x to get

$$u_t = -cf'(x - ct), \qquad u_x = f'(x - ct),$$

Hence

$$u_t + cu_x = 0. \tag{1.2}$$

Equation (1.2) is a first-order linear partial differential equation that, in the sense just described, is the simplest wave equation. It is called the **advection equation** and its general solution is (1.1), where f is an arbitrary differentiable function. As we noted in Section 6.2, the advection equation describes how a quantity is carried along with the bulk motion of a medium. Similarly, a traveling wave of the form $u = f(x + ct)$ is a left–traveling wave and is a solution of the partial differential equation $u_t - cu_x = 0$.

Other waves of interest in many physical problems are periodic, or sinusoidal waves. These traveling waves are represented by expressions of the form

$$u(x, t) = A \cos(kx - \omega t) \tag{1.3}$$

(see Fig. 7.2). The positive number A is the **amplitude**, k is the **wave number** (the number of oscillations in 2π units of space, observed at a fixed time), and ω the **angular frequency** (the number of oscillations in 2π units of time, observed at a fixed location x). The number $\lambda = 2\pi/k$ is the **wavelength** and

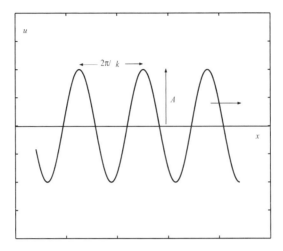

Figure 7.2 Plane wave.

$P = 2\pi/\omega$ is the *time period*. The wavelength measures the distance between successive crests and the time period is the smallest time for an observer located at a fixed position x to see a repeat pattern. If we write (1.3) as

$$u = A\cos k\left(x - \frac{\omega}{k}t\right),$$

then we note that (1.3) represents a traveling wave moving to the right with velocity $c - \omega/k$. This number is called the **phase velocity**, and it is the speed one would have to move to remain at the same point on the traveling wave. For calculations, the complex exponential form

$$u(x,t) = e^{i(kx-\omega t)} \tag{1.4}$$

is often used rather than (1.3). These waves are called **plane waves**. Computations involving differentiations are easier with (1.4) and afterward, making use of Euler's formula $\exp(i\theta) = \cos\theta + i\sin\theta$, the real or imaginary part may be taken to recover real solutions. Again, waves of the type (1.3) [or (1.4)] are characteristic of linear processes and linear equations.

Not every wave propagates so that its profile remains unchanged or undistorted; a surface wave on the ocean is an obvious example. Less familiar perhaps, but just as common, are strong stress or pressure waves that propagate in solids or gases. To fix the idea and to indicate how nonlinearity affects the shape of a wave, let us consider a strong pressure wave propagating in air, caused, for example, by an explosion. (This is not the case of acoustics, where sound waves propagate at relatively small pressure changes.) The distortion of

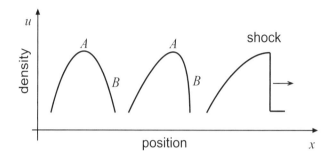

Figure 7.3 Distortion of a wave into a shock.

a wave profile results from the property of most materials to transmit signals at a speed that increases with increasing pressure. Therefore a strong pressure wave that is propagating in a medium will gradually distort and steepen until it propagates as an idealized discontinuous disturbance, or shock wave . This is what also occurs in sonic booms. Figure 7.3 shows various snapshots of a stress wave propagating into a material. The wave steepens as time increases because signals or disturbances travel faster when the pressure is higher. Thus the point A moves to the right faster than the point B. The shock that forms in pressure is accompanied by discontinuous jumps in the other parameters, such as density, particle velocity, temperature, energy, and entropy.

Physically, a shock wave is not a strict discontinuity but rather an extremely thin region where the change in the state is steep. The width of the shock is small, usually of the order of a few mean free paths, or average distance to a collision, of the molecules. In a shock there are two competing effects that cause this thinness, the nonlinearity of the material that is causing the shock to form, and the dissipative effects (e.g., viscosity) that are tending to smear the wave out. Usually these two effects just cancel and the front assumes a shape that does not change in time. The same mechanism that causes pressure waves to steepen into shocks, that is, an increase in signal transmission speed at higher pressures, causes release waves or **rarefaction waves** to form that lower the pressure. Figure 7.4 shows a spreading rarefaction. Again, point A at higher pressure moves to the right faster than point B, thereby flattening the wave. So far we have mentioned two types of waves, those that propagate undistorted at constant velocity and those that distort because the speed of propagation depends on the amplitude of the wave. There is a third interesting phenomenon that is also relevant, namely that of **dispersion**. In this case the speed of propagation depends on the wavelength of the particular wave. So, for example, longer waves can travel faster than shorter ones. Thus an observer of a wave at fixed location x_0 may see a different temporal wave pattern from

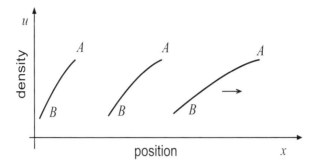

Figure 7.4 A rarefaction wave.

another observer at fixed location x_1. Dispersive wave propagation arises from both linear and nonlinear models.

The behavior of linear partial differential equations with constant coefficients can be characterized by the plane wave solutions they admit. When we substitute a plane wave form (1.4) into such an equation, we obtain a relation between the frequency and the wave number,

$$\omega = \omega(k).$$

This equation is called the **dispersion relation**. We say the equation is **diffusive** if $\omega(k)$ is complex, and **dispersive** if $\omega(k)$ is real and $\omega(k)/k$ is independent of k; if $\omega(k)$ is real and $\omega(k)/k$ depends on k, then we say the equation is **hyperbolic**.

Example 7.1

The reader should check that the advection–diffusion equation $u_t + \gamma u_x = D u_{xx}$ has complex dispersion relation $\omega = \gamma k - i D k^2$. Therefore the advection–diffusion equation is diffusive. Plane wave solutions are given by

$$u(x,t) = e^{-Dk^2 t} e^{i(kx - \gamma t)}.$$

The factor $\exp(ik(x - \gamma t))$ represents a periodic, right-traveling wave with wave number k, and the factor $\exp(-Dk^2 t)$ represents a decaying amplitude. Qualitatively, two conclusions can be drawn:

(i) For waves of constant wavelength (k constant) the attenuation of the wave increases with increasing D; hence D is a measurement of diffusion.

(ii) For constant γ the attenuation increases as k increases; hence smaller wavelengths attenuate faster than longer wavelengths.

From this example the reader should note that the diffusion equation ($\gamma = 0$) has plane wave solutions $u(x,t) = e^{-Dk^2 t}e^{ikx}$, which are not traveling waves, but stationary waves that decay.

Example 7.2

Consider the partial differential equation

$$u_{tt} + \gamma u_{xxxx} = 0, \qquad \gamma > 0,$$

which arises in studying the transverse vibrations of a beam. Substituting (1.4) into this equation gives the dispersion relation

$$\omega = \pm\sqrt{\gamma}k^2,$$

which is real, and $\omega(k)/k$ is not constant. Hence the beam equation is dispersive and the phase velocity is $\pm\sqrt{\gamma}k$, which depends on the wave number. Therefore shorter wavelength vibrations travel faster.   □

Generally, in the dispersive case the plane wave solution takes the form of a right-traveling wave,

$$u(x,t) = e^{i(kx-\omega(k)t)} = e^{ik(x - \frac{\omega(k)}{k}t)}.$$

The wave speed $\omega(k)/k$, or phase velocity, is dependent upon the wave number k. Therefore, wave packets of different wavelength will propagate at different speeds, causing a signal to disperse. Common examples of dispersive waves are water waves, vibrations in solids, and electromagnetic waves in dielectrics.

We may think of plane wave solutions of the form $e^{i(kx-\omega(k)t)}$ as being a one-parameter family of solutions, with $k \in \mathbb{R}$. If we superimpose them over all wave numbers, we obtain

$$u(x,t) = \int_{-\infty}^{\infty} a(k)e^{i(kx-\omega(k)t)}\,dk.$$

Under reasonable conditions we expect this to be a solution of the differential equation under investigation. If

$$u(x,0) = f(x), \qquad x \in \mathbb{R},$$

is an initial condition, then

$$u(x,0) = f(x) = \int_{-\infty}^{\infty} a(k)e^{ikx}\,dk,$$

which is the Fourier transform of the unknown coefficient function $a(k)$. Therefore

$$a(k) = \frac{1}{2\pi} \int_{-\infty}^{\infty} f(x)e^{-ikx}dx,$$

and we have obtained a formal solution to a Cauchy problem for the partial differential equation having the form

$$
\begin{aligned}
u(x,t) &= \frac{1}{2\pi} \int_{-\infty}^{\infty} \left(\int_{-\infty}^{\infty} f(y)e^{-iky}dy \right) e^{i(kx - \omega(k)t)} dk \\
&= \frac{1}{2\pi} \int_{-\infty}^{\infty} \left(\int_{-\infty}^{\infty} e^{-ik(y-x)}e^{-i\omega(k)t}dk \right) f(y)\,dy.
\end{aligned}
$$

So, again, Fourier analysis is at the heart of the problem. In principle, this integral could be calculated to find the solution.

7.1.1 The Advection Equation

We showed that the simplest wave equation, the **advection equation**,

$$u_t + cu_x = 0, \quad x \in \mathbb{R}, \ t > 0, \tag{1.5}$$

has general solution

$$u = f(x - ct), \tag{1.6}$$

which is a right-traveling wave propagating at constant velocity c, where f is an arbitrary function. If we impose the initial condition

$$u(x,0) = \phi(x), \quad x \in \mathbb{R}, \tag{1.7}$$

then it follows that $f(x) = \phi(x)$, and so the solution to the initial value problem (1.5) and (1.7) is

$$u(x,t) = \phi(x - ct). \tag{1.8}$$

The straight lines $x - ct = $ constant, which are called **characteristic curves** (or, **characteristics**), play a special role in this problem. We interpret characteristics as lines in space-time along which signals are carried (see Fig. 7.5), here the signal being a constant value of the initial data. Furthermore, on these characteristics, the partial differential equation (1.5) reduces to the ordinary differential equation $du/dt = 0$. That is to say, if C is a typical characteristic $x = ct + k$ for some constant k, then the directional derivative of u along that curve is

$$
\begin{aligned}
\frac{du}{dt}(x(t),t) &= u_x(x(t),t)\frac{dx}{dt} + u_t(x(t),t) \\
&= u_x(x(t),t)c + u_t(x(t),t),
\end{aligned}
$$

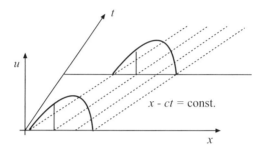

Figure 7.5 The solution surface to $u_t + cu_x = 0$, $c > 0$, and the characteristic curves in the xt-plane. The solution is constant on a characteristic curve and the curve carries that constant value at $t = 0$ into $t > 0$.

which is the left side of (1.5) evaluated along C. Notice that the family of characteristic curves $x - ct = k$ propagates at speed c, which is the reciprocal of their slope in the xt coordinate system.

Now let us complicate the partial differential equation (1.5) by replacing the constant c by a function of the independent variables t and x and consider the initial value problem

$$u_t + c(x,t)u_x = 0, \quad x \in \mathbb{R}, \ t > 0, \tag{1.9}$$
$$u(x,0) = \phi(x), \quad x \in \mathbb{R},$$

where $c(x,t)$ is a given function. Let C be the family of curves defined by the differential equation

$$\frac{dx}{dt} = c(x,t). \tag{1.10}$$

Then along a member of C

$$\frac{du}{dt} = u_x \frac{dx}{dt} + u_t = u_x c(x,t) + u_t = 0.$$

Hence u is constant on each member of C. The curves C defined by (1.10) are the **characteristic curves**. Notice that they propagate at a speed $c(x,t)$, which depends on both time and the position in the medium.

Example 7.3

Consider the initial value problem

$$u_t + 2tu_x = 0, \quad x \in \mathbb{R}, \ t > 0,$$
$$u(x,0) = \exp(-x^2), \quad x \in \mathbb{R}, \ t > 0.$$

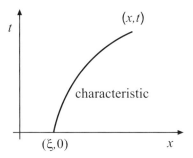

Figure 7.6 Characteristic $x - t^2 + \xi$.

The characteristic curves are defined as solutions of the differential equation $dx/dt = 2t$. We obtain the family of parabolas

$$x = t^2 + k,$$

where k is a constant. Because u is constant on these characteristic curves, we can easily find a solution to the initial value problem. Let (x, t) be an arbitrary point where we want to obtain the solution. The characteristic curve through (x, t) passes through the x axis at $(\xi, 0)$, and it has equation $x = t^2 + \xi$ (see Fig. 7.6). Since u is constant on this curve,

$$u(x, t) = u(\xi, 0) = e^{-\xi^2} = e^{-(x-t^2)^2},$$

which is the unique solution to the initial value problem. The speed of the signal at (x, t) is $2t$, which is dependent upon t. The wave speeds up as time increases, but it retains its initial shape.   □

The preceding equation is defined on an infinite spatial domain $-\infty < x < \infty$. Now, suppose there are boundaries in the problem. To illustrate the basic points it is sufficient to study the simple advection equation

$$u_t + cu_x = 0, \quad 0 < x < 1, \ t > 0,$$

with $c > 0$. Assume an initial condition

$$u(x, 0) = f(x), \quad 0 < x < 1.$$

On the unbounded interval $(-\infty, \infty)$ the solution is a unidirectional right-traveling wave $u = f(x - ct)$. On a bounded interval what kinds of boundary conditions can be imposed at $x = 0$ and $x = 1$? See Fig. 7.7. Because u is given by the initial condition $f(x)$ along the initial line $t = 0$, $0 < x < 1$, data cannot be prescribed arbitrarily on the segment A along the boundary $x = 1$. This is

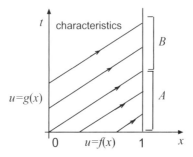

Figure 7.7 Characteristic diagram.

because the characteristics $x - ct = $ constant are moving to the right and they carry the initial data to the segment A. Boundary data can be imposed along the line $x = 0$, since that data is carried along the forward-going characteristics to the segment B along $x = 1$. Then boundary conditions along B cannot be prescribed arbitrarily. Thus, it is clear that the problem

$$u_t + cu_x = 0, \quad 0 < x < 1, \quad t > 0,$$
$$u(x, 0) = f(x), \quad 0 < x < 1,$$
$$u(0, t) = g(t), \quad t > 0,$$

is properly posed. There are no backward-going characteristics in this problem, so there are no left-traveling waves. Thus waves are not reflected from the boundary $x = 1$. To find the solution to this initial boundary value problem we must separate the domain into two portions, $0 < x < ct$ and $x > ct$. The solution in $x > ct$ is determined by the initial data, whereas the solution in $0 < x < ct$ is determined by the boundary data. These two regions are separated by the characteristic $x = ct$; we refer to the region $x > ct$ as the region of spacetime *ahead* of this limiting characteristic, and the region $0 < x < ct$ as the region *behind* the limiting characteristic.

In summary, care must be taken to properly formulate and solve boundary value problems for one-dimensional, unidirectional wave equations. The situation is much different for second-order hyperbolic partial differential equations like $u_{tt} - c^2 u_{xx} = 0$. In this case both forward- and backward-going characteristics exist, and so left-traveling waves are also possible, as well as a mechanism for reflections from boundaries. This seems plausible because we can factor the differential operator into a product of two advection-like operators, $\frac{\partial^2}{\partial t^2} - c^2 \frac{\partial^2}{\partial x^2} = \left(\frac{\partial}{\partial t} - c\frac{\partial}{\partial x}\right)\left(\frac{\partial}{\partial t} + c\frac{\partial}{\partial x}\right)$, one right-moving and one left-moving.

Example 7.4

(**Characteristic coordinates**) This example presents an alternate method for solving advection equations

$$u_t + c(x,t)u_x = 0.$$

Actually, it is the same thing as before, but with a different interpretation and strategy. When we find the characteristic curves $\xi(x,t) = \xi = $ const. by solving the characteristic equation $dx/dt = c(x,t)$, we can define a transformation of spacetime $(x,t) \to (\xi,\tau)$ by

$$\xi = \gamma(x,t), \quad \tau = t.$$

The ξ, τ are called characteristic coordinates, and they are coordinates that ride with the wave, or moving coordinates. If we transform u to U via $U(\xi,\tau) = u(x,t)$, then the chain rule gives

$$u_t = U_\xi \xi_t + U_\tau \tau_t = U_\xi \xi_t + U_\tau, \quad u_x = U_\xi \xi_x + U_\tau \tau_x = U_\xi \xi_x.$$

The PDE becomes, upon substitution,

$$u_t + c(x,t)u_x = U_\xi \xi_t + U_\tau + c(x,t)U_\xi \xi_x = 0.$$

But, differentiating $\xi(x,t) = $ const. with respect to t gives $\xi_x(dx/dt) + \xi_t = \xi_x c(x,t) + \xi_t = 0$. So the PDE reduces to simply $U_\tau = 0$, independent of ξ. This is plausible because 'ξ is riding on the wave.' Hence, $U(\xi,\tau) = F(\xi)$, where F is an arbitrary function. This means $u(x,t) = F(\xi(x,t))$ is the general solution to the original problem. The basic idea is that under a change to characteristic coordinates the advection operator transforms to

$$\frac{\partial}{\partial t} + c(x,t)\frac{\partial}{\partial x} = \frac{\partial}{\partial \tau}. \quad \square$$

Example 7.5

Use the method of the last example to solve a problem with a source,

$$u_t + 2tu_x = -3u.$$

The characteristics are $x - t^2 = $ const., so the characteristic coordinates are $\xi = x - t^2$, $\tau = t$. Then $U(\xi,\tau)$ satisfies $U_\tau = -3U$, which gives

$$U(\xi,\tau) = F(\xi)e^{-3\tau}, \quad F \text{ arbitrary.}$$

Then the general solution is

$$u(x,t) = F(x - t^2)e^{-3t}. \quad \square$$

Example 7.6

(**Linear systems**) We can treat linear systems of advection equations in a similar fashion as linear systems of ordinary differential equations. Let $\mathbf{u}(x,t) = [u_1(x,t), \ldots, u_n(x,t)]^{\mathrm{T}}$ be a vector of n unknowns and consider the system of n equations

$$\mathbf{u}_t + A\mathbf{u}_x = \mathbf{0}$$

where A is a real $n \times n$ matrix with real, distinct eigenvalues. Then, A is diagonalizable, meaning there is an invertible matrix P such that $P^{-1}AP = D$, where D is a diagonal matrix with the eigenvalues on the diagonal; the columns of P are the eigenvectors. Note that the system is coupled, with each equation containing possibly all the $u_k(x,t)$. If we introduce a new dependent variable $\mathbf{v}$ defined by

$$\mathbf{u} = P\mathbf{v},$$

then the system becomes

$$P\mathbf{v}_t + AP\mathbf{v}_x = \mathbf{0} \quad \text{or} \quad \mathbf{v}_t + P^{-1}AP\mathbf{v}_x = \mathbf{0}.$$

Thus,

$$\mathbf{v}_t + D\mathbf{v}_x = \mathbf{0}.$$

Therefore the transformation to a new variable $\mathbf{v}$ decoupled the system. The kth equation in this system is

$$\frac{\partial v_k}{\partial t} + \lambda_k \frac{\partial v_k}{\partial x} = 0,$$

which has general solution $v_k(x,t) = \phi_k(x - \lambda_k t)$, a traveling wave moving at speed λ_k of the kth eigenvalue. Here, $\mathbf{v}$ is an arbitrary function. The solution to the original system is then $\mathbf{u} = P^{-1}\mathbf{v}$. Initial conditions determine the arbitrary functions $\mathbf{v}$. We leave examples for the Exercises. □

EXERCISES

1. For the following equations find the dispersion relation and phase velocity. Classify as diffusive, dispersive, or hyperbolic.

   a) $u_t + cu_x + ku_{xxx} = 0$.

   b) $u_{tt} - c^2 u_{xx} = 0$.

   c) $u_{tt} + u_{xx} = 0$. (Note that this is an elliptic equation.)

2. Verify that the equation $u_t + c(x,t)u_x = 0$ has a solution $u(x,t) = f(a(x,t))$ for an arbitrary differentiable function f, where $a(x,t) = $ constant are the integral curves of $dx/dt = c(x,t)$.

3. Solve the following initial value problems on $t > 0$, $x \in \mathbb{R}$. Sketch the characteristic curves and several time snapshots of the wave in each case.

   a) $u_t + 3u_x = 0$, $\quad u(x,0) = \frac{1}{1+x^2}$.

   b) $u_t + xu_x = 0$, $\quad u(x,0) = \exp(-x)$.

   c) $u_t - x^2 t u_x = 0$, $\quad u(x,0) = x + 1$.

4. On the domain $0 < x < 1$, $t > 0$, consider the problem

$$u_t + cu_x = 0, \quad 0 < x < 1, \ t > 0,$$
$$u(x,0) = 0, \quad 0 < x < 1,$$
$$u(0,t) = te^{-t}, \quad t > 0,$$

   with $c > 0$. Find a formula for the solution, and sketch a graph of $u(1,t)$ for $t > 0$ when $c = 2$.

5. Solve the boundary value problem

$$u_t + 2tu_x = -3u, \quad u(0,t) = t^2,$$

   in the region $x > 0$, $t > 0$, $x < t^2$.

6. Solve the Cauchy problem

$$u_t + xu_x = -tu, \quad u(x,0) = \cos x.$$

7. Solve the Cauchy problem

$$u_t + xu_x = -2u, \quad u(x,1) = \cos x,$$

   where the data is given along $t = 1$.

8. Sometimes data is given along a curve in the xt plane. Solve the problem

$$2u_t + u_x = -2u, \quad u(x,t) = f(x,t) \text{ on } x = t.$$

9. Find the solution and sketch a characteristic diagram

$$u_t + cu_x = 0, \qquad x \in \mathbb{R}, \ t > 0, \ c > 0$$
$$u(x,0) = \begin{cases} 1 - x^2, & |x| \le 1 \\ 0, & |x| > 1 \end{cases}.$$

10. Solve the linear system

$$\frac{\partial u_1}{\partial t} + 8\frac{\partial u_2}{\partial x} = 0,$$

$$\frac{\partial u_2}{\partial t} + 2\frac{\partial u_1}{\partial x} = 0,$$

subject to the initial conditions $u_1(x,0) = f(x)$, $u_2(x,0) = g(x)$, in the region $x \in \mathbb{R}$, $t > 0$.

11. Consider the the *linearized Korteweg–de Vries* equation

$$u_t + u_{xxx} = 0.$$

a) Find the dispersion relation and determine how the phase velocity depends on the wave number.

b) By superimposing the plane wave solutions, find an integral representation for the solution to the Cauchy problem on $\mathbb{R}$. Write your answer in terms of the **Airy function**

$$\mathrm{Ai}(x) = \frac{1}{2\pi}\int_{-\infty}^{\infty} \cos\left(\frac{z^3}{3} + xz\right) dz.$$

c) Take the initial condition to be $f(x) = e^{-x^2}$ and use a computer algebra system to plot solution profiles when $t = 0.5$ and $t = 1$.

7.2 Nonlinear Waves

7.2.1 Nonlinear Advection

In the last section we examined the two simple model wave equations, $u_t + cu_x = 0$ and $u_t + c(x,t)u_x = 0$, that are both first-order and linear. Now we study the same type of equation when a nonlinear advection term is introduced. The resulting equation is often called the kinematic wave equation. In particular we consider

$$u_t + c(u)u_x = 0, \quad x \in \mathbb{R}, \ t > 0, \tag{2.1}$$
$$u(x,0) = \phi(x), \quad x \in \mathbb{R}, \tag{2.2}$$

where $c'(u) > 0$, with initial condition Using the guidance of the earlier examples we define the **characteristic curves** by the differential equation

$$\frac{dx}{dt} = c(u). \tag{2.3}$$

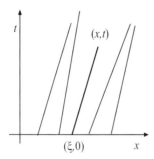

Figure 7.8 A characteristic diagram showing characteristics, or signals, moving at different speeds; each characteristic carries a constant value of u.

Then along a particular characteristic curve $x = x(t)$ we have

$$\frac{du}{dt}(x(t), t) = u_x(x(t), t)c(u(x(t))) + u_t(x(t), t) = 0.$$

Therefore u is constant along the characteristics, and the characteristics are straight lines since

$$\frac{d^2 x}{dt^2} = \frac{d}{dt}\left(\frac{dx}{dt}\right) = \frac{d}{dt}c(u(x(t))) = c'(u)\frac{du}{dt} = 0.$$

In the nonlinear case, however, the speed of the characteristics as defined by (2.3) depends on the value u of the solution at a given point. To find the equation of the characteristic C through (x, t) we note that its speed is

$$\frac{dx}{dt} = c(u(\xi, 0)) = c(\phi(\xi))$$

(see Fig. 7.8). This results from applying (2.3) at $(\xi, 0)$. Hence, after integrating,

$$x = c(\phi(\xi))t + \xi \tag{2.4}$$

gives the equation of the desired characteristic C. Equation (2.4) defines $\xi = \xi(x, t)$ implicitly as a function of x and t, and the solution $u(x, t)$ of the initial value problem (2.1) and (2.2) is given by

$$u(x, t) = \phi(\xi) \tag{2.5}$$

where ξ is defined by (2.4).

In summary, for the nonlinear equation (2.1):

1. Every characteristic is a straight line.

2. The solution is constant on each such line.

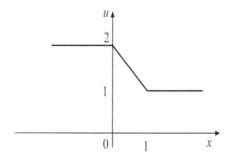

Figure 7.9 Initial wave profile.

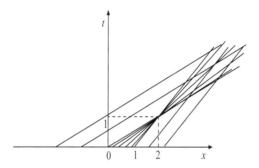

Figure 7.10 Characteristic diagram showing colliding characteristics.

3. The speed of each charactistic is equal to the value of $u(x,t)$ on it.

4. The speed $c(u)$ is the speed that signals, or waves, are propagated in the system.

Example 7.7

Consider the initial value problem

$$u_t + uu_x = 0, \quad x \in \mathbb{R}, \quad t > 0,$$

$$u(x,0) = \phi(x) = \begin{cases} 2, & x < 0, \\ 2 - x, & 0 \leq x \leq 1, \\ 1, & x > 1. \end{cases}$$

The initial curve is sketched in Fig. 7.9. Since $c(u) = u$ the characteristics are straight lines emanating from $(\xi, 0)$ with speed $c(\phi(\xi)) = \phi(\xi)$. These are plotted in Fig. 7.10. For $x < 0$ the lines have speed 2; for $x > 1$ the lines have speed 1; for $0 \leq x \leq 1$ the lines have speed $2 - x$ and these all intersect at (2,

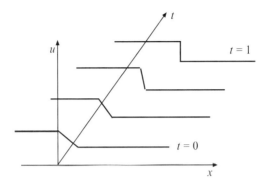

Figure 7.11 Solution surface with time profiles.

1). Immediately one observes that a solution cannot exist for $t > 1$, since the characteristics cross beyond that time and they carry different constant values of u. Figure 7.11 shows several wave profiles that indicate the steepening that is occurring. At $t = 1$ *breaking* of the wave occurs, which is the first instant when the solution becomes multiple valued. To find the solution for $t < 1$ we first note that $u(x,t) = 2$ for $x < 2t$ and $u(x,t) = 1$ for $x > t+1$. For $2t < x < t+1$ equation (2.4) becomes

$$x = (2 - \xi)t + \xi,$$

which gives

$$\xi = \frac{x - 2t}{1 - t}.$$

Equation (2.5) then yields

$$u(x,t) = \frac{2 - x}{1 - t}, \qquad 2t < x < t + 1, \ t < 1.$$

This explicit form of the solution also indicates the difficulty at the breaking time $t = 1$. □

In general the initial value problem (2.1)–(2.2) may have a solution only up to a finite time t_b, which is called the **breaking time**. Let us assume in addition to $c'(u) > 0$ that the initial wave profile satisfies the conditions

$$\phi(x) \geq 0, \qquad \phi'(x) < 0.$$

At the time when breaking occurs the gradient u_x will become infinite. To compute u_x we differentiate (2.4) implicitly with respect to x to obtain

$$\xi_x = \frac{1}{1 + c'(\phi(\xi))\phi'(\xi)t}.$$

Then from (2.5)

$$u_x = \frac{\phi'(\xi)}{1 + c'(\phi(\xi))\phi'(\xi)t}.$$

The gradient catastrophe will occur at the minimum value of t, which makes the denominator zero. Hence

$$t_b = \min_{\xi} \frac{-1}{\phi'(\xi)c'(\phi(\xi))}, \qquad t_b \geq 0.$$

In Example 7.7 we have $c(u) = u$ and $\phi(\xi) = 2 - \xi$. Hence $\phi'(\xi)c'(\phi(\xi)) = (-1)(1) = -1$ and $t_b = 1$ is the time when breaking occurs.

    In summary we have observed that the nonlinear partial differential equation

$$u_t + c(u)u_x = 0, \quad c'(u) > 0$$

propagates the initial wave profile at a speed $c(u)$, which depends on the value of the solution u at a given point. Since $c'(u) > 0$, large values of u are propagated faster than small values and distortion of the wave profile occurs. This is consistent with our earlier remarks of a physical nature, namely that wave distortion and shocks develop in materials because of the property of the medium to transmit signals more rapidly at higher levels of stress or pressure. Mathematically, distortion and the development of shocks or discontinuous solutions are distinctively nonlinear phenomena caused by the term $c(u)u_x$ in the last equation.

Example 7.8

(Implicit solution) We have shown that when the advection speed is constant, that is, we have

$$u_t + cu_x = 0,$$

the general solution is given explicitly by $u = F(x - ct)$, where F is arbitrary. Could such a simple form of the solution occur for the nonlinear equation

$$u_t + c(u)u_x = 0?$$

The answer is yes, and an implicit, general solution to this equation is easily shown to be

$$u = F(x - c(u)t),$$

when it exists. (Use the chain rule and substitute into the equation.) The function F is determined, for example, by an initial condition. For example, consider PDE

$$u_t + u^2 u_x = 0.$$

The general solution is given implicitly by $u = F(x - u^2 t)$, which is easily verified. If $u(x, 0) = x$, then $F(x) = x$ and $u = x - u^2 t$. Solving for u gives

$$u = \frac{1}{2t}\left(-1 \pm \sqrt{1 + 4tx}\right)$$
$$= \frac{1}{2t}\left(-1 + \sqrt{1 + 4tx}\right),$$

where we have taken the positive square root to meet the initial condition. The solution is valid for $t < -1/4x$. (See Exercise 5.) □

Example 7.9

(**Traffic flow**) Everyone who drives has experienced traffic issues, such as jams, poorly timed traffic lights, high road density, etc. In this abbreviated example we suggest how some of these issues can be understood with a simple model. Traffic moving in a single direction x with car density $\rho(x, t)$, given in cars per kilometer, can be modeled by a conservation law

$$\rho_t + (\rho v)_x = 0,$$

where $v = v(x, t)$ is the local car speed (kilometers per hour), and ρv is the the flux, in cars per hour. Importantly, we are making a continuum assumption about cars, surely a questionable one. As always, we need a constitutive assumption to close the system. By experience, the speed of traffic surely depends on traffic density, or, $v = F(\rho)$. The simplest model is to assume that the flux ρv is zero when ρ is zero, and is jammed (no flux) when the density is some maximum value ρ_J. Therefore, we take

$$\rho v = \rho v_M \left(1 - \frac{\rho}{\rho_J}\right),$$

where v_M is the maximum velocity of cars. Note that this flux curve is a parabola, concave down. The conservation law can then be written, after rescaling,

$$u_t + v_M[u(1 - u)]_x = 0, \quad u = \frac{\rho}{\rho_M}.$$

This equation has the same form as the kinematic wave equation and can be expanded to

$$u_t + v_M(1 - 2u)u_x = 0,$$

so $c(u) = v_M(1 - 2u)$ is the speed that traffic waves move in the system. It is different from the speed of cars! Because $c'(u) < 0$, signals are propagated backward into the traffic flow. □

EXERCISES

1. Consider the Cauchy problem

$$u_t = xuu_x, \quad x \in \mathbb{R}, \ t > 0,$$
$$u(x,0) = x, \quad x \in \mathbb{R}.$$

Find the characteristics, and find a formula that determines the solution $u = u(x,t)$ implicitly as a function of x and t. Does a smooth solution exist for all $t > 0$?

2. Consider the initial value problem

$$u_t + uu_x = 0, \quad x \in \mathbb{R}, \ t > 0,$$
$$u(x,0) = \begin{cases} 1 - x^2, & |x| \leq 1, \\ 0, & |x| > 1. \end{cases}$$

Sketch the characteristic diagram. At what time t_b does the wave break? Find a formula for the solution.

3. Consider the initial value problem

$$u_t + uu_x = 0, \quad x \in \mathbb{R}, \ t > 0,$$
$$u(x,0) = \exp(-x^2), \quad x \in \mathbb{R}.$$

Sketch the characteristic diagram and find the point (x_b, t_b) in space–time where the wave breaks.

4. Consider the Cauchy problem

$$u_t + c(u)u_x = 0, \quad x \in \mathbb{R}, \ t > 0,$$
$$u(x,0) = f(x), \quad x \in \mathbb{R}.$$

Show that if the functions $c(u)$ and $f(x)$ are both nonincreasing or both nondecreasing, then no shocks develop for $t \geq 0$.

5. Consider the problem

$$u_t + u^2 u_x = 0, \quad x \in \mathbb{R}, \ t > 0,$$
$$u(x,0) = x, \quad x \in \mathbb{R}.$$

Derive the solution

$$u(x,t) = \begin{cases} x, & t = 0, \\ \frac{\sqrt{1+4xt}-1}{2t}, & t \neq 0, \ 1 + 4tx > 0. \end{cases}$$

When do shocks develop? Verify that $\lim_{t\to 0^+} u(x,t) = x$.

6. Consider the **signaling problem**

$$u_t + c(u)u_x = 0, \quad t > 0, \ x > 0,$$
$$u(x, 0) = u_0, \quad x > 0,$$
$$u(0, t) = g(t), \quad t > 0,$$

where c and g are given functions and u_0 is a positive constant. If $c'(u) > 0$, under what conditions on the signal g will no shocks form? Determine the solution in this case in the domain $x > 0$, $t > 0$.

7. In the traffic flow model in Example 7.9, explain what occurs if the initial car density has each of the following shapes: (a) a density bump in the traffic having the shape of a bell-shaped curve; (b) a density dip in the traffic having the shape of an inverted bell-shaped curve; (c) a density that is jammed for $x < 0$, with no cars ahead for $x > 0$ (a stop light); (d) a density that is shaped like a curve $\pi/2 + \arctan x$ where the traffic ahead has increasing density.

   In each case, sketch a qualitative characteristic diagram and sketch several density profiles. On the characteristic diagram sketch a sample car path.

8. The height $h = h(x, t)$ of a flood wave can be modeled by

$$h_t + (vh)_x = 0,$$

where v, the average stream velocity, is $v = a\sqrt{h}$, $a > 0$ (Chezy's law). Show that flood waves propagate 1.5 times faster than the average stream velocity.

7.2.2 Traveling Wave Solutions

One approach to analysis of partial differential equations is to examine model equations to gain a sense of the role played by various terms in the equation and how those terms relate to fundamental physical processes. In the preceding paragraphs we observed that the advection equation

$$u_t + cu_x = 0 \tag{2.6}$$

propagates an initial disturbance or signal at velocity c, while maintaining the precise form of the signal. On the other hand, the nonlinear equation

$$u_t + uu_x = 0 \tag{2.7}$$

propagates signals that distort the signal profile; that is, the nonlinear advection term uu_x causes a shocking up or rarefaction effect. We know from Chapter 6 that the diffusion equation

$$u_t - Du_{xx} = 0, \qquad (2.8)$$

which contains a diffusion term Du_{xx}, causes spreading of a density profile. Insights can be gained about the nature of the various terms in an evolution equation by attempting to find either traveling wave solutions

$$u = U(x - ct), \qquad (2.9)$$

or plane wave solutions

$$u = A \ \exp[i(kx - \omega t)]. \qquad (2.10)$$

The latter is applicable to linear equations with constant coefficients, whereas traveling wave may be characteristic of linear or nonlinear processes.

In general, **traveling wave solutions** have no conditions at the boundaries (plus or minus infinity). However, there are often boundary conditions. Traveling wave solutions that approach constant states at $\pm\infty$ are called **wave front** solutions. They correspond to a traveling wave moving into and from a constant state. If these two states are equal, then we say the wave front is a **pulse**.

Example 7.10

(**Burgers' equation**) Consider the nonlinear equation

$$u_t + uu_x - Du_{xx} = 0, \quad D > 0, \qquad (2.11)$$

which is known as **Burgers' equation**. The term uu_x will have a shocking up effect that will cause waves to break, and the term Du_{xx} is a diffusion term, like the one occurring in the diffusion equation (2.8). We attempt to find a traveling wave solution of (2.11) of the form

$$u = U(x - ct), \qquad (2.12)$$

where the wave profile U and the wave speed c are to be determined. Substituting (2.12) into (2.11) gives

$$-cU'(s) + U(s)U'(s) - DU''(s) = 0,$$

where

$$s = x - ct.$$

We interpret s as a moving coordinate. Noting that $UU' = \frac{1}{2}\frac{d}{ds}U^2$ and performing an integration gives

$$-cU + \tfrac{1}{2}U^2 - DU' = B,$$

where B is a constant of integration. Hence

$$\frac{dU}{ds} = \frac{1}{2D}(U - f_1)(U - f_2), \qquad (2.13)$$

where

$$f_1 = c - \sqrt{c^2 + 2B}, \qquad f_2 = c + \sqrt{c^2 + 2B}.$$

We assume $c^2 > -2B$, and so $f_2 > f_1$. Separating variables in (2.13) and integrating again gives

$$\frac{s}{2D} = \int \frac{dU}{(U - f_1)(U - f_2)} = \frac{1}{f_2 - f_1} \ln \frac{f_2 - U}{U - f_1}, \qquad f_1 < U < f_2.$$

Solving for U yields

$$U(s) = \frac{f_2 + f_1 e^{Ks}}{1 + e^{Ks}}, \qquad (2.14)$$

where

$$K = \frac{1}{2D}(f_2 - f_1) > 0.$$

For large positive s we have $U(s) \to f_1$, and for large negative s we have $U(s) \to f_2$. It is easy to see that $U'(s) < 0$ for all s and that $U(0) = \frac{1}{2}(f_1 + f_2)$. The graph of U is shown in Fig. 7.12 for different values of D. The weaker the diffusive effect (small D), the sharper the gradient in U. The traveling wave solution to (2.11) is

$$u(x, t) = \frac{f_2 + f_1 e^{K(x-ct)}}{1 + e^{K(x-ct)}},$$

where the speed is determined from the definition of f_1 and f_2 to be

$$c = \tfrac{1}{2}(f_1 + f_2).$$

Graphically, the traveling wave solution is the profile $U(s)$ in Fig. 7.12 moving to the right at speed c. The solution of Burgers' equation, because it resembles the actual profile of a shock wave, is called the *shock structure* solution; it joins the asymptotic states f_1 and f_2. Without the Du_{xx} term the solutions of (2.11) would shock up and tend to break. The presence of the diffusion term prevents this breaking effect by countering the nonlinearity. The result is competition and balance between the nonlinear advection term uu_x and the diffusion term $-Du_{xx}$, much the same as occurs in a real shock wave in the narrow region where the gradient is steep. In this shock wave context the $-Du_{xx}$ term could also be interpreted as a viscosity term, as will become clear in the final chapter. □

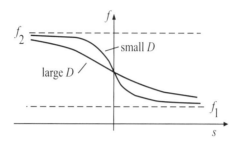

Figure 7.12 Traveling wave front solutions to Burgers' equation.

Example 7.11

(**Korteweg–de Vries equation**) A perturbation analysis of the equations that govern long waves in shallow water leads to another fundamental equation of applied mathematics, the **Korteweg–de Vries** *equation* (**KdV** *equation* for short)

$$u_t + uu_x + ku_{xxx} = 0, \quad k > 0. \tag{2.15}$$

This equation is a model equation for a nonlinear dispersive process. The uu_x term causes a shocking-up effect. In Burgers' equation this effect was balanced by a diffusion term $-Du_{xx}$, creating a shock structure solution. Now we replace this term by ku_{xxx}, which is interpreted as a dispersion term (plane wave solutions of the linear equation $u_t + ku_{xxx} = 0$ are dispersive). The KdV equation and modifications of it arise in many physical contexts (see C. S. Gardner et al., *SIAM Review*, Vol. 18, p. 412 (1976)). We follow the same reasoning as in Burgers' equation. That is, we assume a solution of the form

$$u = U(s), \quad s = x - ct, \tag{2.16}$$

where the waveform U and the wave speed c are to be determined. Substituting (2.16) into (2.15) gives

$$-cU' + \frac{1}{2}\frac{d}{ds}U^2 + kU''' = 0,$$

which when integrated gives

$$-cU + \frac{1}{2}U^2 + kU'' = A.$$

Multiplying by U' and integrating again yields

$$-\frac{c}{2}U^2 + \frac{1}{6}U^3 + \frac{1}{2}kU'^2 = AU + B.$$

One can then write

$$\frac{dU}{ds} = \pm\sqrt{\frac{1}{3k}}\,\phi(U)^{1/2}, \qquad \phi(U) = -U^3 + 3cU^2 + 6AU + 6B. \qquad (2.17)$$

Since $\phi(U)$ is a cubic polynomial, there are five possibilities:

(i) ϕ has one real root α.

(ii) ϕ has three distinct real roots $\gamma < \beta < \alpha$.

(iii) ϕ has three real roots satisfying $\gamma = \beta < \alpha$.

(iv) ϕ has three real roots satisfying $\gamma < \alpha = \beta$.

(v) ϕ has a triple root γ.

Clearly, if α is a real root of ϕ, then $U = \alpha$ is a constant solution. We seek real, nonconstant, bounded solutions that will exist only when $\phi(U) \geq 0$. On examining the direction field it is easy to see that only unbounded solutions of (2.17) exist for cases (i) and (iv). We leave case (v) and case (ii) to the reader. Case (ii) leads to cnoidal waves. We examine case (iii) in which a class of solutions known as *solitons* arise. In this case (2.17) becomes

$$\frac{ds}{\sqrt{3k}} = \frac{dU}{(U-\gamma)\sqrt{\alpha-U}}. \qquad (2.18)$$

Letting $U = \gamma + (\alpha - \gamma)\text{sech}^2 w$ and noting $dU = -2(\alpha-\gamma)\text{sech}^2 w \tanh w \; dw$, it follows that

$$\frac{ds}{\sqrt{3k}} = \frac{-2(\alpha-\gamma)\text{sech}^2 w \tanh w \; dw}{(\alpha-\gamma)\text{sech}^2 w\sqrt{(\alpha-\gamma)-(\alpha-\gamma)\text{sech}^2 w}} = \frac{-2dw}{\sqrt{\alpha-\gamma}}.$$

Integrating gives

$$w = -\sqrt{\frac{\alpha-\gamma}{12k}}\,s,$$

and therefore a solution to (2.18) is given by

$$U(s) = \gamma + (\alpha - \gamma)\text{sech}^2\left(\sqrt{\frac{\alpha-\gamma}{12k}}\,s\right). \qquad (2.19)$$

Clearly $U(s) \to \gamma$ as $s \to \pm\infty$. A graph of the waveform U is shown in Fig. 7.13. It is instructive to write the roots α and γ in terms of the original parameters. To this end

$$\phi(U) = -U^3 + 3cU^2 + 6AU + 6B$$
$$= (U - \gamma)^2(\alpha - U)$$
$$= -U^3 + (\alpha + 2\gamma)U^2 + (-2\alpha\gamma - \gamma^2)U + \gamma^2\alpha.$$

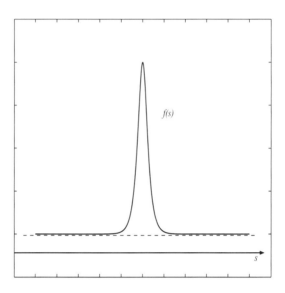

Figure 7.13 Soliton solution to the KdV equation.

Thus the wave speed c is given by

$$c = \frac{\alpha + 2\gamma}{3},$$

and we can write the solution as

$$u(x,t) = \gamma + a\,\mathrm{sech}^2\left(\sqrt{\frac{a}{12k}}\left(x - \left[\gamma + \frac{a}{3}\right]t\right)\right), \qquad a \equiv \alpha - \gamma.$$

Let us note several features of this wave front solution, or pulse. The velocity relative to the asymptotic state at $\pm\infty$ is proportional to the amplitude a. The width of the wave, defined by $\sqrt{12k/a}$, increases as k increases; that is, the wave disperses. Finally, the amplitude a is independent of the asymptotic state at $\pm\infty$. Such a waveform is known as a **soliton**, and many important equations of mathematical physics have soliton-like solutions (for example, the Boussinesq equation, the Sine–Gordon equation, the Born–Infeld equation, and nonlinear Schrödinger-type equations). In applications, the value of such solutions is that if a pulse or signal travels as a soliton, then the information contained in the pulse can be carried over long distances with no distortion or loss of intensity. Solitons, or solitary waves, have been observed in canals and waterways.   □

Reaction–advection–diffusion equations of the form

$$u_t = Du_{xx} + f(u, u_x)$$

form an important class of equations that often admit traveling wave solutions. If we assume $u = U(z)$, $z = x - ct$, then

$$-cU' = DU'' + F(U, U').$$

For wave front solutions we require $F(u_\pm, 0) = 0$, where $u_\pm$ are the constant states at $\pm\infty$. This equation is often examined in the UV phase plane (see Chapter 2), where $V = U'$. Then

$$U' = V, \quad V' = -\frac{1}{D}(F(U, V) + cV).$$

Showing the existence of traveling wave solutions then amounts to showing there is an orbit connecting the critical points $(u_-, 0)$ and $(u_+, 0)$.

7.2.3 Conservation Laws

We observed how nonlinear advection can cause solutions to 'shock up' and cease to exist as a smooth solution. In this section we discover what happens after the gradient catastrophe occurs.

The first-order partial differential equation

$$u_t + J(u)_x = 0, \quad x \in \mathbb{R}, \ t > 0, \tag{2.20}$$

which represents a local conservation law, was derived under the conditions of smoothness of the density and flux. If there are discontinuities, we must take the integral form of the conservation law,

$$\frac{d}{dt} \int_a^b u(x, t) \, dx = J(u(a, t)) - J(u(b, t)), \tag{2.21}$$

which holds in all cases.

Example 7.12

The model nonlinear equation

$$u_t + uu_x = 0$$

can be written in conservation form as

$$u_t + \left(\tfrac{1}{2}u^2\right)_x = 0,$$

where the flux is $J(u) = \tfrac{1}{2}u^2$. The integral form of this law is

$$\frac{d}{dt} \int_a^b u(x, t) \, dx = \frac{1}{2}(u(a, t)^2 - u(b, t)^2). \quad \square$$

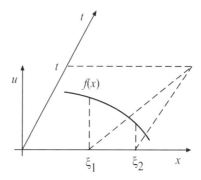

Figure 7.14 Characteristics determined by the initial condition.

In general we can write (2.20) in the form

$$u_t + c(u)u_x = 0, \qquad (2.22)$$

where $c(u) = J'(u)$. We noted earlier that the initial value problem associated with (2.20) does not always have a solution for all $t > 0$. We showed that u is constant on the characteristic curves defined by $dx/dt = c(u)$, and the initial condition

$$u(x, 0) = f(x)$$

is propagated along straight lines with speed $c(u)$. If $c'(u) > 0$ and $f'(x) < 0$, then the characteristics issuing from two points ξ_1 and ξ_2 on the x axis with $\xi_1 < \xi_2$ have speeds $c(f(\xi_1))$ and $c(f(\xi_2))$, respectively. It follows that $c(f(\xi_1)) > c(f(\xi_2))$, and therefore the characteristics must intersect (see Fig. 7.14). Because u is constant on characteristics, the solution is meaningless at the value of t where the intersection occurs. The first such t, denoted earlier by t_b, is the breaking time. A smooth solution exists only up to time t_b; beyond t_b something else must be tried. The key to continuing the solution for $t > t_b$ is to take a hint from the propagation of stress waves in a real physical continuum. There, under certain conditions, a smooth wave profile evolves into a shock wave. Hence, it is reasonable to expect that a discontinuous solution may exist and propagate for $t > t_b$. This discontinuous solution cannot satisfy the differential form (2.22) of the conservation law, but the integral form (2.21) remains valid. Therefore, we turn to a study of the simplest kinds of such solutions in an effort to obtain a condition that must hold at a discontinuity. Let $J = J(u)$ be continuously differentiable, and let $x = s(t)$ be a smooth curve in space-time across which u is discontinuous. Suppose u is smooth on each side of the curve (see Fig. 7.15). Next, select a and b such that the curve $x = s(t)$

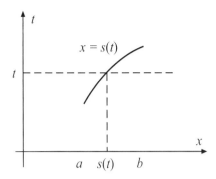

Figure 7.15 Shock path in space–time.

intersects $a \leq x \leq b$ at time t and let u_0 and u_1 denote the right and left limits of u at $s(t)$, respectively. That is,

$$u_0 = \lim_{x \to s(t)+} u(x, t), \quad u_1 = \lim_{x \to s(t)-} u(x, t).$$

From the conservation law (2.21) and Leibniz' rule for differentiating integrals,

$$J(u(a, t)) - J(u(b, t)) = \frac{d}{dt} \int_a^{s(t)} u(x, t) \, dx + \frac{d}{dt} \int_{s(t)}^b u(x, t) \, dx$$

$$= \int_a^{s(t)} u_t(x, t) \, dx + u_1 \frac{ds}{dt} + \int_{s(t)}^b u_t(x, t) \, dx - u_0 \frac{ds}{dt}.$$

Because u_t is bounded in each of the intervals $[a, s(t)]$ and $[s(t), b]$ separately, both integrals tend to zero in the limit as $a \to s(t)^-$ and $b \to s(t)^+$. Therefore

$$J(u_1) - J(u_0) = (u_1 - u_0)\frac{ds}{dt}. \tag{2.23}$$

Equation (2.23) is a **jump condition**, often called the Rankine–Hugoniot condition, that relates the values of u and the flux in front of and behind the discontinuity to the speed ds/dt of the discontinuity. Thus the integral form of the conservation law provides a restriction on possible jumps across a simple discontinuity. A conventional notation for (2.23) is

$$[J(u)] = [u]\frac{ds}{dt},$$

where $[\cdot\cdot]$ indicates the jump in a quantity across the discontinuity.

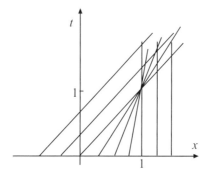

Figure 7.16 Characteristic diagram. Colliding characteristics indicate formation of a gradient catastrophe when the smooth solution ceases to exist.

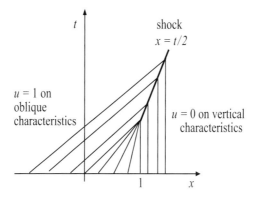

Figure 7.17 Characteristic diagram showing the formation of the shock and its path.

Example 7.13

Consider the Cauchy problem

$$u_t + (\tfrac{1}{2}u^2)_x = 0, \quad x \in \mathbb{R},\ t > 0,$$

$$u(x,0) - f(x) = \begin{cases} 1, & x \le 0, \\ 1 - x, & 0 < x < 1, \\ 0, & x \ge 1. \end{cases}$$

The characteristic diagram is shown in Fig. 7.16. Lines emanating from the x axis have speed $c(f(x)) = f(x)$. It is clear from the geometry that $t_b = 1$ and that a single valued solution exists for $t < 1$. For $t > 1$ we fit a shock or discontinuity beginning at $(1,1)$, separating the state $u_1 = 1$ on the left from the state $u_0 = 0$ on the right. Thus $[u] = 1 - 0 = 1$ and $[F] = \frac{1}{2}u_1^2 - \frac{1}{2}u_0^2 = \frac{1}{2}$. Therefore the discontinuity must have speed $s'(t) = [F]/[u] = \frac{1}{2}$. The resulting characteristic diagram is shown in Fig. 7.17. Therefore we have obtained a solution for all $t > 0$ satisfying the conservation law. $\square$

EXERCISES

1. Consider the problem

$$2u_t + u_x = 0, \quad x > 0, \ t > 0,$$
$$u(0, t) = 0, \quad t > 0,$$
$$u(x, 0) = 1, \quad x \geq 0.$$

In what sense are $u_1(x, t) = 1 - H(t - 2x)$ and $u_2(x, t) = 1 - H(t - x)$ both solutions to the problem? Which one is the appropriate solution? (H is the Heaviside function.)

2. Consider the equation $u_t + c(u)u_x = 0$, $x \in \mathbb{R}$, $t > 0$, with initial data $u(x, 0) = u_1$ for $x < 0$, and $u(x, 0) = u_0$ for $x \geq 0$, with $c'(u) > 0$.

   a) Find a nonsmooth solution in the case $u_0 > u_1$.

   b) Find a discontinuous (shock) solution in the case $u_0 < u_1$.

   c) Discuss (a) and (b) in the context of shock waves and rarefaction waves.

3. In the traveling wave solution to Burgers' equation, the **shock thickness** is defined by $(f_2 - f_1)/ \max |U'(s)|$. Give a geometric interpretation of this condition, and show that the shock thickness is given by $8D/(f_2 - f_1)$.

4. Find all wave front solutions of the reaction–advection equation

$$u_t + u_x = u(1 - u).$$

5. Consider the nonlinear advection–diffusion equation

$$\left(u - \frac{1}{2}u^2 \right)_t = u_{xx} - u_x.$$

   a) What is the speed c of right-traveling wave front solutions $u = U(z)$, $z = x - ct$, that satisfy the conditions $U(-\infty) = 1$, $U(\infty) = 0$?

   b) Determine explicitly the right-traveling wave front in (a) that satisfies the condition $U(0) = \frac{1}{2}$.

6. Show that Burgers' equation $u_t + u u_x = D u_{xx}$ can be reduced to the diffusion equation via the transformation $u = -2D\phi_x/\phi$ (this is the **Cole–Hopf** transformation). If the initial condition $u(x,0) = f(x)$ is imposed, derive the solution

$$u(x,t) = \int_{-\infty}^{\infty} \frac{x-\xi}{t} \exp\left(\frac{-G}{2D}\right) d\xi \int_{-\infty}^{\infty} \exp\left(\frac{-G}{2D}\right) d\xi,$$

where

$$G = G(\xi, x, t) = \int_0^\xi f(\eta)\, d\eta + \frac{(x-\xi)^2}{2t}.$$

7.3 Quasi-linear Equations

A quasi-linear partial differential equation is an equation that is linear in its derivatives. Many of the equations that occur in applications (for example, the nonlinear equations of fluid mechanics that we will develop in a subsequent section) are quasi-linear. In this section we focus on the initial value problem for first-order equations and introduce the idea of the general solution. Finally, as an application, we introduce age-structured models in population dynamics.

In this section we study the Cauchy problem for the quasi-linear equation

$$u_t + c(x,t,u)u_x = f(x,t,u), \quad x \in \mathbb{R}, t > 0, \tag{3.1}$$

where the initial condition has the form

$$u(x,0) = \phi(x), \quad x \in \mathbb{R}. \tag{3.2}$$

Motivated by our earlier approach, we define a family of curves by the differential equation

$$\frac{dx}{dt} = c(x,t,u). \tag{3.3}$$

On these curves u is no longer constant, but rather

$$\frac{du}{dt} = f(x,t,u). \tag{3.4}$$

The pair of differential equations (3.3)–(3.4) is called the **characteristic system** associated with (3.1). The initial data can be represented

$$x = \xi, \quad u = \phi(\xi) \quad \text{on} \quad t = 0, \tag{3.5}$$

where ξ is a parameter representing an arbitrary value on the x axis. We may interpret equations (3.3)–(3.4) as a system of two nonautonomous equations for $x = x(t)$ and $u = u(t)$ (the values of x and u along the characteristic curves), with initial values given in (3.5). The general solution of (3.3)–(3.4) involves two arbitrary constants and has the form

$$x = F(t, c_1, c_2), \quad u = G(t, c_1, c_2).$$

The initial data imply $\xi = F(0, c_1, c_2)$, $\phi(\xi) = G(0, c_1, c_2)$, which gives the constants in terms of ξ, or $c_1 = c_1(\xi)$, $c_2 = c_2(\xi)$. Therefore, x and u are given as functions of t and their initial values,

$$x = F(t, c_1(\xi), c_2(\xi)), \quad u = G(t, c_1(\xi), c_2(\xi)). \tag{3.6}$$

In principle, the first of these equations can be solved to obtain $\xi = \xi(x, t)$, which can then be substituted into the second equation to obtain an explicit representation $u = u(x, t)$ to the solution to (3.1)–(3.2).

Example 7.14

Solve the IVP

$$
\begin{aligned}
u_t + u u_x &= -u, & x \in \mathbb{R}, t > 0, \\
u(x, 0) &= -\frac{x}{2}, & x \in \mathbb{R}.
\end{aligned}
$$

The characteristic system is

$$\frac{dx}{dt} = u, \quad \frac{du}{dt} = -u,$$

and the initial condition gives

$$x = \xi, \quad u = -\frac{\xi}{2} \quad \text{on} \quad t = 0.$$

The general solution of the characteristic system is

$$x = -c_1 e^{-t} + c_2, \quad u = c_1 e^{-t}.$$

Applying the initial data gives $\xi = -c_1 + c_2$ and $-\frac{\xi}{2} = c_1$. Therefore

$$x = \frac{\xi}{2} e^{-t} + \frac{\xi}{2}, \quad u = -\frac{\xi}{2} e^{-t}.$$

Therefore $\xi = \xi(x, t) = 2x/(1 + e^{-t})$, and so

$$u(x, t) = \frac{-x e^{-t}}{1 + e^{-t}}. \quad \square$$

We have carried out this solution process formally. In most problems it will not be possible to solve the characteristic system or the resulting algebraic equations that determine the constants $c_1 = c_1(\xi)$, $c_2 = c_2(\xi)$ as functions of ξ. As a consequence, we often have to be satisfied with the parametric representation (3.6). In practice, numerical methods are usually applied to (3.3)–(3.4) to advance the initial conditions forward in time.

In some contexts the general solution of the quasi-linear equation (3.1) is needed, in terms of a single, arbitrary function. To determine the general solution we proceed with some general definitions and observations. An expression $\psi(x, t, u)$ is called a **first integral** of the characteristic system (3.3)–(3.4) if $\psi(x, t, u) = k$ (constant), on solutions to (3.3)–(3.4). That is, $\psi(X(t), t, U(t)) = k$ for all t in some interval I, if $x = X(t)$, $u = U(t)$, $t \in I$, is a solution to (3.3)–(3.4). Taking the total derivative of this last equation with respect to t and using the chain rule gives

$$\psi_x c + \psi_t + \psi_u f = 0, \quad t \in I, \tag{3.7}$$

where each of the terms in this equation are evaluated at $(X(t), t, U(t))$.

Furthermore, if $\psi(x, t, u)$ is a first integral of the characteristic system, then the equation $\psi(x, t, u) = k$ defines, implicitly, a surface $u = u(x, t)$, on some domain D in the xt plane, provided $\psi_u \neq 0$. Thus

$$\psi(x, t, u(x, t)) = k, \quad (x, t) \in D.$$

Using the chain rule, we can calculate the partial derivatives of u as

$$\psi_t + \psi_u u_t = 0, \quad \psi_x + \psi_u u_x = 0,$$

or

$$u_t = -\frac{\psi_t}{\psi_u}, \quad u_x = \frac{\psi_x}{\psi_u}, \quad (x, t) \in D.$$

Each term on the right is evaluated at $(x, t, u(x, t))$. Observe also that the curve $(X(t), t, U(t))$ lies on this surface because $\psi(X(t), t, U(t)) = k$.

We claim that this surface $u = u(x, t)$ is a solution to the partial differential equation (3.1). To prove this, fix an arbitrary point (ξ, τ) in D. Then

$$u_t(\xi, \tau) + c u_x(\xi, \tau) = -\frac{\psi_t}{\psi_u} - c\frac{\psi_x}{\psi_u}, \tag{3.8}$$

where c and all the terms on the right side are evaluated at (ξ, τ, ω), where $\omega = u(\xi, \tau)$. But a solution curve $(X(t), t, U(t))$ lies on the surface and passes through the point (ξ, τ, ω) at $t = \tau$. That is, $(\xi, \tau, \omega) = (X(\tau), \tau, U(\tau))$. Therefore, the right side of (3.8) may be evaluated at $(X(\tau), \tau, U(\tau))$.

In general, we expect to find two independent first integrals to the characteristic system, $\psi(t, x, u)$ and $\chi(t, x, u)$. The **general solution** of (3.1) is then

$$H(\psi(t, x, u), \chi(t, x, u)) = 0, \qquad (3.9)$$

where H is an arbitrary function. We can formally solve for one of the variables and write this as $\psi(t, x, u) = G(\chi(t, x, u))$, where G is an arbitrary function. We leave it as an exercise using the chain rule to show that if $\psi(t, x, u)$ and $\chi(t, x, u)$ are first integrals of the characteristic system, then (3.9) defines, implicitly, a solution $u = u(x, t)$ to (3.1) for any function H, provided $H_\psi \psi_u + H_\chi \chi_u \neq 0$.

Example 7.15

Consider the equation

$$u_t + 2tu_x = \frac{x}{2u + 1}.$$

The characteristic system is

$$\frac{dx}{dt} = 2t, \qquad \frac{du}{dt} = \frac{x}{2u + 1}.$$

The first equation gives first integral $\psi = x - t^2 = c_1$. Then the second equation becomes

$$\frac{du}{dt} = \frac{t^2 + c_1}{2u + 1}.$$

Separating variables and integrating gives another first integral $\chi = u^2 + u + \frac{2}{3}t^3 - xt = c_2$. Therefore, the general solution is

$$H\left(x - t^2, u^2 + u + \frac{2}{3}t^3 - xt\right) = 0,$$

where H is an arbitrary function. This expression defines implicit solutions. It can also be written

$$u^2 + u + \frac{2}{3}t^3 - xt = G\left(x - t^2\right),$$

where G is an arbitrary function. This quadratic in u can be solved to determine explicit solutions. □

7.3.1 Age-Structured Populations

Elementary population models in biology lump the total population at time t into a single number $p(t)$, regardless of the age of the individuals. We have studied these models, e.g., the Malthus law and the logistics law, in earlier chapters. However, because individuals may differ in their vital statistics (birth- and deathrates, fertility), in some demographic models it is necessary to superimpose an age distribution at each time. These models are called **age-structured** models.

We consider a population of females whose age structure at time $t = 0$ is $f(a)$. That is, $f(a)\,da$ is approximately the initial number of females between age a and $a + da$. Females are usually used in demographic models because they have a definite beginning and end to their reproductive capacity. And, for simplicity, we impose the age range to be $0 \leq a < \infty$, even though the age at death is finite. Given the mortality and fecundity rates of females, the goal is to determine the age structure, or population density, $u(a, t)$ at time t, where $u(a, t)\,da$ is approximately the number of females between age a and $a + da$ at time t. The total number of females at time t is

$$N(t) = \int_0^\infty u(a, t)\,da.$$

We can derive a partial differential equation for the density by accounting for the change in population over a small time dt, where members leave the population only because of death. In the time interval from t to $t+dt$ all females of age a become females of age $a+da$. The number of females at time t between age a and $a + da$ is $u(a, t)\,da$, and the number between $a + da$ and $a + 2da$ at time $t + dt$ is $u(a + da, t + dt)\,da$. The difference between these two populations is the number that die during the time interval. If λ represents the mortality rate, or the fraction that die per unit time, then the number that die from time t to $t + dt$ is $\lambda[u(a, t)\,da]\,dt$. The mortality rate λ could depend upon age a, time t, age density u, or even the total population $N(t)$. Consequently, to leading order we have

$$u(a + da, t + dt)da - u(a, t)\,da = -\lambda u(a, t)\,dadt + \cdots.$$

By Taylor's expansion, $u(a + da, t + dt) = u(a, t) + u_t(a, t)\,dt + u_a(a, t)\,da + \cdots$. Therefore $u_t(a, t)\,dt + u_a(a, t)\,da = -\lambda u(a, t)\,dt + \cdots$. Dividing by dt, using $da = dt$, and taking the limit as $dt \to 0$ gives

$$u_t + u_a = -\lambda u, \tag{3.10}$$

which is the **McKendrick–Von Foerster equation**. The initial condition is

$$u(a, 0) = f(a). \tag{3.11}$$

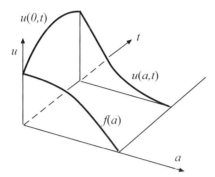

Figure 7.18 Graphical representation of an age-structured model showing age profiles at time $t = 0$ ($f(a)$) and at time t ($u(a,t)$). The boundary condition $u(0,t)$ represents the number of newborns.

The boundary condition at age $a = 0$ is not as simple as it may first appear because $u(0,t) = B(t)$, the number of newborns at time t, is not known *a priori*. Rather, it depends upon the fertility and ages of the reproducing females at time t. To formulate the boundary condition let $b(a,t)$ denote the average number of offspring per female of age a at time t. We call b the fecundity rate or maternity function. We expect b to be zero up until the age of maturity and zero after menopause. The total number of offspring produced by females in the age interval a to $a + da$ is $b(a,t)u(a,t)\,da$. Thus the total number of offspring $B(t)$ produced at time t is

$$B(t) = u(0,t) = \int_0^\infty b(a,t)u(a,t)\,da. \tag{3.12}$$

Therefore the boundary condition at $a = 0$ depends upon the solution to the differential equation. Such boundary conditions are called *nonlocal* because they involve an integration of the unknown in the problem. The initial boundary value problem (3.10)–(3.12) is an example of an age-structured model. These nonlocal, nonlinear models can be complex, and thus it is common to make assumptions that lead to simplification and the possibility of a tractable problem. For example, we assume the mortality rate λ is constant, and the fecundity rate b is dependent only upon age. Then we obtain the simplified the McKendrick–Von Foerster model

$$
\begin{aligned}
u_t + u_a &= -\lambda u, \quad a > 0, t > 0, \\
u(a,0) &= f(a), \quad a > 0, \\
u(0,t) &= \int_0^\infty b(a)u(a,t)\,da, \quad t > 0.
\end{aligned}
$$

The characteristic system for the partial differential equation is

$$\frac{da}{dt} = 1, \quad \frac{du}{dt} = -\lambda u.$$

It is easy to find the first integrals $a - t = c_1$ and $ue^{\lambda t} = c_2$. Therefore the general solution to the pde is

$$u(a,t) = g(a - t)e^{-\lambda t},$$

where g is an arbitrary function. In the domain $a > t$ the solution is determined by the initial condition at $t = 0$. Therefore $g(a) = f(a)$ and we obtain

$$u(a,t) = f(a - t)e^{-\lambda t}, \quad a > t.$$

In the domain $0 < a < t$ the solution is determined by the boundary condition. From the general solution we have

$$u(0,t) = B(t) = g(-t)e^{-\lambda t},$$

which determines the arbitrary function, $g(t) = B(-t)e^{-\lambda t}$. Therefore

$$u(a,t) = B(t - a)e^{-\lambda a}, \quad 0 < a < t.$$

Hence, we will have determined the solution once the function B is found.

To determine B we use the nonlocal boundary condition to obtain

$$\begin{aligned}
B(t) &= \int_0^\infty b(a)u(a,t)\,da \\
&= \int_0^t b(a)u(a,t)da + \int_t^\infty b(a)u(a,t)\,da \\
&= \int_0^t b(a)B(t - a)e^{-\lambda a}\,da + \int_t^\infty b(a)f(a - t)e^{-\lambda t}\,da.
\end{aligned}$$

This equation is a nonhomogeneous Volterra integral equation for B and it can be written in the form

$$B(t) = \int_0^t b(a)B(t - a)e^{-\lambda a}\,da + F(t), \tag{3.13}$$

where $F(t) = \int_t^\infty b(a)f(a - t)e^{-\lambda t}\,da$. Equation (3.13) is called the **renewal equation**. In summary, the solution to McKendrick–Von Foerster model has been reduced to the solution of an integral equation.

We show how an asymptotic solution for large times can be obtained. For large t the nonhomogeneous term $F(t)$ vanishes because the fecundity b is zero at some finite age. Thus, the renewal equation can be approximated by

$$B(t) = \int_0^\infty b(a)B(t - a)e^{-\lambda a}\,da, \quad t \text{ large.}$$

Note that we can replace the upper limit by infinity since the fecundity rate vanishes at a finite age. Let us assume an exponential solution of the form $B(t) = U_0 e^{rt}$, where r is an unknown growth rate. Substituting gives

$$1 = \int_0^\infty b(a) e^{-(r+\lambda)a} \, da. \qquad (3.14)$$

This is the **Euler–Lotka equation**, and it determines the growth rate r. The population density in this asymptotic case is

$$u(a, t) = U_0 e^{rt} e^{-(r+\lambda)a}, \quad t > a, \quad t \text{ large.}$$

The decaying function $U(a) = e^{-(r+\lambda)a}$ is called the **stable age structure**.

More extensive material on age structured problems can be found in Brauer and Castillo-Chavez (2001) and Kot (2001). Age-structured problems are a special case of more general physiological structured problems where the age variable a can be replaced by length, size, weight, development, and so on.

EXERCISES

1. Insofar as possible, solve the following pure initial value problems:

   a) $u_t + u_x = -2tu$, $u(x, 0) = \phi(x)$.

   b) $u_t + u_x = g(x, t)$, $u(x, 0) = \phi(x)$.

   c) $u_t + x u_x = e^t$, $u(x, 0) = \phi(x)$.

2. Use characteristics to find the general solution to the following equations:

   a) $u_t + u_x = -u^2$.

   b) $t^2 u_t + x^2 u_x = 2xt$.

   c) $u u_t = x - t$.

   d) $t u_t + x u_x = xt(1 + u^2)$.

   e) $t u_t + u u_x = x$.

   f) $u(u_t - u_x) = x - t$.

   g) $u u_t + x u_x = t$.

3. Consider the initial value problem

   $$u_t + u u_x + u = 0, \quad x \in \mathbb{R}, \ t > 0,$$
   $$u(x, 0) = -2x, \quad x \in \mathbb{R}.$$

   Find the time t_b that a gradient catastrophe occurs and beyond which there is no continuously differentiable solution.

4. Consider the initial value problem

$$u_t + uu_x + \alpha u^2 = 0, \quad x \in \mathbb{R}, \ t > 0; \ \alpha > 0,$$
$$u(x,0) = 1, \quad x \in \mathbb{R}.$$

Derive the solution $u(x,t) = (1 + \alpha t)^{-1}$.

5. Consider the initial value problem

$$u_t + uu_x = -u, \quad x \in \mathbb{R}, \ t > 0,$$
$$u(x,0) = ke^{-x^2}, \quad x \in \mathbb{R},$$

where $k > 0$. Determine the values of k for which the solution blows up.

6. Consider the initial value problem for the damped advection equation,

$$u_t + uu_x = -\alpha u, \quad x \in \mathbb{R}, \ t > 0,$$
$$u(x,0) = \phi(x), \quad x \in \mathbb{R},$$

where $\alpha > 0$. Show that the solution is given implicitly by

$$x = \phi(\xi)e^{-\alpha t}, \quad u = \frac{\phi(\xi)}{\alpha}(1 - e^{-\alpha t}) + \xi.$$

Determine a condition on ϕ that guarantees no shocks will form. If a shock does form, show that the breaking time is greater than that for the undamped equation when $\alpha = 0$.

7. Consider

$$u_t + uu_x + u = 0, \quad x \in \mathbb{R}, \ t > 0,$$
$$u(x,0) = k\exp(-x^2), \quad x \in \mathbb{R}; \ k > 0.$$

For what values of k does breaking occur?

8. Solve the signaling problem

$$u_t + u_x = u^2, \quad x > 0, \ -\infty < t < \infty,$$
$$u(0,t) = g(t), \quad -\infty < t < \infty.$$

9. Consider an age-structured female population where $u = u(a,t)$ is the age density of females at age a at time t and where no individual survives beyond age δ. The model is

$$u_t + u_a = -\frac{\gamma}{\delta - a}u, \quad 0 < a < \delta, \ t > 0.$$

Find the solution on the given domain subject to the initial and boundary conditions

$$u(a,0) = f(a), \quad u(0,t) = B(t); \quad 0 < a < \delta, \ t > 0,$$

where f and B are given functions.

10. Consider a population of female organisms whose mortality rate is λu, with λ constant, and whose fecundity is given by $b(a) = ae^{-\gamma a}$. From the Euler–Lotka equation determine the growth rate r of the stable age structured.population.

11. Consider the age-structured model

$$
\begin{aligned}
u_t + u_a &= -\lambda(N(t))u, \quad a > 0, t > 0, \\
u(a, 0) &= f(a), \quad a > 0, \\
u(0, t) &= \int_0^\infty b(N(t))u(a, t)da, \quad t > 0,
\end{aligned}
$$

where the mortality and fecundity rates depend upon the total population $N(t)$, with $b'(N) \leq 0$, $\lambda'(N) \geq 0$.

a) Show that $N'(t) = (b(N) - \lambda(N))N$.

b) Assuming the differential equation in part (a) has a nonzero equilibrium N^*, show that there is an equilibrium population density $u = U(a)$ given by

$$
U(a) = N^*\lambda(N^*)e^{-\lambda(N^*)a}.
$$

12. In the McKendrick–Von Foerster model assume the mortality rate λ and the fecundity rate b are constant.

a) Show that $B' = (b - \lambda)B$.

b) Find $B(t)$ and the population density $u(a, t)$.

c) What is the size $N(t)$ of the population at time t?

7.4 The Wave Equation

The science of acoustics is the study of sound propagation in solids, fluids, and gases. The fundamental equation is the wave equation that was obtained as an approximation for small-amplitude signals.

7.4.1 The Acoustic Approximation

As before, we imagine a gas flowing in a tube of constant cross-sectional area A in such a way that the physical parameters are constant in any cross section. We fix two cross sections $x = a$ and $x = b$. The principle of mass conservation states simply that mass is neither created nor destroyed. Therefore, the rate of change of mass inside the interval $a \leq x \leq b$ equals the mass flux in at $x = a$ less the mass flux out at $x = b$. In symbols

$$\frac{d}{dt} \int_a^b A\rho(x,t)\, dx = A\rho(a,t)v(a,t) - A\rho(b,t)v(b,t), \tag{4.1}$$

where $\rho(x,t)$ is the density and $v(x,t)$ is the velocity. The quantity $A\rho(x,t)v(x,t)$ gives the amount of fluid that crosses the location x per unit time and is therefore the mass flux. Equation (4.1) can be written

$$\int_a^b \rho_t(x,t)\, dx = - \int_a^b (\rho(x,t)v(x,t))_x\, dx.$$

Because the interval $a \leq x \leq b$ is arbitrary it follows that

$$\rho_t + (\rho v)_x = 0, \tag{4.2}$$

which is the **continuity equation**, or the partial differential equation expressing conservation of mass.

Momentum balance requires that the rate of change of momentum inside $[a,b]$ must equal the transport of momentum carried into and out of the region at $x = a$ and $x = b$ plus the momentum created inside $[a,b]$ by the force (pressure times area) p acting at a and b. This is, of course, motivated by Newton's second law of motion, which states that the rate of change of momentum is the force, or $\frac{d}{dt}(mv) = F$. In mathematical terms,

$$\frac{d}{dt} \int_a^b A\rho(x,t)v(x,t)\, dx = A\rho(a,t)v^2(a,t) - A\rho(b,t)v^2(b,t) + Ap(a,t) - Ap(b,t). \tag{4.3}$$

The term $A\rho(x,t)v^2(x,t)$ is the momentum times the velocity at x and is therefore the **momentum flux** at x. Bringing the time derivative into the integral and using the fundamental theorem of calculus turns (4.3) into

$$\int_a^b (\rho(x,t)v(x,t))_t\, dx = - \int_a^b (\rho(x,t)v^2(x,t) + p(x,t))_x\, dx.$$

The arbitrariness of the interval $a \leq x \leq b$ forces

$$(\rho v)_t + (\rho v^2 + p)_x = 0, \tag{4.4}$$

which is a partial differential equation expressing **conservation of momentum**. Both (4.2) and (4.4) are of the form

$$\frac{\partial}{\partial t}[\cdots] + \frac{\partial}{\partial x}[\cdots] = 0,$$

and are in **conservation form**.

In the simplest gas dynamical calculations equations (4.2) and (4.4) may be supplemented by the barotropic equation of state

$$p = f(\rho), \quad f'(\rho) > 0, \tag{4.5}$$

which gives pressure as a function of density only. It is convenient to introduce the **local sound speed** c defined by

$$c^2 = f'(\rho). \tag{4.6}$$

Then (4.2) and (4.4) may be written

$$p_t + v p_x + c^2 \rho v_x = 0, \tag{4.7}$$

$$\rho v_t + \rho v v_x + p_x = 0, \tag{4.8}$$

where we have used $p_x = c^2 \rho_x$ and $p_t = c^2 \rho_t$ to rewrite (4.2), and we have used (4.2) to replace the quantity ρ_t in (4.4), thereby giving (4.8). Equations (4.7) and (4.8) along with (4.5) provide a starting point for the study of flow of a gas (Chapter 8).

The equations governing the adiabatic flow of a gas are nonlinear and cannot be solved in general. Rather, only special cases can be resolved. The simplest case, that of acoustics, arises when the deviations from a constant equilibrium state $v = 0$, $\rho = \rho_0$, and $p = p_0$ are assumed to be small. Therefore we consider the isentropic equations

$$\rho_t + v \rho_x + \rho v_x = 0, \tag{4.9}$$

$$\rho v_t + \rho v v_x + p_x = 0, \tag{4.10}$$

with the barotropic equation of state

$$p = F(\rho).$$

It is common in acoustics to introduce the *condensation* δ defined by

$$\delta \equiv \frac{\rho - \rho_0}{\rho_0},$$

which is the relative change in density from the constant state ρ_0. We assume that v, δ, and all their derivatives are small compared to unity. Then the equation of state can be expanded in a Taylor series as

$$p = F(\rho_0) + F'(\rho_0)(\rho - \rho_0) + \tfrac{1}{2} F''(\rho_0)(\rho - \rho_0)^2 + O((\rho - \rho_0)^3)$$
$$= F(\rho_0) + \rho_0 F'(\rho_0)\delta + \tfrac{1}{2} \rho_0^2 F''(\rho_0)\delta^2 + O(\delta^3).$$

Whence

$$p_x = \rho_0 F'(\rho_0)\delta_x + O(\delta^2),$$

and therefore (4.9) and (4.10) become

$$\delta_t + v\delta_x + (1 + \delta)v_x = 0,$$
$$(1 + \delta)v_t + (1 + \delta)vv_x + F'(\rho_0)\delta_x + O(\delta^2) = 0.$$

Discarding products of small terms gives a set of linearized equations for the small deviations δ and v, namely

$$\delta_t + v_x = 0, \quad v_t + F'(\rho_0)\delta_x = 0. \tag{4.11}$$

These equations are known as the **acoustic approximation**. Eliminating v from (4.11) gives

$$\delta_{tt} - c_0^2 \delta_{xx} = 0, \tag{4.12}$$

where

$$c^2 = F'(\rho_0). \tag{4.13}$$

Equation (4.12) is a second-order linear partial differential equation known as the **wave equation**. The quantity c in (4.13) is the propagation speed of acoustic waves. For an ideal gas $p = F(\rho) = k\rho^\gamma$, and therefore

$$c = \sqrt{\frac{\gamma p_0}{\rho_0}}$$

is the speed that small-amplitude signals are propagated in an ideal gas. By eliminating δ from equations (4.11) we note that the small velocity amplitude v satisfies the wave equation as well.

At this point it is worthwhile to test the acoustic approximation to determine its range of validity. Let U and Γ denote scales for the velocity v and condensation δ, respectively. From Chapter 1 we know that such scales can be defined as the maximum value of the quantity, for example, $U = \max|v|$ and $\Gamma = \max|\delta|$. If L and τ denote length and time scales for the problem, then the balance of the terms in (4.11) requires

$$\Gamma\tau^{-1} \approx UL^{-1}, \quad U\tau^{-1} \approx c_0^2 \Gamma L^{-1}.$$

Thus

$$c \approx L\tau^{-1}.$$

In the approximation we assumed that the convective term vv_x was small compared to v_t. Consequently, for acoustics

$$U^2 L^{-1} \ll U\tau^{-1},$$

or

$$U \ll c. \tag{4.14}$$

That is, the acoustic approximation can be justified when the maximum particle velocity is small compared to the speed c that sound waves travel in the medium. The ratio U/c is called the **Mach number** M and therefore acoustics are valid when $M \ll 1$. For phenomena involving ordinary audible sounds the approximation is highly satisfactory, but for flows resulting from the presence of high-speed aircraft the full nonlinear equations (4.9) and (4.10) would be required for an accurate description. In the theory of *nonlinear acoustics* a few, but not all, of the nonlinear terms are retained in the perturbation equations.

Example 7.16

(**Vibrations of a guitar string**) In this example we take an aside to mention the most common application of the wave equation, namely, the vibrations and sounds produced by plucking a guitar string. A detailed derivation of the governing equation, which is a simple application of Newton's second law, can be found in any elementary partial differential equations text (e.g., see Logan, 2006, Strauss 1992.). We envision a tightly stretched, perfectly elastic string attached at two ends, $x = 0$ and $x = L$. This is the undisturbed equilibrium configuration. The wave equation is a model that governs the vertical displacements $u = u(x, t)$ from equilibrium of the string in a plane at each location x in $[0, L]$ and for each $t > 0$. The excitation of the string is induced by initial conditions that specify the initial displacement $u(x, 0)$ and the initial velocity $u_t(x, 0)$. For example, the string could be plucked. The equation of motion plainly states that, locally, the mass times the acceleration is the vertical force. Symbolically,

$$\rho(x)u_{tt} = Tu_{xx}, \qquad \text{(wave equation)}$$

or

$$u_{tt} = c^2(x)u_{xx}, \quad c^2(x) \equiv \frac{T}{\rho(x)}.$$

Here, ρ is the linear density (mass per length) of the string, and T is the constant tension in the string; $c(x)$ is the wave speed, or the speed of propagation on the string. Like acoustics, the wave equation governs small vibrations from equilibrium. Terms representing body forces (like gravity) or damping forces can be added to the right side of the equation.   □

7.4.2 Solutions to the Wave Equation

The one-dimensional wave equation

$$u_{tt} - c^2 u_{xx} = 0 \tag{4.15}$$

arises naturally in acoustics and it describes how small amplitude waves (density and particle velocity) propagate. There are many other physical situations modeled by this important equation, for example, electromagnetism, wave on taut strings (guitar strings), and vibrations in elastic media. From earlier remarks we note that the wave equation is hyperbolic. It is easy to see (Exercise 2) that the general solution of the wave equation is given by

$$u(x,t) = f(x + ct) + g(x - ct), \tag{4.16}$$

where f and g are arbitrary functions; hence solutions are the superposition of right- and left-traveling waves moving at speed c. The functions f and g are determined by the initial and boundary data, although it is difficult to do this in specific cases.

The Cauchy problem for the wave equation is

$$u_{tt} - c^2 u_{xx} = 0, \quad x \in \mathbb{R},\ t > 0, \tag{4.17}$$
$$u(x,0) = F(x), \quad u_t(x,0) = G(x), \quad x \in \mathbb{R},$$

where F and G are given functions, and it can be solved exactly by **D'Alembert's solution**.

Theorem 7.17

In (4.17) let $F \in C^2(\mathbb{R})$ and $G \in C^1(\mathbb{R})$. Then the solution of (4.17) is given by

$$u(x,t) = \frac{1}{2}(F(x+ct) + F(x-ct)) + \frac{1}{2c}\int_{x-ct}^{x+ct} G(y)\,dy. \tag{4.18}$$

Proof

The derivation of (4.18) follows directly by applying the initial conditions to determine the arbitrary functions f and g in (4.16). We have

$$u(x,0) = f(x) + g(x) = F(x), \tag{4.19}$$
$$u_t(x,0) = cf'(x) - cg'(x) = G(x). \tag{4.20}$$

Dividing (4.20) by c and integrating yields

$$f(x) - g(x) = \frac{1}{c}\int_0^x G(y)\,dy + A, \tag{4.21}$$

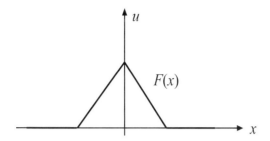

Figure 7.19 Initial signal, a triangular wave.

where A is a constant of integration. Adding and subtracting (4.19) and (4.21) yield the two equations

$$f(x) = \frac{1}{2}F(x) + \frac{1}{2c}\int_0^x G(y)\, dy + \frac{1}{2}A,$$

$$g(x) = \frac{1}{2}F(x) - \frac{1}{2c}\int_0^x G(y)\, dy - \frac{1}{2}A,$$

from which (4.18) follows. Since $F \in C^2$ and $G \in C$, it follows that $u \in C^2$ and satisfies the wave equation. The solution can be verified using Leibniz' rule.

$\square$

Much insight can be gained by examining a simple problem. Therefore let us consider the initial value problem

$$u_{tt} - c^2 u_{xx} = 0, \quad x \in \mathbb{R},\ t > 0,$$
$$u(x,0) = F(x), \quad u_t(x,0) = 0, \quad x \in \mathbb{R}.$$

By (4.18) the solution is

$$u(x,t) = \tfrac{1}{2}(F(x - ct) + F(x + ct)).$$

To fix the idea, suppose $F(x)$ is the initial signal shown in Fig. 7.19. At time $t > 0$ we note that u is the average of $F(x - ct)$ and $F(x + ct)$, which is just the average of the two signals resulting from shifting $F(x)$ to the right ct units and to the left ct units. A sequence of time snapshots at $t_1 < t_2 < t_3$ in Fig. 7.20 shows how the two signals $F(x - ct)$ and $F(x + ct)$ are averaged to produce u. The initial signal splits into two signals moving in opposite directions, each at speed c. The left-moving signal travels along the characteristics $x + ct = $ constant, and the right-moving signal travels along the characteristics $x - ct = $ constant. Therefore, there are two families of characteristic lines along which

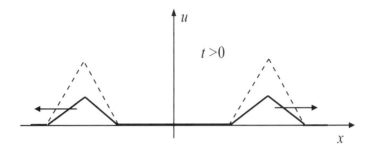

Figure 7.20 Time snapshots showing $u(x,t)$ (solid) as the average of two profiles $F(x+ct)$ and $F(x-ct)$ (dashed).

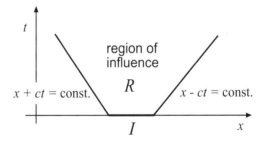

Figure 7.21 Region of influence R of an interval I.

disturbances propagate. If the initial data F and G in (4.17) are supported in an interval I on the x axis (i.e., F and G are zero outside I), then the region R of the xt plane that is affected by the disturbances within I is called the **region of influence** of I (see Fig. 7.21). This region has as its lateral boundaries the two characteristics $x + ct = $ constant and $x - ct = $ constant emanating from the left and right endpoints of I, respectively. A signal in I can never affect the solution outside R; or stated differently, if the initial data are supported in I, then the solution is supported in R. These statements follow directly from (4.18).

We also pose the question of determining which initial values affect the value of u at a given point (x_0, t_0). From D'Alembert's solution,

$$u(x_0, t_0) = \frac{1}{2}(F(x_0 - ct_0) + F(x_0 + ct_0)) + \frac{1}{2c} \int_{x_0 - ct_0}^{x_0 + ct_0} G(y)\, dy.$$

Because of the integrated term, u will depend only on the values in the interval $I \colon [x_0 - ct_0, x_0 + ct_0]$. This interval, called the **interval of dependence**, is found by tracing the characteristics $x - ct = x_0 - ct_0$ and $x + ct = x_0 + ct_0$

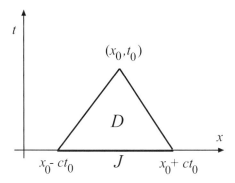

Figure 7.22 Domain D and its interval of dependence J.

back to the initial $t = 0$ line (Fig. 7.22).

When boundary conditions are present the situation is more complicated because waves are reflected from the boundaries. For example, consider the mixed initial boundary value problem

$$u_{tt} - c^2 u_{xx} = 0, \quad 0 < x < l,\ t > 0,$$
$$u(x, 0) = F(x), \quad u_t(x, 0) = G(x), \quad 0 \le x \le l,$$
$$u(0, t) = a(t), \quad u(l, t) = b(t), \quad t > 0,$$

where $F \in C^2[0, l]$, $G \in C^1[0, l]$, and a, $b \in C^2[0, \infty)$. If $u(x, t)$ is a class C^2 solution on $t \ge 0$ and $0 \le x \le l$, then necessarily the compatibility conditions $F(0) = a(0)$, $F(l) = b(0)$, $G(0) = a'(0)$, $g(l) = b(0)$, $a''(0) = c^2 F''(0)$, and $b''(0) = c^2 F(l)$ must hold at the corners $x = 0$, $x = l$, and $t = 0$. This problem can be transformed into one with homogeneous boundary conditions, which can be solved by eigenfunction expansions.

A geometric construction of the solution based on the following theorem gives insight into the nature of the solution to the wave equation.

Theorem 7.18

Let $u(x, t)$ be of class C^3 for $t > 0$ and $x \in \mathbb{R}$. Then u satisfies the wave equation (4.15) if, and only if, u satisfies the difference equation

$$u(x - ck, t - h) + u(x + ck, t + h) = u(x - ch, t - k) + u(x + ch, t + k) \quad (4.22)$$

for all h, $k > 0$.   □

Proof

Necessity follows easily from the fact that the general solution of the wave equation is given by (4.16) and both the forward- and backward-going waves $f(x+ct)$ and $g(x-ct)$ satisfy the difference equation. Conversely, if u satisfies the difference equation, then set $h=0$, add $-2u(x,t)$ to both sides and then divide by $c^2 k^2$ to get

$$\frac{u(x-ck,t)+u(x+ck,t)-2u(x,t)}{c^2 k^2} = \frac{u(x,t-k)+u(x,t+k)-2u(x,t)}{c^2 k^2}.$$
(4.23)

By Taylor's theorem

$$u(x \pm ck, t) = u(x,t) \pm u_x(x,t)ck + \tfrac{1}{2}u_{xx}(x,t)c^2 k^2 + o(k^2),$$
$$u(x, t \pm k) = u(x,t) \pm u_t(x,t)k + \tfrac{1}{2}u_{tt}(x,t)k^2 + o(k^2).$$

Substituting these quantities into (4.23) gives

$$u_{xx}(x,t) = \frac{1}{c^2}u_{tt}(x,t) + \frac{o(k^2)}{k^2}.$$

Taking $k \to 0$ shows that u satisfies (4.15) and completes the proof. $\qquad\square$

Geometrically, the points $A\colon (x-ck, t-h)$, $B\colon (x+ch, t+k)$, $C\colon (x+ck, t+h)$, and $D\colon (x-ch, t-k)$ are the vertices of a **characteristic parallelogram** $ABCD$ formed by two pairs of characteristics, one forward-going pair and one backward-going pair. Then (4.22) may be written

$$u(A) + u(C) = u(B) + u(D).$$

This equation can be used to geometrically construct a solution to the initial boundary value problem as follows. We divide the region $0 < x < l$, $t > 0$ into regions I, II, III ..., as shown in Fig. 7.23, by drawing in the characteristics emanating from the lower corners $(0,0)$ and $(l,0)$ and continually reflecting them from the boundaries. The solution in region I is completely determined by D'Alembert's formula. To find the solution at a point R in region II we sketch the characteristic parallelogram $PQRS$ and use Theorem x to find $u(R) = u(S)+u(Q)-u(P)$. The quantity $u(S)$ is determined by the boundary condition at $x=0$, and $u(P)$ and $u(Q)$ are known from the solution in region I. A similar argument can be made for points in region III. To determine the solution at a point M in region IV we complete the characteristic parallelogram $KLMN$ as shown in the diagram and use Theorem 7.11 to conclude $u(M) = u(L) + u(N) - u(K)$. Again the three quantities on the right are known and we can proceed to the next step and obtain u in the entire region $t > 0$, $0 < x < l$.

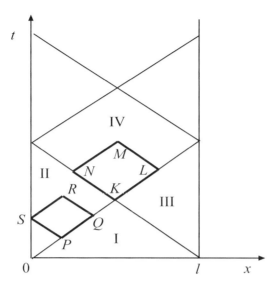

Figure 7.23 Geometrical construction of the solution to the wave equation.

7.4.3 Scattering and Inverse Problems

By a scattering problem we mean the following. Consider a situation where an object, called a **scatterer**, lies in some medium. An incoming *incident wave* impinges on the scatterer and produces a **reflected wave** and a **transmitted wave** (Fig. 7.24). The **direct scattering problem** is to determine the reflected and transmitted waves (amplitude, wave number, frequency, etc.) if the properties of the incident wave and the scatterer are known. The **inverse scattering problem** is to determine the properties of the scatterer, given the incident, reflected, and possibly the transmitted waves. Inverse scattering problems arise in a number of physical situations. For example, in radar or sonar theory a known incident wave and observed reflected wave are used to detect the properties or presence of aircraft or submarine objects. In tomography, X-rays and sound waves are used to determine the presence or properties of tumors by detecting density variations; in this case the incident, reflected, and transmitted waves are all known. In geologic exploration, explosions may be set off on the earth's surface producing waves that, when reflected from underground layers, may indicate the presence of oil or other structures of geologic interest. In nondestructive testing, analysis of reflected and transmitted waves can indicate flaws in a material, for example, the nose cone of a shuttle craft. Inverse scattering theory is an active area of applied mathematics research. In the succeeding paragraphs we set up and solve a simple inverse problem in-

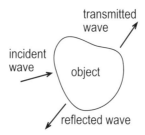

Figure 7.24 Object scattering an incident wave.

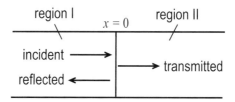

Figure 7.25 The interface between two materials scattering an incident wave.

volving the wave equation. We assume that an infinite bar $-\infty < x < \infty$ is composed of two homogeneous materials separated by an interface at $x = 0$. One material in region I $(x < 0)$ has constant density ρ_1 and stiffness (Young's modulus) y_1 and the other in region II $(x > 0)$ has constant density ρ_2 and stiffness y_2. The sound velocities in region I and region II are c_1 and c_2, respectively. Recall that $c^2 = y/\rho$. The inverse scattering problem can be stated as follows: Suppose an observer at $x = -\infty$ sends out an incident right-traveling wave $\exp(i(x - c_1 t))$ of unit amplitude and unit wave number. When the wave impinges on the interface a wave is reflected back into region I and another wave is transmitted into region II (see Fig. 7.25). If the material properties ρ_1 and y_1 of region I are known, is it possible to determine the properties of ρ_2 and y_2 of region II by measuring the amplitude and wave number of the reflected wave? We answer this question in the context of small displacement theory where the displacements u_1 and u_2 in region I and region II, respectively, satisfy the one-dimensional wave equation

$$u_{tt} - c^2 u_{xx} = 0. \tag{4.24}$$

At the interface $x = 0$ the displacements must be continuous, since we assume that the regions do not separate. Therefore

$$u_1(0^-, t) = u_2(0^+, t), \quad t > 0 \tag{4.25}$$

Here $u_1(0^-, t)$ denotes the limit of $u_1(x, t)$ as $x \to 0^-$ and $u_2(0^+, t)$ denotes the limit of $u_2(x, t)$ as $x \to 0^+$. Further, we require that the force across the interface be continuous or

$$y_1 \frac{\partial u_1}{\partial x}(0^-, t) = y_2 \frac{\partial u_2}{\partial x}(0^+, t), \quad t > 0. \tag{4.26}$$

Equations (4.25) and (4.26) are called the *interface conditions*. We seek u_1 and u_2 satisfying (4.24) subject to (4.25) and (4.26). A reasonable attempt at a solution is to take u_1 to be the incident wave superimposed with the reflected wave and u_2 to be the transmitted wave. Hence we try a solution of the form

$$u_1 = \exp(i(x - c_1 t)) + A \exp(i\alpha(x + c_1 t)),$$
$$u_2 = B \exp(i\beta(x - c_2 t)).$$

The quantities A and α denote the amplitude and wave number of the reflected wave moving at speed c_1 in region I to the left, and B and β denote the amplitude and wave number of the transmitted wave moving at speed c_2 in region II to the right. Clearly u_1 and u_2 satisfy the wave equation with $c_1^2 = y_1/\rho_1$ and $c_2^2 = y_2/\rho_2$, respectively. The inverse scattering problem can now be stated in analytic terms as follows: Given A, α, ρ_1, and y_1, can ρ_2 and y_2 be determined?

The interface conditions (4.25) and (4.26) will yield relations among the constants. Condition (4.25) implies

$$\exp(-ic_1 t) + A \exp(i\alpha c_1 t) = B \exp(-i\beta c_2 t),$$

which, in turn, gives

$$\alpha = -1, \quad \beta = \frac{c_1}{c_2}, \quad B = 1 + A. \tag{4.27}$$

Therefore

$$u_1 = \exp[i(x - c_1 t)] + A \exp[-i(x + c_1 t)],$$
$$u_2 = (1 + A) \exp \left[i \frac{c_1}{c_2}(x - c_2 t) \right].$$

An application of (4.26) forces

$$y_1(1 - A) = y_2(1 + A)\frac{c_1}{c_2}. \tag{4.28}$$

A close examination of (4.27) and (4.28) yields the information we seek. Knowledge of A, y_1, and c_1 permits the calculation of the ratio y_2/c_2 or $\sqrt{y_2 \rho_2}$. Consequently both y_2 and ρ_2 cannot be calculated, only their product. If, however, there is means to measure the speed c_2 of the transmitted wave or determine its wave number β, then the material properties of region II can be found.

The direct scattering problem has a simple solution. If all of the material parameters ρ_1, y_1, ρ_2, and y_2 are known, relations (4.27) and (4.28) uniquely determine the reflected and transmitted waves.

7.4.4 The Schrödinger Equation

In Chapter 3 we introduced the Schrödinger equation, one of the fundamental equations of modern physics. Chapter 5 gave a glimpse of eigenvalue problems associated with the Schrödinger equation, in particular, determining the discrete spectrum (eigenvalues) and eigenfunctions (bound states) for a potential well. Now we have the machinery to discuss some additional mathematical problems arising from quantum mechanics. Our scope is limited; we do not go into the philosophy of quantum mechanics, about which much has been written, and we do not delve into detailed applications and interpretations in atomic physics; these issues are beyond our scope. This section is written for the general applied mathematics reader. There are myriad outstanding texts at all levels that can be consulted for additional understanding. For example, the elementary text by Griffiths (2005) is highly readable; Messiah (1958) has a more advanced perspective.

One problem was mentioned with only a cursory explanation: the *continuous spectrum*. In this section we discuss the meaning in more detail and relate it to scattering phenomena, which we examined in the last section.

We work in one dimension and remind the reader that the fundamental equation of quantum mechanics is the **time-dependent** Schrödinger equation

$$i\hbar \Psi_t = -\frac{\hbar^2}{2m}\Psi_{xx} + V(x)\Psi, \quad x \in \mathbb{R},\ t > 0,$$

where $\Psi = \Psi(x,t)$ is the wave function and $V = V(x)$ is the potential energy. This is a second-order linear PDE that appears, if it does not contain the complex constant i, to be diffusion-like. For the special case of constant V the dispersion relation is

$$\omega = \omega(k) = \frac{-\hbar}{2m}k^2 + \frac{V}{\hbar},$$

which is real; therefore Schrödinger equations are dispersive; we expect wave speeds to depend on wave numbers. Wave numbers, or spatial frequencies, are related to energy levels. A simple separation of variables method, assuming $\Psi(x,t) = g(t)y(x)$, leads to solutions of the form

$$\Psi(x,t) = Ae^{-iEt/\hbar}y(x),$$

where $y = y(x)$ is the solution to the **time-dependent** Schrödinger equation

$$-\frac{\hbar^2}{2m}y'' + V(x)y = Ey,$$

which is a Sturm–Liouville differential equation for the steady-state wave function $y(x)$; the separation constant E is the energy.

There are two interpretations of the wave function Ψ. The first is the *statistical interpretation*. If X_t is a random variable, or random process, for the location of the particle at time t, then the square of the wave function $\Psi^*\Psi = |\Psi|^2$ is the probability distribution[1] for X_t, meaning

$$\Pr(a < X_t \leq b) = \int_a^b |\Psi|^2(x,t)dx = \int_a^b |y(x)|^2 dx.$$

In particular,

$$\Pr(X_t \in \mathbb{R}) = \int_{-\infty}^{\infty} |\Psi|^2 dx = \int_{-\infty}^{\infty} |y(x)|^2 dx = 1.$$

This condition, that the particle is located somewhere, is a normalization condition on wave functions $y(x)$, which are solutions to the singular Sturm–Liouville problem

$$-\frac{\hbar^2}{2m}y'' + V(x)y = Ey, \quad x \in \mathbb{R},$$

$$\int_{-\infty}^{\infty} |y(x)|^2 dx = 1.$$

The discrete spectrum, or set of eigenvalues, determines the allowed energies E. For notational simplicity in the sequel, this model can be written as

$$y'' + U(x)y = \lambda y, \quad x \in \mathbb{R},$$

$$\int_{-\infty}^{\infty} |y(x)|^2 dx = 1,$$

where

$$U(x) = \frac{2m}{\hbar^2}V(x), \quad \lambda = \frac{2m}{\hbar^2}E.$$

It is important to notice that if $\lambda > U(x)$, then we expect oscillatory solutions; if $\lambda < U(x)$ we expect exponential solutions. The first corresponds to a classical particle with energy higher than the potential energy; the second is classically not possible.

Example 7.19

(**Particle in a box**) The simplest case is a free particle (no force and zero potential) of mass m trapped in a finite box $0 < x < a$ with infinite potential

---

[1] In this section we use Ψ^* to denote the complex conjugate of Ψ.

at $x = 0, a$. Therefore the probability of the particle being at $x = 0$ or $x = a$ is zero, and inside the box we have

$$-y'' = \lambda y, \quad x \in (0, a), \quad y(0) = y(a) = 0.$$

This is a regular Sturm-Liouville problem we have solved many times. There are infinitely many positive eigenvalues (energies)

$$\lambda_n = \left(\frac{n\pi}{a}\right)^2, \quad n = 1, 2, 3, ldots$$

with corresponding normalized wave functions (bound states)

$$y_n(x) = \sqrt{\frac{2}{a}} \sin\left(\lambda_n x\right), \quad n = 1, 2, 3, \dots. \quad \square \tag{4.29}$$

This example well illustrates the common statistical interpretation. The particle can exist with only certain discrete energies E_n in bound states y_n. The lowest state corresponding to $n = 1$ is called the **ground state**, and the states for which $n > 1$ are **excited states**. With this interpretation, we can answer the question of the nature of the wave function—it resides in a Hilbert space, $L^2(0, a)$. The probability density functions

$$y_n(x) = \frac{2}{a} \sin^2\left(\lambda_n x\right), \quad n = 1, 2, 3\dots$$

determine the probability of finding the particle in the nth state. Because the wave functions form an orthonormal basis for the Hilbert space, any time-dependent wave function $\Psi(x, t)$ can be written as a linear combination of the basis functions, or

$$\Psi(x, t) = \sqrt{\frac{2}{a}} \sum_{n=1}^{\infty} c_n e^{-iE_n t/\hbar} \sin\left(\lambda_n x\right).$$

If $\Psi(x, 0) = f(x)$ is the wave function at time $t = 0$, then

$$\Psi(x, 0) = f(x) = \sqrt{\frac{2}{a}} \sum_{n=1}^{\infty} c_n \sin\left(\lambda_n x\right),$$

and c_n are Fourier coefficients, or the orthogonal projections of f on the nth eigenspace generated by $y_n(x)$; that is, $c_n = \sqrt{\frac{2}{a}} \int_0^a f(x) \sin\left(\lambda_n x\right) dx$. Thus, the wave function $\Psi(x, t)$, or the wave function-squared, defines how the initial probability density $|f(x)|^2$ for position evolves over time.

Let's briefly review one aspect of the dynamics of a classical particle. Newton's second law for a particle of mass m in a conservative force field having potential $V = V(x)$ leads easily, as we have seen in Chapter 1, to the conservation law

$$y = \pm\sqrt{2/m}\sqrt{E - V(x)}, \quad y = \frac{dx}{dt},$$

where E is the energy. Thus, the particle can only exist in the region where $E \geq V(x)$. Classically, the region $E < V(x)$ is forbidden. The contrast with a quantum mechanical particle is both revolutionary and exciting.

With all this in mind, we now we take up the concept of the continuous spectrum for the Schrödinger equation. This leads to the second interpretation of the wave function. As preparation, we look at the discrete spectrum for an infinite potential well.

Example 7.20

(**Infinite potential well—discrete spectrum**) Assume the potential is given by a delta distribution with pole at zero, $V(x) = -q\delta(x)$. Thus there is an infinitely deep potential well at $x = 0$. We have $V(x) = 0$ for $x \neq 0$, and using the notation in (4.29), the Schrödinger equation becomes

$$y'' + (q\delta(x) + \lambda)\, y = 0, \quad x \in \mathbb{R}. \tag{4.30}$$

We impose the condition $y \in L^2(\mathbb{R})$, which implies

$$\lim_{x \to \pm\infty} y(x) = 0.$$

This is a distributional differential equation and therefore we use the same method as we used in Chapter 5 for constructing a Green's function. There are two cases, $\lambda < 0$ and $\lambda > 0$. To examine the negative energy case we set $\lambda = -k^2$. Then, if $x \neq 0$ we have $y'' - k^2 y = 0$, $x \in \mathbb{R}$. Bounded solutions y are given in $x < 0$ and $x > 0$ by

$$y_L(x) = Be^{kx}, \ x < 0; \quad y_R(x) = Ae^{-kx}, \ x > 0.$$

Continuity at $x = 0$ gives $A = B$. These two decaying solutions connect at the origin with a discontinuity in their first derivatives. To derive the condition we integrate (4.30) over an interval $(-\eta, \eta)$ about the origin and then take the limit as $\eta \to 0$. We obtain

$$\int_{-\eta}^{\eta} y''(x)dx + q \int_{-\eta}^{\eta} \delta(x)y(x)dx - k^2 \int_{-\eta}^{\eta} y(x)dx = 0.$$

By continuity of y the last integral will be zero; the second integral can be calculated by the sifting property of the delta distribution, and the first integral can be found by the fundamental theorem of calculus. We obtain

$$y_R'(0^+) - y_L'(0^-) + qy(0) = -kA - kB + qy(0) = -2k + q = 0.$$

Therefore, $k = \frac{q}{2}$, which gives a *single* eigenvalue, or energy level,

$$\lambda = -\frac{q^2}{4}.$$

We can easily find A by normalization. Here again we get a discrete spectrum, the surprise being that there is only one eigenvalue and one bound state. □

Example 7.21

(**Potential well—continuous spectrum**) We continue with the preceding example, but now assume $\lambda = k^2 > 0$. With the same notation, we obtain

$$y'' + \left(q\delta(x) + k^2\right)y = 0, \quad x \in \mathbb{R}.$$

For $x \neq 0$ we obtain $y'' + k^2 y = 0$, which has sines and cosines as independent solutions. This signals trouble for eigenvalues and bound states, as well as the statistical interpretation of eigenfunctions. If these are the solutions, there can be no way to normalize the wave functions. We can give up now, or we could proceed. Let's proceed and write the solution in complex exponential form as

$$y_L(x) = Ae^{ikx} + Be^{-ikx}, \; x < 0; \quad y_R(x) = Ce^{ikx} + De^{-ikx}, \; x > 0.$$

The solutions are oscillatory and there are no boundary conditions to force any of the constants to be zero. Going ahead, let's apply continuity at $x = 0$, giving

$$A + B = C + D.$$

Now a jump condition. As in the preceding example we integrate the distributional equation over an interval $(-\eta, \eta)$ and take the limit as $\eta \to 0$. We obtain

$$y_R'(0^+) - y_L'(0^-) + qy(0) = 0.$$

Therefore

$$ik(C - D - A + B) = -q(A + B).$$

The result is two equations for the four unknown coefficients and the number k, which is a possible eigenvalue. In summary, we arrived at results that seem to lead nowhere.

Let us step back and ask what we have in terms of scattering theory. From this perspective, the spatial solution $y_L(x)$, for example, in $x > 0$, arises from separating variables and leads to the time-dependent solution

$$\Psi(x, t) = e^{-iEt/\hbar}\left(Ae^{ikx} + Be^{-ikx}\right)$$
$$= Ae^{ik[x - (k\hbar/2m)t]} + Be^{-ik[x + (k\hbar/2m)t]}.$$

This expression is the sum of a right-traveling wave and a left-traveling wave; so, as in scattering, we can think of Ae^{ikx} as an incident right traveling wave and Be^{-ikx} as a reflected wave. Similar comments apply to the solution for $x > 0$. Thus, we can regard the solution as a scattering experiment where a 'stream' of particles coming in from the left $(-\infty)$ is scattered by the presence of the potential well into a reflected wave and a transmitted wave. This means there is no incoming wave from the right $(+\infty)$ and we can set $D = 0$. It makes sense therefore to set $A = 1$ (fix the amplitude of the incoming wave), and so the preceding relations yield

$$1 + B = C, \quad ik(C - 1 + B) = -q(1 + B).$$

It follows that

$$B = -\frac{q}{q + 2ik}, \quad C = \frac{2ik}{q + 2ik}.$$

Our calculation is valid for each value of $k > 0$. For each k the wave function is (see Fig. 7.26)

$$y_k(x) = \begin{cases} e^{ikx} - \frac{q}{q+2ik}e^{-ikx}, & x < 0, \\ \frac{2ik}{q+2ik}e^{ikx}, & x > 0 \end{cases}$$

The spectrum, $\lambda = k^2$, is continuous, and the wave functions do not represent

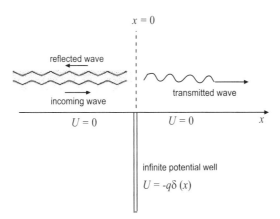

Figure 7.26 Schematic of a scattering solution showing an incoming incident wave and the transmitted and reflected waves from an infinite potential well at the origin.

bound states, but rather scattering states. At this point the whole concept of the actual states of real particles and their probability densities is now in question. □

Example 7.22

(**Reflection–transmission coefficients**) There is an interesting addendum to the result of the last example. If we define

$$\mathcal{R} = |B|^2 = \frac{1}{1 + 4k^2/q^2}, \quad \mathcal{T} = |C|^2 = \frac{1}{1 + q^2/4k^2},$$

then it is easily to show that $\mathcal{R} + \mathcal{T} = 1$, and so $\mathcal{R}$ and $\mathcal{T}$ represent probabilities. These are the reflection and transmission probabilities. A unit amplitude incoming wave has a probability $\mathcal{R}$ of being reflected and a probability $\mathcal{T}$ of being transmitted. Note that the probability of transmission, or penetration, increases with energy $E\ (k)$, and reflection decreases with the energy of the wave. □

We have considered two interpretations of the wave functions: the statistical interpretation and the scattering interpretation. In the latter case we make a further intuitive argument for a particle interpretation in spite of the non-normalizable wave functions and the lack of bound states. We noted earlier that a particle is not localized in space. One way to think about it is in terms of a wave packet, which is a superposition of a group of closely packed plane waves, or scattering solutions, that have nearly the same frequency (energy). In such a case, the wave packet can be written as

$$\Psi(x,t) = \int_{-\infty}^{+\infty} A(k) e^{ik(x - c(k)t)} \, dk,$$

where $A(k)$ represents a narrow envelop of amplitudes depending also on the frequencies, and $c(k)$ is the *phase velocity*. With further analysis (see the references), it can be shown that this entire wave packet travels at a *group velocity* that is twice the phase velocity, which is consistent with the classical particle speed.

In the exercises below we use the notation given in (4.29).

EXERCISES

1. Maxwell's equations for a nonconducting medium with permeability μ and permittivity ε are

$$\frac{\partial \mathbf{B}}{\partial t} + \nabla \times \mathbf{E} = 0, \qquad \varepsilon \frac{\partial \mathbf{E}}{\partial t} = \frac{1}{\mu} \nabla \times \mathbf{B},$$

$$\nabla \cdot \mathbf{B} = 0, \qquad \nabla \cdot \mathbf{E} = 0,$$

where $\mathbf{B}$ is the magnetic induction and $\mathbf{E}$ is the electric field. Show that the components of $\mathbf{B}$ and $\mathbf{E}$ satisfy the three-dimensional wave equation $u_{tt} - c^2(u_{xx} + u_{yy} + u_{zz}) = 0$ with propagation speed $c = (\varepsilon\mu)^{-1/2}$, where $u = u(t,x,y,z)$.

2. Show that the transformation

$$\xi = x - ct, \quad \eta = x + ct,$$

transforms the wave equation (4.15) into the partial differential equation

$$\varphi_{\xi\eta} = 0, \quad \varphi(\xi, \eta) = u(x, t).$$

As a consequence, show that the general solution to (4.15) is given by (4.16).

3. Solve

$$u_{tt} - u_{xx} = 0, \quad 0 < x < \pi, \ t > 0,$$
$$u(0, t) = u(\pi, t) = 0, \quad t > 0,$$
$$u(x, 0) = 0, \quad u_t(x, 0) = 4 \sin x, \quad 0 < x < \pi.$$

4. Solve

$$u_{tt} - u_{xx} = 0, \quad x \in \mathbb{R}, \ t > 0,$$
$$u(x, 0) = \sin x, \quad u_t(x, 0) = 1, \quad x \in \mathbb{R}.$$

5. Let $v \in C^2$ satisfy the boundary value problem

$$v_{tt} - \frac{y_0}{\rho_0} v_{xx} = 0, \quad 0 < x < l, \ t > 0,$$
$$v(x, 0) = v_t(x, 0) = 0, \quad 0 < x < l,$$
$$v(0, t) = v(l, t) = 0, \quad t > 0.$$

a) Prove that $v = 0$ by introducing the **energy integral**

$$E(t) = \int_0^l \left(\tfrac{1}{2} \rho_0 v_t^2 + \tfrac{1}{2} y_0 v_x^2 \right) dx$$

and showing that $E(t) \equiv 0$ for all $t \geq 0$.

b) Prove that solutions to the boundary value problem

$$u_{tt} - c^2 u_{xx} = 0, \quad 0 < x < l, \ t > 0,$$
$$u(x, 0) = F(x), \quad u_t(x, 0) = G(x), \quad 0 < x < l,$$
$$u(0, t) = H(t), \quad u(l, t) = K(t), \quad t > 0,$$

are unique.

6. Let $u = u(x, y)$ be a solution to the boundary value problem in the unit square $\Omega = \{(x, y) : 0 < x, \ y < 1\}$:

$$u_{xy} = 0 \text{ on } \Omega, \qquad u = f \text{ on } \partial\Omega.$$

Show that f must satisfy the relation $f(O) + f(P) = f(Q) + f(R)$, where $O = (0, 0)$, $P = (1, 1)$, $Q = (1, 0)$, and $R = (0, 1)$. Discuss well-posedness for this hyperbolic equation.

7. Use Fourier transforms to find the solution to

$$u_{tt} - c^2 u_{xx} = 0, \qquad x \in \mathbb{R}, \quad t > 0$$
$$u(x, 0) = 0, \qquad u_t(x, 0) = G(x), \qquad x \in \mathbb{R},$$

where G vanishes outside some closed, bounded interval.

8. Consider the boundary value problem

$$u_{tt} - c^2 u_{xx} = 0, \quad t > 0, \quad 0 < x < 1,$$
$$u(x, 0) = F(x), \qquad u_t(x, 0) = G(x), \quad 0 < x < 1,$$
$$u(0, t) = a(t), \qquad u(1, t) = b(t), \quad t > 0.$$

Transform this problem to one with a nonhomogeneous partial differential equation and homogeneous boundary conditions.

9. Consider the nonhomogeneous problem

$$u_{tt} - c^2 u_{xx} = f(x, t), \qquad x \in \mathbb{R}, \ t > 0,$$
$$u(x, 0) = 0, \qquad u_t(x, 0) = 0, \qquad x \in \mathbb{R}.$$

By integrating over the characteristic triangle T bounded by characteristics emanating backward from the point (x_0, t_0) to the x axis, show that

$$u(x_0, t_0) = \frac{1}{2c} \int\!\!\int_T f(x, t) \, dx \, dt.$$

10. Consider a bar with constant stiffness y_0 and density ρ_0 occupying the region $x \leq 0$. An incident small displacement wave $\exp(i(x - c_0 t))$ impinges on $x = 0$ from $x = -\infty$.

   a) Find the reflected wave if the end at $x = 0$ is fixed.

   b) Find the reflected wave if the end at $x = 0$ is free.

   c) Find the reflected wave if a mass M is attached to the end at $x = 0$. (Hint: Show that the boundary condition is $M u_{tt} + y_0 u_x = 0$.)

11. Find the solution to the *outgoing signaling problem*

$$u_{tt} - c^2 u_{xx} = 0, \quad x > 0, \; -\infty < t < \infty.$$
$$u_x(0, t) = Q(t), \quad -\infty < t < \infty.$$

(Hint: There are no boundaries to the right, so attempt a right-traveling wave solution.)

12. A bar of constant cross-sectional area is composed of two materials. For $-\infty < x < 0$ and $1 < x < \infty$ the material parameters are y_0 and ρ_0 and for $0 < x < 1$ the parameters are y_1 and ρ_1. An incident displacement wave $\exp(i(x - c_0 t))$ impinges on the system from $-\infty$. Assuming the linearized theory and the fact that displacements are continuous across the discontinuities, what are the frequency and wave number of the transmitted wave in the region $x > 1$?

13. In three dimensions the wave equation is

$$u_{tt} - c^2 \Delta u = 0,$$

where Δ is the Laplacian. For waves with spherical symmetry $u = u(r, t)$ and $\Delta u = u_{rr} + (2/r)u_r$. By introducing the variable $U = ru$, show that the general solution for the spherically symmetric wave equation is

$$u = \frac{1}{r} f(r - ct) + \frac{1}{r} g(r + ct).$$

14. Show that $u(x, t) = f(x - ct)$ is a weak solution to the wave equation $u_{tt} - c^2 u_{xx} = 0$, $(x, t) \in \mathbb{R}^2$, for any locally integrable function f on $\mathbb{R}$.

15. Use D'Alembert's formula to formally derive the solution to

$$u_{tt} - c^2 u_{xx} = 0, \quad x \in \mathbb{R}, t > 0,$$
$$u(x, 0) = 0, \; u_t(x, 0) = \delta(x - \xi), \quad x, \xi \in \mathbb{R},$$

in the form

$$u(x, t) = \frac{1}{2c}(H(x + ct) - H(x - ct)),$$

where H is the Heaviside function. Comment on the physical meaning.

16. Find an integral representation for the solution to the initial boundary value problem for $u = u(x, y, t)$:

$$u_t = u_{yy}, \quad x, \; y > 0, \; t > 0,$$
$$u(x, y, 0) = 0, \quad x, \; y > 0,$$
$$u(0, 0, t) = u_0(t), \quad t > 0,$$
$$u_t(x, 0, t) = -u_x(x, 0, t) + u_y(x, 0, t), \quad x > 0, \; t > 0$$
$$\lim_{y \to \infty} u(x, y, t) = 0.$$

17. Show that the nonlinear wave equation

$$u_{tt} - u_{xx} + 4u^3 = 0$$

has solutions of the form $u(x, t) = F(\theta)$, where $\theta = kx - \omega t$, $\omega^2 > k^2$, and where F is a periodic function. (Hint: Transform to the FF'-phase plane.) Show that the period is given by

$$P = \sqrt{2(\omega^2 - k^2)} \int_{-a^{1/4}}^{a^{1/4}} \frac{d\vartheta}{\sqrt{a - \vartheta^4}},$$

where a is a constant characterizing an orbit.

18. For a particle in a box (Example 7.19) find the expected value and variance of the position when the particle is in the nth bound state. Plot the density $|y_n(x)|^2$ for the first two states $n = 1, 2$.

19. (Potential barrier) An incoming, right-traveling wave from $-\infty$ encounters, with positive energy E, encounters an infinite potential hill given by a delta function $U(x) = q\delta(x)$. (a) Find the reflected and transmitted wave scattered by the potential, and compute the reflection and transmission coefficients. (b) What is the probability of penetration of the barrier? [This quantum scattering phenomenon is called *tunneling*, which cannot occur for a classical particle with energy less than the maximum potential of the hill, and its application to the electromagnetic properties of materials is far-reaching.]

20. Consider a step potential $U(x) = q > 0$ for $x > 0$ and $U(x) = 0$ for $x < 0$. For positive energies, do you expect the system to have bound states and a discrete spectrum? For energies between 0 and q, find the scattering solution for the Schrödinger equation. What is the continuous spectrum? What is the probability that an incoming wave from $-\infty$ penetrates the barrier?

21. On the region $x \geq 0$ a potential function is given by $U(x) = 0$ for $0 < x \leq 1$, and $U(x) = q > 0$ for $x > 1$. At $x = 0$ the potential is infinite, giving the boundary condition $y(0) = 0$.

   a) Find the discrete spectrum and bound states. [Obtain an equation giving the eigenvalues and show them graphically.] How many eigenvalues are there?

   b) Find the scattering states for an incoming wave from $+\infty$.

22. (Finite potential well) Consider a finite potential well $U(x) = -q$ for $-1 < x < 1$ and $U(x) = 0$ otherwise. Find the discrete spectrum and *even* $(y(-x) = y(x))$ bound states.

REFERENCES AND NOTES

Most books on partial differential equations discuss wave phenomena. The texts listed below are only a small, specialized sample. Whitham (1974) and Logan (2008) contain a large amount of material on linear and nonlinear waves. Brauer and Castillo-Chavez (2001), Murray (2001), and Kot (2001) discuss age structure and partial differential equations. A very readable treatment of quantum mechanics is in Griffiths (2005); Messiah (1958) is a more advanced work. Strauss (1992) is an overall outstanding textbook on partial differential equations, and in particular, wave phenomena. [See Logan (2000) for a review of several books on partial differential equations.]

Brauer F. & Castillo-Chavez, C. 2001. *Mathematical Models in Population Biology and Epidemiology*, Springer-Verlag, New York.

Griffiths, D. J. 2005. *Introduction to Quantum Mechanics*, Pearson Prentice-Hall, Englewood Cliffs, NJ.

Kot, M. 2001. *Elements of Mathematical Ecology*, Cambridge University Press, Cambridge.

Logan, J. D. 2000. Featured review: PDE Books: Present and Future, *SIAM Review* **42**(3), 515-522.

Logan, J. D. 2006. *Applied Partial Differential Equations*, 2nd ed., Springer-Verlag, NY.

Logan, J. D. 2008. *An Introduction to Nonlinear Partial Differential Equations*, 2nd ed., Wiley-Interscience, New York.

Messiah, A. 1958. *Quantum Mechanics*, Vol. I, John Wiley & Sons, New York.

Murray, J. D. 2001. *Mathematical Biology*, Vol 1, Springer–Verlag, New York.

Strauss, W. A. 1992. *Partial Differential Equations*, John Wiley & Sons, New York.

Whitham, G. B. 1974. *Linear and Nonlinear Waves*, Wiley-Interscience, New York.

8
Mathematical Models of Continua

Continuum mechanics (fluid dynamics, elasticity, gasdynamics, etc.) is the study of motion, kinematics, and dynamics of continuous systems of particles such as fluids, solids, and gases. Out of these subjects evolve some of the most important partial differential equations of mathematical physics and engineering and some of the most important techniques for solving applied problems. For example, the origins of singular perturbation theory lie in the study of the flow of air around a wing or airfoil, and many of the problems of bifurcation theory have their beginnings in fluid mechanics. More than any other area of engineering or physics, continuum mechanics remains a paradigm for the development of techniques and examples in applied mathematics.

The field equations, or governing equations, of continuum mechanics are a set nonlinear partial differential equations for unknown quantities such as density, pressure, displacement, particle velocity, and so on, which describe the state of the continuum at any instant. They arise, as in the case of the heat conduction equation from (a) conservation laws such as conservation of mass, momentum, and energy and (b) assumptions regarding the makeup of the continuum, called constitutive relations or equations of state. In certain cases, say, when the amplitude of waves is small, the equations reduce to some of the simpler equations of applied mathematics, like the one-dimensional wave equation.

The equations are developed from the concept that the material under investigation is a continuum; that is, it exhibits no structure however finely it is divided. This development gives rise to a model in which the fluid parameters are defined at all points as continuous functions of space and time. The molec-

Applied Mathematics 4th ed.,
By J. David Logan

ular aggregation of matter is completely extraneous to the continuum model. The model may be expected to be invalid when the size of the continuum region of interest is the same order as the characteristic dimension of the molecular structure. For gases this dimension is on the order of 10^{-7} meters, which is the mean free path, and for liquids it is on the order of 10^{-10} meters, which is a few intermolecular spacings. Thus the continuum model is violated only in extreme cases.

In many ways, as we will see, describing the motion of solids requires a different strategy from that in fluids and gases. For example, we often work with displacement in solids, but velocity in fluids and gases. Moreover, models of the forces and stresses in solids lead to complicated constitutive equations that take the subject to a higher level and into subjects such as elasticity, viscoelastic materials, material with memory, and so on. Our main focus in this text is the kinematics and dynamics of gases and fluids. Finally, continuum mechanics, perhaps more than other areas of applied mathematics, requires a good knowledge of several areas of physics, particularly mechanics and thermodynamics. And, the properties of materials and their behavior, now coined *material science*, spans into electrodynamics, quantum and atomic physics, and other areas of modern physics.

8.1 Kinematics and Mass Conservation

8.1.1 Description of Flow

The continuum assumption. The underlying physical model to which we apply the general ideas is a one-dimensional continuum that can be thought of as a solid cylindrical tube, or a cylinder through which a fluid is flowing. The fluid may be a liquid or gas and the lateral wall of the pipe is assumed to have no effect on the flow parameters. Throughout the motion we assume each cross-section remains planar and moves longitudinally down the cylinder with no variation of any of the flow parameters in any cross section. It is this assumption, that the only variation is longitudinal, that gives the description a one-dimensional character. The situation is the same as in the one-dimensional models studied in Chapters 6 and 7.

To understand how physical quantities are defined we consider a thought experiment to observe how the density of a fluid is related to its molecular structure. At time t we consider a disk D of fluid of width α centered at x_0 and having cross-sectional area A. See Fig. 8.1. The average density of the fluid in D is $\rho_\alpha = M_\alpha / \alpha A$, where M_α is the mass of fluid in D. To define the

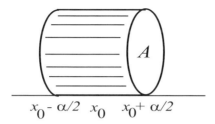

Figure 8.1 A fluid disk D with width α, centered at x_0.

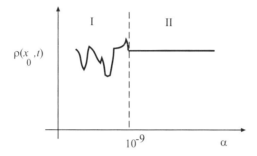

Figure 8.2 Plot of the average density ρ_α vs. the disk of width α (meters).

density $\rho(x_0, t)$ we examine what happens as α approaches zero. The graph in Fig. 8.2 records the results of the experiment. In region II, because there are many particles (molecules) inside D, one would expect that the average density ρ_α would vary very little. If, however, α were on the order of molecular distances, say $\alpha = 10^{-9}$ meters, there may be only a few molecules in D and we might expect large fluctuations in ρ_α even for small changes in α. Such rapid fluctuations are depicted in region I. It seems unreasonable, therefore, to define $\rho(x_0, t)$ as the limiting value of ρ_α as $\alpha \to 0$. Rather, $\rho(x_0, t)$ should be defined as

$$\rho(x_0, t) = \lim_{\alpha \downarrow \alpha^*} \rho_\alpha,$$

where α^* is the value of α where density nonuniformities begin to occur; here, for example, $\alpha^* = 10^{-9}$ meters. In a similar fashion other physical quantities can be considered as point functions in continuously distributed matter without regard to its molecular or atomistic structure. This continuum assumption is often phrased as follows—the fluid is composed of small regions or **fluid elements** that can be idealized as points at which the flow variables become continuous functions of position and time, but they are not so small that discernable fluctuations of these quantities exist within the element.

Kinematics of motion. We now discuss the *kinematics* of fluid motion, meaning how motion is described. *Dynamics*, or the cause of motion, comes later. In a one-dimensional moving fluid or solid there are two coordinate systems used to keep track of the motion. For example, suppose water is flowing in a stream and one wishes to measure the temperature. One can stand on the bank at a fixed location x, measured from some reference position $x = 0$, and insert a thermometer, thereby measuring the temperature $\theta(x, t)$ as a function of time and the *fixed* position x. The coordinate x is called a spatial (laboratory) or **Eulerian coordinate**. On the other hand, one can measure the temperature from a boat riding with the flow. In this case, at time $t = 0$ each particle (section) is labeled with a particle label h, and each particle always retains its label as it moves downstream. The result of the measurement is the temperature $\Theta(h, t)$ as a function of time t and the **Lagrangian** or **material coordinate** h. The variable x is a fixed spatial coordinate, and the representation of the field functions or physical variables (temperature, density, pressure, etc.) as functions of t and x gives a **Eulerian description** of the flow. A representation in terms of t and the material variable h gives a *Lagrangian description* of the flow. For physical variables we reserve lowercase letters for Eulerian quantities and capital letters for Lagrangian quantities. Thus, $f(x, t)$ denotes the measurement of a physical quantity in Eulerian coordinates, and $F(h, t)$ denotes the corresponding description in terms of Lagrangian coordinates.

Now let I be an interval along the axis of the cylinder where the fluid begins its motion. At time $t = 0$ we label all of the particles.[1] with a Lagrangian coordinate h such that

$$x = h \quad \text{at} \quad t = 0,$$

where x is a fixed Eulerian coordinate in I. By a **fluid motion** or **flow**, we mean a twice continuously differentiable mapping $\phi \colon I \times [0, t_1] \to \mathbb{R}$ defined by

$$x = \phi(h, t), \tag{1.1}$$

which is invertible for each $t \in [0, t_1]$. Rather than write ϕ in (1.1), we use the symbol x and write

$$x = x(h, t). \tag{1.2}$$

The convention of using x to denote both a spatial coordinate and a function is common and often preferred over (1.1). In words, (1.2) gives the position x at time t of the particle, or cross section, labeled h. For a fixed $h = h_0$ the curve

$$x = x(h_0, t), \quad 0 \le t \le t_1$$

defines the **particle path** of the particle labeled h_0.

---

[1] In a one-dimensional continuum the word *particle* is synonymous with *cross section*, and we use the terms interchangeably.

The invertibility assumption on ϕ guarantees that (1.2) can be solved for h to obtain

$$h = h(x, t). \tag{1.3}$$

This equation determines the particle, or cross section, at position x at time t. Since (1.2) and (1.3) are inverses,

$$x = x(h(x, t), t), \tag{1.4}$$

and

$$h = h(x(h, t), t). \tag{1.5}$$

Consequently, if $f(x, t)$ is a physical quantity in Eulerian form, then

$$F(h, t) = f(x(h, t), t) \tag{1.6}$$

gives the Lagrangian description. Conversely, if $F(h, t)$ is a Lagrangian quantity, then

$$f(x, t) = F(h(x, t), t) \tag{1.7}$$

gives its description in Eulerian form. The **duality principle** expressed by (1.6) and (1.7) is physically meaningful; for example, (1.7) states that the Eulerian measurement f made by an individual at x at time t coincides with the measurement F made by one moving on the particle h, at the instant t, when h is at x.

The chain rule shows that the derivatives are related by

$$F_h(h, t) = f_x(x(h, t), t) x_h(h, t) \tag{1.8}$$

and

$$F_t(h, t) = f_t(x(h, t), t) + f_x(x(h, t), t) x_t(h, t), \tag{1.9}$$

where $f_x(x(h, t), t)$ means $f_x(x, t)$ evaluated at $x = x(h, t)$, for example. Good notation in fluid dynamics is essential. Here we have indicated explicitly the points at which the derivatives are evaluated. Shortcuts can sometimes lead to confusing and little understood expressions and equations, and writing out the general formulas in detail at least once can dismiss much of the confusion.

We are now prepared to introduce displacement, velocity, and acceleration, the key kinematical descriptors in classical mechanics. If $x = h$ at $t = 0$, the **displacement** $U = U(h, t)$ of the section h is defined by

$$U(h, t) = x(h, t) - h.$$

That is, displacement is where it is now minus where it was at $t = 0$. This is a Lagrangian definition. The Eulerian form of displacement can be found by substituting $h = h(x, t)$:

$$u(x, t) = x - h(x, t).$$

As in classical mechanics we define the **velocity** of a particle or cross section h as the time rate of change of its position, or

$$V(h,t) = x_t(h,t),$$

which is the same as $V(h,t) = U_t(h,t)$. The Eulerian form of the velocity is defined by

$$v(x,t) = V(h(x,t),t),$$

which gives the velocity of the cross section now at location x. The **acceleration** of the cross section labeled h is

$$A(h,t) = V_t(h,t) = x_{tt}(h,t).$$

Therefore, the acceleration of the cross section now at location x is given in Eulerian form by

$$a(x,t) = V_t(h(x,t),t) = V_t(h,t)|_{h(x,t)}. \tag{1.10}$$

From (1.9) it follows that

$$a(x,t) = v_t(x,t) + v(x,t)v_x(x,t).$$

The time rate of change of a physical quantity following a cross section or particle h that is now located at x, such as occurs on the right side of (1.10), is called the material derivative of that quantity. Precisely, the **material derivative** of $f(x,t)$ is defined by

$$\frac{Df}{Dt}(x,t) \equiv F_t(h(x,t),t) = F_t(h,t)|_{h=h(x,t)}, \tag{1.11}$$

where on the right we have explicitly noted that the derivative is taken before the evaluation at $h(x,t)$ is made. The quantity Df/Dt is interpreted as the time derivative of F following the cross section h, frozen at the instant h is located at x. From (1.9) it easily follows that

$$\frac{Df}{Dt} = f_t + vf_x. \tag{1.12}$$

Therefore, in Eulerian coordinates, the acceleration can be written

$$a = \frac{Dv}{Dt}.$$

Another kinematic relation that plays an important role in the derivation of the governing equations is the time rate of change of the Jacobian of the mapping ϕ defining the flow. The **Jacobian** is defined by

$$J(h,t) \equiv x_h(h,t).$$

Hence

$$J_t(h,t) = \frac{\partial}{\partial h} x_t(h,t) = V_h(h,t) = v_x(x(h,t),t)x_h(h,t),$$

or

$$J_t(h,t) = v_x(x(h,t),t)J(h,t). \tag{1.13}$$

Equation (1.13) is the **Euler expansion formula**. To obtain its Eulerian form, we let $h = h(x,t)$, so

$$\frac{Dj}{Dt}(x,t) = v_x(x,t)j(x,t),$$

where $j(x,t) = J(h(x,t),t)$.

Example 8.1

At $t = 0$ let $0 \le h \le 1$ and define a fluid motion $x(h,t) = (1+t^2)h$. The inverse is $h(x,t) = x/(1+t^2)$. The displacements are

$$U(h,t) = x(h,t) - h = ht^2, \quad u(x,t) = x - h(x,t) = \frac{xt^2}{1+t^2}.$$

Where are the cross sections when $t = 2$? Clearly, $0 \le x \le 5$. The velocities are

$$V(x,t) = x_t(h,t) = 2th, \quad v(x,t) = V(h(x,t),t) = 2t\frac{x}{1+t^2}.$$

If $\theta(x,t) = xt^3$ is the laboratory temperature measured at time t, then the Lagrangian temperature, riding along on section h, is

$$\Theta(h,t) = \theta(x(h,t),t) = x(h,t)t^3 = (1+t^2)t^3 h.$$

The Eulerian acceleration can be computed, for example, by the material derivative, that is,

$$a(x,t) = \frac{Dv}{Dt} = v_t(x,t) + v(x,t)v_x(x,t)$$

$$= \frac{1-t^2}{(1+t^2)^2}2x + \frac{2xt}{1+t^2}\frac{2t}{1+t^2} = \frac{2x}{1+t^2}.$$

Of course, the latter could be computed easier using $A(h,t) = V_t(h,t) = 2h$, so that

$$a(x,t) = A(h(x,t),t) = 2\frac{x}{1+t^2}.$$

Finally, the particle paths are the family of parabolas $x = (1+t^2)h$ in the tx plane, where h is a parameter with $0 \le h \le 1$.   □

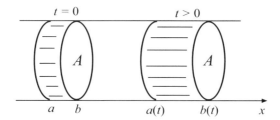

Figure 8.3 A material quantity of fluid at $t = 0$ and at $t > 0$.

Remark 8.2

Finally, for reference we record a few more kinematical relationships that are useful in the sequel. Let $f(x, t)$ be a Eulerian measurement with corresponding Lagrangian measurement $F(h, t)$. Then

1. $F_h = f_x x_h = f_x (1 + U_h)$.

2. $U_h = \frac{u_x}{1 - u_x}$.

3. $F_h = \frac{1}{1 - u_x} f_x$.

4. $F_t = \frac{V}{1 + U_h} F_h = f_x$.

5. $J(h, t) = 1 + U_h$.

6. $v = \frac{u_t}{1 - u_x}$.

The proofs of these identities can be carried out by the chain rule and are left as exercises. $\square$

8.1.2 Mass Conservation

The **field equations** that govern the motion of a one-dimensional continuous medium express conservation of mass, momentum, and energy, and they are universal in that they are valid for any medium. In this section we derive the first of these, mass conservation.

The derivation is based on a Lagrangian approach. First, we find the analytic implications of the fact that the mass in an arbitrary material portion of the cylinder at $t = 0$ does not change as that portion of material moves in time. At time $t = 0$ we consider an arbitrary portion of fluid between $x = a$ and $x = b$ (see Fig. 8.3), and after time t we suppose that this portion of fluid has moved to the region between $x = a(t) \equiv x(a, t)$ and $x = b(t) \equiv x(b, t)$. If $\rho(x, t)$ denotes the density of the fluid, then the amount of fluid between $a(t)$ and

$b(t)$ is

$$\int_{a(t)}^{b(t)} \rho(x,t) A \, dx.$$

Mass conservation requires

$$\frac{d}{dt} \int_{a(t)}^{b(t)} \rho(x,t) \, dx = 0. \tag{1.14}$$

Using Leibniz's formula for differentiating integrals with variable limits, we could proceed directly and compute the left side of (1.14) to get a local form of the law; but we save this as an exercise. We prefer instead a method that easily generalizes to higher dimensions and is more in the spirit of techniques in fluid dynamics. We change variables in (1.14) according to $x = x(h,t)$. Then $dx = J(h,t) \, dh$, and (1.14) becomes

$$\frac{d}{dt} \int_a^b R(h,t) J(h,t) \, dh = 0, \tag{1.15}$$

where $R(h,t)$ (read as an uppercase ρ) is the Lagrangian density defined by $R(h,t) = \rho(x(h,t),t)$. Note that (1.14) has been changed to an integral with constant limits so that the derivative may be brought directly under the integral because R and J are continuously differentiable. Consequently,

$$\int_a^b (RJ_t + JR_t) \, dh = \int_a^b [R(h,t) v_x(x(h,t),t) + R_t(h,t)] J \, dh = 0,$$

where we used the Euler expansion formula (1.13). Because a and b are arbitrary and J is nonzero, we conclude that

$$R_t(h,t) + R(h,t) v_x(x(h,t),t) = 0, \tag{1.16}$$

from which we get

$$R_t + \frac{R}{J} V_h = 0,$$

or

$$R_t + \frac{R}{1 + U_h} V_h = 0. \tag{1.17}$$

This is the **conservation of mass** law in Lagrangian form. The Eulerian form can be obtained directly from (1.16) by substituting $h = h(x,t)$. We get

$$\frac{D\rho}{Dt} + \rho v_x = 0, \tag{1.18}$$

or

$$\rho_t + v \rho_x + \rho v_x = 0. \tag{1.19}$$

Equation (1.18), or (1.19), is known as the **continuity equation**, and it is one of the fundamental equations of one-dimensional flow. It is a first-order nonlinear partial differential equation in terms of the density ρ and velocity v. The corresponding Lagrangian form (1.17) is an equation involving the density R and displacement U. (Note $V = V_t$.) To close the system we need an additional equation, and that comes from momentum balance. We record the main result:

$$\left. \begin{array}{l} \rho_t + v\rho_x + \rho v_x = 0 \\ R_t + \frac{R}{1+U_h} V_h = 0 \end{array} \right\} \quad \text{(Mass conservation)}$$

Remark 8.3

We can easily derive an alternate form of mass conservation in material coordinates by writing (1.17) in the form

$$(1 + U_h)R_t + RU_{ht} = [(1 + U_h)R]_t = 0,$$

and integrating to get

$$(1 + U_h)R = R(h, 0) \quad \square.$$

Special types of flows can be identified with corresponding simplifying assumptions.

Definition 8.4

A fluid motion is said to be **incompressible** if

$$\frac{D\rho}{Dt} = 0 \quad \text{or} \quad R_t = 0.$$

A fluid motion is **steady** if ρ and v are independent of t. $\quad \square$

If the flow is incompressible, then it easily follows that $v_x = 0$, or $V_h = 0$. For steady flows, $(\rho v)_x = 0$, or $V_h = 0$.

EXERCISES

1. A one-dimensional flow is defined by

$$x = (h - t)(1 + ht)^{-1}, \quad 0 < t < t_1.$$

a) Sketch the particle paths on xt and ht diagrams.

b) Find $V(h, t)$ and $v(x, t)$.

c) Verify the Eulerian expansion formula.

d) Verify that $Dv/Dt = v_t + vv_x$.

e) If the density is $R(h,t) = h^2$, find the density that an observer would measure at $x = 1$.

2. Repeat (a) through (e) for a fluid motion defined by $x = \frac{1}{2}h(1 + e^t)$.

3. Repeat (a) through (e) for a fluid motion defined by $x = \frac{2t+1}{t+1}h$.

4. Use the chain rule to verify (1.8) and (1.9).

5. Show that $R_t = \frac{R(h,0)}{1+U_h}$.

6. For a valid flow, state why $u_x < 1$ and $U_h > -1$.

7. Show that the material derivative operator D/Dt is additive and satisfies the product rule. Show that

$$\frac{D}{Dt}\psi(f) = \psi'(f)\frac{Df}{Dt}.$$

8. The reciprocal W of the Lagrangian density R is called the **specific volume**, or volume per mass. If at time $t = 0$ the density is a constant, $R(h,0) = \rho 0$, prove that conservation of mass and momentum can be expressed as

$$W_t - \rho_0^{-1}V_h = 0,$$
$$V_t + \rho_0^{-1}P_h = 0.$$

9. Show that the Jacobian $J(h,t)$ is given by

$$J(h,t) = \frac{R(h,0)}{R(h,t)},$$

and therefore measures the ratio of the initial density to the density at time t. Interpret the meaning of the Jacobian when it is written in terms of the specific volume,

$$J(h,t) = \frac{W(h,t)}{W(h,0)}.$$

10. Derive the continuity equation (1.19) directly by calculating

$$\frac{d}{dt}\int_{a(t)}^{b(t)} \rho(x,t)\,dx$$

using Leibniz' formula.

11. Prove that

$$\frac{d}{dt} \int_{a(t)}^{b(t)} \rho(x,t) g(x,t)\, dx = \int_{a(t)}^{b(t)} \rho(x,t) \frac{Dg}{Dt}(x,t)\, dx$$

for any sufficiently smooth function g, where $a(t) \leq x \leq b(t)$ is a material region.

12. Write the differential equation $F_t + F^2 F_h = 0$, where $F = F(h,t)$, in Eulerian form.

8.2 Momentum and Energy

8.2.1 Momentum Conservation

In classical mechanics, a particle of mass m having velocity $\mathbf{v}$ has linear momentum $m\mathbf{v}$. Newton's second law asserts that the time rate of change of momentum of the particle is equal to the net external force acting upon it. To generalize this law to one-dimensional continuous media we assume the *balance of linear momentum principle*, which states that the time rate of change of linear momentum of any portion of the fluid equals the sum of the external forces acting upon it. The **linear momentum** at time t of material in the material region $a(t) \leq x \leq b(t)$ is defined by

$$\mathbf{i}A \int_{a(t)}^{b(t)} \rho(x,t) v(x,t)\, dx,$$

where $\mathbf{i}$ is the unit vector in the positive x direction, $v(x,t)$ is the velocity, and A is the cross-sectional area.

The precise characterization of the forces acting on a material region in a continuous medium was the result of ideas evolving from works of Newton, Euler, and Cauchy. Basically, there are two types of forces that act on the material region, body forces and surface forces. **Body forces** are forces such as gravity, or an electric or magnetic field. Such a force is assumed to act on each cross section of the region and it is represented by

$$\mathbf{f}(x,t) = f(x,t)\mathbf{i}.$$

The units of f are force per unit mass, and the total body force acting on the region $a(t) \leq x \leq b(t)$ is therefore

$$\mathbf{i}A \int_{a(t)}^{b(t)} \rho(x,t) f(x,t)\, dx.$$

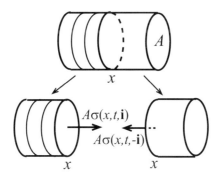

Figure 8.4 Stress vectors.

Surface forces are forces like pressure that act across sections in the fluid medium. More specifically, consider a cross section at time t located at x. By $\sigma(x,t,\mathbf{i})$ we denote the force per unit area on the material *on* the negative (left) side of the cross section *due to* the material on the positive (right) side of the section. Similarly $\sigma(x,t,-\mathbf{i})$ denotes the force per unit area exerted on the material *on* the right side of the section *by* the material on the left side of the section. By convention, the third argument in σ, here either $\mathbf{i}$ or $-\mathbf{i}$, is a vector that points outward and normal from the surface on which the force is acting. The vectors $\sigma(x,t,\mathbf{i})$ and $\sigma(x,t,-\mathbf{i})$ are called surface tractions or **stress vectors**. Figure 8.4 depicts a geometric description of these notions. Before a cross section is indicated, no stress is defined; however, once a cross section x is indicated, the forces $A\sigma(x,t,\mathbf{i})$ and $A\sigma(x,t,-\mathbf{i})$, which are exerted on the shaded and unshaded portions, respectively, are defined. In anticipation of what we prove later, we have drawn these forces opposite and equal in Fig. 8.4. At present it is not known in which direction these point. And we emphasize that the third argument in σ, either $\mathbf{i}$ or $-\mathbf{i}$, does not define the direction of the stress but serves only to indicate the orientation, or normal direction, of the surface of the section. We can now write a quantitative description of the balance of the linear momentum principle for a material region $a(t) \leq x \leq b(t)$,

$$\mathbf{i}\frac{d}{dt}A\int_{a(t)}^{b(t)} \rho(x,t)v(x,t)\ dx$$

$$= \mathbf{i}A\int_{a(t)}^{b(t)} \rho(x,t)f(x,t)\ dx + A\sigma(b(t),t,\mathbf{i}) + A\sigma(a(t),t,-\mathbf{i}). \qquad (2.1)$$

In words, the time rate of change of momentum of the material region $[a(t),b(t)]$ equals the total body force plus the surface forces on $[a(t),b(t)]$. A schematic of the forces is shown in Fig. 8.5. To put (2.1) in a more workable

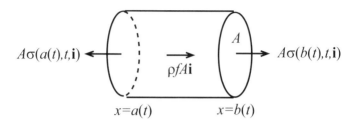

Figure 8.5   Forces on a fluid element.

form we calculate the left side in a similar manner as was done for the integral
form of the conservation of mass equation in the preceding section. Changing
variables to the Lagrangian coordinate h and using the Euler expansion formula
gives

$$\frac{d}{dt}\int_{a(t)}^{b(t)} \rho v \, dx = \frac{d}{dt}\int_{a}^{b} RVJ \, dh = \int_{a}^{b}\left(J\frac{\partial}{\partial t}(RV) + RVJ_t\right) dh$$

$$= \int_{a}^{b}\left[\frac{\partial}{\partial t}(RV) + RVv_x\right]J \, dh = \int_{a(t)}^{b(t)}\left[\frac{D}{Dt}(\rho v) + \rho v v_x\right] dx.$$

From the mass conservation equation (1.18) it follows that

$$\frac{D}{Dt}(\rho v) = \rho\frac{Dv}{Dt} + v\frac{D\rho}{Dt} = \rho\frac{Dv}{Dt} - v\rho v_x.$$

Hence,

$$\frac{d}{dt}\int_{a(t)}^{b(t)} \rho v \, dx = \int_{a(t)}^{b(t)} \rho\frac{Dv}{Dt} \, dx, \tag{2.2}$$

and (2.1) can be written

$$\mathbf{i}\int_{a(t)}^{b(t)}\left(\rho\frac{Dv}{Dt} - \rho f\right) dx = \sigma(b(t), t, \mathbf{i}) + \sigma(a(t), t, -\mathbf{i}). \tag{2.3}$$

This equation can be further simplified using the following result, which is
the continuum mechanical version of the action–reaction principle expressed in
Newton's third law.

Theorem 8.5

The balance of linear momentum principle expressed by (2.3) implies

$$\sigma(x, t, -\mathbf{i}) = -\sigma(x, t, \mathbf{i}) \tag{2.4}$$

for any cross section x. $\square$

Proof

Let x_0 be arbitrary with $a(t) < x_0 < b(t)$. Applying (2.3) to the region between $a(t)$ and x_0 and to the region between x_0 and $b(t)$, we get

$$\mathbf{i} \int_{a(t)}^{x_0} \left[\rho \frac{Dv}{Dt} - \rho f \right] dx = \sigma(x_0, t, \mathbf{i}) + \sigma(a(t), t, -\mathbf{i}),$$

and

$$\mathbf{i} \int_{x_0}^{b(t)} \left[\rho \frac{Dv}{Dt} - \rho f \right] dx = \sigma(b(t), t, \mathbf{i}) + \sigma(x_0, t, -\mathbf{i}).$$

Adding these two equations and then subtracting (2.3) from the result gives

$$\sigma(x_0, t, \mathbf{i}) + \sigma(x_0, t, -\mathbf{i}) = 0,$$

which, because of the arbitrariness of x_0, proves (2.4). □

This result permits us to define the scalar **stress** component $\sigma(x, t)$ by the equation

$$\sigma(x, t, \mathbf{i}) = \sigma(x, t)\mathbf{i}.$$

Therefore (2.3) becomes

$$\int_{a(t)}^{b(t)} \left(\rho \frac{Dv}{Dt} - \rho f \right) dx = \sigma(b(t), t) - \sigma(a(t), t)$$

$$= \int_{a(t)}^{b(t)} \sigma_x(x, t) \, dx.$$

Because the interval $[a(t), b(t)]$ is arbitrary, we have

$$\rho \frac{Dv}{Dt} = \rho f + \sigma_x, \tag{2.5}$$

which is the Eulerian form expressing balance of linear momentum.

It is straightforward to write the momentum equation in Lagrangian coordinates. We use $T(h, t)$ to denote the Lagrangian stress; so, $T(h, t) = \sigma(x(h, t), t)$. With appropriate evaluation of the terms, equation (2.5) is clearly

$$RV_t = RF + T_h h_x,$$

or

$$RV_t = RF + T_h \frac{1}{1 + U_h}.$$

Now we use mass conservation in the form (8.3) to get

$$R(h, 0)U_{tt} = R(h, 0)F + T_h. \tag{2.6}$$

In summary, the **momentum equations** are:

$$\left.\begin{array}{l} \rho\frac{Dv}{Dt} = \rho f + \sigma_x \\ R(h,0)U_{tt} = R(h,0)F + T_h \end{array}\right\} \qquad \text{(Momentum conservation)}$$

Conservation of mass and momentum are valid in a general sense, but they do not account for the differences in materials, e.g., gases, fluids, solids, etc. And, as yet, they are under-determined with more variables than equations. This brings us to a discussion of constitutive relations, or equations that specify the stresses in a material.

Example 8.6

(**Barotropic gas**) A familiar and simple form of the equations can be written if we take the stress to be $\sigma(x,t) = -p(x,t)$, where p is the **pressure**. Then the momentum equation (2.5) with no body forces becomes

$$v_t + vv_x + \frac{1}{\rho}p_x = 0. \tag{2.7}$$

If we further impose that the pressure is a function of density alone, $p = F(\rho)$, the mass and momentum equations become

$$\rho_t + v\rho_x + \rho v_x = 0$$
$$v_t + vv_x + \frac{1}{\rho}F'(\rho)\rho_x = 0,$$

which are the the the equations for **barotropic flow**. Upon linearization about a uniform state, these equations lead to the acoustical equations, and ultimately the wave equation. □

8.2.2 Stress Waves in Solids

The general conservation laws for mass and momentum in continuous media also hold for solid mechanics. Our physical model is a solid cylindrical bar where we seek to describe longitudinal vibrations, that is, the motion of the planar cross sections of the bar. Because we want to keep track of the displacements of these sections, we write the governing equations in Lagrangian form with displacement as a dependent quantitiy, or

$$(1 + U_h)R = R(h,0), \tag{2.8}$$
$$R(h,0)U_{tt} = R(h,0)F + T_h. \tag{2.9}$$

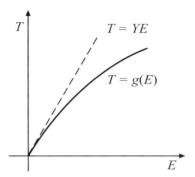

Figure 8.6 Stress T vs. strain E. The strain is $E = U_h$ and Y is Young's modulus.

To close the system we need a constitutive relation for the stress $T(h,t)$ in terms of the remaining dependent variables. Unfortunately, this greatly depends on the type of material we are considering, e.g., a metal, a crystal lattice, an elastomer with interwoven polymer chains, a viscoelastic material, and so on. The internal forces differ tremendously in these materials.

Here we focus on elastic materials and attempt to formulate a constitutive assumption relating the distortion (compression or elongation) the bar undergoes subject to an applied stress. To define the distortion we consider at time $t = 0$ a small portion of the bar between h and $h + \Delta h$. At time $t > 0$ this material is located between $x(h,t)$ and $x(h + \Delta h, t)$. The **distortion** is the fractional change given by

$$
\text{distortion} = \frac{\text{new length} - \text{original length}}{\text{original length}}
$$
$$
= \frac{x(h + \Delta h, t) - x(h,t) - \Delta h}{\Delta h}
$$
$$
= \frac{\partial x}{\partial h}(h,t) - 1 + O(\Delta h).
$$

We define the **strain** E to be the lowest-order approximation of the distortion, that is,

$$
E \equiv \frac{\partial x}{\partial h}(h,t) - 1.
$$

If $U(h,t)$ denotes the displacement of a cross section h at time t, then $x(h,t) = h + U(h,t)$ and

$$
E = \frac{\partial U}{\partial h}(h,t). \tag{2.10}
$$

The constitutive relation is an equation that gives the Lagrangian stress $T(h,t)$ as a definite function of the strain $E(h,t)$. Quite generally, the graph may look like the stress–strain curve $T = g(E)$ shown in Fig. 8.6. If, however, only small strains are of interest, then the stress–strain curve can be approximated by a straight line by a linear equation

$$T(h,t) = Y(h)E(h,t). \tag{2.11}$$

The constant proportionality factor Y is called **Young's modulus**[2] or the *stiffness*, and the linear **stress–strain relation** (2.11) is called **Hooke's law**,[3]. We are assuming that Y does not depend on time t; time-dependent materials are said to have *memory*. With this assumption, the momentum equation (2.9) becomes

$$R_0(h)U_{tt} = R_0(h)F + (Y(h)U_h)_h, \quad R_0(h) = R(h,0), \tag{2.12}$$

which is a one-dimensional wave equation for the displacement U. Once this equation is solved for U, then the mass balance equation (2.8) gives the density $R(h,t)$.

Generally, a rod of finite extent, $0 \le h \le l$, is subject to boundary conditions at its ends $h = 0$, $h = l$. For definiteness we focus on the end $h = l$. Clearly if the end is held fixed, then

$$U(l,t) = 0, \quad t > 0 \qquad \text{(fixed end)}.$$

If the end is free, or no force acts on the face at $h = l$, then $T(l,t) = 0$ or $Y(l)E(l,t) = 0$, or

$$Y(l)\frac{\partial U}{\partial h}(l,t) = 0, \quad t > 0, \quad \text{(free end)}.$$

Other conditions are possible. For example, if the left end is fixed, for example, attached to a rigid wall, and the right end is attached to a Hookean spring that is attached to a wall, having spring constant k, then the boundary conditions are

$$U(0,t) = 0, \quad Y(l)U_h(l,t) = -\frac{k}{A}U(l,t),$$

where A is the area of the face, and k is given in force per length.

---

[2] Thomas Young (1733–1829) was an English medical doctor and scientist.
[3] Robert Hooke (1635–1703), FRS, was an English natural philosopher, architect, polyscientist, and mathematician who, in his time, was considered one of the premier English scientists.

Remark 8.7

(**Units**) From a practical viewpoint, we note that Young's modulus is given in GPa (giga Pascals), an mks unit for stress. A Pascal is one Newton per meter-squared, and the prefix *giga* means 10^9. So, Young's modulus is large. For example, common materials are diamond (1000 GPa), stainless steel (200 GPa), and rubber (0.007 GPa). □

Example 8.8

(**Elongation of an elastic cord**) Consider a long, thin elastic cord (say, a bungee cord) that has natural length L_0, constant modulus Y, and constant density R_0. Then the cord is hung vertically and stretches under the force of gravity g. How much does it stretch? We wait until the system is in steady state, so the governing equation is

$$T_h = -R_0 g.$$

Using $T = 0$ at $h = L_0$ after integration gives

$$T(h) = -R_0 g h + R_0 g L_0.$$

Finally, substituting $T = Y U_h$ and then integrating again gives

$$U(h) = \frac{R_0 g}{2Y}(2L_0 - h)h.$$

Setting $h = L_0$ gives the final hanging length

$$L = L_0 \left(1 + \frac{R_0 g L_0}{2Y}\right). \quad \square$$

Eulerian formulation. In Eulerian coordinates, the constitutive assumption becomes

$$\sigma(x,t) = y(x,t)\epsilon(x,t), \qquad (2.13)$$

where $\sigma(x,t) = T(h(x,t),t)$, $y(x,t) = Y(h(x,t))$, and $\epsilon(x,t) = E(h(x,t),t)$. Using the relation

$$\left.\frac{\partial U}{\partial h}\right|_{h=h(x,t)} = \frac{u_x}{1-u_x},$$

where u is the Eulerian displacement, Hooke's law (2.13) can be written

$$\sigma(x,t) = y(x,t)\frac{u_x(x,t)}{1-u_x(x,t)}. \qquad (2.14)$$

Then, σ can be eliminated from the momentum equation (2.8) to obtain

$$\rho(v_t + vv_x) = \rho f + \frac{\partial}{\partial x}\left(\frac{yu_x}{1 - u_x}\right). \tag{2.15}$$

In summary, this momentum equation (2.15) and the mass conservation equation

$$\rho_t + v\rho_x + \rho v_x = 0, \tag{2.16}$$

give two equations for the three unknowns v, u, and ρ. The third equation is the defining relationship between u and v,

$$v = \frac{Du}{Dt} = u_t + vu_x. \tag{2.17}$$

The Eulerian equations (2.15), (2.16), and (2.17) are nonlinear equations, and the constitutive equation (2.14) is nonlinear as well. This is in stark contrast to the Lagrangian equations. Thus, linearity in one formulation does not imply linearity in the other.

Further, equations (2.15), (2.16), and (2.17) cannot be resolved analytically. However, as was the case in the acoustic approximation, we can obtain a linearized theory for small displacements u. To this end, consider a bar with initial density $\rho_0(x)$ and stiffness (Young's modulus) $y_0(x)$. For a small displacement theory assume u and its derivatives are small compared to unity. From (2.17),

$$v = u_t(1 - u_x)^{-1} = u_t(1 + u_x + u_x^2 - \cdots) = u_t + \mathrm{O}(u^2),$$

and

$$v_t = u_{tt} + \mathrm{O}(u^2).$$

Let $\bar{\rho}(x,t)$ denote a small deviation from the initial density $\rho_0(x)$; that is,

$$\rho(x,t) = \rho_0(x) + \bar{\rho}(x,t).$$

Then, using the mean value theorem,

$$\begin{aligned} |\bar{\rho}(x,t)| &= |\rho(x,t) - \rho_0(x)| \\ &= |R(x - u, t) - R(x, 0)| \\ &\leq |R_h(\tilde{x}, \tilde{t})u| + |R_t(\tilde{x}, \tilde{t})|t, \end{aligned}$$

where $0 < \tilde{t} < t$ and $x < \tilde{x} < x - u$. Hence the deviation $\bar{\rho}$ is small provided that u *and* t are small and that the derivatives R_h and R_t are bounded. A similar calculation shows that if $y(x,t) = y_0(x) + \bar{y}(x,t)$, then the deviation $\bar{y}$ satisfies the equality $|\bar{y}(x,t)| = |Y_h(\tilde{x})u|$, and therefore $\bar{y}$ is small if u is small and Y_h is bounded.

With this information, the momentum equation can be written (taking $f = 0$)

$$(\rho_0 + \overline{\rho})(u_{tt} + O(u^2)) + (\rho_0 + \overline{\rho})(vv_x) = \frac{\partial}{\partial x}[(y_0 + \overline{y})(u_x)(1 + u_x + O(u^2))].$$

Retaining the lowest-order terms gives the **linearized, small displacement equation**

$$\rho_0(x)u_{tt} = \frac{\partial}{\partial x}(y_0(x)u_x). \tag{2.18}$$

Subject to the assumptions in the last paragraph we may expect (2.18) to govern small longitudinal vibrations of a bar of density $\rho_0(x)$ and stiffness $y_0(x)$. Notice that the unknown density dropped out of (2.15) and could be replaced by the initial density $\rho_0(x)$. Therefore (2.18) is one equation in the single unknown u; the conservation of mass equation is then a consistency relation.

If $\rho_0(x) = \rho_0$ and $y_0(x) = y_0$, where ρ_0 and y_0 are constants, then (2.18) reduces to the classical wave equation

$$u_{tt} - c^2 u_{xx} = 0, \quad c^2 = \frac{y_0}{\rho_0}.$$

If the extent of the rod is $0 \leq x \leq l$, then the wave equation is accompanied by boundary conditions. For a fixed endpoint, say at $x = l$, then

$$u(l,t) = 0, \quad t > 0 \quad \text{(fixed end)}.$$

If the end is free (no stress), then

$$y(l,t)\frac{u_x(l,t)}{1 - u_x(l,t)} = 0, \quad t > 0 \quad \text{(free end)}.$$

A properly posed problem for determining the small displacements $u(x,t)$ of a bar of length l with stiffness $y_0(x)$ and density $\rho_0(x)$ consists of the partial differential equation (2.18) along with boundary conditions at $x = 0$ and $x = l$, and with initial conditions of the form

$$u(x,0) = u_0(x), \quad u_t(x,0) = v_0(x), \quad 0 < x < l,$$

where $u_0(x)$ and $v_0(x)$ are the given initial displacement and velocity, respectively.

EXERCISES

1. Derive the conservation of mass and momentum equations for one-dimensional flow in a cylinder of variable (Lagrangian) cross-sectional area $A(h,t)$. Assume $A(h,t) = A(h,0)$ for all $t > 0$. Specifically, show

$$\frac{D(a\rho)}{Dt} + a\rho v_x = 0, \qquad \rho a \frac{Dv}{Dt} = \rho a f + \frac{\partial}{\partial x}(a\sigma),$$

where $a(x,t) = A(h(x,t),0)$.

2. A rubber cord of natural length 15 meters is hung vertically from a bridge. Its density is 1200 kg per cubic meter, and its modulus is 0.007 GPa. What is it hanging length?

3. A time-periodic solution of the wave equation

$$\rho_0(x)u_{tt} = \frac{\partial}{\partial x}(y_0(x)u_x), \quad 0 < x < l,$$

of the form

$$u(x,t) = w(x)\exp(i\omega t),$$

subject to boundary conditions at $x = 0$ and $x = l$, is called a standing wave. The set of values ω and the corresponding spatial distributions $w(x)$ are called the **fundamental frequencies** and **normal modes** of vibration, respectively. Determine the fundamental frequencies and normal modes for the following wave equations.

a) $u_{tt} - c^2 u_{xx} = 0, \quad 0 < x < \pi,$

   $u(0,t) = 0, \quad u_x(\pi,t) = 0.$

b) $u_{tt} - c^2 u_{xx} = 0, \quad 0 < x < \frac{\pi}{2},$

   $u(0,t) = 0, \quad u_x\left(\frac{\pi}{2},t\right) + \frac{3}{2}u\left(\frac{\pi}{2},t\right) = 0.$

c) $u_{tt} = \frac{\partial}{\partial x}(x^2 u_x), \quad 1 < x < e,$

   $u(1,t) = u(e,t) = 0.$

d) $x u_{tt} = (x u_x), \quad 0 < x < 2,$

   $\lim_{x \to 0^+} x u_x(x,t) = 0, \quad u(2,t) = 0.$

8.2.3 Thermodynamics and Energy Conservation

The Euler equations of motion for a one-dimensional continuum with no body forces are

$$\rho_t + v\rho_x + \rho v_x = 0, \tag{2.19}$$

$$v_t + vv_x + \frac{1}{\rho}p_x = 0. \tag{2.20}$$

This is a pair of first-order nonlinear partial differential equations for the three unknowns ρ, v, and p. Intuition tells us that a third equation is needed. Further reflection dictates that particular physical properties of the medium must play a role, and such properties are not included in (2.19)–(2.20), which are completely general and hold for any continuum. Equations that specify properties of the medium are known as equations of state or **constitutive relations**. Such equations give relations between observable effects and the internal constitution of the material, and they are generally expressible in terms of thermodynamic variables such as density, pressure, energy, entropy, temperature, and so on.

Example 8.9

(**Barotropic gas**) In Example 8.6 we introduced a simple barotropic equation of state where the pressure depends only upon the density, or

$$p = F(\rho), \tag{2.21}$$

where F is a given differentiable function and $F'(\rho) > 0$. For example, the barotropic equation of state

$$p = k\rho^\gamma, \quad \gamma > 1,$$

is a law used for many gases, called γ–law gases (e.g., air is a gamma law gas). Implicit in (2.21) is the assumption that other thermodynamic variables, e.g., temperature, are not involved. The barotropic gas describes only certain classes of flows. □

Therefore, in contrast to (2.21), it is more often the case that the equation of state introduces yet another unknown variable in the problem, such as the temperature. If that is the case, then still another equation is required. Such an equation must come from energy conservation. For general fluid motions a complete set of field equations usually consists of conservation equations for mass, momentum, and energy, as well as one or more constitutive relations. A discussion of energy conservation naturally entails development of some basic concepts in equilibrium thermodynamics.

Classic thermodynamics deals with relations between equilibrium states of uniform matter and the laws that govern changes in those states. It is implicit that the changes occur through a sequence of uniform equilibrium states, and the usual results of equilibrium thermodynamics apply. (It is possible to imagine processes that take place so rapidly that local fluid elements are incapable of establishing instantaneous equilibrium. Such processes belong to the study of nonequilibrium fluid dynamics.) Each thermodynamic quantity, such as temperature, pressure, and so on, is assumed to be a function of position and time, and thermodynamical relations are assumed to hold locally. To illustrate this principle, consider a hypothetical gas in a container and let P, R, and Θ denote its pressure, density, and temperature, respectively. It is observed experimentally that the ratio of pressure to density is proportional to the temperature, or

$$\frac{P}{R} = \mathcal{R}\Theta, \tag{2.22}$$

where $\mathcal{R}$ is the constant of proportionality, characteristic of the gas. Equation (2.22) is a Lagrangian statement because it holds for a material volume of gas. The assumption of local thermodynamic equilibrium allows us to assume that

$$\frac{p}{\rho} = \mathcal{R}\theta,$$

where p, ρ, and the temperature θ are the local Eulerian quantities that are functions of x and t.

Example 8.10

(**Ideal gas**) Under normal conditions most gases obey the ideal gas law

$$p = \mathcal{R}\rho\theta, \tag{2.23}$$
$$e = c_v\theta + \text{constant}, \tag{2.24}$$

where p is pressure, ρ is density, θ is temperature, and e is the internal energy per unit mass (or specific internal energy). The constant c_v is the specific heat at constant volume and $\mathcal{R}$ is the *gas constant* for the particular gas. A gas satisfying (2.23) and (2.24) is called an **ideal gas** and many compressible fluids of practical interest can be treated approximately as ideal gases. Hence (2.23) and (2.24) describe a wide range of phenomena in fluid dynamics. These equations coupled with (2.19) and (2.20) give only four equations for the five unknowns p, v, ρ, θ, and e, so it is again evident that another equation is required.   □

A common statement of the first law of thermodynamics is that energy is conserved if heat is taken into account. Thus, the first law of thermodynamics provides a concise statement of energy conservation. It comes from considering the consequences of adding a small amount of heat to a unit mass of a material substance in a way that equilibrium conditions are established at each step. Some of the energy will go into the work $pd(1/\rho)$ done in expanding the specific volume $1/\rho$ by $d(1/\rho)$, and the remainder will go into increasing the specific internal energy e by de. The precise relationship is

$$q = de + pd\rho^{-1}, \tag{2.25}$$

where the differential form q is the heat added. For an equilibrium process equation (2.25) is the **first law of thermodynamics**. In general, q is not an exact differential, that is, there does not exist a state function Q for which $q = dQ$. If $q = 0$, then the process is called **adiabatic**.

Example 8.11

(**Equations of state**) For an ideal gas described by equations (2.23) and (2.24), we have

$$q = de + pd\rho^{-1} = c_v\, d\theta + \mathcal{R}\theta\, d(\ln\rho^{-1}). \tag{2.26}$$

From (2.26) we notice that θ^{-1} is an integrating factor for the differential form q. Therefore

$$\theta^{-1}q = d(c_v\ln\theta + \mathcal{R}\ln\rho^{-1}) = ds,$$

where

$$s \equiv c_v\ln\theta + \mathcal{R}\ln\rho^{-1} + \text{constant} \tag{2.27}$$

is the **entropy**. Consequently, for an ideal gas the first law (2.25) takes the form

$$de = \theta\, ds - pd\rho^{-1}. \tag{2.28}$$

Equation (2.27) may also be written in terms of the pressure as

$$p = k\rho^{\gamma} \exp\left(\frac{s}{c_v}\right), \tag{2.29}$$

where

$$\gamma \equiv 1 + \frac{\mathcal{R}}{c_v}$$

and k is a constant. Combining (2.23) and (2.24) gives yet another form of the equation of state, namely

$$e = \frac{p}{(\gamma - 1)\rho} + \text{constant} \tag{2.30}$$

Finally, in some contexts, particularly in studying chemical reactions, it is useful to introduce the **enthalpy** h defined by $h = e + p/\rho$. For an ideal gas $h = c_p\theta$, where $c_p = \mathcal{R} + c_v$ is the specific heat at constant pressure. Hence $\gamma = c_p/c_v$ is the ratio of specific heats. For air, $\gamma = 1.4$, and for a monatomic gas, $\gamma = \frac{5}{3}$. Generally, $\gamma > 1$. ⊔

Example 8.12

Other equations of state have been proposed to include various effects. The **Abel** or **Clausius** equation of state,

$$p\left(\frac{1}{\rho} - \alpha\right) = \mathcal{R}\theta, \qquad \alpha \text{ constant}$$

introduces a constant covolume α to account for the size of the molecules. The **van der Waals equation of state**

$$(p + \beta\rho^2)\left(\frac{1}{\rho} - \alpha\right) = \mathcal{R}\theta, \qquad \alpha, \beta \text{ constant}.$$

contains the term $\beta\rho^2$ to further account for intermolecular forces. The **Tait equation**

$$\frac{p + B}{p} = \left(\frac{\rho}{\rho_0}\right)^{\bar{\gamma}},$$

where $\bar{\gamma}$ and B are constants, has been used to model the behavior of liquids at high pressures. $\qquad \square$

We now formulate a partial differential equation governing the flow of energy in a system. The approach is consistent with the earlier development of balance laws in integral form. Let $[a(t), b(t)]$ be a one-dimensional material region with cross-sectional area A. We define the **kinetic energy** of the fluid in the region by

$$A \int_{a(t)}^{b(t)} \tfrac{1}{2}\rho v^2 \, dx \qquad \text{(Kinetic energy)}$$

and the **internal energy** in the region by

$$A \int_{a(t)}^{b(t)} \rho e \, dx. \qquad \text{(Internal energy)}$$

By the general conservation of energy principle *the time rate of change of the total energy equals the rate that the forces do work on the region plus the rate that heat flows into the region.* There are two forces, the body force $f(x,t)$ acting at each cross section of the region, and the stress $\sigma(x,t)$ acting at the two ends. Since force times velocity equals the rate work is done, the total **rate that work is done** on the region is

$$A \int_{a(t)}^{b(t)} \rho f v \, dx - A\sigma(a(t), t)v(a(t), t) + A\sigma(b(t), t)v(b(t), t).$$

The **rate that heat flows** into the region is $AJ(a(t), t) - AJ(b(t), t)$, where $J(x, t)$ is the heat flux in energy units per unit area per time. Therefore, we *postulate* the **balance of energy law**

$$\frac{d}{dt} \int_{a(t)}^{b(t)} \left(\tfrac{1}{2} \rho v^2 + \rho e \right) \, dx$$

$$= \int_{a(t)}^{b(t)} \rho f v \, dx - \sigma(a(t), t) v(a(t), t) + \sigma(b(t), t) v(b(t), t)$$

$$+ J(a(t), t) - J(b(t), t),$$

or

$$\frac{d}{dt} \int_{a(t)}^{b(t)} \left(\tfrac{1}{2} \rho v^2 + \rho e \right) \, dx = \int_{a(t)}^{b(t)} \left(\rho f v + (\sigma v)_x - J_x \right) \, dx.$$

Assuming sufficient smoothness of the state variables and using the arbitrariness of the interval $[a(t), b(t)]$, we have the following Eulerian differential form of the conservation of energy law:

$$\frac{1}{2} \rho \frac{D v^2}{Dt} + \rho \frac{De}{Dt} = \rho f v + (\sigma v)_x - J_x. \tag{2.31}$$

Equation (2.31) can be recast into various forms. First, multiplication of the momentum balance equation (2.5) by v gives

$$\frac{1}{2} \rho \frac{D v^2}{Dt} = \rho f v + v \sigma_x.$$

Subtracting this from (2.31) gives an equation for the rate of change of internal energy

$$\rho \frac{De}{Dt} = \sigma v_x - J_x. \tag{2.32}$$

An alternate approach to energy conservation comes from an examination of the first law of thermodynamics (2.25). The second law of thermodynamics states that the differential form q in general has integrating factor θ^{-1} and $\theta^{-1} q = ds$, where s is the **entropy** and θ is the absolute temperature. Thus

$$\theta \, ds = de + p d\rho^{-1}.$$

This combined form of the first and second laws of thermodynamics can be reformulated as a partial differential equation. Because it refers to a given material region, we *postulate*

$$\theta \frac{Ds}{Dt} = \frac{De}{Dt} + p \frac{D(1/\rho)}{Dt}, \tag{2.33}$$

which is another local form of the conservation of energy principle. Noting that $D\rho^{-1}/Dt = -(1/\rho^2)D\rho/Dt = -(1/\rho)v_x$ and putting $\sigma = -p$, equation (2.33) may be subtracted from (2.32) to obtain

$$\theta\frac{Ds}{Dt} = -\frac{1}{\rho}\phi_x,$$

which relates the entropy change to the heat flux. If we assume the *constitutive relation* $J = -K\theta_x$, which is Fourier's law, then the energy equation (2.33) can be expressed as

$$\frac{De}{Dt} + p\frac{D(1/\rho)}{Dt} = \frac{1}{\rho}(K\theta_x)_x. \tag{2.34}$$

Example 8.13

(**Adiabatic flow**) In adiabatic flow $\theta Ds/Dt = 0$, and the energy equation (2.33) becomes

$$\frac{De}{Dt} + p\frac{D(1/\rho)}{Dt} = 0. \tag{2.35}$$

Equations (2.19), (2.20), and (2.35), along with a **thermal equation of state** $p = p(\rho, \theta)$ and a **caloric equation of state** $e = e(\rho, \theta)$, give a set of five equations for the unknowns ρ, v, p, e, and θ. □

Example 8.14

(**Ideal gas**) For adiabatic flow of an ideal gas, expression (2.30) for the energy can be substituted into (2.35) to obtain

$$\frac{Dp}{Dt} - \frac{\gamma p}{\rho}\frac{D\rho}{Dt} = 0. \tag{2.36}$$

Equations (2.19), (2.20), and (2.36) give three equations for ρ, v, and p. Because $Ds/Dt = 0$, the entropy is constant for a given fluid particle (cross section); in Lagrangian form $S = S(h)$. In this case the equations governing **adiabatic flow** are often written

$$\frac{D\rho}{Dt} + \rho v_x = 0, \quad \rho\frac{Dv}{Dt} + p_x = 0, \quad \frac{Ds}{Dt} = 0,$$

along with the equation of state (2.29), giving $p = p(s, \rho)$. If the entropy is constant initially, that is, $S(h,0) = s_0$ for all h, then $s(x,t) = s_0$ for all x and t, and the resulting adiabatic flow is called **isentropic**. This case reduces to barotropic flow as discussed earlier. □

In general, for nonadiabatic flow involving heat conduction, equations (2.19), (2.20), and (2.34), along with thermal and caloric equations of state give five equations for ρ, v, p, e, and θ. If there is no motion ($v = 0$) and $s = c_v \ln \theta +$ const., then (2.34) reduces to the classic **diffusion equation**

$$\rho c_v \theta_t = (K\theta_x)_x.$$

EXERCISES

1. By introducing the enthalpy $h = e + p/\rho$ into the energy equation $De/Dt + pD(\rho^{-1})/Dt = 0$, show that for steady flow

$$h_x - \frac{1}{\rho}p_x = 0.$$

   Prove that

$$h + \tfrac{1}{2}v^2 = \text{constant}.$$

2. Derive (2.29) and (2.33).

3. Show that the equations governing the one-dimensional, time-dependent flow of a compressible fluid under constant pressure are

$$v_t + vv_x = 0,$$
$$\rho_t + (\rho v)_x = 0,$$
$$e_t + (ev)_x + pv_x = 0.$$

   Subject to initial conditions $v(x,0) = f(x)$, $\rho(x,0) = g(x)$, and $e(x,0) = h(x)$, where f, g, and h are given smooth functions with nonnegative derivatives, derive the solution $v = f(x - vt)$ and

$$\rho = \frac{g(x - vt)}{1 + tf'(x - vt)}, \qquad e = \frac{h(x - vt) + p}{1 + tf'(x - vt)} - p.$$

   Here, e is measured in energy per unit volume.

8.3 Gas Dynamics

8.3.1 Riemann's Method

Heretofore there has been little discussion of how fluid motion begins. It is common in gas dynamics, or compressible fluid dynamics, to imagine the flow being induced by the motion of a piston inside the cylindrical pipe. Such a

Figure 8.7 Piston problem. A receding, decelerating piston

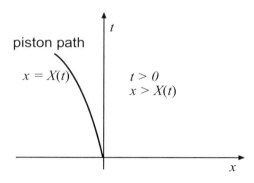

Figure 8.8 Space-time diagram of the piston problem.

device is not as unrealistic as it may first appear. The piston may represent the fluid on one side of the valve after it is opened or it may represent a detonator in an explosion process; while in aerodynamics it may represent a blunt object moving into a gas.

Accordingly, we set up and solve a simple problem using **Riemann's method**, which is used in more general problems in gas dynamics. We consider a gas in a tube initially in the constant state

$$v = 0, \quad \rho = \rho_0, \quad p = p_0, \quad c = c_0, \tag{3.1}$$

with equation of state

$$p = k\rho^\gamma, \quad k, \gamma > 0. \tag{3.2}$$

To initiate the motion a piston located initially at $x = 0$ is withdrawn slowly according to

$$x = X(t), \quad X'(t) < 0, \quad X''(t) < 0, \tag{3.3}$$

where X is given function (see Fig. 8.7). The problem is to determine v, p, c, and ρ for all $t > 0$ and $X(t) < x < \infty$. We may regard this as a boundary value problem where initial conditions are given along the positive x axis and v is given by the piston velocity along the piston path (see Fig. 8.8). The method of solution is motivated by the study of the simple nonlinear model equation $u_t + c(u)u_x = 0$ from Chapter 7. There we were able to define a family of curves called characteristics along which signals propagated and along which the partial differential equation reduced to an ordinary differential equation. In the case of this model equation $du/dt = 0$ along $dx/dt = c(u)$. We follow a similar strategy for the nonlinear acoustic equations in Chapter 7, equations (4.7) and (4.8), and attempt to find characteristic curves along which the partial differential equations reduce to simpler equations. To this end we multiply (4.7) by c and then add and subtract it from (4.8) to obtain

$$p_t + (v + c)p_x + \rho c(v_t + (v + c)v_x) = 0, \tag{3.4}$$

$$p_t + (v - c)p_x - \rho c(v_t + (v - c)v_x) = 0. \tag{3.5}$$

Hence, along the families of curves C^+ and C^- defined by

$$C^+ : \frac{dx}{dt} = v + c, \tag{3.6}$$

$$C^- : \frac{dx}{dt} = v - c, \tag{3.7}$$

we have

$$\frac{dp}{dt} + \rho c \frac{dv}{dt} = 0, \quad \text{on } C^+, \tag{3.8}$$

$$\frac{dp}{dt} - \rho c \frac{dv}{dt} = 0, \quad \text{on } C^-. \tag{3.9}$$

Equations (3.8) and (3.9) may be rewritten as

$$\frac{c}{\rho} \frac{dp}{dt} \pm \frac{dv}{dt} = 0, \quad \text{on } C^+, C^-.$$

Integrating gives

$$\int \frac{c(\rho)}{\rho} d\rho + v = \text{constant} \quad \text{on } C^+, \tag{3.10}$$

$$\int \frac{c(\rho)}{\rho} d\rho - v = \text{constant} \quad \text{on } C^-. \tag{3.11}$$

The left sides of (3.10) and (3.11) are called the **Riemann invariants**; they are quantities that are constant along the characteristic curves C^+ and C^- defined by (3.6) and (3.7). If the equation of state is defined by (3.2), then

$$c^2 = k\gamma\rho^{\gamma-1},$$

and

$$\int \frac{c(\rho)}{\rho} d\rho = \frac{2c}{\gamma - 1}.$$

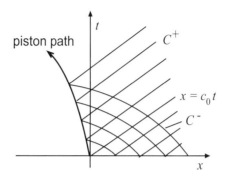

Figure 8.9 C^+ and C^- characteristics. The C^+ characteristics are slowly fanning out.

Therefore, the Riemann invariants are

$$r_+ \equiv \frac{2c}{\gamma - 1} + v = \text{constant} \quad \text{on } C^+, \tag{3.12}$$

$$r_- \equiv \frac{2c}{\gamma - 1} - v = \text{constant} \quad \text{on } C^-. \tag{3.13}$$

We have enough information from the Riemann invariants to determine the solution of the piston withdrawal problem. First consider the C^- characteristics. Since $c = c_0$ when $v = 0$ the characteristics that begin on the x axis leave with speed $-c_0$ and the constant in (3.13) has valued $2c_0/(\gamma - 1)$. Hence on the C^- characteristics

$$\frac{2c}{\gamma - 1} - v = \frac{2c_0}{\gamma - 1}. \tag{3.14}$$

Since the last equation must hold along every C^- characteristic it must hold everywhere and therefore r_- is constant in the entire region $t > 0$, $X(t) < x < \infty$. The C^- characteristics must end on the piston path, since their speed $v_p - c_p$ at the piston is more negative than the speed v_p of the piston (see Fig. 8.9). It is easy to see that the C^+ characteristics are straight lines. Adding and subtracting (3.14) and (3.12) shows

$$c = \text{constant} \quad \text{on a } C^+ \text{ characteristic},$$

$$v = \text{constant} \quad \text{on a } C^+ \text{ characteristic}.$$

We have used the fact that (3.14) holds everywhere. The speed $v + c$ of a C^+ characteristic is therefore constant and thus the characteristic is a straight line. The C^+ characteristics emanating from the x axis have speed c_0 (since $v = 0$ on the x axis) and carry the constant state $v = 0$, $c = c_0$, $p = p_0$, $\rho = \rho_0$ into the region $x > c_0 t$. This is just the uniform state ahead of the signal $x = c_0 t$

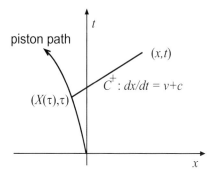

Figure 8.10 Positive characteristic carrying information from the piston path.

beginning at the origin traveling at speed c_0 into the constant state. This signal indicates the initial motion of the piston.

A C^+ characteristic beginning on the piston at $(X(\tau), \tau)$ and passing through (x, t) has equation

$$x - X(\tau) = (c + v)(t - \tau). \tag{3.15}$$

The speed $v + c$ can be calculated as follows. Clearly $v = X'(t)$, which is the velocity of the piston. From (3.14)

$$\frac{2c}{\gamma - 1} - X'(\tau) = \frac{2c_0}{\gamma - 1},$$

and hence

$$c = c_0 + \frac{\gamma - 1}{2} X'(\tau).$$

Thus

$$x - X(\tau) = \left(c_0 + \frac{\gamma + 1}{2} X'(\tau) \right) (t - \tau) \tag{3.16}$$

is the equation of the desired C^+ characteristic (see Fig. 8.10). The fact that v, c, p, and ρ are constant on the C^+ characteristics gives the solution

$$v(x, t) = X'(\tau),$$

where τ is given implicitly by (3.16). Obviously

$$c(x, t) = c_0 + \frac{\gamma - 1}{2} v(x, t).$$

The quantities p and ρ may be calculated from the equation of state and the definition of c^2. Qualitatively we think of C^- characteristics as carrying the values

$$r_- = \frac{2c}{\gamma - 1} - v$$

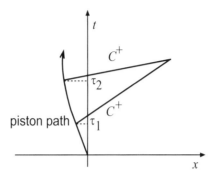

Figure 8.11 Colliding positive characteristics.

from the constant state back into the flow. The C^+ characteristics carry information from the piston forward into the flow. Whenever one of the Riemann invariants is constant throughout the flow, we say that the solution in the nonuniform region is a **simple wave**. It can be shown that a simple wave solution always exists adjacent to a uniform state provided that the solution remains smooth. A complete analysis is given in Courant and Friedrichs (1948).

In the piston withdrawal problem previously discussed we assumed that the piston path $x = X(t)$ satisfied the conditions $X'(t) < 0$ and $X''(t) < 0$, which means the piston is always accelerating backward. If there is ever an instant of time when the piston slows down, that is, $X''(t) > 0$, then a smooth solution cannot exist for all $t > 0$, for two distinct C^+ characteristics emanating from the piston will cross. Suppose, for example, that τ_1 and τ_2 are two values of t in a time interval where $X''(t) > 0$. The speed of characteristics leaving the piston is

$$v + c = c_0 + \frac{\gamma + 1}{2} X'(\tau).$$

If $\tau_1 < \tau_2$, then $X'(\tau_1) < X'(\tau_2)$ and it follows that $v + c$ at τ_1 is smaller than $v + c$ at τ_2. Therefore the characteristic emanating from $(X(\tau_2), \tau_2)$ is faster than the characteristic emanating from $(X(\tau_1), \tau_1)$. Thus the two characteristics must cross (see Fig. 8.11).

Example 8.15

Beginning at time $t = 0$ a piston located at $x = 0$ moves forward into a gas under uniform conditions with equation of state given by (3.2). Its path is given by $X(t) = at^2$, $a > 0$. We determine the first instant of time that two characteristics cross and the wave breaks. For this problem the preceding analysis remains valid and the C^+ characteristics emanating from the piston

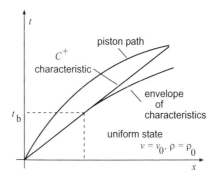

Figure 8.12 The positive characteristics emanating from the piston path and forming an envelope. The characteristics intersect in the region between the envelope and the curve $x = c_0 t$. The shock begins at time t_b

have equation (3.16) or

$$x - a\tau^2 = (c_0 + (\gamma + 1)a\tau)(t - \tau),$$

or

$$F(x, t, \tau) \equiv \gamma a\tau^2 + (c_0 - (\gamma + 1)at)\tau + x - c_0 t = 0. \qquad (3.17)$$

Along such a characteristic v is constant and hence

$$v(x, t) = X'(\tau) = 2a\tau,$$

where $\tau = \tau(x, t)$ is to be determined from (3.17). To solve (3.17) for τ requires by the implicit function theorem that $F_\tau(x, t, \tau) \neq 0$. Since $F_\tau = 2\gamma a\tau + (c_0 - (\gamma + 1)at)$, the first instant of time t that (3.17) cannot be solved for τ is

$$t_b = \min_{t \geq 0} \frac{c_0 + 2\gamma a\tau}{(\gamma + 1)a} = \frac{c_0}{(\gamma + 1)a}.$$

That is, the breaking time t_b will occur along the first characteristic (indexed by τ) where $F_\tau = 0$. The characteristic diagram is shown in Fig. 8.12 in the special case $a = c_0 = 1$ and $\gamma = 3$. $\quad\square$

8.3.2 Rankine–Hugoniot Conditions

As we observed earlier smooth solutions break down when characteristics intersect, since constant values are carried along the characteristics. The solution that develops is a discontinuous one and it propagates as a shock wave. We now determine what conditions hold across such a discontinuity. We proceed as in Section 8.1 where conservation laws are discussed.

The integral form of the conservation of mass law, namely

$$\frac{d}{dt} \int_a^b \rho(x,t)\,dx = -\rho(x,t)v(x,t)|_a^b \tag{3.18}$$

holds in all cases, even when the functions ρ and v are not smooth. In Chapter 7 we derived the jump condition across a shock, which we briefly review. Let $x = s(t)$ be a smooth curve in space–time that intersects the interval $a \le x \le b$ at time t, and suppose v, ρ, and p suffer simple discontinuities along the curve. Otherwise v, ρ, and p are assumed to be C^1 functions with finite limits on each side of $x = s(t)$. Then, by Leibniz' rule,

$$
\begin{aligned}
\frac{d}{dt} \int_a^b \rho(x,t)\,dx &= \frac{d}{dt} \int_a^{s(t)} \rho(x,t)\,dx + \frac{d}{dt} \int_{s(t)}^b \rho(x,t)\,dx \\
&= \int_a^{s(t)} \rho_t(x,t)\,dx + \rho(s(t)^-,t)s'(t) \\
&\quad + \int_{s(t)}^b \rho_t(x,t)\,dx - \rho(s(t)^+,t)s'(t).
\end{aligned}
$$

Both integrals on the right side approach zero as $a \to s(t)^-$ and $b \to s(t)^+$ and therefore from (3.18) it follows that

$$(\rho(s(t)^-,t) - \rho(s(t)^+,t))s'(t) = -\rho(x,t)v(x,t)|_{s(t)^-}^{s(t)^+}. \tag{3.19}$$

If the values (one-sided limits) of ρ and v on the right and left sides of $x = s(t)$ are denoted by the subscripts zero and one, respectively, for example,

$$\rho_0 = \lim_{x \to s(t)^+} \rho(x,t), \qquad \rho_1 = \lim_{x \to s(t)^-} \rho(x,t),$$

then (3.19) can be written

$$(\rho_1 - \rho_0)s'(t) = \rho_1 v_1 - \rho_0 v_0. \tag{3.20}$$

In a similar fashion the condition

$$(\rho_1 v_1 - \rho_0 v_0)s'(t) = \rho_1 v_1^2 + p_1 - \rho_0 v_0^2 - p_0 \tag{3.21}$$

can be obtained from the integral form of the conservation of momentum law. The two conditions (3.20) and (3.21) are known as the **Rankine–Hugoniot jump conditions**; they relate the values ρ_0, v_0, p_0 ahead of the discontinuity

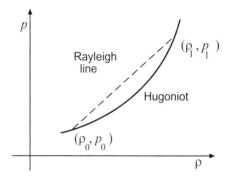

Figure 8.13 The Hugoniot diagram.

to the speed s' of the discontinuity and to the values ρ_1, v_1, and p_1 behind it. If the state ahead is at rest, that is, $v_0 = 0$, then the conditions become

$$(\rho_1 - \rho_0)s' = \rho_1 v_1, \tag{3.22}$$
$$\rho_1 v_1 s' = \rho_1 v_1^2 + p_1 - p_0. \tag{3.23}$$

Supplemented with the equation of state

$$p_1 = f(\rho_1), \tag{3.24}$$

the equations can be regarded as three equations in the four unknowns ρ_1, v_1, p_1, and s'. If any one of these quantities is known, the remaining three may be determined.

It is helpful to picture the information contained in (3.22)–(3.24) on a $p\rho$ diagram as shown in Fig. 8.13. The solid curve is the graph of the equation of state $p = f(\rho)$. It is known as the **Hugoniot curve**; and all states, both initial and final, must lie on this curve. The straight line connecting the state ahead (ρ_0, p_0) to the state behind (ρ_1, p_1) is the **Rayleigh line**. If (3.23) is rewritten using equation (3.22) as

$$p_1 - p_0 = \rho_0 v_1 s',$$

then it is clear that the slope of the Rayleigh line is given by

$$\frac{p_1 - p_0}{\rho_1 - \rho_0} = \frac{\rho_0}{\rho_1} s'^2.$$

These ideas extend beyond this simple case to more general problems in gas-dynamics (see Courant and Friedrichs (1948)).

EXERCISES

1. Derive the jump condition (3.21).

2. Discuss the piston withdrawal problem when the piston path is given by $x = -V_0 t$, where V_0 is a positive constant. Assume an equation of state $p = k\rho^\gamma$ with uniform conditions $v = 0$, $p = p_0$, $\rho = \rho_0$, $c = c_0$ ahead of the piston at $t = 0$.

3. At $t = 0$ the gas in a tube $x \geq 0$ is at rest. For $t > 0$ a piston initially located at $x = 0$ moves according to the law $x = X(t)$, where $X(t)$ is small. Show that in the acoustic approximation the motion induced in the gas is

$$v = X'\left(t - \frac{x}{c_0}\right),$$

and find the corresponding density variation.

4. A sphere of initial radius r_0 pulsates according to $r(t) = r_0 + A\sin\omega t$. In the linearized theory find the motion of the gas outside the sphere.

8.4 Fluid Motions in $\mathbb{R}^3$

In the early sections the equations governing fluid motion were developed in one spatial dimension. This limitation is severe, however, because most interesting fluid phenomena occur in higher dimensions. In this section we derive the field equations for fluid dynamics in three dimensions, with the two-dimensional case being an obvious corollary. Our use of vector notation makes this higher dimensional case appear very clean compared to the detail written out earlier in the one-dimensional case; moreover, much of the road work has been done and the key ideas have been presented. The result is a section that may appear more like a formal mathematical treatment, rather than a physical one.

8.4.1 Kinematics

Let Ω_0 be an arbitrary closed bounded set in $\mathbb{R}^3$. We imagine that Ω_0 is a region occupied by a fluid at time $t = 0$. By a **fluid motion** we mean a mapping ϕ_t: $\Omega_0 \to \Omega_t$ defined for all t in an interval I containing the origin, which maps the region Ω_0 into $\Omega_t = \phi_t(\Omega_0)$, the latter being the region occupied by the same fluid at time t (see Fig. 8.14). We assume that ϕ_t is represented by the formula

$$\mathbf{x} = \mathbf{x}(\mathbf{h}, t), \tag{4.1}$$

where $\mathbf{h} = (h_1, h_2, h_3)$ is in Ω_0 and $\mathbf{x} = (x_1, x_2, x_3) = (x, y, z)$, with

$$\mathbf{x}(\mathbf{h}, 0) = \mathbf{h}.$$

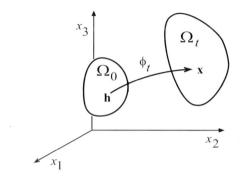

Figure 8.14 Fluid motion.

Thus $\mathbf{h}$ is a Lagrangian coordinate or particle label given to each fluid particle at $t = 0$, and $\mathbf{x}$ is the Eulerian coordinate representing the laboratory position of the particle $\mathbf{h}$ at time t. We further assume that the function $\mathbf{x}(\mathbf{h}, t)$ is twice continuously differentiable on its domain $\Omega_0 \times I$ and that for each $t \in I$ the mapping ϕ_t has a unique inverse defined by

$$\mathbf{h} = \mathbf{h}(\mathbf{x}, t).$$

Hence

$$\mathbf{x}(\mathbf{h}(\mathbf{x}, t), t) = \mathbf{x}, \qquad \mathbf{h}(\mathbf{x}(\mathbf{h}, t), t) = \mathbf{h}.$$

By convention, functions of $\mathbf{h}$ and t that represent the results of Lagrangian measurements will be denoted by capital letters, and functions of $\mathbf{x}$ and t that represent the results of a fixed laboratory measurement will be denoted by lowercase letters. The measurements are connected by the formulas

$$f(\mathbf{x}, t) = F(\mathbf{h}(\mathbf{x}, t), t), \qquad F(\mathbf{h}, t) = f(\mathbf{x}(\mathbf{h}, t), t).$$

In the sequel we often use the notation $F|_{\mathbf{x}}$ and $f|_{\mathbf{h}}$ to denote the right sides of these equations, respectively.

For a given fluid motion (4.1), the **fluid velocity** is defined by

$$\mathbf{V}(\mathbf{h}, t) = \frac{\partial \mathbf{x}}{\partial t}(\mathbf{h}, t).$$

Then

$$\mathbf{v}(\mathbf{x}, t) = \mathbf{V}(\mathbf{h}(\mathbf{x}, t), t).$$

$\mathbf{V}(\mathbf{h}, t) = \langle V_1(\mathbf{h}, t), V_2(\mathbf{h}, t), V_3(\mathbf{h}, t) \rangle$ is the actual velocity vector of the fluid particle $\mathbf{h}$ at time t, whereas $\mathbf{v}(\mathbf{x}, t) = \langle v_1(\mathbf{x}, t), v_2(\mathbf{x}, t), v_3(\mathbf{x}, t) \rangle$ is the velocity vector measured by a fixed observer at location $\mathbf{x}$. If $\mathbf{h}_0$ is a fixed point in Ω_0, then the curve $\mathbf{x} = \mathbf{x}(\mathbf{h}_0, t)$, $t \in I$, is called the **particle path** of the particle $\mathbf{h}_0$. The following theorem states that knowledge of the Eulerian velocity field $\mathbf{v}(\mathbf{x}, t)$ is equivalent to knowledge of all the particle paths.

Theorem 8.16

If the velocity vector field $\mathbf{v}(\mathbf{x}, t)$ is known, then all of the particle paths can be determined, and conversely. $\quad\square$

Proof

First we prove the converse. If each particle path is given, then the function $\mathbf{x} = \mathbf{x}(\mathbf{h}, t)$ is known for all $h \in \Omega_0$. Consequently, $\mathbf{h} = \mathbf{h}(\mathbf{x}, t)$ can be determined and $\mathbf{v}$ can be calculated from

$$\mathbf{v}(\mathbf{x}, t) = \mathbf{V}(\mathbf{h}(\mathbf{x}, t), t) = \mathbf{x}_t(\mathbf{h}(\mathbf{x}, t), t).$$

Now assume $\mathbf{v}(\mathbf{x}, t)$ is given. Then

$$\frac{\partial \mathbf{x}}{\partial t}(\mathbf{h}, t) = \mathbf{v}(\mathbf{x}(\mathbf{h}, t), t),$$

or

$$\frac{\partial \mathbf{x}}{\partial t} = \mathbf{v}(\mathbf{x}, t), \tag{4.2}$$

where explicit dependence on $\mathbf{h}$ has been dropped. Equation (4.2) is a first-order system of differential equations for $\mathbf{x}$. Its solution subject to the initial condition

$$\mathbf{x} = \mathbf{h} \quad \text{at} \quad t = 0$$

gives the particle paths $\mathbf{x} = \mathbf{x}(\mathbf{h}, t)$. $\quad\square$

It is interesting to contrast the particle paths with the so-called **streamlines** of the flow. The latter are integral curves of the vector field $\mathbf{v}(\mathbf{x}, t_0)$ frozen at some fixed but arbitrary instant t_0 of time. Thus the streamlines are found by solving the system

$$\frac{d\mathbf{x}}{ds} = \mathbf{v}(\mathbf{x}, t_0), \tag{4.3}$$

where $\mathbf{x} = \mathbf{x}(s)$ and s is a parameter along the curves. If $\mathbf{v}$ is independent of t, then the flow is called **steady** and the streamlines coincide with the particle paths. In a time-dependent flow they need not coincide.

Example 8.17

Consider the fluid motion

$$x_1 = t + h_1, \quad x_2 = h_2 e^t, \quad x_3 = th_1 + h_3, \quad t > 0.$$

Then

$$\mathbf{V} = \left\langle \frac{\partial x_1}{\partial t}, \frac{\partial x_2}{\partial t}, \frac{\partial x_3}{\partial t} \right\rangle = \langle 1, h_2 e^t, h_1 \rangle.$$

Inverting the motion,

$$h_1 = x_1 - t, \quad h_2 = x_2 e^{-t}, \quad h_3 = x_3 - tx_1 + t^2.$$

Therefore

$$\mathbf{v} = \langle 1, x_2, x_1 - t \rangle,$$

and thus the motion is not steady. The streamlines at time t_0 are given by solutions to (4.3), or

$$\frac{dx_1}{ds} = 1, \quad \frac{dx_2}{ds} = x_2, \quad \frac{dx_3}{ds} = x_1 - t_0.$$

Hence

$$x_1 = s + c_1, \quad x_2 = c_2 e^s, \quad x_3 = \tfrac{1}{2}s^2 + (c_1 - t_0)s + c_3,$$

where c_1, c_2, and c_3 are constants. $\quad\square$

As in the one-dimensional case we define the **material derivative** by

$$\frac{Df}{Dt}(\mathbf{x}, t) \equiv F_t(\mathbf{h}, t)|_{\mathbf{h}=\mathbf{h}(\mathbf{x},t)},$$

where f and F are the Eulerian and Lagrangian representations of a given measurement, respectively. A straightforward calculation using the chain rule shows that

$$F_t(\mathbf{h}, t) = \frac{\partial}{\partial t} f(\mathbf{x}(\mathbf{h}, t), t)$$

$$= f_t(\mathbf{x}(\mathbf{h}, t), t) + \sum_{j=1}^{3} f_{x_j}(\mathbf{x}(\mathbf{h}, t), t) V_j(\mathbf{h}, t)$$

$$= f_t(\mathbf{x}(\mathbf{h}, t), t) + \mathbf{V} \cdot \nabla f|_h.$$

Evaluating at $\mathbf{h} = \mathbf{h}(\mathbf{x}, t)$ gives

$$\frac{Df}{Dt}(\mathbf{x}, t) = f_t(\mathbf{x}, t) + \mathbf{v}(\mathbf{x}, t) \cdot \nabla f(\mathbf{x}, t). \tag{4.4}$$

If $\mathbf{f} = \langle f_1, f_2, f_3 \rangle$ is a vector function, then componentwise,

$$\frac{Df_i}{Dt} = \frac{\partial f_i}{\partial t} + \sum_{j=1}^{3} v_j \frac{\partial f_i}{\partial x_j}, \quad i = 1, 2, 3,$$

and we write

$$\frac{D\mathbf{f}}{Dt} = \mathbf{f}_t + \sum_{j=1}^{3} v_j \frac{\partial \mathbf{f}}{\partial x_j} \equiv \mathbf{f}_t + (\mathbf{v} \cdot \nabla)\mathbf{f}.$$

Here $\mathbf{v} \cdot \nabla$ is a special notation for the operator $v_1 \partial/\partial x_1 + v_2 \partial/\partial x_2 + v_3 \partial/\partial x_3$. The material derivative Df/Dt at $(\mathbf{x}, t)$ is a measure of the time rate of change of the quantity F moving with the particle $\mathbf{h}$ frozen at the instant the particle is at $\mathbf{x}$.

Another kinematical result is the three-dimensional analog of the **Euler expansion theorem**. It gives the time rate of change of the **Jacobian**

$$J(\mathbf{h}, t) = \det\left(\frac{\partial x_i}{\partial h_j}\right)$$

of the transformation ϕ_t. We record this as a theorem.

Theorem 8.18

In Lagrangian and Eulerian form, respectively,

$$\begin{aligned} J_t(\mathbf{h}, t) &= J(\mathbf{h}, t)\nabla \cdot \mathbf{v}(\mathbf{x}, t)|_{\mathbf{h}}, \\ \frac{Dj}{Dt}(\mathbf{x}, t) &= j(\mathbf{x}, t)\nabla \cdot v(\mathbf{x}, t), \end{aligned}$$

where $j(\mathbf{x}, t) = J(\mathbf{h}(\mathbf{x}, t), t)$. $\square$

Details of the proof, based on the formula for differentiating a determinant, are left as an exercise. The reader will be asked to show the result for two dimensions.

In the one-dimensional case the derivation of many of the results depends on knowing the time derivative of an integral over the moving material region. For three dimensions we state this as a fundamental theorem.

Theorem 8.19

(**The convection theorem**) Let $g = g(\mathbf{x}, t)$ be a continuously differentiable function. Then

$$\frac{d}{dt}\int_{\Omega_t} g\, d\mathbf{x} = \int_{\Omega_t}\left(\frac{Dg}{Dt} + g\nabla \cdot \mathbf{v}\right) d\mathbf{x}. \quad\square \tag{4.5}$$

Proof

The region of integration Ω_t of the integral on the left side of (4.5) depends on t, and therefore the time derivative cannot be brought under the integral sign directly. A change of variables in the integral from Eulerian coordinates $\mathbf{x}$ to Lagrangian coordinates $\mathbf{h}$ will transform the integral to one over the fixed, time-independent volume Ω_0. The derivative may then be brought under the integral sign and the calculation can proceed. Letting $\mathbf{x} = \mathbf{x}(\mathbf{h}, t)$, we have

$$\int_{\Omega_t} g(\mathbf{x}, t)\, d\mathbf{x} = \int_{\Omega_0} G(\mathbf{h}, t)J(\mathbf{h}, t)\, d\mathbf{h}.$$

Here we have used the familiar formula for changing variables in a multiple integral where the Jacobian enters as a factor in the new volume element. Then, also using Theorem 8.10,

$$
\begin{aligned}
\frac{d}{dt} \int_{\Omega_t} g \, d\mathbf{x} &= \int_{\Omega_0} \frac{\partial}{\partial t}(GJ) \, d\mathbf{h} \\
&= \int_{\Omega_0} (G_t + G\nabla \cdot \mathbf{v}|_{\mathbf{x}=\mathbf{x}(\mathbf{h},t)}) J \, d\mathbf{h} \\
&= \int_{\Omega_t} \left(\frac{Dg}{Dt} + g\nabla \cdot \mathbf{v} \right) d\mathbf{x},
\end{aligned}
$$

where in the last step we transformed back to Eulerian coordinates and used (4.4). $\qquad\square$

The following important corollary follows immediately.

Corollary 8.20

(**Reynold's transport theorem**) We have

$$
\frac{d}{dt} \int_{\Omega_t} g \, dx = \int_{\Omega_t} g_t \, dx + \int_{\partial \Omega_t} g\mathbf{v} \cdot \mathbf{n} \, d\tau, \tag{4.6}
$$

where $\mathbf{n}$ is the outer unit normal to the surface $\partial \Omega_t$. $\qquad\square$

Proof

The integrand on the right side of (4.5) is

$$
\frac{Dg}{Dt} + g\nabla \cdot \mathbf{v} = g_t + \nabla \cdot g\mathbf{v}.
$$

Thus

$$
\int_{\Omega_t} \left(\frac{Dg}{Dt} + g\nabla \cdot \mathbf{v} \right) dx = \int_{\Omega_t} g_t \, dx + \int_{\Omega_t} \nabla \cdot g\mathbf{v} \, dx.
$$

An application of the divergence theorem yields

$$
\int_{\Omega_t} \nabla \cdot g\mathbf{v} \, dx = \int_{\partial \Omega_t} g\mathbf{v} \cdot \mathbf{n} \, dA,
$$

and the corollary follows. $\qquad\square$

The interpretation is straightforward. From calculus we recall that the integral $\int_{\partial \Omega_t} g\mathbf{v} \cdot \mathbf{n} \, dA$ represents the flux of the vector field $g\mathbf{v}$ through the surface $\partial \Omega_t$. Hence (4.6) states that the time rate of change of the quantity $\int_{\Omega_t} g$, where both the integrand and the region of integration depend on t, equals the integral, frozen in time, of the change in g plus the flux or convection of g through the boundary $\partial\Omega_t$. Equation (4.6) is the three-dimensional version of Leibniz' formula.

Mass Conservation. With all this machinery, it is a simple matter to obtain a mathematical expression of mass conservation. We take it as a physical axiom that a given material volume of fluid has the same mass as it evolves in time. Symbolically

$$\int_{\Omega_t} \rho(\mathbf{x}, t)\, d\mathbf{x} = \int_{\Omega_0} \rho(\mathbf{x}, 0)\, d\mathbf{x},$$

where $\rho(\mathbf{x}, t)$ is the Eulerian density. Thus

$$\frac{d}{dt} \int_{\Omega_t} \rho(\mathbf{x}, t)\, d\mathbf{x} = 0,$$

and Theorem 8.11 gives, upon taking $g = \rho$,

$$\frac{D\rho}{Dt} + \rho \nabla \cdot \mathbf{v} = 0, \tag{4.7}$$

because the volume Ω_t is arbitrary. The first-order nonlinear partial differential equation (4.7) is the **continuity equation,** and it is a mathematical expression for **conservation of mass**.

An important class of fluid motions are those in which a material region maintains the same volume. Thus we say a fluid motion $\phi_t \colon \Omega_0 \to \Omega_t$ is **incompressible** if

$$\frac{d}{dt} \int_{\Omega_t} d x = 0, \quad \text{for all } \Omega_0.$$

As earlier, we denote the Lagrangian density by $R(\mathbf{h}, t) = \rho(\mathbf{x}(\mathbf{h}, t), t)$.

Theorem 8.21

The following are equivalent:

(i) ϕ_t *is incompressible,*

(ii) $\nabla \cdot \mathbf{v} = 0$,

(iii) $R_t = 0$,

(iv) $\frac{D\rho}{Dt} = 0$.   □

Proof

The proof is straightforward. By definition, (iii) and (iv) are equivalent, while (ii) and (iv) are equivalent by conservation of mass (4.7). Finally (i) and (ii) are equivalent by setting $g = 1$ in the convection theorem.   □

8.4.2 Dynamics

The nature of the forces on a fluid element was presented in Section 8.1 for one-dimensional flows. The idea in three dimensions is again to generalize Newton's second law, which states for a particle of mass m that $d\mathbf{p}/dt = \mathbf{F}$, or the time rate of change of momentum is the total force. For a material region Ω_t of fluid we define the **momentum** by

$$\int_{\Omega_t} \rho \mathbf{v}\, d\mathbf{x}.$$

The forces on the fluid region are of two types, body forces, which act at each point of the region, and surface forces or tractions, which act on the boundary of the region. Body forces will be denoted by $\mathbf{f}(\mathbf{x}, t)$, which represents the force per unit mass acting at the point $\mathbf{x}$ at time t. Thus the total body force on the region Ω_t is

$$\int_{\Omega_t} \rho \mathbf{f}\, d\mathbf{x}.$$

At each point $\mathbf{x}$ on the boundary $\partial\Omega_t$ of the given region Ω_t we assume that there exists a vector $\sigma(\mathbf{x}, t; \mathbf{n})$ called the **stress vector**, which represents the force per unit area acting at $\mathbf{x}$ at time t on the surface $\partial\Omega_t$ by the material exterior to Ω_t. We note that the stress at $\mathbf{x}$ depends on the orientation of the surface through $\mathbf{x}$. Simply put, different surfaces will have different stresses (see Fig. 8.15). The dependence of σ upon the orientation of the surface is denoted by the argument $\mathbf{n}$ in $\sigma(\mathbf{x}, t; \mathbf{n})$, where $\mathbf{n}$ is the outward unit normal to the surface. We think of $\mathbf{n}$ as pointing toward the material that is causing the stress. Therefore the total surface force or traction on the region Ω_t is given by

$$\int_{\partial\Omega_t} \sigma(x, t; \mathbf{n})\, dA.$$

We now postulate the **balance of momentum** principle, also called Cauchy's stress principle, which states that for all material regions Ω_t

$$\frac{d}{dt}\int_{\Omega_t} \rho \mathbf{v}\, d\mathbf{x} = \int_{\Omega_t} \rho \mathbf{f}\, d\mathbf{x} + \int_{\partial\Omega_t} \sigma(\mathbf{x}, t; \mathbf{n})\, dA. \tag{4.8}$$

In words, the time rate of change of momentum equals the total force. Because

$$\frac{d}{dt}\int_{\Omega_t} \rho \mathbf{v}\, d\mathbf{x} = \int_{\Omega_t} \rho \frac{D\mathbf{v}}{Dt}\, d\mathbf{x},$$

we may write (4.8) as

$$\int_{\Omega_t} \rho \left(\frac{D\mathbf{v}}{Dt} - \mathbf{f} \right) d\mathbf{x} = \int_{\partial\Omega_t} \sigma(\mathbf{x}, t; \mathbf{n})\, dA. \tag{4.9}$$

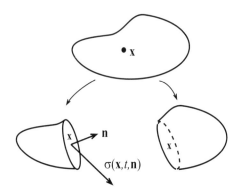

Figure 8.15 Diagram showing the dependence of the stress vector σ on the unit normal $\mathbf{n}$. At $\mathbf{x}$, the material on the right is exerting a force on the material to the left; the stress vector $\sigma(x, t; \mathbf{n})$ indicates the direction of the force.

At a given point $\mathbf{x}$ in a fluid on a surface, defined by the orientation $\mathbf{n}$, the stress on the fluid exterior to Ω_t caused by the fluid interior to Ω_t is equal and opposite to the stress on the interior caused by the fluid exterior to Ω_t. More precisely, see the following theorem.

Theorem 8.22

(**Action–reaction**) The stress vector satisfies the condition

$$\sigma(\mathbf{x}, t; -\mathbf{n}) = -\sigma(\mathbf{x}, t; \mathbf{n}). \quad \square \tag{4.10}$$

Proof

Let Ω_t be a material region and divide it into two regions Ω_t^1 and Ω_t^2 by a surface S that passes through an arbitrary point $\mathbf{x}$ in Ω_t (refer to Fig. 8.16). Applying (4.9) to Ω_t^1 and Ω_t^2, we obtain

$$\int_{\Omega_t^1} \rho \left(\frac{D\mathbf{v}}{Dt} - \mathbf{f} \right) d\mathbf{x} = \int_{\partial\Omega_t^1} \sigma \, dA, \tag{4.11}$$

$$\int_{\Omega_t^2} \rho \left(\frac{D\mathbf{v}}{Dt} - \mathbf{f} \right) d\mathbf{x} = \int_{\partial\Omega_t^2} \sigma \, dA. \tag{4.12}$$

Subtracting (4.9) from the sum of (4.11) and (4.12) gives

$$\mathbf{0} = \int_S \left(\sigma(\mathbf{x}, t; \mathbf{n}) + \sigma(\mathbf{x}, t; -\mathbf{n}) \right) dA,$$

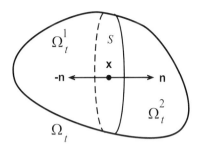

Figure 8.16 Action-reaction principle.

where $\mathbf{n}$ is the outward unit normal to Ω_t^1. The integral mean value theorem implies

$$[\sigma(\mathbf{z}, t; \mathbf{n}_1) + \sigma(\mathbf{z}, t; -\mathbf{n}_1)] \cdot \text{Area}(S) = \mathbf{0}, \tag{4.13}$$

where $\mathbf{z}$ is some point on S and $\mathbf{n}_1$ is the outward unit normal to Ω_t^1 at $\mathbf{z}$. Taking the limit as the volume of Ω_t goes to zero in such a way that the area of S goes to zero and $\mathbf{x}$ remains on S we get $\mathbf{z} \to \mathbf{x}$ and $\mathbf{n}_1 \to \mathbf{n}$, and so (4.13) implies the result.   □

As it turns out, the stress vector $\sigma(\mathbf{x}, t; \mathbf{n})$ depends in a very special way on the unit normal $\mathbf{n}$. The next theorem is one of the fundamental results in fluid mechanics.

Theorem 8.23

(**Cauchy theorem**) Let $\sigma_i(\mathbf{x}, t; \mathbf{n})$, $i = 1, 2, 3$, denote the three components of the stress vector $\sigma(\mathbf{x}, t; \mathbf{n})$, and let $\mathbf{n} = (n_1, n_2, n_3)$. Then there exists a matrix function $\sigma_{ji}(\mathbf{x}, t)$ such that

$$\sigma_i = \sum_{j=1}^{3} \sigma_{ji} n_j. \tag{4.14}$$

In fact

$$\sigma_{ji}(\mathbf{x}, t) = \sigma_i(\mathbf{x}, t; \mathbf{e}_j), \tag{4.15}$$

where $\mathbf{e}_j$, $j = 1, 2, 3$, denote the unit vectors in the direction of the coordinate axes x_1, x_2, and x_3, respectively.   □

Proof

Let t be a fixed instant of time and let Ω_t be a tetrahedron at $\mathbf{x}$ as shown in Fig. 8.17 with the three faces S_1, S_2, and S_3 parallel to the coordinate planes and the face S is the oblique face. Using the integral mean value theorem, we have

$$
\begin{aligned}
\int_{\partial \Omega_t} \sigma \, dA &= \sum_{j=1}^{3} \int_{S_j} \sigma \, dA + \int_{S} \sigma \, dA \\
&= A(S_1)\sigma(\mathbf{z}_1, t; -\mathbf{e}_1) + A(S_2)\sigma(\mathbf{z}_2, t; -\mathbf{e}_2) \\
&\quad + A(S_3)\sigma(\mathbf{z}_3, t; -\mathbf{e}_3) + A(S)\sigma(\mathbf{z}, t; \mathbf{n}),
\end{aligned}
$$

where $\mathbf{z}_i$ is a point on S_i and $\mathbf{z}$ is a point on S. Here A denotes the area function and $\mathbf{n}$ is the outer normal to S. Because

$$
A(S_j) = A(S)n_j \equiv l^2 n_j
$$

(we have defined l to be $\sqrt{A(S)}$), it follows that

$$
\frac{1}{l^2} \int_{\partial \Omega_t} \sigma \, dA = -\sum_{j=1}^{3} \sigma(\mathbf{z}_j, t; \mathbf{e}_j)n_j + \sigma(\mathbf{z}, t; \mathbf{n}),
$$

where Theorem 8.22 has been applied. Taking the limit as $l \to 0$ gives

$$
\lim_{l \to 0} \frac{1}{l^2} \int_{\partial \Omega_t} \sigma \, dA = -\sum_{j=1}^{3} \sigma(\mathbf{x}, t; \mathbf{e}_j)n_j + \sigma(\mathbf{x}, t; \mathbf{n}).
$$

We now show that

$$
\lim_{l \to 0} \frac{1}{l^2} \int_{\partial \Omega_2} \sigma(\mathbf{x}, t; \mathbf{n}) \, dA = \mathbf{0}. \tag{4.16}
$$

To this end, it follows from (4.9) that

$$
\left| \int_{\partial \Omega_t} \sigma(\mathbf{x}, t; \mathbf{n}) \, dA \right| \leq \int_{\Omega_t} \left| \rho \left(\frac{D\mathbf{v}}{Dt} - \mathbf{f} \right) \right| \, d\mathbf{x} \leq M l^3,
$$

where the last inequality results from the boundedness of the integrand. Dividing by l^2 and taking the limit as $l \to 0$ proves (4.16). Consequently

$$
\sigma(\mathbf{x}, t; \mathbf{n}) = \sum_{j=1}^{3} \sigma(\mathbf{x}, t; \mathbf{e}_j)n_j, \tag{4.17}
$$

or in component form

$$
\sigma_i(\mathbf{x}, t; \mathbf{n}) = \sum_{j=1}^{3} \sigma_i(\mathbf{x}, t; \mathbf{e}_j)n_j, \qquad i = 1, 2, 3.
$$

Defining the σ_{ji} by

$$
\sigma_{ji}(\mathbf{x}, t) \equiv \sigma_i(\mathbf{x}, t; \mathbf{e}_j)
$$

gives the final result. $\square$

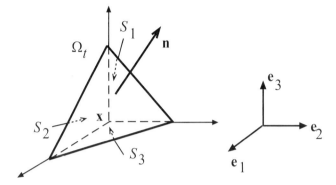

Figure 8.17 Proof of Cauchy's theorem.

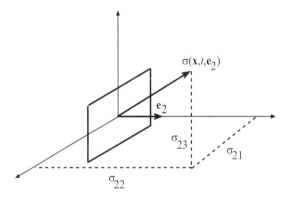

Figure 8.18 The stress vector $\sigma(\mathbf{x}, t, \mathbf{e}_2)$ and its three stress components.

The nine quantities σ_{ji} are the components of the **stress tensor**. By definition, σ_{ji} is the ith component of the stress vector on the face whose normal is $\mathbf{e}_j$. For example, σ_{21}, σ_{22}, and σ_{23} are shown in Fig. 8.18. Equation (4.17) shows that $\sigma(\mathbf{x}, t; \mathbf{n})$ can be resolved into a linear combination of $\sigma(\mathbf{x}, t; \mathbf{e}_1)$, $\sigma(\mathbf{x}, t; \mathbf{e}_2)$, and $\sigma(\mathbf{x}, t, \mathbf{e}_3)$, that is, into three stresses that are the stresses on the coordinate planes at $\mathbf{x}$. These stresses are shown in Fig. 8.19. The components of the three vectors shown are the σ_{ji}.

With the aid of Theorem 8.23 a vector partial differential equation can be formulated that expresses the momentum balance law (4.9). This calculation is facilitated by adopting a notational convention that saves in writing summation signs in complicated expressions. This practice is called the *summation convention* and it assumes that a sum is taken over any repeated index in a

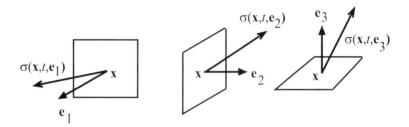

Figure 8.19 Stress tensor components.

given term. Thus

$$a_i b_{ij} \quad \text{means} \quad \sum_i a_i b_{ij},$$

and

$$a_i b_{ij} c_j \quad \text{means} \quad \sum_i \sum_j a_i b_{ij} c_j.$$

The range of the summation index (or indices) is determined from context. For example (4.14) may be written $\sigma_i = \sigma_{ji} n_j$ with a sum over $j = 1$ to $j = 3$ assumed on the right side. Any index not summed in a given expression is called a free index. Free indices vary over their appropriate ranges; for example, in (4.14) the index i is free and ranges over $i = 1, 2, 3$. In (4.15) both i and j are free with $i, j = 1, 2, 3$. In terms of this convention the divergence theorem may be expressed

$$\int_\Omega \frac{\partial}{\partial x_i} f_i \, dx = \int_{\partial \Omega} f_i n_i \, dA, \tag{4.18}$$

where $\mathbf{f} = (f_1, f_2, f_3)$ and $\mathbf{n} = (n_1, n_2, n_3)$. On both sides of the equation i is assumed to be summed from $i = 1$ to $i = 3$.

In component form (4.9) is

$$\int_{\Omega_t} \rho \left(\frac{Dv_i}{Dt} - f_i \right) dx = \int_{\partial \Omega_t} \sigma_i \, dA. \tag{4.19}$$

But the right side is

$$\int_{\partial \Omega_t} \sigma_i \, dA = \int_{\partial \Omega_t} \sigma_{ji} n_j \, dA = \int_{\Omega_t} \frac{\partial}{\partial x_j} \sigma_{ji} \, dx,$$

where we have used (4.14) and (4.18). Then (4.19) becomes

$$\int_{\Omega_t} \left(\rho \frac{Dv_i}{Dt} - \rho f_i - \frac{\partial}{\partial x_j} \sigma_{ji} \right) dx = 0, \qquad i = 1, 2, 3.$$

Because of the arbitrariness of the region of integration

$$\rho \frac{Dv_i}{Dt} = \rho f_i + \frac{\partial}{\partial x_j} \sigma_{ji}, \qquad i = 1, 2, 3. \tag{4.20}$$

Equations (4.20) are the **Cauchy equations** or **equations of motion**, and they represent balance of linear momentum. Along with the continuity equation (4.7) we have four equations for ρ, the three components v_i of the velocity, and the nine components σ_{ji} of the stress tensor. What remains, which is the most difficult issue of all, is specification of the stress tensor.

There is a common way to write the Cauchy stress principle in terms of a fixed laboratory volume Ω_1. It is expressed by the following theorem.

Theorem 8.24

Cauchy's stress principle (4.8) is equivalent to

$$\frac{d}{dt} \int_{\Omega_1} \rho v_i \, d\mathbf{x} + \int_{\partial \Omega_1} \rho v_i v_j n_j \, dA = \int_{\Omega_1} \rho f_i \, d\mathbf{x} + \int_{\partial \Omega_1} \sigma_i \, dA, \tag{4.21}$$

where Ω_1 is any fixed region in $\mathbb{R}^3$. $\square$

Proof

Using Corollary 8.12 with $g \equiv \rho v_i$ gives

$$\frac{d}{dt} \int_{\Omega_t} \rho v_i \, d\mathbf{x} = \int_{\Omega_t} (\rho v_i)_t \, d\mathbf{x} + \int_{\partial \Omega_t} \rho v_i v_j n_j \, dA.$$

Therefore

$$\int_{\Omega_t} \rho \frac{Dv_i}{Dt} \, d\mathbf{x} = \int_{\Omega_t} (\rho v_i)_t \, d\mathbf{x} + \int_{\partial \Omega_t} \rho v_i v_j n_j \, dA.$$

Now let Ω_1 be the volume that coincides with Ω_t at $t = t_1$. Then

$$\int_{\Omega_1} \rho \frac{Dv_i}{Dt} \, d\mathbf{x} = \frac{d}{dt} \int_{\Omega_1} \rho v_i \, d\mathbf{x} + \int_{\partial \Omega_1} \rho v_i v_j n_j \, dA.$$

Applying (4.9) gives (4.21). $\sqcup$

Example 8.25

(**Force on an object**) As an application of (4.21) we compute the force on a stationary object in a steady fluid flow. Suppose S_1 is the boundary of the object and let S be an imaginary surface that contains the object (see Fig. 8.20).

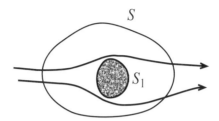

Figure 8.20 Flow around an obstacle with boundary S_1. S is the boundary of any region containing the obstacle.

Assume the flow is steady and that there is no body force ($\mathbf{f} = \mathbf{0}$). Letting Ω_1 be the region in between S and S_1 and applying the vector form of (4.21) gives

$$\int_{\partial\Omega_1} (\rho\mathbf{v}(\mathbf{v}\cdot\mathbf{n}) - \sigma)\, dA = \mathbf{0}, \qquad \partial\Omega_1 = S_1 \cup S,$$

or

$$\int_S (\rho\mathbf{v}(\mathbf{v}\cdot\mathbf{n}) - \sigma)\, dA = \int_{S_1} (\sigma - \rho\mathbf{v}(\mathbf{v}\cdot\mathbf{n}))\, dA = \int_{S_1} \sigma\, dA,$$

where the fact that $\mathbf{v}\cdot\mathbf{n} = 0$ on S_1 has been used. But the force $\mathbf{F}$ on the obstacle is

$$
\begin{aligned}
\mathbf{F} &= -\int_{S_1} \sigma\, dA \\
&= \int_S (\sigma - \rho\mathbf{v}(\mathbf{v}\cdot\mathbf{n}))\, dA.
\end{aligned}
$$

This is a useful result since it permits the calculation of $\mathbf{F}$ at a control surface at a distance from the obstacle; it may be virtually impossible to obtain the stresses on the surface of the obstacle itself. □

8.4.3 Energy and Constitutive Theory

The four equations, mass and momentum conservation exhibited in (4.7) and (4.20), clearly do not represent a complete determined system for all of the unknowns. At this point in the one-dimensional case we introduced an equation of state or constitutive relation. In the present case specification of the stress tensor σ_{ij} is required to close the system.

For a fluid (gas or liquid) in a motionless state it is clear that the stress vector σ at a point on a surface in the fluid is always normal to the surface, that is, it is in the direction of the unit normal $\mathbf{n}$ (or $-\mathbf{n}$). This is not true for solids, since they can sustain complicated shear or tangential stresses and not undergo motion. This normality property extends to fluids in uniform motion where the velocity field is constant over some time interval.

Example 8.26

(**Inviscid fluid**) One of the simplest classes of general nonuniform fluid motions is that class of motions termed **inviscid** where the normality assumption holds, namely

$$\sigma(\mathbf{x}, t; \mathbf{n}) = -p(\mathbf{x}, t)\mathbf{n}. \tag{4.22}$$

Hence, for inviscid flow the stress across a surface is proportional to the normal vector where the proportionality factor p is the **pressure**. The concept of an inviscid fluid is highly useful in technology, since many real fluids are actually modeled by (4.22) and the calculations are far simpler using (4.22) than for more complicated constitutive relations. From (4.22) it is easy to calculate the components σ_{ij} of the stress tensor. We have

$$\sigma_{ij} = -p(\mathbf{x}, t)\delta_{ij}, \tag{4.23}$$

where δ_{ij} is the Kronecker delta symbol defined by $\delta_{ij} = 1$ if $i = j$ and $\delta_{ij} = 0$ if $i \neq j$. Finally, if we denote the stress matrix $T = z(\sigma_{ij})$, then (4.22) can be written simply as

$$T = -p(\mathbf{x}, t)I,$$

where I is the identity matrix. ⊔

From this last example, it follows that

$$\frac{\partial}{\partial x_j}\sigma_{ji} = -\frac{\partial}{\partial x_j}\left(p(\mathbf{x}, t)\delta_{ij}\right) = -\frac{\partial}{\partial x_i}p(\mathbf{x}, t),$$

we may write the Cauchy equations (4.20) as

$$\rho\frac{Dv_i}{Dt} = \rho f_i - \frac{\partial p}{\partial x_i}, \qquad i = 1, 2, 3,$$

or in vector form as

$$\rho\frac{D\mathbf{v}}{Dt} = \rho\mathbf{f} - \nabla p. \tag{4.24}$$

Equation (4.24) is called the **Euler equation**. Equations (4.7) and (4.24) represent four equations for the five unknowns ρ, $\mathbf{v}$, and p. Supplemented by equations of state and an energy conservation equation, a complete set of equations can be found. For inviscid fluids the discussion now follows the one-dimensional analysis presented in Section 8.2.

Example 8.27

(**Ideal fluid**) An **ideal fluid** is an inviscid fluid undergoing an incompressible motion, and much of the literature in fluid mechanics deals with ideal fluids. The ideal fluid equations consist of the Euler equations (4.24) and $\nabla \cdot \mathbf{v} = 0$.
□

Equation (4.23) suggests that in the most general case the stress tensor takes the form

$$\sigma_{ij} = -p\delta_{ij} + \tau_{ij}, \tag{4.25}$$

where p is identified with the thermodynamic pressure and τ_{ij} define the components of the oblique, or non-normal, stresses across a fluid surface. These oblique stress are partly caused by viscous forces where adjacent fluid elements undergo shear stresses across surfaces separating them. The τ_{ij} define the components of the **viscous stress tensor**. . As one may imagine, it is a difficult task to specify how these components relate to the other flow variables.

Energy balance. As in the one-dimensional case, we postulate a **balance of energy law**:

$$\frac{d}{dt}\int_{\Omega_t}\left(\frac{1}{2}\rho\mathbf{v}\cdot\mathbf{v}+\rho e\right)d\mathbf{x}=\int_{\Omega_t}\rho\mathbf{f}\cdot\mathbf{v}\,d\mathbf{x}+\int_{\partial\Omega_t}\sigma\cdot\mathbf{v}\,dA-\int_{\partial\Omega_t}\mathbf{q}\cdot\mathbf{n}\,dA, \tag{4.26}$$

where Ω_t is an arbitrary material region with exterior unit normal $\mathbf{n}$ and $\mathbf{q}$ represents the heat flux density. In words, the time rate of change of the kinetic plus internal energy in Ω_t equals the rate work is done by the body forces $\mathbf{f}$, plus the rate work is done by the surface stresses σ, plus the rate heat flows into the region. A differential form of the energy balance law can be easily obtained by appealing to the arbitrariness of Ω_t after an application of the divergence theorem and Cauchy's theorem. We record the result, leaving the standard argument for the reader.

Theorem 8.28

If the functions are sufficiently smooth, equation (4.26) implies

$$\rho\frac{D}{Dt}\left(\frac{1}{2}\mathbf{v}\cdot\mathbf{v}+e\right)=\rho\mathbf{f}\cdot\mathbf{v}+\frac{\partial}{\partial x_j}(\sigma_{ji}v_i)-\nabla\cdot\mathbf{q}. \quad \square \tag{4.27}$$

Theorem 8.29

The change in internal energy is given by

$$\rho\frac{De}{Dt}=-\nabla\cdot\mathbf{q}+\sigma_{ji}\frac{\partial v_i}{\partial x_j}. \quad \square \tag{4.28}$$

Proof

Multiplying the momentum balance law (4.20) by v_i and summing over $i = 1, 2, 3$ gives

$$\frac{\rho}{2} \frac{D}{Dt} (\mathbf{v} \cdot \mathbf{v}) = \rho f_i v_i + \frac{\partial}{\partial x_j} (\sigma_{ji} v_i) - \sigma_{ji} \frac{\partial v_i}{\partial x_j}.$$

Subtracting from (4.27) yields (4.28) and hence the result. □

From the combined form of the first and second laws of thermodynamics,

$$\theta \frac{Ds}{Dt} = \frac{De}{Dt} + p \frac{D(1/\rho)}{Dt}, \tag{4.29}$$

we can obtain an expression for how the entropy of a fluid particle changes. Combining (4.28) and (4.29) and using $D\rho^{-1}/Dt = -\rho^{-2}D\rho/Dt = \rho^{-1}\nabla \cdot \mathbf{v}$, we get the following theorem.

Theorem 8.30

The energy balance equation is equivalent to

$$\rho\theta \frac{Ds}{Dt} = p\nabla \cdot \mathbf{v} - \nabla \cdot \mathbf{q} + \sigma_{ji} \frac{\partial v_i}{\partial x_j}. \quad \Box \tag{4.30}$$

We can interpret the preceding energy equation as follows. The left side is the heat added; the term $-\nabla \cdot \mathbf{q}$ on the right represents the heat flux into the system, and therefore the sum of the other two terms, $\Psi \equiv p\nabla \cdot \mathbf{v} + \sigma_{ji}\partial v_i/\partial x_j = \tau_{ji}\partial v_i/\partial x_j$ must represent the heat generated due to work done by the forces to deform the system. The function Ψ is called the **dissipation function**, and, more precisely, it is the rate per unit volume at which mechanical energy is dissipated into heat.

With this notation, the energy equation (4.30) may be written

$$\rho \frac{Ds}{Dt} = -\frac{1}{\theta}\nabla \cdot \mathbf{q} + \frac{1}{\theta} \Psi. \tag{4.31}$$

We can obtain some important conclusions from these results. Thermodynamics also requires that the entropy increase equal or exceed the heat added divided by the absolute temperature. Thus we *postulate* the inequality

$$\frac{d}{dt} \int_{\Omega_t} \rho s \, d\mathbf{x} \geq -\int_{\partial\Omega_t} \frac{1}{\theta} \mathbf{q} \cdot \mathbf{n} \, dA.$$

The differential form of this axiom is called the **Clausius–Duhem inequality**:

$$\rho \frac{Ds}{Dt} \geq -\nabla \cdot (\theta^{-1}\mathbf{q}). \tag{4.32}$$

Equations (4.31) and (4.32) may be combined to get

$$\theta^{-1}\Psi - \theta^{-2}\mathbf{q} \cdot \nabla \theta \geq 0. \tag{4.33}$$

Sufficient conditions for (4.33) are

$$\Psi \geq 0, \qquad -\mathbf{q} \cdot \nabla \theta \geq 0.$$

That is, (i) *deformation does not convert heat into mechanical energy*, and (ii) *heat flows against the temperature gradient*.

Viscous stress. To proceed further, some assumption is required regarding the form of the viscous stress tensor τ_{ij}. A **Newtonian fluid** is one in which there is a linear dependence of τ_{ij} on the **rate of deformation**

$$D_{ij} \equiv \tfrac{1}{2}(\partial v_i/\partial x_j + \partial v_j/\partial x_i).$$

That is,

$$\tau_{ij} = C_{ijrs}D_{rs},$$

where the coefficients C_{ijrs} may depend on the local thermodynamic states θ and ρ. Arguments of symmetry and invariance of the stress tensor under translations and rotations lead to (see, e.g., Segel and Handelman (1977) for an enlightening treatment)

$$\tau_{ij} = 2\mu D_{ij} + \lambda\delta_{ij}\nabla \cdot \mathbf{v},$$

where μ and λ are coefficients of viscosity. Generally μ and λ may depend on temperature or density, but here we assume they are constant. The previous assumption forces the stress tensor to have the form

$$\sigma_{ij} = (-p + \lambda\nabla \cdot \mathbf{v})\delta_{ij} + 2\mu D_{ij},$$

and therefore the momentum law (4.20) becomes

$$\rho \frac{Dv_i}{Dt} = \rho f_i - \frac{\partial}{\partial x_i}(p - (\lambda + \mu)\nabla \cdot \mathbf{v}) + \mu\nabla^2 v_i \tag{4.34}$$

for $i = 1, 2, 3$.

In the incompressible case where

$$\nabla \cdot \mathbf{v} = 0$$

(4.34) reduces in vector form to

$$\rho \frac{D\mathbf{v}}{Dt} = \rho\mathbf{f} - \nabla p + \mu\nabla^2\mathbf{v}, \tag{4.35}$$

which are the **Navier–Stokes equations**. These last two equations are among the most studied equations in fluid mechanics, and still many unanswered questions regarding them are yet to be proved. Exact solutions of (4.35) in special cases are given in the exercises.

Finally, for viscous flows it is generally assumed that the fluid adheres to a rigid boundary. This leads to the **adherence boundary condition**

$$\lim_{\mathbf{x} \to \mathbf{x}_b} \mathbf{v}(\mathbf{x}) = \mathbf{v}(\mathbf{x}_b),$$

where $\mathbf{x}_b$ is a point on a rigid boundary and $\mathbf{v}(\mathbf{x}_b)$ is the known velocity of the boundary.

EXERCISES

1. Prove the Euler expansion when $n = 2$. Hint: Recall the rule for differentiating a determinant:

$$\frac{d}{dt} \begin{vmatrix} A & B \\ C & D \end{vmatrix} = \begin{vmatrix} \frac{dA}{dt} & B \\ \frac{dC}{dt} & D \end{vmatrix} + \begin{vmatrix} A & \frac{dB}{dt} \\ C & \frac{dD}{dt} \end{vmatrix}.$$

2. Let $R(\mathbf{h}, t)$ be the Lagrangian density. Show that

$$J(\mathbf{h}, t) = \frac{R(\mathbf{h}, 0)}{R(\mathbf{h}, t)}.$$

3. Derive the Lagrangian form of the mass conservation law

$$\Delta_t + (\nabla \cdot \mathbf{v}|_{\mathbf{h}})R = 0.$$

4. Prove that if mass conservation holds, then

$$\frac{d}{dt} \int_{\Omega_t} \rho g \, d\mathbf{x} = \int_{\Omega_t} \rho \frac{Dg}{Dt} \, d\mathbf{x}$$

for any continuously differentiable function g. Show that the scalar function g can be replaced by a vector function $\mathbf{g}$.

5. If $\sigma(\mathbf{x}, t; \mathbf{n}) = -p(\mathbf{x}, t)\mathbf{n}$, prove that $\sigma_{ij} = -p(\mathbf{x}, t)\delta_{ij}$.

6. A two-dimensional fluid motion is given by

$$x_1 = h_1 e^t, \quad x_2 = h_2 e^{-t}, \quad t > 0.$$

a) Find $\mathbf{V}$ and $\mathbf{v}$.

b) Find the streamlines and show they coincide with the particle paths.

c) If $\rho(\mathbf{x}, t) = x_1 x_2$, show that the motion is incompressible.

7. The Eulerian velocity of a two-dimensional fluid flow is

$$\mathbf{v} = \langle t, x_1 + 1 \rangle, \quad t > 0.$$

Find the streamlines passing through the fixed point (a_0, b_0) and find the particle path of the particle $(h_1, h_2) = (a_0, b_0)$. Does the particle path ever coincide with a streamline at any time t_0?

8. A steady flow in two dimensions is defined by

$$\mathbf{v} = \langle 2x_1 + 3x_2, x_1 - x_2 \rangle, \quad t > 0.$$

Find the particle paths. Compute the acceleration measured by an observer located at the fixed point $(3, 4)$.

9. A flow is called **potential** on a region D if there exists a continuously differentiable function $\phi(t, \mathbf{x})$ on D for which $\mathbf{v} = \nabla \phi$, and a flow is called **irrotational** on D if $\nabla \times \mathbf{v} = \mathbf{0}$.

a) Prove that a potential flow is irrotational.

b) Show that an irrotational flow is not always potential by considering

$$\mathbf{v} = \left\langle \frac{-y}{x^2 + y^2}, \frac{x}{x^2 + y^2} \right\rangle \quad \text{on } \mathbb{R}^2 - \{\mathbf{0}\}.$$

c) Show that if D is simply connected then irrotational flows are potential.

10. a) Show that
$$(\mathbf{v} \cdot \nabla)\mathbf{v} = \tfrac{1}{2} \nabla |\mathbf{v}|^2 - \mathbf{v} \times (\nabla \times \mathbf{v}).$$

b) Prove **Bernoulli's theorem**: For a potential flow of an inviscid fluid, suppose there exists a function $\psi(t, \mathbf{x})$ for which $\mathbf{f} = -\nabla \psi$. Then for any path C connecting two points $\mathbf{a}$ in $\mathbf{b}$ in a region where $\mathbf{v} = \nabla \phi$,

$$\left(\phi_t + \frac{1}{2} |\mathbf{v}|^2 + \psi \right)\Big|_a^b + \int_C \frac{dp}{\rho} = 0.$$

(Hint: Integrate the equation of motion and use (a).)

11. In a region D where a flow is steady, irrotational, and incompressible with constant density ρ, prove that

$$\frac{1}{2} |\mathbf{v}|^2 + \frac{1}{\rho} p + \psi = \text{constant},$$

where $\mathbf{f} = -\nabla \psi$.

12. For a fluid motion with $\sigma_{ij} = \sigma_{ji}$ prove that

$$\frac{dK}{dt} = \int_{\Omega_t} \rho \mathbf{f} \cdot \mathbf{v} \, dx + \int_{\partial\Omega_t} \sigma \cdot \mathbf{v} \, dA - \int_{\Omega_t} \sigma_{ij} D_{ij} \, dx$$

where $K(\Omega_t) \equiv \frac{1}{2} \int_{\Omega_t} \rho \mathbf{v} \cdot \mathbf{v} \, dx$ is the kinetic energy of Ω_t. This result is the **energy transport theorem**.

13. For an ideal fluid prove that $K(\Omega) = $ constant, where Ω is a fixed region in $\mathbb{R}^3$ and $\mathbf{v}$ is parallel to $\partial\Omega$ (see Exercise 12).

14. The alternating symbol ε_{ijk} is defined by

$$\varepsilon_{ijk} = \begin{cases} 0, & \text{if any two indices are equal,} \\ 1, & \text{if } (ijk) \text{ is an even permutation of } 123, \\ -1, & \text{if } (ijk) \text{ is an odd permutation of } 123. \end{cases}$$

A permutation (ijk) is an even (odd) permutation of (123) if it takes an even (odd) number of switches of adjacent elements to get it to the form (123). For example,

$$321 \rightarrow 231 \rightarrow 213 \rightarrow 123,$$

so (321) is an odd permutation of (123). Show that if $\mathbf{a} = \langle a_1, a_2, a_3 \rangle$ and $\mathbf{b} = \langle b_1, b_2, b_3 \rangle$, then the kth component of the cross product $\mathbf{a} \times \mathbf{b}$ is given by

$$(\mathbf{a} \times \mathbf{b})_k = \varepsilon_{ijk} a_i b_j, \qquad \text{sum on } i, j = 1, 2, 3.$$

15. (**Conservation of Angular Momentum**) A particle of mass m moving in $\mathbb{R}$ with velocity vector $\mathbf{v}$ has linear momentum $\mathbf{p} = m\mathbf{v}$ and angular momentum about the origin $\mathbf{0}$ given by $\mathbf{L} = \mathbf{x} \times \mathbf{p}$, where $\mathbf{x}$ is its position vector. If $\mathbf{F}$ is a force on the particle, then $\mathbf{F} = d\mathbf{p}/dt$ by Newton's second law. Show that $\mathbf{N} = d\mathbf{L}/dt$, where $\mathbf{N} = \mathbf{x} \times \mathbf{F}$ is the torque (or moment of force) about $\mathbf{0}$. For a continuum argue that one should postulate

$$\frac{d}{dt} \int_{\Omega_t} (\mathbf{x} \times \rho \mathbf{v}) \, dx = \int_{\Omega_t} (\mathbf{x} \times \rho \mathbf{f}) \, dx + \int_{\partial\Omega_t} (\mathbf{x} \times \sigma) \, dA, \qquad (4.36)$$

where Ω_t is any material region. Write (4.36) in component form using the alternating symbol ε_{ijk}. Note that (4.36) does not account for any internal angular momentum of the fluid particles that would occur in a fluid where the particles are rotating rods; such fluids are called *polar*.

16. Prove that (4.36) of Exercise 15 is equivalent to the symmetry of the stress tensor σ_{ij} (i.e., $\sigma_{ij} = \sigma_{ji}$). Hint:

$$\frac{d}{dt} \int_{\Omega_t} \varepsilon_{ijk} x_i \rho v_j \, dx = \int_{\Omega_t} \varepsilon^{ijk} x_i \rho \frac{Dv_j}{Dt} \, dx.$$

Use the equation of motion and

$$\frac{\partial}{\partial x_l}\,(x_i\sigma_{lj}) = x_i\,\frac{\partial}{\partial x_l}\,\sigma_{lj} + \sigma_{ij},$$

followed by an application of the divergence theorem. This theorem shows that balance of angular momentum is equivalent to symmetry of the stress tensor. Thus, of the nine components of σ_{ij} only six are independent.

17. Let $\Gamma \subseteq \Omega_0$ be a simple closed curve and let $\Gamma_t = \phi_t(\Gamma)$, where ϕ_t is an isentropic fluid motion. The *circulation* about Γ_t is defined by the line integral $C_{\Gamma_t} \equiv \int_{\Gamma_t} \mathbf{v}\cdot d\mathbf{l}$. Prove **Kelvin's theorem,** which asserts that C_{Γ_t} is a constant in time.

18. The **vorticity** ω of a flow is defined by $\omega = \nabla\times\mathbf{v}$. Determine the vorticity of the following flows and sketch the streamlines in each case.

   a) $\mathbf{v} = v_0\exp(-x_2^2)\mathbf{j}$.

   b) $\mathbf{v} = v_0\exp(-x_2^2)\mathbf{j}$.

   c) $\mathbf{v} = \omega_0(-x_2\mathbf{i} + x_1\mathbf{j})$.

19. If ρ is constant and $\mathbf{f} = \nabla\psi$ for some ψ, prove that (4.35) implies the *vorticity equation*

$$\frac{D\omega}{Dt} = (\omega\cdot\mathbf{grad})\mathbf{v} + \frac{\mu}{\rho}\,\Delta\omega.$$

20. Consider an incompressible viscous flow of constant density ρ_0 under the influence of no body forces governed by the Navier–Stokes equation (4.35).

   a) If l, ν/l^2, U, and $\rho_0 vU/l$, where $v \equiv \mu/\rho_0$ are length, time, velocity, and pressure scales, show that in dimensionless form the governing equations can be written

$$\nabla\cdot\mathbf{v} = 0, \qquad \mathbf{v}_t + (\mathbf{v}\cdot\nabla)\mathbf{v} = -\nabla p + \frac{1}{\mathrm{Re}}\Delta\mathbf{v},$$

   where $\mathrm{Re} \equiv Ul/\nu$ is a constant called the **Reynolds number**. The constant ν is the kinematic viscosity.

   b) For flows with small Reynolds number, that is, $\mathrm{Re} \ll 1$, show that

$$\Delta p = 0, \qquad \frac{\partial}{\partial t}\,(\Delta\mathbf{v}) = \Delta(\Delta\mathbf{v}).$$

   (Hint: $\nabla\times(\nabla\times\mathbf{v}) = \nabla(\nabla\cdot\mathbf{v}) - \Delta\mathbf{v}$.)

21. (**Plane Couette flow**) An incompressible viscous fluid of constant density under no body forces is confined to lie between two infinite flat plates at $z = 0$ and $z = d$ in $\mathbb{R}^3$. The lower plate $z = 0$ is stationary and the upper plate $z = d$ is moved at constant velocity U in the x direction. That is, if $\mathbf{v} = (u, v, w)$, the boundary conditions $\mathbf{v} = \mathbf{0}$ on $z = 0$ and $\mathbf{v} = \langle U, 0, 0 \rangle$ on $z = d$. Write out the governing equations and show that the velocity field and pressure are given by

$$u = \frac{U}{d}\, z, \quad v = w = 0, \quad p = p_0 = \text{constant}.$$

Show that the stress on the lower plate is $\mu U/d$.

22. (**Plane Poiseuille flow**) An incompressible viscous fluid of constant density under no body forces is confined to lie between two stationary infinite planes $z = -d$ and $z = d$. A flow is forced by a constant pressure gradient $\nabla p = \langle -C, 0, 0 \rangle$, $C > 0$. Show that the fluid velocity is given by

$$u = \frac{C}{2\mu}\,(d^2 - z^2), \quad v = w = 0.$$

What is the pressure? Show that the mass flow per unit width in the x direction is given by $2Cd^3/3\mu$.

23. (**Poiseuille flow in a pipe**) Consider a cylindrical pipe of length L and radius R where an incompressible, viscous fluid is flowing (e.g., blood flow in an artery). In cylindrical coordinates (r, θ, z), denote the velocity field by $\mathbf{v} = \langle v_r, v_\theta, v_z \rangle$. Then assume boundary conditions $p = p_0$ at $z = 0$, $p = p_L$ at $z = L$, and $\mathbf{v} = 0$ on $r = R$, and $v_r = v_\theta = 0$ at $z = 0, L$.

a) Write out the governing partial differential equations in cylindrical form. (You may have to look up forms for the divergence and gradient operations in cylindrical coordinates.) Physically, justify the boundary conditions.

b) Assuming a solution of the form $v_z = v_z(r, z)$, $p = p(r, z)$, with $v_\theta = v_r = 0$, show that the pressure is linear along the length of the tube and

$$v_z = \frac{p_0 - p_L}{4\mu L}\,(R^2 - r^2).$$

c) Show that the velocity profile in the tube has the shape of a paraboloid.

REFERENCES AND NOTES

Continuum mechanics has long been a central part of applied mathematics, and it is an entity within itself. If fact, not many years ago continuum mechanics was considered an area of applied mathematics as much as engineering is today. The literature is enormous. As such, it is not possible to reference a large selection of texts or articles. We mention a few elementary introductions that relate to our material and to various extensions.

A concise, readable, elementary introduction is Chorin and Marsden (1993). A thorough and outstanding presentation is given in Segel and Handelman (1987). Classic texts for gas dynamics and shock waves include Courant and Friedrichs (1948) and Whitham (1974). Holmes (2009), very similar to Segel and Handelman and earlier editions of the present text, has a very readable introduction to constitutive relations in solid mechanics. Hydrogeological systems are discussed in de Marsily (1986); see also Logan (2001) for emphasis on contaminant transport and for additional references.

Chorin, A. J. & Marsden, J. 1993. *A Mathematical Introduction to Fluid Mechanics*, 3rd ed., Springer-Verlag, New York.

Courant, R. & K. O. Friedrichs, K. O. 1948. *Supersonic Flow and Shock Waves*, Wiley-Interscience, New York (reprinted by Springer-Verlag, New York, 1976).

Holmes, M. 2009. *Introduction to the Foundations of Applied Mathematics*, Springer-Verlag, New York.

Logan, J. D. 2001. *Transport Modeling in Hydrogeochemical Systems*, Springer-Verlag, New York.

de Marsily, G. 1986. *Quantitative Hydrogeology*, Academic Press, Inc., San Diego.

Segel, L. A. & Handelman, G. H. 1977. *Mathematics Applied to Continuum Mechanics*, Macmillan, New York (reprinted by Dover Publications, New York 1987).

Whitham, G. B., 1974. *Linear and Nonlinear Waves*, Wiley-Interscience, New York.

9
Discrete Models

From elementary courses we know that differential equations are often approximated by discrete, or, finite difference formulas; for example, the Euler and the Runge–Kutta numerical algorithms lead to difference equations. But, as well, discrete equations arise naturally without a continuous counterpart. The recent emphasis on quantitative methods in the biological sciences has brought discrete models to the forefront, not only in classical areas like population dynamics and epidemiology, but in newer applications in genomics arising from the accumulation of DNA sequence data, bioinformatics, and other phylogenetic processes. The digital revolution in electrical engineering has made discrete models and discrete transforms central in the development of advanced technological devices. Digital signal processing is a major area of study. Discrete models are conceptually simpler than their differential equation counterparts. However, discrete models are less amenable to analytic solution techniques and their dynamics can be more complicated, often exhibiting cycling and chaotic behavior.

Another important endeavor is to understand how stochasticity, or randomness, enters and affects various systems. For example, there is always stochasticity in the environment that affects populations, noise in electrical circuits and devices that affect their responses, and random forces on structures that affect their vibrations. We consider some aspects of randomness in this chapter as well.

Applied Mathematics 4th ed.,
By J. David Logan
Copyright © 2013 John Wiley & Sons, Inc.

9.1 One-Dimensional Models

9.1.1 Linear and Nonlinear Models

In differential equations the time t runs continuously, and so differential equations are often referred to as continuous time models. However, some processes are better formulated as **discrete time models** where time ticks off in discrete units $t = 0, 1, 2, 3, ...$ (say, in days, months, or years, etc.). For example, if the money in a savings account is compounded monthly, then we need only compute the principal each month. In a fisheries model, the number of fish may be estimated once a year. Or, a wildlife conservationist may census a deer population in the spring and fall to estimate their numbers and make decisions on allowable harvesting rates. Data is usually collected at discrete times. In electrical engineering, continuous data can be generated graphically on an oscilloscope, but digital data is equally common and is collected, for example, in a sampling process, in discrete time steps.

In a discrete time model the state function is a sequence x_t, i.e., x_0, x_1, x_2, x_3, ..., rather than a continuous function. The subscripts denote the time; e.g., in a weekly census of mosquitos grown in a laboratory, x_5 would denote the number of mosquitos at the fifth week. We graph the states, or sequence, x_t as a set of points (t, x_t) in a tx plane, often connecting them by straight line segments.

A one-stage, or first-order, **discrete time model**, the analog of a first-order differential equation, is an equation of the form

$$x_{t+1} = f(t, x_t), \quad t = 0, 1, 2, 3, ..., \tag{1.1}$$

where f is a given function. Such equations are called **difference equations** or **recursion relations**. Knowledge of the initial state x_0 allows us to compute the subsequent states recursively, in terms of the previously computed states. Thus, (1.1) is a deterministic update rule that tells us how to compute the next value in terms of the previous one. A sequence x_t that satisfies the model is a solution to the equation.

If $f(t, x_t) = a_t x_t + b_t$, where a_t and b_t are given, fixed sequences, then the model is linear; otherwise, (1.1) is nonlinear. In the sequel we mostly examine the autonomous equation

$$x_{t+1} = f(x_t), \quad t = 0, 1, 2, 3, ..., \tag{1.2}$$

where the right side does not depend explicitly on t. Discrete models are also defined in terms of changes $\Delta x_t = x_{t+1} - x_t$ of the state x_t. Thus,

$$\Delta x_t = g(x_t)$$

defines a discrete model, where g is the given change. Finally, some models are also defined in terms of the relative change, or *per capita* change, $\Delta x_t/x_t$. Thus,

$$\frac{\Delta x_t}{x_t} = h(x_t),$$

where h is the given *per capita* change. Any form can be obtained easily from another by simple algebra.

A second-order, discrete time model has the form

$$x_{t+2} = f(x_{t+1}, x_t), \quad t = 0, 1, 2, 3, \dots.$$

Now a state depends upon two preceding states, and both x_0 and x_1 are required to start the process.

Difference equations are recursion formulas, and programs for computing the values x_t are easily composed on computer algebra systems and graphing calculators.

Example 9.1

The simplest discrete model, the **growth-decay** process, is linear and has the form

$$x_{t+1} = x_t + rx_t = (1+r)x_t. \tag{1.3}$$

For example, if x_t is the principal at month t in a savings account that earns 0.3% per month, then the principal at the $(t+1)$st month is $x_{t+1} = x_t + 0.003x_t$. We can perform successive iterations to obtain

$$\begin{aligned}
x_1 &= (1+r)x_0, \\
x_2 &= (1+r)x_1 = (1+r)^2 x_0, \\
x_3 &= (1+r)x_2 = (1+r)^3 x_0,
\end{aligned}$$

and so on. By induction, the solution to (1.3) is

$$x_n = (1+r)^t x_0.$$

If $r > 0$ then x_t grows geometrically and we have a **growth model**. If $-1 < r < 0$, then $1+r$ is a proper fraction and x_t goes to zero geometrically; this is a **decay model**. If $-2 < r < -1$, then the factor $1+r$ is negative and the solution x_t will oscillate between negative and positive values as it converges to zero. Finally, if $r < -2$ the solution oscillates without bound. It is also easily checked by direct substitution that the sequence

$$x_t = C(1+r)^t$$

is a solution to the equation for any value of C:

$$x_{t+1} = C(1+r)^{t+1} = (1+r)C(1+r)^t = (1+r)x_t.$$

If x_0 is fixed, then $C = x_0$. Discrete growth-decay models are commonplace in finance, in ecology, and in other areas. For example, in ecology we often write the model (1.3) as

$$\frac{\Delta x_t}{x_t} = r.$$

Then we can recognize r as the constant *per capita* growth rate. For populations in general, the constant r is given by $r = b - d + i - \epsilon$, where b, d, i, and ϵ are the birth, death, immigration, and emigration rates, respectively. When $r > 0$ the model (1.3) is called the **Malthus model** of population growth. □

Example 9.2

The last example showed that the difference equation $x_{t+1} = \lambda x_t$ has general solution $x_t = C\lambda^t$, where C is any constant. Let us modify the equation and consider the linear model

$$x_{t+1} = \lambda x_t + p, \tag{1.4}$$

where p is a constant. We can consider this model to be, say, the monthly growth of principal x_t in a bank account where r $(\lambda = 1+r)$ is the monthly interest rate and p is a constant monthly addition to the account. In an ecological setting, λ may be the growth rate of a population and p a constant recruitment rate. We can solve this equation by recursion to find a formula for x_t. If x_0 is the initial value, then iteration yields

$$
\begin{aligned}
x_1 &= \lambda x_0 + p, \\
x_2 &= \lambda x_1 + p = \lambda^2 x_0 + \lambda p + p, \\
x_3 &= \lambda x_2 + b = \lambda^3 x_0 + \lambda^2 p + \lambda p + p, \\
&\cdots \\
x_t &= \lambda^t x_0 + \lambda^{t-1} p + \lambda^{t-2} p + \cdots + \lambda p + p.
\end{aligned}
$$

By the formula for the sum of a geometric sequence,

$$\lambda^{t-1} + \lambda^{t-2} + \cdots + \lambda + 1 = \frac{1 - \lambda^t}{1 - \lambda},$$

and consequently

$$x_t = \lambda^t x_0 + p\frac{1 - \lambda^t}{1 - \lambda}, \tag{1.5}$$

which is the solution to (1.4). □

Example 9.3

(**Logistic model**) In the Malthus model of population growth,

$$\frac{\Delta x_t}{x_t} = r \qquad \text{or} \qquad x_{t+1} = (1 + r)x_t,$$

the population x_t grows unboundedly if the *per capita* growth rate r, or the change in population per individual, is positive. Such a prediction cannot be accurate over a long time. If, for example, $r = b - d$, where b is the *per capita* birth rate and d is the *per capita* death rate, then

$$\frac{\Delta x_t}{x_t} = b - d.$$

We might expect that for early times, when the population is small, there are ample environmental resources to support a high birth rate, with the death rate being small. But for later times, as the population grows, there is a higher death rate as individuals compete for space and food (intraspecific competition). Thus, we should argue for a decreasing *per capita* growth rate r as the population increases. The simplest such assumption is to take a linearly decreasing per capita rate, that is, $r(1 - x_t/K)$, where K is the *carrying capacity*, or the population where the growth rate is zero. See Fig. 9.1. Then

$$\frac{\Delta x_t}{x_t} = r\left(1 - \frac{x_t}{K}\right). \tag{1.6}$$

In the standard form

$$x_{t+1} = x_t\left(1 + r - \frac{r}{K}x_t\right), \tag{1.7}$$

which is nonlinear. The discrete logistics model, quadratic in the population, is the simplest nonlinear model that we can develop. It is tempting to identify $b = r$ as a constant birth rate and $d = \frac{r}{K}x_t$ as the death rate depending upon the population. But an alternative is to take $b = r - \frac{r}{2K}x_t$ and $d = \frac{r}{2K}x_t$, with both depending upon population. A plot of the population is shown in Fig. 9.2 with $r = 0.5$, $K = 100$, and $x_0 = 20$. It shows a steady increase up to the carrying capacity, where it levels off. The accompanying MATLAB *m-file*, or script, produces the sequence in Fig. 9.2 for $t = 1, ..., 20$. In the next section we observe that this is not the whole story; different values of the parameters can lead to interesting, unusual, and complex behavior. □

```
function logistic
r=0.5; K=100;
x=15; xhistory=x;
for t=1:50;
    x=x.*(1+r-r*x/K);
```

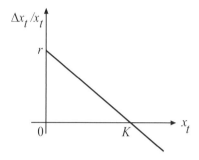

Figure 9.1 *Per capita* growth rate for the logistic model.

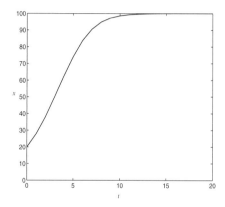

Figure 9.2 Logistic population growth.

```
        xhistory=[xhistory, x];
    end
    plot(xhistory)
```

Example 9.4

(**The Ricker model**). The Ricker model is a nonlinear, discrete time ecological model for the yearly population x_t of a fish stock where adult fish cannibalize the young. The dynamical equation is

$$x_{t+1} = bx_t e^{-cx_t},$$

where $b > 1$ and c is positive. In a heuristic manner, one can think of the model in the following way. In year t there are x_t adult fish, and they would normally give rise to bx_t adult fish the next year, where b is the number of fish produced per adult. However, if adults eat the younger fish, only a fraction of

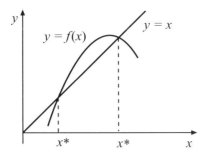

Figure 9.3 Graphical method for finding equilibrium as the intersection points of the graphs of $y = x$ and $y = f(x)$.

those will survive to the next year to be adults. We assume the probability of a fish surviving cannibalism is e^{-cx_t}, which decreases as the number of adults increases. Thus, $bx_t e^{-cx_t}$ is the number of adults the next year. One cannot solve this model to obtain a formula for the fish stock x_t, so we must be satisfied to plot its solution using iteration, as was done for the logistic model in the last example. □

9.1.2 Equilibria, Stability, and Chaos

An important question for discrete models, as it is for continuous models, is whether the state of the system approaches an equilibrium as time gets large (i.e., as t approaches infinity). Equilibrium solutions are constant solutions, or constant sequences. We say $x_t = x^*$ is an **equilibrium solution** of $x_{t+1} = f(x_t)$ if

$$x^* = f(x^*) \tag{1.8}$$

Equivalently, x^* is a value that makes the change Δx_t zero. Graphically, we can find equilibria x^* as the intersection points of the graph of $y = f(x)$ and $y = x$ in an xy plane. See Fig. 9.3.

Example 9.5

(**Logistic model**) Setting $\Delta x_t = 0$ in the logistics model gives $rx^* \left(1 - \frac{x^*}{K}\right) = 0$ or, $x^* = 0$ and $x^* = K$. These two equilibria represent extinction and the population carrying capacity, respectively. Graphically, we can plot $y = x$ vs. $y = f(x) = x\left(1 + r - \frac{r}{K}x\right)$; the equilibria are at the intersections points of the two curves. See Fig. 9.4. □

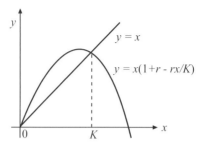

Figure 9.4 Graphical method for finding equilibrium 0 and K for the logistics model.

Example 9.6

(**Ricker model**) An equilibrium state for the Ricker model must satisfy

$$x^* = bx^* e^{-cx^*},$$

or

$$x^*(1 - be^{-cx^*}) = 0.$$

Therefore one equilibrium state is $x^* = 0$, which corresponds to extinction. Setting the other factor equal to zero gives

$$be^{-cx^*} = 1$$

or

$$x^* = \frac{\ln b}{c}.$$

If $b > 1$ we obtain a positive, viable equilibrium population. □

If there is an equilibrium solution x^* to a discrete model (1.8), we always ask about its permanence, or stability. For example, suppose the system is in equilibrium and we perturb it by a small amount (natural perturbations are present in all physical and biological systems). Does the system return to that state or does it do something else (e.g., go to another equilibrium state, or blow up)? We say an equilibrium state is **locally asymptotically stable** if small perturbations decay and the system returns to the equilibrium state. If small perturbations of the equilibrium do not cause the system to deviate too far from the equilibrium, we say the equilibrium is stable. If a perturbation grows, then we say the equilibrium state is **unstable**. In the next paragraph we use a familiar argument to determine the stability of an equilibrium population.

Let x^* be an equilibrium state for (1.8), and assume y_0 represents a small deviation from x^* at $t = 0$. Then this perturbation will be propagated in time, having value y_t at time t. Does y_t decay, or does it grow? The dynamics of the state $x^* + y_t$ (the equilibrium state plus the deviation) must still satisfy the dynamical equation. Substituting $x_t = x^* + y_t$ into (1.8) gives

$$x^* + y_{t+1} = f(x^* + y_t).$$

We can simplify this equation using the assumption that the deviations y_t are small. We can expand the right side in a Taylor series centered about the value x^* to obtain

$$x^* + y_{t+1} = f(x^*) + f'(x^*)y_t + \frac{1}{2!}f''(x^*)y_t^2 + \frac{1}{3!}f''(x^*)y_t^3 + \cdots.$$

Because the deviations are small we can discard the higher powers of y_t and retain only the linear term. Moreover, $x^* = f(x^*)$ because x^* is an equilibrium. Therefore, small perturbations are governed by the linearized perturbation equation, or **linearization**

$$y_{t+1} = f'(x^*)y_t.$$

This difference equation is the growth-decay model (note that $f'(x^*)$ is a constant). For conciseness, let $\lambda = f'(x^*)$. Then the difference equation has solution

$$y_t = y_0 \lambda^t.$$

If $|\lambda| < 1$ then the perturbations y_t decay to zero and x^* is locally asymptotically stable; if $|\lambda| > 1$ then the perturbations y_t grow and x^* is unstable. If $\lambda = 1$, then the linearization gives no information about stability and further calculations are required, for example, examining the higher order-terms in the series above. The stability indicator λ is called the **eigenvalue**. Therefore, if the absolute value of the slope of the tangent line to $f(x)$ at the intersection with $y = x$ is less than 1, then the equilibrium is asymptotically stable; if the absolute value of the slope is greater than 1, the equilibrium is unstable. In other words, we get stability if f is not too steep at the intersection point.

Example 9.7

(**Chaotic behavior**) Consider the Ricker model

$$x_{t+1} = bx_t e^{-cx_t},$$

where $c > 0$ and $b > 1$. Let us check the equilibrium

$$x^* = \frac{\ln b}{c}$$

for stability. Here $f(x) = bxe^{-cx}$ and we must calculate the eigenvalue, or the derivative of f evaluated at equilibrium. To this end $f'(x) = (-cx + 1)be^{-cx}$; evaluating at x^* gives

$$\lambda = f'(x^*) = (-cx^* + 1)be^{-cx^*} = 1 - \ln b.$$

We get asymptotic stability when

$$-1 < 1 - \ln b < 1,$$

or

$$1 < b < e^2 \approx 7.39.$$

If $b < e^2$, then the perturbations decay and the equilibrium is asymptotically stable. Let us follow up on this by performing some simulations with different values of b and observe the effect. Fix $c = 0.001$. Figure 9.5 shows the results for $b = 6.5, 9, 13, 18$. The equilibrium is at $x^* = 1000 \ln b$. For $b = 6.5$, within the stability range, the solution does indeed represent an asymptotically stable, decaying oscillation. For $b = 9$, however, a periodic, 2-cycle appears, alternating between high and low population values. When b is increased further, there is some value of b where suddenly the 2-cycle bifurcates into a 4-cycle. The plot shows the periodic 4-cycle when $b = 13$. This period doubling continues as b increases further, giving 8-cycles, 16-cycles, and so on. It can be shown that these cycles are stable limit cycles. But there is a critical value of b where the patterns disappear, and there are seemingly random fluctuations in the population ($b = 18$ shows this case). When b exceeds this critical value these non-periodic fluctuations become highly sensitive to initial conditions, a behavior known as **chaos**. The frames in Fig. 9.6 show the solution for conditions $x_0 = 99$ and $x_0 = 101$ with $b = 18$. We observe much different solutions. In the chaotic regime, the solution is deterministic (rather than stochastic), but highly unstable with respect to initial data. □

Chaotic behavior is common in discrete models, even in one dimension. Such behavior does not occur in autonomous differential equations until dimension 3.

Next we illustrate a graphical procedure to determine the stability of an equilibrium solution of $x_{t+1} = f(x_t)$. The discussion refers to Figs. 9.7 and 9.8. We first sketch the curves $y = x$ and $y = f(x)$ on a set of axes, as in Fig. 9.4. We then mark the beginning value x_0 on the x axis. To find x_1 we move vertically to the graph of $f(x)$, because $x_1 = f(x_0)$, and mark x_1 on the y axis. To find x_2 we reflect x_1 back to the x axis through the line $y = x$. Now we have x_1 on the x axis. We repeat the process by moving vertically to the curve $f(x)$, which gives x_2. We reflect it back to the x axis, and so on. This

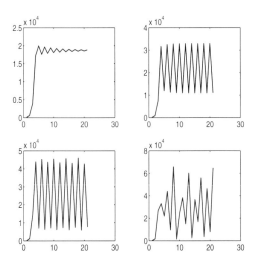

Figure 9.5 Population dynamics for the Ricker model for $c = 0.001$ and $b = 6.5, 9, 13, 18$, showing a decaying oscillation, 2-cycle, 4-cycle, and chaos, as b increases.

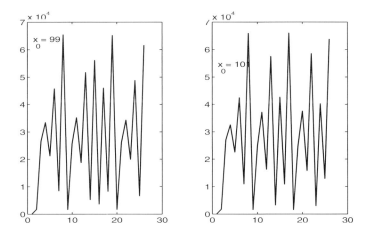

Figure 9.6 Plots of discrete solutions of the Ricker equation in the case the initial populations ($x_0 = 99$, $x_0 = 100$) are close, in the chaos regime.

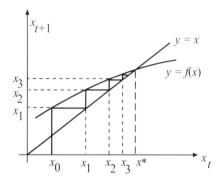

Figure 9.7 Cobweb diagram for a stable equilibrium.

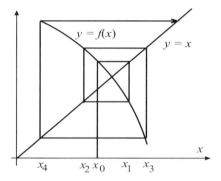

Figure 9.8 Cobweb diagram for an unstable equilibrium.

sequence of moves is accomplished by starting at x_0, going vertically to $f(x)$, going horizontally to the diagonal, vertically to the curve, horizontally to the diagonal, alternating back and forth. The plot of these vertical and horizontal moves that give the segments between the curve $f(x)$ and the diagonal is called the **cobweb diagram**. If the equilibrium is asymptotically stable, the cobweb will converge to the equilibrium represented by the intersection point. If the equilibrium is unstable, the cobweb will diverge from the intersection point. Figure 9.8 shows a divergent cobweb.

EXERCISES

1. Solve the following linear difference equations and plot their solutions:

   a) $x_{t+1} = \frac{1}{2}x_t$, $x_0 = 100$.

   b) $x_{t+1} = -1.3x_t$, $x_0 = 25$.

c) $x_{t+1} = -0.2x_t + 0.7$, $x_0 = 50$.

d) $x_{t+1} = 4x_t + 1$, $x_0 = 1$.

2. If you borrow $15,000 to purchase a new car at a monthly rate of 0.9% and you pay off your loan in 4 years, what is your monthly payment?

3. Use iteration to show that the solution to the difference equation $x_{t+1} = \lambda x_t + p_t$ can be written

$$x_t - x_0\lambda^t + \sum_{k=0}^{t-1} p_k \lambda^{t-k-1}.$$

4. The next three exercises investigate second-order, linear, discrete models, which are analogs of the differential equation $u'' = pu' - qu$. Consider a model of the form

$$x_{t+2} = -px_{t+1} - qx_t$$

Assume that the solution sequence takes the form $x_t = r^t$ for some r to be determined. Show that r satisfies the quadratic equation (called the **characteristic equation**)

$$r^2 + pr + q = 0.$$

In the case that there are two distinct real roots r_1 and r_2 we have two distinct solutions r_1^t and r_2^t. Show that for any two constants A and B, the discrete function $x_t = Ar_1^t + Br_2^t$ is a solution to the difference equation. This expression is called the general solution, and all solutions have this form. Use this method to find the general solution of

$$x_{t+2} = x_{t+1} + 6x_t.$$

Find a specific solution that satisfies the conditions $x_0 = 1$ and $x_1 = 5$.

5. Referring to Exercise 4, show that if the characteristic equation has two equal real roots $(r = r_1 = r_2)$, then $x_n = r^n$ and $x_n = nr^n$ are both solutions to the equation. Thus, the general solution is $x_t = Ar^t + Btr^t = r^t(A + Bt)$, where A and B are arbitrary constants. Find the general solution of the discrete model

$$x_{t+2} = 4x_{t+1} - 4x_t.$$

Find a specific solution that satisfies the conditions $x_0 = 1$ and $x_1 = 0$.

6. Investigate the case when the roots of the characteristic equation in Exercise 4 are complex conjugates, that is, $r = a \pm bi$. Find two real solutions. Follow these steps. (i) Use the exponential representation of a complex number, $a + bi = \rho e^{i\theta}$, to write

$$x_t = e^{rt} = \rho^t \cos \theta t + i\rho^t \sin \theta t.$$

(ii) Show that the real and imaginary parts of this complex solution are real solutions. (iii) What is the general solution? Then solve the following equation:

$$x_{t+2} = x_{t+1} - x_t.$$

Find and plot the solution if the initial states are $x_0 = 1$ and $x_1 = \frac{1}{2}$.

7. Find the general solution of the **Beverton–Holt** model equation

$$x_{t+1} = \frac{ax_t}{b + x_t}$$

by making the substitution $w_t = \frac{1}{x_t}$.

8. Draw the cobweb diagram for the model

$$x_{t+1} = \frac{8x_t}{1 + 2x_t}$$

when $x_0 = 0.5$.

9. A density-limited population model is

$$x_{t+1} = \frac{ax_t}{(1 + bx_t)^c},$$

where a, b, and c are positive.

a) Use scaling to reduce the number of parameters in the model to two.

b) Using the scaled equation, find the equilibria and determine conditions for stability for each.

c) Plot, in two-dimensional parameter space, the region of stability in the case there is a positive equilibrium.

10. The *Gilpin–Ayala* population model is given by

$$\Delta P_t = rP_t \left(1 - \left(\frac{P_t}{K} \right)^\theta \right),$$

where r, K, and θ are positive constants. (a) Find the equilibria. (b) Find a formula for the *per capita* population growth. (c) Write the model in the form $P_{t+1} = f(P_t)$ and determine the stability of the equilibria. (d) Sketch the region in the $r\theta$ plane that gives values for which the nonzero equilibrium is stable.

9.2 Systems of Difference Equations

9.2.1 Linear Models

A linear model is a system of difference equations of the form

$$\mathbf{x}_{t+1} = A\mathbf{x}_t, \quad t = 0, 1, 2, ..., \tag{2.1}$$

where $\mathbf{x}_t$ is an n-vector of quantities at time t and A is a constant $n \times n$ matrix. Such models arise naturally in Markov[1] processes (discussed below), where A is a Markov matrix with nonnegative entries and whose columns sum to 1. Linear models occurs in other areas, for example, for stage-structured models in biology where A is a Leslie matrix, or a population projection, matrix. By iteration, if $\mathbf{x}_0$ is an initial vector, then

$$\mathbf{x}_t = A^t \mathbf{x}_0 \tag{2.2}$$

is the solution to (2.1). We can get a more usable representation of the solution in terms of the eigenvalues and eigenvectors of A. Suppose that A has n distinct, real eigenvalues λ_k with associated independent eigenvectors $\mathbf{v}_k$, $k = 1, 2, ..., n$. Then the eigenvectors form a basis for $\mathbb{R}^n$ and we can express the initial vector

$$\mathbf{x}_0 = c_1 \mathbf{v}_1 + \cdots + c_n \mathbf{v}_n,$$

where the c_k are the coordinates of $\mathbf{x}_0$ in the basis of eigenvectors. Then, using linearity,

$$\mathbf{x}_t = A^t \mathbf{x}_0 = c_1 A^t \mathbf{v}_1 + \cdots + c_n A^t \mathbf{v}_n = c_1 \lambda_1^t \mathbf{v}_1 + \cdots + c_n \lambda_n^t \mathbf{v}_n, \tag{2.3}$$

which is the solution.

We recall that we can find the power A^t by diagonalizing the matrix A. Let D be the diagonal matrix with the eigenvalues on the diagonal, and let Q be the matrix whose columns are the associated eigenvectors. Then, from linear algebra, $Q^{-1}AQ = D$ (we say the matrix Q diagonalizes A). Thus $A = QDQ^{-1}$ and $A^t = (QDQ^{-1})^t = QD^tQ^{-1}$, where D^t is a diagonal matrix with the entries λ_i^t on the diagonal. This leads to an alternate solution form

$$\mathbf{x}_t = QD^tQ^{-1}\mathbf{x}_0.$$

An important case occurs when there is a positive, real **dominant eigenvalue**, say λ_1. This means $\lambda_1 > |\lambda_2| \geq |\lambda_3| \geq \cdots \geq |\lambda_n|$. Then

$$
\begin{aligned}
\mathbf{x}_t &= c_1 \lambda_1^t \mathbf{v}_1 + c_2 \lambda_2^t \mathbf{v}_2 + \cdots + c_n \lambda_n^t \mathbf{v}_n \\
&= \lambda_1^t \left[c_1 \mathbf{v}_1 + c_2 \left(\frac{\lambda_2}{\lambda_1} \right)^t \mathbf{v}_2 + \cdots + c_n \left(\frac{\lambda_n}{\lambda_1} \right)^t \mathbf{v}_n \right].
\end{aligned}
\tag{2.4}
$$

---

[1] A. Markov (1856–1922) was a Russian mathematician who made significant contributions to analysis, number theory, approximation theory, and probability.

We have $|\lambda_i/\lambda_1| < 1$ for $i = 2, 3, ..., n$, and therefore all the terms in parentheses decay to zero as $t \to \infty$. Consequently, for large t the solution to the linear model behaves approximately as

$$\mathbf{x}_t \approx c_1 \lambda_1^t \mathbf{v}_1 \quad \text{as} \quad t \to \infty.$$

Asymptotically, if $\lambda_1 > 0$, the solution exhibits geometric growth or decay, depending upon the magnitude of λ_1. For some matrices A, to be defined later, there is a dominant eigenvalue and its associated eigenvector $\mathbf{v}_1$ is positive, having positive entries. In this case, the dominant eigenvalue λ_1 is called the **growth rate** and the vector $\mathbf{v}_1$ is the longtime structure of the solution. In the context of population models, $\mathbf{v}_1$, appropriately normalized, is the **stable age structure**.

If some of the eigenvalues of A are complex, but the set of eigenvalues is still distinct, the argument above can be repeated. Using Euler's formula, $e^{i\theta} = \cos\theta + i\sin\theta$, every complex term in solution (2.3) can be written in terms of real and imaginary parts, each of which is a real solution. For example, if $\mathbf{v} = \mathbf{u} + i\mathbf{w}$ and $\lambda = \alpha + i\beta = re^{i\theta}$ is a complex eigenpair, where r is the modulus of λ and θ is its argument, then

$$
\begin{aligned}
\mathbf{v}\lambda^t &= (\mathbf{u} + i\mathbf{w})(\alpha + i\beta)^t \\
&= r^t(\mathbf{u} + i\mathbf{w})(\cos\theta t + i\sin\theta t) \\
&= r^t(\mathbf{u}\cos\theta t - \mathbf{w}\sin\theta t) + ir^t(\mathbf{u}\sin\theta t + \mathbf{w}\cos\theta t).
\end{aligned}
$$

The real and imaginary parts of this vector are two linearly independent, real solutions. The complex conjugate eigenpair $\mathbf{v} = \mathbf{u} - i\mathbf{w}$ and $\lambda = \alpha - i\beta$ gives the same two independent solutions. In summary, the two complex terms in (2.3) associated with complex conjugate eigenpairs may be replaced by the two real, independent vector solutions. Therefore, we may always write the solution as a real solution. Complex eigenvalues clearly lead to oscillations in the system. The analysis of the case of a repeated eigenvalue of A is left to the Exercises.

An **equilibrium solution** $\mathbf{x}^*$ of (2.1) satisfies

$$\mathbf{x}^* = A\mathbf{x}^*.$$

Clearly, $\mathbf{x}^* = 0$ is always an equilibrium solution; there are nontrivial equilibria if, and only if, $\lambda = 1$ is an eigenvalue of A. Notice that (2.4) also implies that the solution to the linear model (2.1) satisfies the condition $\mathbf{x}_t \to \mathbf{0}$ if, and only if, all the eigenvalues have modulus less than one, or $|\lambda_i| < 1$. In this case we say $\mathbf{x} = \mathbf{0}$ is an **asymptotically stable** equilibrium. If there is an eigenvalue with $|\lambda| > 1$, then the equilibrium $\mathbf{x} = \mathbf{0}$ is **unstable**.

In Chapters 1 and 2, where we discussed continuous population models, we had in mind tracking one or more species and we lumped all the individuals of a given species into a single quantity, without regard to development stage, size, age, or any other structured quantity. However, some animals, such as insects, have well-defined life stages, such as egg–larva–pupa–adult. Still other species' populations can be divided into juveniles and adults. We want to develop dynamics of how that animal might go through its life stages, keeping track of the population of each stage.

Before working through an example, we remark that many population models in ecology track only the females of the species. They are the important ones that give birth; males may sire many offspring from many females. Sorry, men!

Example 9.8

(**Leslie model**) Consider a hypothetical animal that has two life stages, juvenile and adult. At time t (say, given in weeks) we let J_t denote the number of juveniles and A_t denote the number of adults. Over each time step of one week we assume that new juveniles are born, some die, and some graduate to adulthood. For adults, some die and some are recruited from the graduating juvenile stage. In symbols, for the juvenile population we may write

$$J_{t+1} = J_t - mJ_t - gJ_t + fA_t. \tag{2.5}$$

In words, the number of juveniles at the next week (J_{t+1}) equals the number at the last week (J_t), minus the number that died (mJ_t), minus the number that became adults (gJ_t), plus the number of births (fA_t). The constant m is the weekly juvenile mortality rate, g the weekly graduation rate, and f the average, weekly fecundity (fertility) of adults. Because adults reproduce, we assume the number of juveniles produced each week is proportional to the number of adults present. For example, if each adult on the average gives rise to three juveniles each week, then $f = 3$. Then, 100 adults would give rise to 300 juveniles per week. In the same way, for the adult population,

$$A_{t+1} = A_t - \mu A_t + gJ_t, \tag{2.6}$$

where μ is the adult weekly mortality rate. The last term represents those juveniles that became adults. Equations (2.5)–(2.6) represent a system of two coupled, first-order difference equations. They provide a way to take information from time t and project it forward to the next time $t + 1$. Therefore, if we know the initial numbers in each stage, we can recursively calculate the population numbers for all times t. The stage-based population model in the last example can be formulated in matrix form as

$$\begin{pmatrix} J_{t+1} \\ A_{t+1} \end{pmatrix} = \begin{pmatrix} 1 - m - g & f \\ g & 1 - \mu \end{pmatrix} \begin{pmatrix} J_t \\ A_t \end{pmatrix},$$

and can be formulated in vector notation as

$$\mathbf{x}_{t+1} = P\mathbf{x}_t, \tag{2.7}$$

where

$$\mathbf{x}_t = \begin{pmatrix} J_t \\ A_t \end{pmatrix}, \quad P = \begin{pmatrix} 1-m-g & f \\ g & 1-\mu \end{pmatrix}.$$

The matrix P is called the **population projection matrix**, or **Leslie matrix**[2] (after P. H. Leslie, a biologist who developed the ideas in the 1940s). □

Often we express the model in terms of survivorships, rather than mortality rates. The fraction s that survive in a given stage is

$$s = 1- \text{ mortality rate } - \text{ graduation rate.}$$

Then, in our model for juveniles and adults, $s_J = 1 - m - g$ and $s_A - 1 - \mu$, because no adults graduate to the next class. So the Leslie matrix is

$$P = \begin{pmatrix} S_J & f \\ g & S_A \end{pmatrix}.$$

By iteration we see immediately that the solution to (2.7) is

$$\mathbf{x}_t = P^t\mathbf{x}_0,$$

where $\mathbf{x}_0$ is a vector containing the initial stage populations. □

Example 9.9

In the preceding example let us choose $g = m = 0.5$, $f = 2$, and $\mu = 0.9$. Then the model is

$$\mathbf{x}_{t+1} = \begin{pmatrix} 0 & 2 \\ 0.5 & 0.1 \end{pmatrix} \mathbf{x}_t.$$

The eigenpairs are

$$\lambda_1 = 1.051, \quad \mathbf{v}_1 = \begin{pmatrix} 0.885 \\ 0.465 \end{pmatrix}; \qquad \lambda_2 = -0.951, \quad \mathbf{v}_2 = \begin{pmatrix} 0.903 \\ -0.429 \end{pmatrix}.$$

The general solution is

$$\mathbf{x}(t) = c_1(1.051)^t \begin{pmatrix} 0.885 \\ 0.465 \end{pmatrix} + c_2(-0.951)^t \begin{pmatrix} 0.903 \\ -0.429 \end{pmatrix}.$$

---

[2] Actually, the Leslie matrix was introduced to study age-structured models. Later, L. P. Lefkovitch in 1965 generalized the models to stage-structured population models. We use the term Leslie matrix for any population projection matrix.

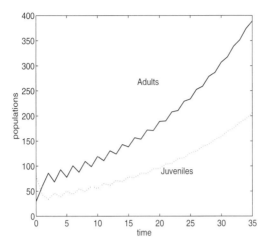

Figure 9.9 Population of adults (solid) and juveniles (dashed).

The dominant eigenvalue is $\lambda_1 = 1.051$ and the first term produces growth of about 5%. The second term will give a decaying oscillation. This is our observation in Fig. 9.9. Over a long time, the juvenile-adult population will approach a stable age structure defined by the eigenvector $(0.885, 0.465)^{\mathrm{T}}$, which, when normalized to make the sum of the entries one, gives $(0.656, 0.344)^{\mathrm{T}}$. Therefore, in the long run, 65.6% of the population will be juveniles, and 34.4% of the population will be adults. In yet different words, there will be about 1.9 juveniles for every adult. □

Doing calculations by hand to get plots is tedious, so we use software. A MATLAB script produces the time series plots shown in Fig. 9.9. □

```
function stagepopulation
clear
m=0.5; g=0.5; f=2; mu=0.9; timesteps=35;
J=80; Jhistory=J; A=30; Ahistory=A;
for t=1:timesteps;
x=(1-m-g)*J+f*A;
y=g*J+(1-mu)*A;
J=x; A=y;
Ahistory=[Ahistory J];
Jhistory=[Jhistory A];
end
time=0:timesteps;
```

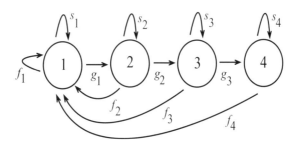

Figure 9.10 Leslie diagram.

```
plot(time,Jhistory), hold on
plot(time,Ahistory), hold off
```

To get a phase plane plot of J_t vs. A_t, just replace the plot commands in the script above by 'plot(Jhistory, Ahistory).'

In general, $\mathbf{x}_t$ may be an n-vector whose entries represent the populations of each of n stages at time t. If $n = 4$, for example, $\mathbf{x}_t = (x_{1t}, x_{2t}, x_{3t}, x_{4t})^T$, and the Leslie matrix (generally, we use the letter A) has the form

$$A = \begin{pmatrix} s_1 + f_1 & f_2 & f_3 & f_4 \\ g_1 & s_2 & 0 & 0 \\ 0 & g_2 & s_3 & 0 \\ 0 & 0 & g_3 & s_4 \end{pmatrix}, \tag{2.8}$$

where f_i is the fecundity of the ith stage, s_i is survival rate (one minus the mortality rate minus the graduation rate) of the ith stage, and g_i is the fraction of the ith stage that passes to the jth stage during the fixed time step. The **Leslie diagram** in Fig. 9.10 shows pictorially how the rates and the stages are connected. In still a more general case, the entries in the Leslie matrix may depend upon t, or they may depend upon the populations themselves, making the model nonlinear. Nonlinear models are discussed later.

When there is a positive, dominant eigenvalue, say λ_1, of the Leslie matrix A, we have $\mathbf{x}_t \sim c_1 \mathbf{v}_1 \lambda_1^t$ for large t. Therefore λ_1 is the long-term growth rate and $c_1 \mathbf{v}_1$ represents the stable stage structure. Because c_1 is a constant, we can just use $\mathbf{v}_1$ to determine the proportion, or percentage, of each stage to the whole population. For example, if $\mathbf{v}_1 = (1, 2, 3)^T$ is an eigenvector associated with a dominant eigenvalue, then the vector $(\frac{1}{6}, \frac{1}{3}\frac{1}{2})^T$ shows the fraction of each stage in the stable age structure.

Finally, we ask when is there a dominant eigenvalue for a given model and a corresponding positive eigenvector. A simple version of the **Perron–Frobenius** theorem, can be stated as follows.

Theorem 9.10

(**Perron–Frobenius**) If A is a non-negative matrix, i.e., its entries are non-negative, and if for some positive integer k the matrix A^k has all positive entries, then A has a simple, positive dominant eigenvalue with a positive eigenvector.
□

There are other results that guarantee the same conclusion under less strenuous hypotheses. (See Caswell 2001.)

One interesting issue in a stage-structured population model is to determine which values in the Leslie matrix, which represent fecundities, graduation, and survivorship probabilities, most affect the dominant eigenvalue λ, which is the long-term growth rate. This is particularly important in making management decisions about ecosystems. We can measure the sensitivity of λ to changes in an entry a_{ij} by calculating the derivative $\partial\lambda/\partial a_{ij}$. (Note that the dominant eigenvalue is a function of the entries in A.) The matrix of these values,

$$S = \left(\frac{\partial\lambda}{\partial a_{ij}}\right),$$

is called the **sensitivity matrix**.

Example 9.11

For a 2×2 matrix

$$A = \begin{pmatrix} a & b \\ c & d \end{pmatrix}$$

with non-negative entries the dominant eigenvalue is easily found from the roots of the characteristic equation,

$$\lambda = \frac{1}{2}\left(\operatorname{tr} A + \sqrt{D}\right), \quad D = (\operatorname{tr} A)^2 - 4\det A.$$

Then the sensitivity matrix is

$$S = \begin{pmatrix} \frac{\partial\lambda}{\partial a} & \frac{\partial\lambda}{\partial b} \\ \frac{\partial\lambda}{\partial c} & \frac{\partial\lambda}{\partial d} \end{pmatrix} = \begin{pmatrix} \frac{1}{2}\left(1 + \frac{a-d}{\sqrt{D}}\right) & \frac{c}{\sqrt{D}} \\ \frac{b}{\sqrt{D}} & \frac{1}{2}\left(1 - \frac{a-d}{\sqrt{D}}\right) \end{pmatrix}.$$

For example, consider the two-stage (juveniles and adults) model

$$x_{t+1} = 1.7y_t, \quad y_{t+1} = 0.6x_t + 0.1y_t$$

with matrix

$$A = \begin{pmatrix} 0 & 1.7 \\ 0.6 & 0.1 \end{pmatrix}.$$

The value 1.7 is the fecundity of adults, and 0.1 is their survival rate; 0.6 is the fraction of juveniles that survive and pass to the adult stage. The dominant eigenvalue is $\lambda = 1.06$, which means that the long-time growth rate of the population is 6% per season. The sensitivity matrix is

$$S = \begin{pmatrix} 0.475 & 0.297 \\ 0.841 & 0.525 \end{pmatrix}.$$

Therefore, we see that the $2, 1$ entry, the survivorship of juveniles affects the dominant eigenvalue most. We can estimate the change $\Delta\lambda$ in the long-term growth rate when the survivorship is changed by Δc with the linear approximation

$$\Delta\lambda = \frac{\partial\lambda}{\partial c}\Delta c = 0.841 \cdot \Delta c. \quad \square$$

Now we proceed in general. Consider the n-dimensional model

$$\mathbf{x}_{t+1} = A\mathbf{x}_t$$

with $\lambda, \mathbf{v}$ the dominant eigenpair. Then

$$A\mathbf{v} = \lambda\mathbf{v}.$$

We know that λ is also an eigenvalue of A^{T} having some eigenvector $\mathbf{w}$. Therefore $A^{\mathrm{T}}\mathbf{w} = \lambda\mathbf{w}$, or, taking the transpose,

$$\mathbf{w}^{\mathrm{T}}A = \lambda\mathbf{w}^{\mathrm{T}}.$$

Therefore $\mathbf{w}$ is a **left eigenvector** of A.

To calculate the change in the dominant eigenvalue with respect to the change of an entry in the matrix, we change the ij-entry of A by a small amount Δa_{ij}, while holding the remaining entries fixed. Let ΔA be the matrix with Δa_{ij} in the ij-position and zeros elsewhere. Then the perturbed matrix $A + \Delta A$ has dominant eigenvalue $\lambda + \Delta\lambda$ and eigenvector $\mathbf{v} + \Delta\mathbf{v}$, or

$$(A + \Delta A)(\mathbf{v} + \Delta\mathbf{v}) = (\lambda + \Delta\lambda)(\mathbf{v} + \Delta\mathbf{v}).$$

Multiplying out and discarding the higher-order terms gives, to leading order,

$$A(\Delta\mathbf{v}) + (\Delta A)\mathbf{v} = \lambda(\Delta\mathbf{v}) + (\Delta\lambda)\mathbf{v}.$$

Multiplying by $\mathbf{w}^{\mathrm{T}}$ on the left gives

$$\mathbf{w}^{\mathrm{T}}(\Delta A)\mathbf{v} = (\Delta\lambda)\mathbf{w}^{\mathrm{T}}\mathbf{v},$$

or

$$\Delta\lambda = \frac{\mathbf{w}^{\mathrm{T}}(\Delta A)\mathbf{v}}{\mathbf{w}^{\mathrm{T}}\mathbf{v}} = \frac{w_i v_j \Delta a_{ij}}{\mathbf{w}^{\mathrm{T}}\mathbf{v}}.$$

In the limit we have

$$S = \left(\frac{\partial \lambda}{\partial a_{ij}}\right) = \frac{1}{\mathbf{w}^T \mathbf{v}}(w_i v_j) = \frac{1}{\mathbf{w}^T \mathbf{v}}\mathbf{w}\mathbf{v}^T, \tag{2.9}$$

which is a simple representation of the sensitivity matrix. As a final note, it is always possible to normalize $\mathbf{v}$ and $\mathbf{w}$ so that $\mathbf{w}^T \mathbf{v} = 1$.

MATLAB is useful to calculate the sensitivity matrix S. The following command line statements calculate S for the model in the last example.

```
A=[0  1.7; 0.6  0.1]
[V,D]=eig(A)
v=V(:,2)
[W,d]=eig(A')
w=W(:,2)
S=(1/(w'*v))*(w*v')
```

Unfortunately, the sensitivity matrix may give information that is hard to interpret when entries in A have large differences. For example, fecundities are usually high compared to survivorships. This fact may confound the calculation. Therefore asking how the dominant eigenvalue changes with respect to a unit change in an entry is not always practical, and it is better to compute the *relative* changes, or the elasticity. We define the **elasticity matrix** E by

$$E = \left(\frac{a_{ij}}{\lambda}\frac{\partial \lambda}{\partial a_{ij}}\right) \simeq \left(\frac{\Delta\lambda/\lambda}{\Delta a_{ij}/a_{ij}}\right).$$

In MATLAB this is simply E = A.*S/max(eig(A)). When an entry of E is large, we say the entry is *elastic*, and when it is small, we say it is *inelastic*. These terms are familiar terms in economics with regard to the elasticity of demand for a product with respect to price changes; there,

$$\text{elasticity} = \frac{\% \text{ change in demand}}{\% \text{ change in price}} \simeq \frac{p}{D}\frac{dD}{dp},$$

where the demand D is a function of price p.

For reference or study, the definitive book by Caswell (2001) has a thorough discussion of matrix models in the biological sciences.

EXERCISES

1. Find the solution to the discrete model

$$\begin{aligned}
X_{t+1} &= X_t + 4Y_t \\
Y_{t+1} &= \frac{1}{2}X_t,
\end{aligned}$$

if $X_0 = 6$ and $Y_0 = 0$.

2. For each of the following Leslie matrices find the dominant eigenvalue (if any), the corresponding eigenvector, and the stable age structure. Interpret each entry in context of a three-stage population model.

a) $A = \begin{pmatrix} 0 & 1 & 5 \\ 0.3 & 0 & 0 \\ 0 & .5 & 0 \end{pmatrix}$.

b) $A = \begin{pmatrix} 0 & 1 & 1 \\ \frac{2}{3} & 0 & 0 \\ 0 & \frac{1}{3} & 0 \end{pmatrix}$.

3. Describe the dynamics of the linear model

$$\mathbf{x}_{t+1} = \begin{pmatrix} 0 & 0 & f \\ a & 0 & 0 \\ 0 & b & 0 \end{pmatrix} \mathbf{x}_t,$$

where a, b, and f are positive constants. Discuss the two cases $abf > 1$ and $abf < 1$.

4. Consider the linear model

$$\mathbf{x}_{t+1} = \begin{pmatrix} ac & bc \\ s(1-a) & 0 \end{pmatrix} \mathbf{x}_t,$$

where s, b, $c > 0$ and $0 < a < 1$.

   a) Determine conditions (if any) for which there are infinitely many equilibria.

   b) In the case where there is a single equilibrium, determine conditions on the parameters that ensure all solutions converge to zero as $t \to \infty$.

5. For many bird species the fecundity and survivorship of adults is independent of the age of the adult bird. So, we can think of the population as composed of two classes, juveniles and adults.

   a) Set up a general Leslie model to describe the dynamics of the female bird population, assuming juveniles do not reproduce and that the juvenile stage lasts only one year. Adult females may live more than one year.

   b) Find the eigenvalues and eigenvectors of the Leslie matrix for this model.

   c) Identify the growth rate and dominant eigenvalue, and determine the long time fraction of the total population that are juveniles. Examine all cases.

6. A population is divided into three age classes, ages 0, 1, and 2. During each time period 20% of the females of age 0 and 70% of the females of age 1 survive until the end of the following breeding season. Females of age 1 have an average fecundity of 3.2 offspring per female and the average fecundity of age 2 females is 1.7. No female lives beyond three years.

   a) Set up the Leslie matrix.

   b) If initially the population consists of 2000 females of age 0, 800 females of age 1, and 200 females of age 2, find the age distribution after three years.

   c) Determine the asymptotic behavior of the population.

7. Show that if $\lambda = 0$ is an eigenvalue of a matrix A, then A^{-1} does not exist.

8. Consider a linear model $\mathbf{x}_{t+1} = A\mathbf{x}_t$. Under what condition(s) on A does the system have a single equilibrium?

9. Consider the stage-structured model

$$\mathbf{x}_{t+1} = \begin{pmatrix} \frac{1}{2} & 1 & \frac{3}{4} \\ \frac{2}{3} & 0 & 0 \\ 0 & \frac{1}{3} & 0 \end{pmatrix} \mathbf{x}_t.$$

   a) Find the eigenvalues and eigenvectors, and identify the growth rate and stable age structure.

   b) Generate the sensitivity matrix S and determine which entry in A most strongly affects the growth rate.

   c) Compute the elasticity matrix E. Do the entries of E change your conclusion in part (b)?

10. A three-stage model of coyote populations (pups, yearlings, and adults) has a Leslie matrix given by

$$\begin{pmatrix} 0.11 & 0.15 & 0.15 \\ 0.3 & 0 & 0 \\ 0 & 0.6 & 0.6 \end{pmatrix}.$$

   Repeat parts (a) and (b) of Exercise 10.

11. A female population with three stages, juveniles J, youth Y, and adults A, have the following vital statistics. In a time step (one season), no animal remains in the same stage; but, on the average, during each time step 50% of the juveniles become youth, and 60% of the youth survive to become adults. On the average, each youth produces 0.8 offspring (juveniles) per

time step, and each adult produces 2.2 offspring per time step. Juveniles do not reproduce.

a) Draw a Leslie diagram with arrows indicating how the stages interact.

b) Write down the discrete model and identify the population projection matrix L.

c) If $J_0 = 25$, $Y_0 = 50$, and $A_0 = 25$, draw plots (on the same set of axes) of the populations of each stage for the first 30 seasons.

d) What is the the growth rate of the total population? Over a long time, what percentage of the population is in each stage? Explain your observations in part (c) based on the eigenvalues and eigenvectors of the population projection matrix L. What are the absolute values of the eigenvalues?

e) Under the same initial conditions, plot the natural logarithms of the populations versus time? Explain the difference observed from part (d).

f) Compute the sensitivity matrix S.

g) The dominant eigenvalue is most sensitive to a change in what nonzero entry in L?

h) Compute the elasticity of the dominant eigenvalue to adult fecundity. Interpret your result in words.

i) If measures are taken to increase the survivorship of youths by 10%, what would be the approximate population growth rate?

j) How much would the fecundity of adults have to decrease to drive the population to extinction?

9.2.2 Nonlinear Interactions

Next we set up and analyze two-dimensional nonlinear discrete systems

$$x_{t+1} = f(x_t, y_t), \tag{2.10}$$
$$y_{t+1} = g(x_t, y_t), \tag{2.11}$$

where f and g are given functions, and $t = 0, 1, 2, \dots$. As in the case of differential equations, we can visualize a solution geometrically as time series, where the sequences x_t and y_t are plotted against t, or as a sequence of points (x_t, y_t) in the xy plane, or phase plane. We often connect the points of the sequence to make the plots continuous.

The importance of equilibria again arises in understanding the behavior of nonlinear systems. An equilibrium solution is a constant solution $x_t = x^*$, $y_t = y^*$. It plots as a single point, sometimes called a **critical point**, in the phase plane; as in the case of differential equations, the behavior of the system is determined by its local behavior near the critical points. Therefore, stability, or permanence of an equilibrium, becomes an issue.

An equilibrium solution satisfies the equations

$$x^* = f(x^*, y^*), \quad y^* = g(x^*, y^*).$$

To analyze the orbital behavior near the equilibrium we consider small perturbations u_t and v_t from x^* and y^*, respectively, or

$$x_t = x^* + u_t, \quad y_t = y^* + v_t.$$

Substituting into the system (2.10)–(2.11) and linearizing leads, in the usual way, to the linearized perturbation equations

$$\mathbf{u}_{t+1} = J\mathbf{u}_t, \tag{2.12}$$

where $\mathbf{u}_t = (u_t, v_t)^{\mathrm{T}}$ and $J = J(x^*, y^*)$ is the **Jacobian matrix** (or the community matrix in ecology),

$$J = \begin{pmatrix} f_x(x^*, y^*) & f_y(x^*, y^*) \\ g_x(x^*, y^*) & g_y(x^*, y^*) \end{pmatrix}. \tag{2.13}$$

Let λ_1 and λ_2 denote the eigenvalues of J. From the section on linear systems, we have the following stability result:

(a). If $|\lambda_1| < 1$ and $|\lambda_2| < 1$, then the perturbations decay and (x^*, y^*) is locally asymptotically stable.

(b). If either $|\lambda_1| > 1$ or $|\lambda_2| > 1$, then some of the perturbations will grow and (x^*, y^*) is unstable.

This result translates into the following general condition for asymptotic stability in terms of the characteristic equation

$$\det(J - \lambda I) = \lambda^2 - (\operatorname{tr} J)\lambda + \det J = 0$$

for the Jacobian matrix. Necessary and sufficient conditions for local asymptotic stability, $|\lambda_1| < 1$ and $|\lambda_2| < 1$, are

$$|\operatorname{tr} J| < 1 + \det J, \quad \det J < 1. \tag{2.14}$$

Conditions (2.14) are called the **Jury conditions**. A proof is requested in the Exercises.

To plot orbits in the phase plane it is useful to find the nullclines where x_t or y_t does not change. Therefore the nullclines are defined by $\Delta x_t = 0$, $\Delta y_t = 0$, or the loci

$$f(x, y) = x, \quad g(x, y) = y,$$

respectively. The x- and y-nullclines intersect at the critical points. The behavior of the orbits near the critical points is similar to that encountered in differential equations: nodal structure, saddle structure, and rotational structure (spirals and ellipses).

Example 9.12

(**Nicholson–Bailey model**) Many insect predators are parasitoids. A parasitoid finds prey, called hosts, and lays its eggs inside or on the victim. The eggs hatch into larvae, which then consume the host. The host then dies and the larvae metamorphose and emerge as adults. The population dynamics is based upon the probability of the host avoiding detection and has the general form

$$\begin{aligned} x_{t+1} &= rx_t F(y_t), \\ y_{t+1} &= cx_t \left(1 - F(y_t)\right), \end{aligned}$$

where x_t is the host population and y_t is the predator population; F is the escapement function, or probability of avoiding detection in one season, r is the geometric growth factor, and c is the number of new parasitoids per host discovered. Think of F as the fraction of hosts that survive to the next time step. The dynamics is discrete because both host and parasitoid usually have life cycles synchronized with seasons. The model was developed in the 1930s by A. J. Nicholson and V. A. Bailey, and it is fundamental in population ecology because of its historical importance, in spite of some difficulties with its predictions. There are various forms for the escapement function F. The usual model is to base it upon a **Poisson random variable** (see Section 9.3). A discrete random variable X is Poisson distributed if

$$\Pr(X = k) = \frac{\lambda^k}{k!} e^{-\lambda}, \quad k = 0, 1, 2, ...,$$

where λ is a positive constant. A Poisson distribution describes the occurrence of discrete, random events, such as encounters between a predator and its prey, and X is the probability of k encounters in a given time period, say over the lifespan of the prey. The constant λ is the average number of encounters during the period. In terms of hosts and parasitoids, the average encounter rate for a given host should be proportional to the number of predators searching, or

$\lambda = ay_t$, during that period. So, the probability of no encounters is the zeroth-order term $(k = 0)$, or

$$F(y_t) = e^{-ay_t}.$$

Therefore

$$x_{t+1} = rx_t e^{-ay_t}, \qquad (2.15)$$
$$y_{t+1} = cx_t \left(1 - e^{-ay_t}\right). \qquad (2.16)$$

The equilibria are found from the equations

$$x = rxe^{-ay}, \qquad y = cx \left(1 - e^{-ay}\right),$$

which give two equilibria

$$x^* = y^* = 0; \qquad x^* = \frac{r \ln r}{(r - 1)ac}, \qquad y^* = \frac{\ln r}{a}.$$

Note that $r > 1$ must be an implicit assumption to the model. To determine stability we must compute the Jacobian matrix (2.13). This tedious calculation is left to the reader. We obtain

$$J(x^*, y^*) = \begin{pmatrix} re^{-ay^*} & -ax^*e^{-ay^*} \\ c(1 - e^{-ay^*}) & acx^*e^{-ay^*} \end{pmatrix}.$$

At extinction,

$$J(0,0) = \begin{pmatrix} r & 0 \\ 0 & 0 \end{pmatrix},$$

which has eigenvalues $r > 1$ and 0. Hence, the extinct population is unstable. The nonzero equilibrium has Jacobian matrix

$$J(x^*, y^*) = \begin{pmatrix} 1 & -\frac{r \ln r}{c(r-1)} \\ \frac{c(r-1)}{r} & \frac{\ln r}{r-1} \end{pmatrix}.$$

To determine stability we apply the Jury conditions (2.14). The trace and determinant are

$$\operatorname{tr} J = 1 + \frac{\ln r}{r - 1}, \qquad \det J = \frac{r \ln r}{r - 1}.$$

We show that $\det J > 1$, and so the equilibrium is unstable. Equivalently, we show $g(r) = r - 1 - r \ln r < 0$ for $r > 1$. But $g(1) = 0$ and $g'(r) = -\ln r < 0$. Therefore we must have $g(r) < 0$. □

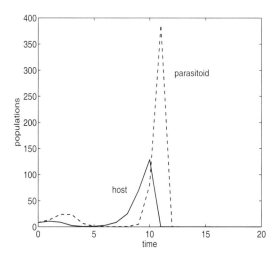

Figure 9.11 Simulations of the Nicholson–Bailey model. The dashed curve is the parasitoid abundance and the solid curve is the host abundance.

The results of the Nicholson–Bailey model are ecologically perplexing. There are no stable equilibria, yet in nature hosts and parasitoids endure, co-evolving together. Does this mean it is an invalid model? To get a sense of the dynamics we simulate the solution with $x_0 = 10$, $y_0 = 10$, taking $a = 1$, $c = 1$, and $r = 2$. Fig. 9.11 shows the unstable oscillations that result. If time is continued, the oscillations continue and become greater. There have been explanations that attempt to resolve this issue, but more to the point, the Nicholson–Bailey model has become the basis of many other models with modifications that attempt to stabilize the dynamics.

Example 9.13

(**Beddington model**) In the Beddington model,

$$
\begin{aligned}
x_{t+1} &= e^{b(1-x_t/K)} x_t e^{-ay_t}, \\
y_{t+1} &= c x_t \left(1 - e^{-ay_t}\right),
\end{aligned}
$$

the growth rate r is replaced by a saturating density dependent rate $e^{b(1-x_t/K)}$. This model has a positive stable equilibrium for a large set of parameter values. Texts in mathematical ecology investigate some of these models in detail. The monograph by Hassell (1978), and many of his papers, offer deep insights into these and other arthropod predator–prey systems. □

Example 9.14

(**Clumped prey**) Instead of searching among randomly distributed hosts, as described by the Poisson distribution, parasitoids may search using a more clumped prey model described by the negative binomial distribution, with escapement function

$$F(x, y) = \left(1 + \frac{ay}{k}\right)^{-k},$$

where k is a *clumping parameter*. As $k \to +\infty$, this distribution approaches the Poisson distribution. Analysis of this model is requested in the exercises. $\quad\square$

Example 9.15

(**Epidemics**) Earlier we investigated a continuous SIR model of the course of an infection. Actually, because health data is collected on a discrete scale, e.g., weekly, it reasonable to develop a discrete model. Consider a disease in a population of fixed size N, and let S_t, I_t, and R_t be the susceptible, infective, and removed classes at week t, respectively. To review, susceptibles are those who can catch the disease but are currently not infected, infectives are those who are infected with the disease and are contagious, and removed individuals have either recovered permanently, are naturally immune, or have died. Clearly $N = S_t + I_t + R_t$. With homogeneous population mixing, the dynamics is given by

$$
\begin{aligned}
S_{t+1} &= S_t - \alpha S_t I_t, \\
I_{t+1} &= I_t + \alpha S_t I_t - \gamma I_t, \\
R_{t+1} &= R_t + \gamma I_t,
\end{aligned}
\tag{2.17}
$$

where the individuals in a given class at week $t + 1$ equals the number at week t, plus or minus those added or subtracted during that week. The constant α is the **transmission coefficient** and γ is the **removal rate**. This model is the **discrete SIR model**. To interpret the constants we argue as follows. If S_0 and I_0 are the number susceptible and infected initially, then $\alpha S_0 I_0$ measures the number that become infected at the outset; thus αS_0 is the number that become infected by a contact with a single infected individual during the initial week. Therefore, α represents the probability that one susceptible person becomes infected. Looked at in a different way, α^{-1} is the average time that it takes for an individual to get infected. The removal rate can be interpreted similarly; γ^{-1} is the average duration of an individual's infectious period. To understand

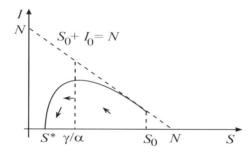

Figure 9.12 Orbit in the discrete SIR model. One infective is introduced into a susceptible population of 500. There is an epidemic with $R_0 = 2.5$ and about 50 individuals do not get the disease.

how the disease progresses we observe that

$$\Delta I_t \; > \; 0 \quad \text{if} \quad S_t > \frac{\gamma}{\alpha};$$

$$\Delta I_t \; < \; 0 \quad \text{if} \quad S_t < \frac{\gamma}{\alpha};$$

$$\Delta I_t \; = \; 0 \quad \text{if} \quad S_t = \frac{\gamma}{\alpha}.$$

Because $\Delta S_t < 0$, we know S_t is always decreasing. Therefore, if $S_0 < \frac{\gamma}{\alpha}$ then $S_t < \frac{\gamma}{\alpha}$ for all t, and the disease decreases. On the other hand, if $S_0 > \frac{\gamma}{\alpha}$ then the disease increases in the population and an epidemic occurs. If we define $R = \frac{\alpha}{\gamma} S_0$, then an epidemic occurs if $R > 1$. Epidemiologists refer to R as the **basic reproduction number** of the infection. Notice that the epidemic peaks when $S_t = \frac{\gamma}{\alpha}$, where ΔI_t changes sign. Simulations of the SIR equations (2.17) are investigated in the Exercises. Both time series plots and orbits in the SI phase plane illustrate the dynamics. Fig. 9.12 shows a typical orbit in the phase plane when there is an epidemic. The initial point is on the line $S_0 + I_0 = N$, and the orbit increases up to a maximum, which lies on the vertical line $S_t = \frac{\gamma}{\alpha}$. Then the disease dies out, leaving a number S^* of infectives who never get the disease. $\square$

The Exercises show the variety and scope of discrete, nonlinear problems in biology and other areas of science. An unsurpassed, elementary treatment of discrete models is contained in Allman and Rhodes (2000).

EXERCISES

1. Consider the linear model

$$\Delta P_t = P_t - 2Q_t, \quad \Delta Q_t = -P_t.$$

(a) Find the equilibria. (b) Draw the nullclines and the arrows indicating the direction of the "flow" on the nullclines and in the regions bounded by the nullclines. (c) On the line $Q_t = P_t$ in the phase plane, draw arrows indicating changes in P and Q. (d) Do the equilibria appear stable or unstable, or can you tell? (e) Construct a rough phase plane diagram indicating solution paths. (f) Find the equations of these curves by finding the solution P_t and Q_t in terms of t. (Hint: Reduce the system to a single, second-order equation.)

2. Prove that the quadratic equation $\lambda^2 - p\lambda + q = 0$ has roots satisfying $|\lambda| < 1$ if, and only if,

$$|p| < 1 + q < 2.$$

Here, $p = \operatorname{tr} J$ and $q = \det J$. The condition is called the **Jury condition**. Hint: treat the real case and complex case separately; in the complex case write the roots in complex exponential form, $\lambda = \rho \exp(\pm i\theta)$.

3. Consider the following model connecting the cinnabar moth and ragweed. The life cycle of the moth is as follows. It lives one year, lays it eggs on the plant, and then dies. The eggs hatch the next spring. The number of eggs laid is proportional to the plant biomass in the preceding year, or

$$E_{t+1} = aB_t.$$

The plant biomass the next year depends upon the biomass in the current year and the number of eggs according to

$$B_{t+1} = ke^{-cE_t/B_t}.$$

a) Explain why this last equation is a reasonable model (plot biomass vs eggs and biomass vs biomass).

b) What are the dimensions of k, c, and a?

c) Rewrite the model equations in terms of the scaled variables $X_t = B_t/k$ and $Y_t = E_t/ka$. Call $b = ac$. What are the dimensions of X_t, Y_t, and b?

d) Find the equilibrium and determine a condition on b for which the equilibrium is stable.

e) If the system is perturbed from equilibrium, describe how it returns to equilibrium in the stable case. That is, do the perturbations oscillate, decay, or other?

4. Explain and analyze the nonlinear model

$$\Delta x_t = rx_t \left(1 - \frac{x_t}{K}\right) - sx_ty_t,$$
$$\Delta y_t = -dy_t + \epsilon x_ty_t$$

with respect to equilibria and their stability. All the parameters are positive.

5. Consider a nonlinear fisheries model

$$\begin{pmatrix} x_{t+1} \\ y_{t+1} \end{pmatrix} = \begin{pmatrix} ame^{-\beta x_t} & \frac{3}{2}mb \\ ae^{-\beta x_t} & 0 \end{pmatrix} \begin{pmatrix} x_t \\ y_t \end{pmatrix},$$

where a, m, b, $\beta > 0$, and x_t and y_t are juveniles and adults, respectively.

a) Find the equilibria (use the notation $R = am\left(1 + \frac{3b}{2}\right)$). Find a condition on R that guarantees a nontrivial equilibrium lies in the first quadrant.

b) For a positive equilibrium, determine conditions on R, a, and b under which the equilibrium is asymptotically stable.

c) For a fixed a, plot the stability region in the b, $\ln R$ parameter plane and identify the stability boundary that separates stability from instability.

6. Analyze the Nicholson–Bailey model if the escapement function is

$$F(y) = \left[1 + \left(\frac{ay}{k}\right)\right]^{-k}.$$

7. Suppose one infected individual is introduced into a population of 500 susceptible individuals. The likelihood that an individual becomes infected is 0.1%, and an infective is contagious for an average of 10 days.

a) Compute the basic reproduction number of the disease and determine if an epidemic occurs.

b) Plot the time series S_t, I_t, and R_t over a 60-day period. At what time does the disease reach a maximum? Over a long time, how many do not get the disease?

c) Plot the orbit in an SI phase plane.

8. Modify a nonfatal SIR disease model to include vital dynamics. That is, assume susceptibles are born at a constant rate m, and at the same time individuals in each class die at a per capita rate m. Assume the population is constant. Working in the SI phase plane, investigate the model thoroughly, finding equilibria (if any) and their stability, and sketching the orbits. Explain the long-term dynamics of the disease.

9. A nonlinear juvenile-adult model where the fecundity depends upon the adult population is given by

$$
\begin{aligned}
J_{t+1} &= cbA_t e^{-A_t} J_t + be^{-A_t} A_t, \\
A_{t+1} &= \tau J_t - \sigma A_t.
\end{aligned}
$$

Here, $b, c > 0$ and $0 < \tau, \sigma < 1$. Find the equilibria and examine their stability.

9.3 Stochastic Models

Many dynamical processes are not deterministic, but rather contain elements of uncertainty. This is particularly the case in the life sciences, especially in ecology, which provide a context for the discussion in much of this section. We show how randomness arises and how it can be introduced into discrete dynamical models.

It is assumed the reader has studied elementary probability. Some of the concepts and definitions are presented in the first subsection, and the notation is established.

9.3.1 Elementary Probability

A **random variable** (often denoted **RV**) is a variable whose numerical value (a real number) is determined by the outcome of a random experiment, and it is characterized by a rule that dictates what the probabilities of various outcomes are. The outcome may be the number in a population, the charge on a capacitor, the length of a random phone call, the total time it takes to log onto the Internet, the IQ of an individual, the time it takes a forager to find a food item, and so on. Typically we use capital letters to denote random variables and lowercase letters to denote real variables or specific values of random variables. To know the random variable X means to know the *probability* that X takes a value less than or equal to a number b, or symbolically,

$$
\Pr(X \leq b).
$$

There are basically two types of random variables, discrete and continuous. A discrete RV takes on discrete values, either finitely or infinity many, and a continuous RV takes on a continuum of values, in an interval.

Continuous RVs. A **continuous random variable** X is characterized by its **probability density function** (pdf) denoted by $f_X(x)$. Pdfs are also called probability distributions. We assume the pdf exists for all real x. The subscript X on f reminds us that it is associated with the random variable X. The pdf has the properties:

1. $f_X(x) \geq 0$ for all x.

2. $\int_{-\infty}^{\infty} f_X(x)\, dx = 1$.

3. $\Pr(X \leq x) = \int_{-\infty}^{x} f_X(y)\, dy$.

   Condition (1) says that the pdf is nonnegative; condition (2) requires that the total area under the graph of the pdf is unity, and condition (3) dictates the value of the probability of X taking a value smaller than x . The values of the random variable itself are usually not of interest. It is the density function that contains all the information about the random variable X and allows us to compute probabilities. The pdf need not be a continuous function. If there is no confusion about what the random variable is, we drop the subscript on $f_X(x)$ and just write $f(x)$. The following are important continuous distributions. (Other distributions are introduced in the Exercises.) As an exercise, the reader should plot the pdf in each case.

Example 9.16

The most common, and most important, pdf is the **normal** density

$$f(x) = \frac{1}{\sqrt{2\pi\sigma^2}} e^{-(x-\mu)^2/2\sigma^2}, \quad x \in \mathbb{R}. \tag{3.1}$$

Its graph is the standard bell-shaped curve. The number μ is the average, or mean, of the density, and σ is the standard variation, which measures the width of the curve at its inflection points, and hence the spread of the distribution (see below). If a random variable X has the normal density we say that X is normally distributed and we write $X \sim N(\mu, \sigma)$. The distribution $N(0, 1)$ is called the **standard normal.**   □

Example 9.17

The **uniform** probability density on an interval $[c, d]$ is the step function

$$f(x) = \frac{1}{d - c} \mathbf{1}_{[c,d]}(x) = \begin{cases} \frac{1}{d-c}, & c \leq x \leq d \\ 0, & \text{otherwise} . \end{cases}$$

The uniform density models situations where all values in an interval are equally probable.   □

Example 9.18

The **exponential** probability density with parameter λ is defined by

$$f_X(x) = \begin{cases} \lambda e^{-\lambda x}, & x \geq 0 \\ 0, & x < 0 \end{cases}.$$

The exponential distribution models, for example, the length X of a phone call, or the time X of survival of a cancer patient after treatment. $\square$

The function $F_X(x)$, which measures the probability of X taking on a value less than x, is called the **cumulative distribution function (cdf)**; this should not be confused with the probability distribution, or pdf. Thus

$$F_X(x) = \Pr(X \leq x) = \int_{-\infty}^{x} f_X(x)\, dx.$$

The cdf has the properties

$$F_X(x) \geq 0, \quad F_X(-\infty) = 0, \quad F_X(+\infty) = 1,$$

and

$$F_X(x) \quad \text{is a nondecreasing function of } x.$$

The cdf is continuous from the right. That is, for each real number a,

$$\lim_{x \to a^+} F_X(x) = F_X(a).$$

If the pdf $f_X(x)$ is continuous at x, then the fundamental theorem of calculus guarantees that

$$f_X(x) = \frac{dF_X(x)}{dx}.$$

Therefore, the pdf is the derivative of the cdf and the cdf is the integral of the pdf. Note that

$$\Pr(x < X \leq x+h) = \Pr(X \leq x+h) - \Pr(X \leq x) = \int_{x}^{x+h} f_X(y)\, dy \simeq f_X(x)h,$$

or

$$f_X(x) \simeq \frac{1}{h} \Pr(x < X \leq x+h).$$

This equation explains what the pdf $f_X(x)$ actually is—it is the probability per unit length, and hence the name "probability density." The probability of the random variable taking a value between x and $x+h$ is approximately the area under the density curve between x and $x+h$. Also observe that

$$\Pr(a < X \leq b) = F_X(b) - F_X(a).$$

In summary, the probability law characterizing a random variable can be defined by the pdf or the cdf; both carry all the probabilistic and statistical information about X.

Usually we are uninterested in the entire distribution. For example, when shopping for a new car we do not necessarily want to see the entire set of gas mileage data for all cars of the type we wish to buy; we are content to know the average gas mileage. Or, we may want to know the average height of a person in a population, or the average IQ of an entering class of students. A measure of the average, or central tendency, of a random variable X is its **mean**, or **expected value,**

$$\mu = \mathrm{E}(X) = \int_{-\infty}^{\infty} x f_X(x)\, dx.$$

The expected value is the coarsest probabilistic information about X. Other common measures of the central tendency are the mode and the median. A measure of how the possible values of the random variable X are spread about its mean is called the **variance,** which is defined by

$$\mathrm{Var}(X) = \mathrm{E}((X-\mu)^2) = \int_{-\infty}^{\infty} (x-\mu)^2 f_X(x)\, dx.$$

The **standard deviation** σ is the square root of the variance, or

$$\sigma = \sqrt{\mathrm{Var}(X)}.$$

Example 9.19

The normal distribution $X \sim \mathrm{N}(\mu, \sigma)$ has expected value $\mathrm{E}(X) = \mu$ and $\mathrm{Var}(X) = \sigma^2$. Expected values and variances for other random variables are contained in the Exercises.   □

If X is a random variable then we can form combinations of X to create new random variables. That is, we can define $Y = g(X)$, where g is a given real-valued function. This equation defines a transformation of a random variable. Let us assume that g is a strictly increasing function whose first derivative is continuous on all of the real line, and let us denote the inverse g^{-1} of g by φ. That is, $g^{-1} = \varphi$. Thus $y = g(x)$ if, and only if, $x = \varphi(y)$, or $g(\varphi(y)) = y$ and $\varphi(g(x)) = x$. We can determine the pdf for Y in terms of the pdf for X by calculating

$$
\begin{aligned}
\mathrm{Pr}(a < Y \le b) &= \mathrm{Pr}(a < g(X) \le b) \\
&= \mathrm{Pr}(\varphi(a) < X \le \varphi(a)) \\
&= \int_{\varphi(a)}^{\varphi(b)} f_X(x)\, dx.
\end{aligned}
$$

In this last integral we can perform a change variables via $x = \varphi(y)$, $dx = \varphi'(y)\,dy$. Then

$$\Pr(a < Y \le b) = \int_a^b f_X(\varphi(y))\varphi'(y)\,dy.$$

By definition of the probability law, therefore, the pdf for Y is

$$f_Y(y) = f_X(\varphi(y))\varphi'(y). \qquad (3.2)$$

If the function g is not strictly increasing, then the density can be found by breaking up the problem into intervals over which g is monotone increasing or monotone decreasing. As an aside, we recall from calculus that the relationship between the derivatives of φ and g is $\varphi'(y) = 1/g'(x)$.

There is an interesting corollary of this result. Nearly the same calculation can be performed to obtain

$$\Pr(c < X \le d) = \Pr(c < \varphi(Y) \le b) = \Pr(g(c) < Y \le g(d)).$$

In terms of integrals,

$$\int_c^d f_X(x)\,dx = \int_{g(c)}^{g(d)} f_Y(y)\,dy.$$

In words, "probability is conserved" by a strictly increasing, continuously differentiable transformation of the random variable.

Next we inquire about the expected value of $Y = g(X)$. We have

$$
\begin{aligned}
\mathrm{E}(g(X)) &= \mathrm{E}(Y) = \int_{-\infty}^{\infty} y f_Y(y)\,dy \\
&= \int_{-\infty}^{\infty} g(x) f_Y(g(x)) g'(x)\,dx \\
&= \int_{-\infty}^{\infty} g(x) f_X(\varphi(g(x))) \varphi'(g(x)) g'(x)\,dx \\
&= \int_{-\infty}^{\infty} g(x) f_X(x)\,dx,
\end{aligned}
$$

where we have used (3.2) and the fact that g and φ are inverses. Using these facts, we can easily show

$$\mathrm{E}(aX + b) = a\mathrm{E}(X) + b, \quad a, b \in \mathbb{R}, \qquad (3.3)$$

and

$$\mathrm{Var}(X) = \mathrm{E}(X^2) - \mathrm{E}(X)^2. \qquad (3.4)$$

This latter equation is commonly used instead of the integral formula to calculate the variance.

One of the key results in probability theory is the central limit theorem. The statement uses the notion of independent random variables, which are defined later. Intuitively, two RVs are independent if the values of one does not depend on values of the other.

Theorem 9.20

(**Central Limit Theorem**) If Y_1, Y_2, $Y_3, ldots$ is a sequence of independent and identically distributed random variables with finite mean μ and variance σ^2, then the limiting distribution of the random variable

$$S = \frac{\sum_{i=1}^{n} \frac{Y_i}{n} - \mu}{\sigma/\sqrt{n}}$$

is the standard normal distribution $N(0, 1)$.   □

Discrete RVs. The preceding definitions and results go over to **discrete random variables** in an obvious way. Let X be a discrete RV that assumes the values x_1, x_2,...,x_k, ... The analog of the pdf is the discrete **probability mass function**

$$p_k = \Pr(X = x_k), \quad k = 1, 2, 3, ...,$$

which defines the probabilities of X taking the various values. Clearly $\sum_k p_k = 1$. The **expected value** of X is the weighted average

$$\mu = \mathrm{E}(X) = \sum_k x_k p_k,$$

and more generally,

$$\mathrm{E}(g(X)) = \sum_k g(x_k) p_k.$$

The **variance** is

$$\mathrm{Var}(X) = E((X - \mu)^2),$$

and relations (3.3) and (3.4) still hold. Sometimes the **coefficient of variation** CV, defined by $CV = \sigma/\mu$, is used as measure of variation; for example, $CV = 0.40$ carries the rough interpretation that on average X varies 40% from its mean value. Some important discrete random variables are presented below, and the reader should plot the pmf in each case for several values of the parameters.

Example 9.21

Some experiments have only two outcomes, success S and failure F. If N independent trials of a given experiment are performed, we can let the random variable X be the number of successes. If $\Pr(S) = p$ and $\Pr(F) = 1 - p$, then the probability mass function is

$$\Pr(X = k) = \binom{N}{k} p^k (1 - p)^{N-k},$$

which is the probability of obtaining k successes in the N trials. This pmf is the **binomial distribution**. We write $X \sim \mathrm{bin}(N, p)$ to say X is binomial distributed. $\square$

Example 9.22

Again consider successive independent trials where an event A occurs with probability p. Let X be the number of trials until A occurs the first time. Then

$$\Pr(X = k) = p(1 - p)^k, \ k = 1, 2, 3, \ldots$$

is called the **geometric distribution** with parameter p. There is an important generalization of the geometric random variable. Again in the binomial setting, let X be the number of trials required until the rth success occurs. Then

$$\Pr(X = k) = \binom{k-1}{r-1} p^r (1 - p)^{k-r}, \quad k = r, r+1, \ldots$$

is the **negative binomial distribution**. $\square$

The negative binomial distribution is an important, frequently used model for spatial distributions (e.g., of plants or animals) which have a clumped character. This is in contrast to the Poisson distribution, discussed next, which models completely random distributions.

Example 9.23

The **Poisson distribution** with parameter $\lambda > 0$ is

$$\Pr(X = k) = \frac{\lambda^k e^{-\lambda}}{k!}, \quad k = 0, 1, 2, \ldots.$$

The Poisson RV models the probability of unpredictable, rare events over a short time scale, where λ is the average occurrence of the event. $\square$

Finally, we review some ideas in **conditional probability**. If A and B are two events with $\Pr B > 0$, then the probability that A occurs, given that B has occurred, is

$$\Pr(A|B) = \frac{\Pr(A \cap B)}{\Pr B}.$$

A key result, the **total probability theorem**, is used in the sequel. It states that if A is composed of mutually disjoint events $A_1, \ldots, A_n$, that is, $A = \cup_{k=1}^{n} A_k$ with $A_i \cap A_j \neq \phi$, then

$$\Pr A = \sum_{k=1}^{n} P(A|A_k) \Pr A_k.$$

Finally, we say any set of events $A_1, \ldots, A_n$ is **independent** if

$$\Pr(A_{i_1} \cap A_{i_2} \cap \cdots \cap A_{i_r}) = \Pr(A_{i_1}) \Pr(A_{i_2}) \cdots \Pr(A_{i_r})$$

for all sequences of integers $i_1 < i_2 < \cdots < i_k$, $\ k = 2, 3, \ldots$. A set of events is **dependent** if they are not independent.

9.3.2 Stochastic Processes

A **stochastic process** (abbreviated **SP**) is an indexed family of random variables X_t (or $X(t)$), where t belongs to some index set T. That is, for each assigned $t \in T$, the quantity X_t is a random variable. Usually the index set is the time interval $T = [0, \infty)$ or the discrete set of integers $T = \{0, 1, 2, 3, \ldots\}$. When the latter is the index set, the SP is often called a random sequence. When the index set is the finite set $T = \{1, 2, 3, \ldots, n\}$, then the SP is a random vector. There are actually four types of real, stochastic processes X_t that can be considered. Two characterizations are based on whether the index set is discrete or is continuous (an interval), and two more are defined by whether X_t itself takes on discrete values or a continuum of values. As an example of a discrete process, let X_t denote the result of a roll of a single die at times $t \in T = \{0, 1, 2, 3, \ldots\}$. At each instant of time t, X_t is a random variable that takes on one of the values 1,2,3,4,5,6, each with equal probability (one-sixth). A typical **realization** of the process is $1, 1, 3, 6, 1, 4, 5, 2, 5, \ldots$. In this case of a die, the X_t are statistically independent, which is generally not the case for stochastic processes. In fact, statistical dependence is usually what makes stochastic processes interesting and stochastic processes are frequently characterized by their memory; that is, they are defined by how a random variable X_t depends on earlier random variables X_s, $s < t$. An example of a continuous SP is a Poisson process. For this process $T = [0, \infty)$ and X_t is discrete; X_t counts the number of times a certain event occurs from time $t = 0$ to time t. The event could be an accident at an intersection, the number of customers arriving at a bank, the number of breakdowns of machines in a plant, or discoveries of prey items by a predator.

Generally, to define a SP we must specify the index set T, the space where the X_t lie, and relations between the X_t for different t (for example, joint distributions, correlations, or other dependencies). In fact, the probability law for a SP is a specification of all the joint distributions; but simplifications are often introduced to make the process tractable. The types of SPs studied most often in applied probability theory include Markov processes, Poisson processes, Gaussian processes, Wiener processes, and martingales—all of these are defined by certain dependencies, or memory. Therefore, definitions of these processes depend on the notion of conditional probability.

Because a stochastic process X_t is defined as a random variable for each assigned time t, there is a probability density function $f(x, t)$ for X_t. Observe that the density has *both* spatial and temporal dependence. From the density function we can calculate the expected value and variance at each time t via

$$\mu(t) = \mathrm{E}(X_t) = \int_{-\infty}^{\infty} x f(x, t) \, dx$$

and

$$\sigma^2(t) = \mathrm{Var}(X_t) = \int_{-\infty}^{\infty} (x - \mu(t))^2 f(x, t) \, dx.$$

Similar definitions hold for discrete random variables.

EXERCISES

1. Calculate the mean, standard deviation, and cumulative distribution function of a uniform random variable X with density $\frac{1}{d-c} \mathbf{1}_{[c,d]}(x)$.

2. Verify that the standard deviation of a normal random variable with density $N(0, \sigma)$ is, in fact, σ.

3. Show that the expected value of the exponential distribution with parameter λ is $\mathrm{E}(X) = \frac{1}{\lambda}$, and the variance is $\mathrm{Var}(X) = \frac{1}{\lambda^2}$. Show that the cumulative distribution function is

$$F(x) = 1 - e^{-\lambda x}, \quad x > 0.$$

4. If X is a continuous RV, find the pdf of the random variable $Y = X^2$. (Hint: Start with $\mathrm{Pr}(Y \leq y)$.)

5. The famous Scottish physicist J. C. Maxwell showed that molecular collisions in a gas result in a speed distribution where the fraction of molecules

in the speed range $[v, v + dv]$ is given by the **Maxwell–Boltzmann** distribution

$$f(v) = \sqrt{\frac{m}{2\pi kT}}\, v^2 e^{-mv^2/2kT}, \quad v \geq 0,$$

where n is the total number of particles, m is the molecular mass of each particle, T is the temperature, and k is Boltzmann's constant ($k = 1.38 \times 10^{-23}$ joules/deg K). Sketch a generic graph of the distribution and compute the average speed of a molecule in a gas.

6. Verify formulas (3.3) and (3.4), and show that

$$\text{Var}(a + bX) = b^2 \text{Var}(X), \quad a, b \in \mathbb{R}.$$

7. If $X \sim \text{N}(\mu, \sigma)$, show that $Z = \frac{X - \mu}{\sigma} \sim \text{N}(0, 1)$. The RV Z is called the z–score.

8. Find the mean and variance of a Poisson random variable.

9. A Cauchy random variable has density defined by

$$f(x) = \frac{a}{\pi}\frac{1}{a^2 + x^2}.$$

Compute the cumulative distribution function and draw a graph. What is $E(X)$ and $\text{Var}(X)$? What is unusual about your answer?

10. On the average there are about 15 cyclones formed off the U.S. Pacific coast each year. In a given year, what is the probability of there being at most 6 cyclones?

11. A random variable X is **gamma distributed** with parameters a and λ if its pdf is

$$f(x) = \frac{\lambda(\lambda x)^{a-1} e^{-\lambda x}}{\Gamma(a)}, \quad x \geq 0,$$

and $f(x) = 0$ for $x < 0$, where

$$\Gamma(a) = \int_0^\infty z^{a-1} e^{-z} dz$$

is the gamma function. Sketch the pdf for $a = 3.5$ and $\lambda = 2$.

12. Let X be a normal random variable with mean μ and variance σ^2. The random variable Y defined by $Y = e^X$ is said to be a **log-normal** random variable. Show that the pdf of Y is given by

$$f_Y(y) = \frac{1}{y\sqrt{2\pi\sigma^2}} e^{-(\ln y - \mu)^2/2\sigma^2},$$

and plot the pdf for various values of σ. What is the mean and variance of Y? (The log-normal density is often associated with phenomena that involve multiplicative effects of a large number of independent random events.)

13. Let $X \sim N(0,1)$. Find the pdf of the random variable X^2. (This pdf is called the **chi-squared** distribution.)

14. Let X be a discrete RV with $\Pr(X = x_j) = p_j$, $j = 0, 1, 2, ...$. The **probability generating function** is defined by

$$G(t) = \mathrm{E}(t^X) = \sum_{j=0}^{\infty} p_j t^j.$$

a) Show that $E(X) = G'(1)$ and $\mathrm{Var}(X) = G''(1) + G'(1) - G'(1)^2$.

b) Use the probability generating function to find the mean and variance of a binomial random variable $X \sim \mathrm{bin}(N, p)$.

15. The **probability generating function** for a continuous RV X is defined by

$$G(t) = \mathrm{E}(t^X) = \int_{-\infty}^{\infty} t^x f(x) \, dx.$$

a) Show that $\mathrm{E}(X) = G'(1)$ and $\mathrm{Var}(X) = G''(1) + G'(1) - G'(1)^2$.

b) Use the probability generating function to find the mean and variance of an exponential distribution X with parameter λ.

16. Let X be a discrete RV with $\Pr(X = x_j) = p_j$, $j = 0, 1, 2, ...$. The **moment generating function** is defined by

$$M(t) = \mathrm{E}(e^{tX}) = \sum_{j=0}^{\infty} p_j e^{jt}.$$

a) Show that $\mathrm{E}(X) = M'(0)$ and $\mathrm{Var}(X) = M''(0) - M'(0)^2$.

b) Use the moment generating function to find the mean and variance of a binomial random variable $X \sim \mathrm{bin}(N, p)$.

17. The **moment generating function** for a continuous RV X is defined by

$$M(t) = \mathrm{E}(e^{tX}) = \int_{-\infty}^{\infty} e^{tx} f(x) \, dx.$$

a) Show that $\mathrm{E}(X) = M'(0)$ and $\mathrm{Var}(X) = M''(0) - M'(0)^2$.

b) Use the moment generating function to find the mean and variance of an exponential distribution X with parameter λ.

18. Let X be a continuous RV. Prove Chebyshev's inequality:

$$\Pr(|X - \mu| \geq k\sigma) \leq \frac{1}{k^2},$$

where μ and σ are the mean and standard variation, and $k > 0$.

19. Let X be a continuous RV with mean μ and standard variation σ. If g is a sufficiently smooth function, show that $\mathrm{Var}(g(X)) \cong g'(\mu)\sigma^2$.

20. If X is geometrically distributed, show that $\mathrm{E}(X) = \frac{1}{p}$ and $\mathrm{Var}(X) = \frac{1-p}{p^2}$.

21. Show that if X is Poisson distributed with parameter λ, then $\mathrm{E}(X) = \lambda$ and $\mathrm{Var}(X) = \lambda$.

22. Consider a set of N objects of which r of them are of a special type. Choosing n objects, without replacement, show that the probability of getting exactly k of the special objects is

$$\Pr(X = k) = \frac{\binom{r}{k}\binom{N-r}{n-k}}{\binom{N}{n}}, \quad k = 0, 1, ..., n.$$

This distribution is called the **hypergeometric distribution**. Show that $\mathrm{E}(X) = \frac{nr}{N}$.

9.3.3 Environmental and Demographic Models

There are myriad reasons for random variations in population growth and death rates. Ecologists often categorize these random effects as **environmental stochasticity** and **demographic stochasticity**. The former includes random weather patterns and other external sources where all individuals are affected equally and there is no individual variation; the latter involves natural variability in behavior, growth, vital rates, and other genetic factors that occur in all populations, even though the environment may be constant. Demographic stochasticity is commonly modeled by a Gaussian (normal) process, but environmental variations can take many forms. For example, an environment may suddenly have a catastrophic event, such a flood, or a bonanza year where the conditions for reproduction are unusually high.

In engineering one can identify similar processes, those that have external noise, akin to environmental stochasticity, and those that have internal or system noise, which is like demographic stochasticity.

There are multiple approaches to creating stochastic discrete models. One is to add, in one way or another, stochasticity to a known deterministic model. Another is to construct probabilistic modela directly. In the sequel we illustrate some of these constructions.

Example 9.24

To fix the idea of environmental stochasticity and adding stochasticity to a known deterministic model, consider the Ricker population model

$$x_{t+1} = bx_t e^{-cx_t},$$

where b is the average yearly birthrate. Suppose, on the average, once every 25 years there is a weather event that lowers the birth rate that year by 30%. We create a fixed, uniformly distributed random process Z_t having range $[0, 1]$. At each time step we modify the birthrate according to the rule: *if* $Z_t < 0.04$ *then* adjust the birthrate to $0.7b$; *else* make no adjustment. In this manner, the birthrate becomes a random variable B_t and we can write the dynamics as

$$X_{t+1} = B_t X_t e^{-cX_t}, \tag{3.5}$$

where X_t, the population, is now a random process. It is interesting to simulate this process and the Exercises will guide the reader through this activity. □

A word about notation—on the right side of (3.5) we mean the specific value $X_t = x_t$ of the random variable, or population, calculated at the time step t; some authors prefer to write the stochastic model (3.5) as $X_{t+1} = B_t x_t e^{-cx_t}$.

To model demographic changes we might assume the birthrate is normally distributed. This assumption follows from the observation that birthrates do not vary much from the average, with large deviations rare. Then we have

$$X_{t+1} = G_t X_t e^{-cX_t},$$

where for each time t we choose the birthrate from the distribution $G_t \sim N(b, \sigma)$, a normal distribution with mean b and standard deviation σ (or variance σ^2). Computer algebra systems, as well as calculators, have commands that generate different types of random variables. (For example, in MATLAB, *rand* generates a uniform random variable on $[0, 1]$, and *randn* generates a normal random variable $N(0, 1)$.)

Generally, a discrete model has the form

$$x_{t+1} = f(x_t, r), \tag{3.6}$$

where r is a parameter. (The analysis can clearly be extended to vector functions and sets of parameters.) In the two examples above, stochasticity was added to the model by redefining the parameter r to be random process R_t, giving a stochastic process X_t satisfying

$$X_{t+1} = f(X_t, R_t).$$

The values of R_t are usually chosen from a fixed distribution with mean μ and variance σ^2. For example, we may choose $R_t = \mu + \sigma W_t$, where $W_t \sim N(0,1)$. Alternately, the values of R_t may be chosen themselves to satisfy a linear autoregressive process, for example, $R_t = aR_{t-1} + b + W_t$, where $W_t \sim N(0,1)$, and a and b are fixed constants. How stochasticity is put into a model depends upon what one seeks and is therefore *strictly an issue of modeling*. Different stochastic mechanisms produce different patterns of variability.

In some models stochasticity may enter only through the initial condition. These situations are modeled by the process

$$X_{t+1} = f(X_t, r), \quad X_0 = C,$$

where C is a fixed random variable. The solution to this random equation can be obtained by solving the deterministic problem, if possible; in this case, the random initial condition is evolved deterministically. The solution has the form

$$X_t = G(t, C, r),$$

which is a transformation of a random variable C (treating t and r as parameters) and therefore the statistical properties of X_t can, in principle, be determined exactly. Simulations, or realizations of these processes, are smooth curves with random initial conditions.

Later in this chapter we examine stochastic models that are obtained directly from probability considerations.

Example 9.25

Consider the random difference equation

$$
\begin{aligned}
X_{t+1} &= aX_t, \quad t = 0, 1, 2, ..., \\
X(0) &= X_0,
\end{aligned}
$$

where a is a positive constant and X_0 is a random variable. Treating the equation deterministically, we obtain

$$X_t = X_0 a^t.$$

Then

$$
\begin{aligned}
\mathrm{E}(X_t) &= \mathrm{E}(X_0 a^t) = a^t \mathrm{E}(X_0), \\
\mathrm{Var}(X_t) &= \mathrm{Var}(X_0 a^t) = a^{2t} \mathrm{Var}(X_0).
\end{aligned}
$$

For example, if $0 < a < 1$, then the expected value of X_t goes to zero as $t \to \infty$, and the variance goes to zero as well; this means the spread of the distribution gets narrower and narrower. In fact, we can determine the pdf for X_t as follows. Letting $f_{X_0}(x)$ denote the pdf of X_0, we have

$$\Pr(X_t \leq x) = \Pr(X_0 a^t \leq x) = \Pr(X_0 \leq xa^{-t})$$

$$= \int_{-\infty}^{xa^{-t}} f_{X_0}(y)\,dy = \int_{-\infty}^{x} f_{X_0}(za^{-t})a^{-t}\,dz.$$

Therefore the pdf of the process is

$$f_{X_t}(x,t) = f_{X_0}(xa^{-t})a^{-t}.$$

For a specific case let X_0 be uniformly distributed on the interval $[0, b]$. That is,

$$f(x) = \frac{1}{b}\mathbf{1}_{[0,b]}(x).$$

Then

$$f_{X_t}(x,t) = \frac{a^{-t}}{b}\mathbf{1}_{[0,b]}(xa^{-t}) = \frac{a^{-t}}{b}\mathbf{1}_{[0,ba^t]}(x),$$

which is uniform. As an exercise, the reader should plot this density for $t = 1, 2, 3$ for $a > 1$ and $0 < a < 1$. This method is successful because this simple discrete model can be solved exactly.   □

EXERCISES

1. A deterministic model for the yearly population of Florida sandhill cranes is

$$x_{t+1} = (1 + b - d)x_t,$$

where $b = 0.5$ is the birthrate and $d = 0.1$ is the deathrate. Initially the population is 100. (Adapted from Mooney and Swift, 1999.)

a) Find a formula for the population after t years and plot the population for the first 5 years.

b) On average, a flood occurs once every 25 years, lowering the birthrate 40% and raising the deathrate 25%. Set up a stochastic model and perform 20 simulations over a 5-year period. Plot the simulations on the same set of axes and compare to the exact solution. Draw a frequency histogram for the number of ending populations for the ranges (bins) 200–250, 250–300, and so on, continuing in steps of 50.

c) Assume the birthrates and deathrates are normal random variables, $N(0.5, 0.03)$ and $N(0.1, 0.08)$, respectively. Perform 20 simulations of the population over a 5-year period and plot them on the same set of axes. On a separate plot, for each year draw side-by-side box plots indicating the population quartiles. Would you say the population in year 5 is normally distributed?

2. A "patch" has area a, perimeter s, and a strip (band) of width w inside the boundary of a from which animals disperse. Only those in the strip disperse. Let u_t be the number of animals in a at any time t. The *per capita* growth rate of all the animals in a is r. The rate that animals disperse from the strip is proportional to the fraction of the animals the strip, with proportionality constant ϵ, which is the emigration rate for those in the strip.

a) Argue that the dynamics is ruled by

$$u_{t+1} = ru_t - \epsilon \left(\frac{ws}{a} u_t \right).$$

b) Determine conditions on the parameters r, w, s, ϵ, and a under which the population is growing.

c) Why do you expect $s = k\sqrt{a}$ for some constant k? Write down the model in the case that the region is a circle.

d) Suppose at each time step the emigration rate is chosen from a normal distribution with mean 0.35 and standard deviation 0.5. With a growth rate of 1.06 in a circle of radius 100 m, with width $w = 10$ m, simulate the dynamics of the process. Take $u_0 = 100$. Sketch three simulations (realizations) on the same set of axes. Does the population grow or become extinct?

e) In a given time step, suppose the animals disperse with $\epsilon = 0.25$ or $\epsilon = 0.45$, each with equal probability. Simulate the dynamics of the process and sketch three realizations.

3. Consider the population growth law $x_{t+1} = r_t x_t$, where the growth rate r_t is a fixed positive sequence and x_0 is given.

a) Let x_T be the population at time T. Which quantity is a better measure of an *average* growth rate over the time interval $t = 0$ to $t = T$, the arithmetic mean $r_A = \frac{1}{T}(r_0 + r_1 + \cdots + r_{T-1})$, or the geometric mean $r_G = (r_0 r_1 \cdots r_{T-1})^{1/T}$?

b) Consider the stochastic model process $X_{t+1} = R_t X_t$, where $R_t = 0.86$ or $R_t = 1.16$, each with probability one-half. Does the population eventually grow, die out, or remain the same? If $X_0 = 100$ is fixed, what is the expected value of the population after 500 generations? Run some simulations to confirm your answer.

4. A stochastic model is defined by

$$X_{t+1} = R_t X_t,$$

where the initial population $X_0 = x_0$ is fixed, and the growth rate R_t at each time is chosen from a fixed distribution with mean μ and variance σ^2. Use the central limit theorem to prove that X_t approaches a log-normal random variable for large t. What is the mean and variance of X_t for large t?

5. Suppose you are playing a casino game where on each play you win or lose one dollar with equal probability. Beginning with \$100, plot several realizations of a process where you play 200 times. Perform the same task if the casino has the probability 0.52 of winning. In the latter case, perform several simulations and estimate the average length of a game.

6. In a generation, suppose each animal in a population produces two offspring with probability $\frac{1}{4}$, one offspring with probability $\frac{1}{2}$, and no offspring with probability $\frac{1}{4}$. Assume an animal itself does not survive over the generation. Illustrate five realizations of the population history over 200 generations when the initial population is 8, 16, 32, 64, and 128. Do your results say anything about extinction of populations?

7. Suppose μ is the probability that each individual in a population will die during a given year. If X_t is the population at year t, then the probability there will be n individuals alive the next year can be modeled by a binomial random variable,

$$\Pr(X_{t+1} = n | X_t = x_t) = \binom{x_t}{n}(1 - \mu)^n \mu^{x_t - n}.$$

In other words, the population at the next time step is the probability of having n successes out of x_t trials, where $1 - \mu$ is the probability of success (living).

a) At time $t = 0$ assume a population has 50 individuals, each having mortality probability $\mu = 0.7$. Use a computer algebra system to perform 10 simulations of the population dynamics over 7 years and plot the results.

b) For the general model, find the expected value $E(X_{t+1})$ and the variance $Var(X_{t+1})$.

8. Consider a random process X_t governed by the discrete model

$$X_{t+1} = \frac{2X_t}{1 + X_t},$$

where X_0 is a random variable with density $f_{X_0}(x)$. Find a formula for the density $f(x, t)$ of X_t.

9. Experimental data for the yearly population of Serengeti wildebeest can be fit by the Beverton–Holt model

$$X_{t+1} = \frac{(1-h)rKX_t}{K + (r-1)X_t},$$

where K is the carrying capacity, r is a growth factor, and h is the annual harvesting rate. The carrying capacity is a function of rainfall and is modeled by $K = 20748R$, where R is the annual rainfall. Actual measured rainfalls over a 25-year period are given in the vector

$$\mathbf{R} = (100, 36, 100, 104, 167, 107, 165, 71, 91, 77, 134, 192, 235,$$
$$159, 211, 257, 204, 300, 187, 84, 99, 163, 97, 228, 208).$$

Using $X_0 = 250,000$ and $r = 1.1323$, simulate the population dynamics over 50 years for different harvesting rates, $h = 0.05, 0.075, 0.1, 0.125, 0.15$, and each year randomly draw the annual rainfall from the vector $\mathbf{R}$. Run enough simulations to make some conclusions about how the fraction of populations that collapsed depends upon the harvesting rate. Assume the threshold value of 150,000 wildebeest as representing a collapsed herd. (What has been described is a *bootstrap method* where simulations are run by randomizing a single data set.)

10. Develop a demographic stochastic model based upon the discrete model

$$x_{t+1} = \frac{bx_t}{1 + cx_t}.$$

Take $c = 1$ and $b = 6$. With $x_0 = 200$, run simulations of the model with noise $E_t \sim \mathrm{N}(0, \sigma)$ for several different values of σ, and comment upon the results.

9.4 Probability-Based Models

9.4.1 Markov Processes

Stochastic processes X_t are characterized by the dependencies among the random variables $X_0, X_1, X_2, \dots$. A Markov process is a special stochastic where only the current value of the random variable is relevant for future predictions. The past history, or how the present emerged from the past, is irrelevant. Many processes in economics, biology, games of chance, engineering, and other areas are of this type—for example, stock prices, a board position in the game of monopoly, the population of a plant species, and so on.

We illustrate the Markov model by considering a specific problem in phylogenetics.

We may recall from elementary biology that the double helix DNA molecule is the genetic material that contains information on coding for and synthesizing proteins. DNA is constructed from four bases, the purines A and G, and the pyrimidines C and T. In the DNA molecule, across the rungs of the double helix molecular ladder, A and T are paired and G and C are paired. Therefore, to represent a DNA molecule mathematically all we need consider is a single sequence, e.g., $ATCCGTAGATGG$, along one side of the ladder. As time evolves, DNA must be copied from generation to generation. During this process there are natural mutations, and our goal is to set up a simple model for determining how these mutations evolve to change an ancestral DNA sequence. Such models are important, for example, in determining phylogenetic distances between different species and to understand the evolution of the genome in organisms, both plant and animal. Phylogentics helps identify related species and to construct evolutionary trees; understanding how viruses evolve, for example, aids in developing strategies for disease prevention and treatment.

The simplest mutations in copying a DNA sequence are called base *substitutions*—a base at a given location is simply replaced by another one at that same location over one generation. There are other types of mutations, but they are not considered in our introductory discussion. If a purine is replaced by a purine or a pyrimidine with a pyrimidine, then the substitution is a *transition*; if the base class is interchanged, then the substitution is a *transversion*.

To fix the idea, consider mutations of a sequence over three generations:

$$\begin{aligned}
\mathrm{S}_0 &: \quad ATCC\mathbf{GT}AGATGG \\
\mathrm{S}_1 &: \quad ATCC\mathbf{AT}AG\mathbf{T}TGG \\
\mathrm{S}_2 &: \quad ATCC\mathbf{GT}AG\mathbf{T}TGG
\end{aligned}$$

The changes are highlighted. Comparing the first two generations, in the fifth position there is a transition and in the ninth position there is a transversion.

Consider an arbitrary sequence. We let P_A, P_G, P_C, and P_T denote the probabilities that A, G, C, and T appear in at a given location. In the sequence S_0, for example, $P_A = 3/12$. Next we fix an arbitrary location at some generation t. During the next generation, $t+1$, the base at that location may stay the same, or it may mutate to another base. We let the sixteen numbers $P(X|Y)$, where X and Y represent the four bases, denote the conditional probabilities that the location has base X at time $t+1$, given that it was at Y at time t. We can arrange these numbers in a *transition matrix*

$$M = \begin{pmatrix} P(A,A) & P(A,G) & P(A,C) & P(A,T) \\ P(G,A) & P(G,G) & P(G,C) & P(G,T) \\ P(C,A) & P(C,G) & P(C,C) & P(C,T) \\ P(T,A) & P(T,G) & P(T,C) & P(T,T) \end{pmatrix}. \tag{4.1}$$

Notice the order of the entries. Columns and rows are labeled in the order A, G, C, T, and the column denotes the state of the ancestral base, and the row is the state of the descendent base. In the transition matrix M, the entries are nonnegative and the columns sum to one because one of the transitions must occur. Such a matrix is called a *Markov matrix*. We assume that, over many generations of equal time steps, the same transition matrix characterizes the transitions at each time step. Now, if $\mathbf{p}_t = (P_A, P_G, P_C, P_T)^T$ denotes the *probability vector* at time t, the model

$$\mathbf{p}_{t+1} = M\mathbf{p}_t, \tag{4.2}$$

$t = 0, 1, 2, ...$, is an example of a *discrete* **Markov process**. It dictates how the probabilities of bases occurring in a generation change over time. Therefore, if $\mathbf{p}_0$ is given, then $\mathbf{p}_t$ can be calculated by iteration.

We recognize (4.2) as a linear, Leslie-type model that we examined in an earlier section. The solution is

$$\mathbf{p}_t = M^t \mathbf{p}_0,$$

where the entries in the power matrix M^t denote the probabilities of being in a given state (base) after t time steps, given the initial state. For example, for the transition matrix (4.1) for a DNA sequence, the $4, 2$ element in M^t would be the probability that the base is T after t time steps, given that it was G at time $t = 0$.

In general, a discrete Markov process, or Markov chain, can be defined as follows. Consider a sequence of random variables $X_0, X_1, X_2, ...$, where each random variable is defined on a finite or countably infinite state space, which we denote by $S = \{0, 1, 2, 3, ...\}$. The stochastic process X_t is **Markov,** if for any $t = 0, 1, 2, 3, ...$, it has the property

$$\Pr(X_{t+1} = a_{t+1} | X_t = a_t, ..., X_0 = a_0) = \Pr(X_{t+1} = a_{t+1} | X_t = a_t), \quad a_t \in S.$$

In other words, the future behavior depends only upon the present state and not on the previous history. The **transition matrix** M for the process is the matrix whose entries are $m_{ij} = \Pr(X_{t+1} = i | X_t = j)$. Thus, m_{ij} is the probability of moving from state j at time t to state i at time $t + 1$. Notice that M may be an infinite matrix if the state space is infinite. Clearly, by definition, the entries in each column of M are nonnegative and sum to one. We sometimes graphically show the states and elements of the transition matrix on a Leslie type diagram, as in Fig. 9.10. The entries of M^t are denoted by $m_{ij}^{(t)}$ and represent the probabilities of moving from the jth state to the ith state in t time steps. A vector $\mathbf{p}$ with nonnegative entries that sum to one is called a **probability vector**. There are many examples of Markov processes in all areas of science, engineering, and economics.

Example 9.26

(**Jukes–Cantor model**) In the Jukes–Cantor model of base substitution,

$$
M = \begin{pmatrix}
1-a & \frac{a}{3} & \frac{a}{3} & \frac{a}{3} \\
\frac{a}{3} & 1-a & \frac{a}{3} & \frac{a}{3} \\
\frac{a}{3} & \frac{a}{3} & 1-a & \frac{a}{3} \\
\frac{a}{3} & \frac{a}{3} & \frac{a}{3} & 1-a
\end{pmatrix},
$$

where $1-a$ is the probability that there is no change in a base, while $a/3$ is the probability of change to any other base. The number a is called the *mutation rate per generation, or the number of substitutions per site per time step;* values of a range from 10^{-9} for some plants, 10^{-8} for mammals, to 10^{-2} for viruses, which mutate quickly. We are assuming a is a small constant and does not depend upon time or upon location in the sequence.

It is straightforward to verify that $\lambda = 1$ is an eigenvalue of M with eigenvector $(1,1,1,1)^{\mathrm{T}}$, and $\lambda = 1 - \frac{4}{3}a$ is an eigenvalue of multiplicity 3 with independent eigenvectors $(1,1,-1,-1)^{\mathrm{T}}, (1,-1,1,-1)^{\mathrm{T}}, (1,-1,-1,1)^{\mathrm{T}}$. Then, if Q is the matrix whose columns are the eigenvectors, then Q diagonalizes M, which means $M = QDQ^{-1}$, where D is the diagonal matrix

$$
D = \begin{pmatrix}
1 & 0 & 0 & 0 \\
0 & 1-\frac{4}{3}a & 0 & 0 \\
0 & 0 & 1-\frac{4}{3}a & 0 \\
0 & 0 & 0 & 1-\frac{4}{3}a
\end{pmatrix}.
$$

Therefore

$$
M^t = QD^tQ^{-1} = Q \begin{pmatrix}
1 & 0 & 0 & 0 \\
0 & \left(1-\frac{4}{3}a\right)^t & 0 & 0 \\
0 & 0 & \left(1-\frac{4}{3}a\right)^t & 0 \\
0 & 0 & 0 & \left(1-\frac{4}{3}a\right)^t
\end{pmatrix} Q^{-1}.
$$

These matrices can be multiplied out to obtain M^t. One can check that the diagonal entries of M^t are $q(t) = \frac{1}{4} + \frac{3}{4}\left(1 - \frac{4}{3}a\right)^t$, and the off-diagonal entries are all given by $\frac{1}{4} - \frac{1}{4}\left(1 - \frac{4}{3}a\right)^t$. Therefore $q(t)$ is the expected fraction of sites in the sequence that are observed to remain unchanged after t time steps. It follows that $1 - q(t)$ is the fraction of sites expected to differ from the original ancestral sequence after t time steps. $\square$

It is not hard to prove the following important theorem (a proof for a two-dimensional matrix is requested in Exercise 1).

Theorem 9.27

Let $M = (m_{ij})$ be an $n \times n$ Markov matrix having nonnegative entries m_{ij} and column sums equal to one. Then $\lambda = 1$ is an eigenvalue of M and the corresponding eigenvectors have nonnegative entries; the remaining eigenvalues λ_i satisfy $|\lambda_i| \leq 1$. Moreover, if the entries of M are strictly positive, then $\lambda = 1$ is a dominant eigenvalue with a single eigenvector. $\square$

If the state space is finite, then over a long time the system may approach a stationary probability distribution defined by a probability vector $\mathbf{p}$ that is an eigenvector of $\lambda = 1$. That is, $M\mathbf{p} = \mathbf{p}$. Depending upon the geometric multiplicity of $\lambda = 1$, the stationary distribution may or may not be unique. A stationary distribution $\mathbf{p}$ is said to be **limiting** if $M^t\mathbf{p}_0 \to \mathbf{p}$ for every initial probability vector $\mathbf{p}_0$. There are general conditions on a Markov matrix that guarantee a unique limiting stationary distribution (e.g., see Allen, 2003).

There are many interesting questions about processes that satisfy the Markov conditions. For example: Which states can lead to some other state? Are there states that are visited infinitely often, or are there states in which the process becomes trapped? Are there states where the process will always end up? Figure 9.14 shows a Leslie-type diagram for a six-state system, and it illustrates different types of states. The mutually communicating states $\{5, 6\}$ and the state $\{4\}$ are trapping states, and regardless of the initial probability vector, the process eventually will migrate to one of these states. States $\{1\}$, $\{2\}$, and $\{3\}$ are transient. Note that 1 communicates with 2, but 2 does not communicate with 1. There is a technical language that precisely characterizes the states of a process, and we refer to Allen (2003) for a full discussion. Other questions regarding the probabilities and expected times for moving from one state to another are discussed in the next section.

EXERCISES

1. Let M be a 2×2 transition matrix for a two-state Markov process. Prove Theorem 9.27 in this special case.

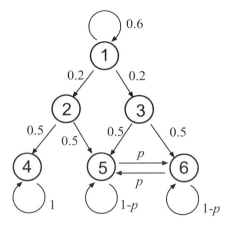

Figure 9.13 Six-state process showing the transition probabilities.

2. For the Markov matrix

$$M = \begin{pmatrix} p & 0 & q \\ q & p & 0 \\ 0 & q & p \end{pmatrix},$$

with $p, q > 0$, find the unique stationary probability distribution. Find a 3×3 Markov matrix with a non-unique stationary probability distribution.

3. Find the transition matrix for the process shown in Fig. 9.14.

4. A regional rent-a-car company has offices located in Lincoln, Omaha, and ten other small cities. Their records show that each week, 55% of the cars rented in Lincoln are returned to Lincoln, 35% are returned to Omaha, and the remaining cars are returned elsewhere. Of those rented in Omaha, 75% are returned to Omaha, 5% to Lincoln, and 20% elsewhere. Of the cars rented in the ten other cities, 25% are returned Omaha and 5% to Lincoln. Draw a Leslie-type diagram illustrating the states and the transition probabilities. Over the long run, if conditions remain the same, what percentage of the cars in their fleet will end up in Lincoln?

5. Consider the matrix

$$M = \begin{pmatrix} 0 & 1 \\ 1 & 0 \end{pmatrix}.$$

Show that there is a stationary probability distribution, but it is not limiting.

6. At each discrete instant of time, measured in minutes, a molecule is either inside a cell membrane, or outside. If it is inside the cell, it has probability

0.2 of passing outside, and if it is outside it has probability 0.1 of entering the cell.

a) If it is inside the cell initially, what is the probability that the molecule is inside the cell at time t? Over the long run, what is the probability that it is inside the cell?

b) If there are 100 molecules inside the cell initially, what is the probability that there are exactly 80 of the molecules remaining in the cell after 1 minute? What is the probability that there are 10 or fewer molecules in the cell after 20 minutes? If N_t is the RV representing the number of molecules in the cell at time t, plot the probability mass function for N_{20}.

7. In a small ecosystem there are two food patches X and Y, and foraging animals move week to week, from patch to patch, with the following probabilities: $P(X, X) = 0.6$, $P(X, Y) = 0.7$, $P(Y, X) = 0.4$, $P(Y, Y) = 0.3$. Write down the transition matrix.

a) What is the limiting distribution of animals in the two patches?

b) If an animal is in patch X at time $t = 5$ weeks, what is the probability that it is in patch Y at $t = 10$ weeks?

8. In the Jukes–Cantor model of base substitution a given base changed to another base with equal probability. A generalization of this model is the two-parameter **Kimura model** where it is assumed that transitions and transversions have different probabilities, b and c, respectively.

a) Write down the Markov matrix M for the Kimura model.

b) Find a limiting stationary distribution.

9.4.2 Random Walks

A random walk is a special type of Markov process where the transitions can only occur between adjacent states. We begin the discussion with a classic example where there are $N + 1$ states $\{0, 1, 2, ..., N\}$, and if the system is in state j, then it moves to $j + 1$ with probability p and to $j - 1$ with probability $q = 1 - p$. Figuratively, we think of an individual walking on a lattice of points, which are the states. At each instant the individual tosses a loaded coin with $\Pr(\text{heads}) = p$ and $\Pr(\text{tails}) = 1 - p$. If heads comes up he moves to the right, and if tails shows he moves to the left—thus the terminology **random walk**. We assume the boundary states 0 and N are **absorbing**, that is, $m_{00} = m_{NN} = 1$.

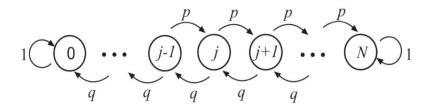

Figure 9.14 Graph for a random walk with absorbing end states.

In other words, if the individual reaches one of these states, then he can never get out. A boundary state that is not absorbing is called a reflecting boundary. Fig. 9.14 shows the graph. With a little imagination the reader can understand why random walks arise in many areas, including gambling (the states represent a player's fortune and p is the probability of winning on a given play), diffusion (the states are geometrical positions and a particle or an animal moves to the left or right with given probabilities), or population models (the states are population levels and there are probabilities of births and deaths in any time period).

We address only one important question in the context of a particle executing a random walk. If a particle is located at position j at time, then what is the probability it will reach position 0 before it reaches N? Let

$$u_j \quad = \quad \text{the probability that the particle will}$$
$$\text{reach 0 before } N \text{ when it starts at } j,$$

for $1 \le j \le N - 1$. To obtain an analytic expression for this probability, we argue as follows. Beginning at j, there are two ways to proceed. In the first step the particle can jump to $j + 1$ with probability p and then have probability u_{j+1} of reaching 0 before N; or, it can jump to $j - 1$ with probability q and then have probability u_{j-1} of reaching 0 before N. According to the law of total probability,

$$u_j = p u_{j+1} + q u_{j-1}, \quad j = 1, ..., N - 1. \tag{4.3}$$

Also, at the boundaries,

$$u_0 = 1, \quad u_N = 0. \tag{4.4}$$

Equation (4.3) is a second-order difference equation that can be solved by techniques introduced Section 9.1. Solutions have the form $u_j = r^j$. Substituting into (4.3) we obtain the characteristic equation

$$pr^2 - r + q = 0.$$

In the case $p \neq q$ the roots are $r = 1$, $r = q/p$, and in the case $p = q = \frac{1}{2}$, the roots are $1, 1$. Therefore, the general solution to (4.3) is

$$u_j = c_1 + c_2 \left(\frac{q}{p}\right)^j, \quad p \neq q, \tag{4.5}$$

$$u_j = c_1 + c_2 j, \quad p = q. \tag{4.6}$$

The arbitrary constants may be found from the boundary conditions (4.4). We obtain

$$u_j = \frac{\left(\frac{q}{p}\right)^j - \left(\frac{q}{p}\right)^N}{1 - \left(\frac{q}{p}\right)^N}, \quad p \neq q,$$

$$u_j = \frac{N - j}{N}, \quad p = q.$$

By exactly the same argument we can calculate the probability

$$v_j = \text{the probability that the particle will}$$
$$\text{reach } N \text{ before } 0 \text{ when it starts at } j.$$

The boundary conditions are now $v_0 = 0$, $v_N = 1$. We get

$$v_j = \frac{1 - \left(\frac{q}{p}\right)^j}{1 - \left(\frac{q}{p}\right)^N}, \quad p \neq q,$$

$$v_j = \frac{j}{N}, \quad p = q.$$

Clearly we have $u_j + v_j = 1$. Therefore, the probability that the particle hits a boundary is 1.

We may not be surprised by the last conclusion. But we can ask what is the expected time to hit a boundary. Let T_j be the RV

$$T_j = \text{the first time a particle arrives at}$$
$$0 \text{ or } N \text{ when it starts at } j.$$

Further, let $\tau_j = \mathrm{E}(T_j)$. Then

$$\tau_j = p(1 + \tau_{j+1}) + q(1 + \tau_{j-1}).$$

The reasoning underlying this difference relation is if the particle begins at j, it may move to $j + 1$ or $j - 1$ with probability p and q, respectively. If it moves to $j+1$ the expected time it hits a boundary is $1+\tau_{j+1}$, counting the jump just taken. If it moves to $j-1$ the expected time is $1+\tau_{j-1}$. Therefore, rearranging,

$$p\tau_{j+1} - \tau_j + q\tau_{j-1} = -1, \tag{4.7}$$

which is a second-order nonhomogeneous difference equation. The boundary conditions are $\tau_0 = \tau_N = 0$. The homogeneous equation is the same as (4.3), and we get the same solution (4.5)–(4.6). To find a particular solution we use the method of undetermined coefficients, as in differential equations, and we find particular solutions $\frac{j}{q-p}$ and $-j^2$ in the cases $p \neq q$ and $p = q$, respectively. In summary, forming the general solution of (4.7) (composed of the sum of the solution to the homogeneous equation and a particular solution) and evaluating the constants using the boundary conditions, we find the expected time for absorption at a boundary is

$$\tau_j = \frac{1}{q-p}\left[j - N\left(\frac{1-\left(\frac{q}{p}\right)^j}{1-\left(\frac{q}{p}\right)^N}\right)\right], \quad p \neq q,$$

$$\tau_j = j(N-j), \quad p = q,$$

where j is the initial position. The exercises request calculations and plots of these times.

Next we describe a combinatorial approach to a random walk problem. Consider a migrating animal, or perhaps a dispersing plant species, that moves in a linear habitat such as a river, or along the bank of a river. As before we zone the habitat into a lattice of sites $\ldots, -2, -1, 0, 1, 2, \ldots$. For this example we assume the habitat is long and has no boundaries. The animal's location at time t $(t = 0, 1, 2, 3, \ldots)$ is the random variable X_t. Starting at $X_0 = 0$, at each time step the animal moves to the right with probability p or to the left with probability $q = 1 - p$. We are interested in calculating the probability mass function, or distribution, of X_t. In other words, what is the probability of the animal being at location k after t time steps? A specific example will lead us to the correct reasoning.

Example 9.28

Consider the probability of being at location $k = 2$ after $t = 10$ time steps. To accomplish this, the animal must move to the right $r = 6$ steps and to the left $l = 4$ steps. A possible path is $RLRRLRLLRR$. The probability that this single path will occur is clearly p^6q^4. How many such paths can be formed with $r = 6$ and $l = 4$? It is clearly "10 choose 6", or $\binom{10}{6} = 210$ paths. Therefore the probability of arriving at $k = 2$ after 10 steps is $\binom{10}{6}p^6q^4$.   □

We can argue in general that if there are t steps terminating at k, then $t = r + l$, $k = r - l$, and

$$\Pr(X_t = k) = \binom{t}{r} p^r q^{t-r}, \quad r = \frac{t+k}{2}.$$

This holds for $t \geq |k|$. If $t < |k|$, then it is impossible to reach k and $\Pr(X_t = k) = 0$. In summary, we have shown that X_t is binomially distributed.

Example 9.29

Assuming $p = q = \frac{1}{2}$, after $2n$ time steps what is the probability that the animal will be back at the origin? We have

$$\Pr(X_{2n} = 0) = \binom{2n}{n} \left(\frac{1}{2}\right)^n \left(\frac{1}{2}\right)^n = \frac{(2n)!}{n!^2 2^{2n}}.$$

With some work we can calculate the right side, but if n is large it will be cumbersome. Fortunately, there is an elegant approximation for $n!$ for large n, and it works well even for n values as low as 6. **Stirling's approximation** is

$$n! \sim \sqrt{2\pi} n^{n+\frac{1}{2}} e^{-n}.$$

Using this formula, we find

$$\Pr(X_{2n} = 0) \sim \frac{1}{\sqrt{n\pi}}. \qquad \square$$

For large t, there is a relationship between the binomial distribution and the normal distribution. Although it is out of our scope to prove it, we present the key result. If t is large and neither p nor q are too close to zero, the binomial $X \sim \text{bin}(t, p)$ can be closely approximated with the standardized normal random variable given by

$$Z = \frac{X - tp}{\sqrt{tpq}}.$$

The approximation is good if tp and tq are greater than 5. In symbols,

$$\lim_{t \to \infty} \Pr\left(a < \frac{X - tp}{\sqrt{tpq}} \leq b\right) = \frac{1}{\sqrt{2\pi}} \int_a^b e^{-s^2} ds.$$

This approximation enables us to easily calculate probabilities that an animal executing a random walk on the integers will lie in some given interval after many steps.

EXERCISES

1. Consider the random walk on the lattice $\{0, 1, 2, ..., 100\}$. If $p = 0.45$, find the expected time τ_j for absorption at a boundary, $j = 0, 1, 2, ..., 100$. Plot τ_j vs. j.

2. A gambler is playing a casino game, and on each play he wins $1 with probability $p = 0.46$, and he loses $1 with probability $q = 0.54$. He starts with $50 and decides to play the game until he goes broke, or until he wins $100. What is the probability of going broke before winning $100? How long should he expect to play? What if the odds are even?

3. Referring to Exercise 2, when $p = 0.46$, simulate three games and plot realizations on the same set of axes.

4. Compute both 6! and 10! exactly and by Stirling's approximation.

5. An animal, starting at the origin, is executing a random walk on the integers with $p = 0.6$. After 40 steps what is the probability $\Pr(X_{40} \leq 10)$? If $p = 0.5$, show that after a million steps the animal will almost certainly lie within 4000 units of the origin. If $p = 0.5$, what is the probability that its position after 50 time steps lies in the interval $5 \leq k \leq 10$?

6. Write a computer program that simulates a random walk on the integers, with $X_0 = 0$ and $p = 0.5$. Sketch three realizations on the same set of axes.

9.4.3 The Poisson Process

Consider a stochastic process taking place in continuous time in which events of a certain type can occur at various instants. For example: phone calls arriving at a relay station, discoveries of food items by a forager, alpha particles emitted from a radioactive source, or customers arriving at bank. These processes can be modeled by a **Poisson process**, defined by the conditions:

$$\Pr(\text{one event in } dt) = \lambda dt + o(dt),$$
$$\Pr(\text{no event in } dt) = 1 - \lambda dt + o(dt),$$
$$\Pr(\text{two or more events in } dt) = o(dt),$$

for small intervals of time dt. The parameter λ is the **rate parameter** of the process. Now, let

$$X_t = \text{ the number of events in } [0, t],$$

and denote the distribution function by

$$p_n(t) = \Pr(X_t = n).$$

We can calculate the probabilities recursively in the way we now describe. We have

$$
\begin{aligned}
p_0(t + dt) &= \Pr(\text{no events in } [0, t + dt]) \\
&= \Pr(\text{no events in } [0, t]) \cdot \Pr(\text{no events in } [t, t + dt]) \\
&= p_0(t)(1 - \lambda dt + \mathrm{o}(dt)).
\end{aligned}
$$

Therefore
$$
\frac{p_0(t + dt) - p_0(t)}{dt} = -\lambda p_0(t) + \frac{\mathrm{o}(dt)}{dt}.
$$

Taking the limit as $dt \to 0$ gives

$$
p_0'(t) = -\lambda p_0(t),
$$

with $p_0(0) = 1$. Therefore the probability that no event occurs up to time t is an exponentially decreasing function

$$
p_0(t) = e^{-\lambda t}.
$$

We use this same method to compute $p_1(t)$. Observe

$$
\begin{aligned}
p_1(t + dt) &= \Pr(\text{one event in } [0, t + dt]) \\
&= \Pr(\text{no events in } [0, t]) \cdot \Pr(\text{one event in } dt) \\
&\quad + \Pr(\text{one event in } [0, t]) \cdot \Pr(\text{no events in } dt) \\
&= p_0(t)\lambda dt + p_1(t)(1 - \lambda dt) + \mathrm{o}(dt)).
\end{aligned}
$$

Therefore,
$$
p_1'(t) = -\lambda p_1(t) + \lambda p_0(t),
$$

with $p_1(0) = 0$.

This method can be extended indefinitely to the other distributions, and we can obtain the system of differential equations

$$
p_n'(t) = -\lambda p_n(t) + \lambda p_{n-1}(t), \quad n = 0, 1, 2, ..., \tag{4.8}
$$

with $p_n(0) = 0$. These are called the **Kolmogorov equations**[3] for the process. They can be solved recursively to get

$$
p_n(t) = \frac{(\lambda t)^n}{n!} e^{-\lambda t}, \tag{4.9}
$$

for $n = 0, 1, 2, \dots$. From this it easily follows that

$$
p_n(t) = \frac{\lambda t}{n} p_{n-1}(t).
$$

---

[3] A. Kolmogorov (1903–1987) was instrumental in developing the modern, axiomatic theory of probability.

Beginning with $p_0(t)$, one can easily calculate the remaining distributions.

The probability distribution (4.9) is the Poisson distribution with parameter λt.

To simulate a Poisson process we need to know the inter-event period, or the time between events. But this, too, is a random variable T. Without loss of generality, we can assume that T is the waiting time from 0 to the first event. To determine how T is distributed, we compute the cumulative distribution function $F_T(t)$. We have

$$
\begin{aligned}
F_T(t) &= \Pr(T \leq t) = 1 - \Pr(T > t) \\
&= 1 - \Pr(\text{no events in } [0, t]) \\
&= 1 - p_0(t) = 1 - e^{-\lambda t}.
\end{aligned}
$$

The pdf is the derivative, or

$$
f_T(t) = \lambda e^{-\lambda t}, \quad t \geq 0,
$$

which is the exponential distribution. Therefore, the inter-event time is exponentially distributed and

$$
E(T) = \frac{1}{\lambda}, \quad \text{Var}(T) = \frac{1}{\lambda^2}.
$$

This means the average waiting time is $1/\lambda$. An important assumption underlying a Poisson process is that it is memoryless. That is, occurrence of an event at time $t = t_0$ has no influence on the future occurrence of events. Symbolically, $\Pr(T > t + t_0 | T > t_0) = \Pr(T > t)$, which is easily obtained from the definition of conditional probability.

In applications to predation events in ecology, the Poisson model describes a search for randomly distributed prey, where success does not deplete the number of prey. This may not be valid in some situations, for example, when the prey are clumped or depleted.

Now we return to the question of simulation of a Poisson process. The key idea is to note that if T is Poisson distributed, then $T = -\frac{\ln U}{\lambda}$, where U is a uniformly distributed random variable on $[0, 1]$. In MATLAB, for example, a sequence of event times $t(k)$ can be calculated recursively in a loop $t(k + 1) = t(k) - \frac{\ln(\text{rand})}{\lambda}$, where $t(0) = 0$.

Example 9.30

Assume a simple birth model where no individual dies, and each has probability of giving birth in a small interval dt to a single offspring with probability $b\,dt$. Then, if n is the population at time t, the population at time $t + dt$ is about

$n + bndt + o(dt)$, where $bndt$ is the probability of n individuals giving birth to a single individual (one event) in the interval dt. We can model the population with a Poisson process with rate parameter $\lambda = \lambda_n = bn$, now depending on n. The random variable X_t represents the number of individuals in the population at time t. That is, $p_n(t) = \Pr(X_t = n)$. The Kolmogorov equations (4.8) become

$$p'_n(t) = -bn p_n(t) + b(n - 1) p_{n-1}(t), \quad n = 0, 1, 2, ... \qquad (4.10)$$

It is shown in Exercise 2 that X_t is negatively binomial distributed.   □

In this chapter we only scratched the surface of a large body of theory and applications of stochastic processes. In particular, we did not discuss continuous random walks, where each time step is infinitesimally small. These processes, called Wiener processes or Brownian motion, are the basis of diffusion theory and play a large role in the development of stochastic differential equations. The references at the end of the chapter may be consulted for readable treatments of these important concepts.

EXERCISES

1. Show that if T is Poisson distributed with parameter λ, then $T = -\frac{\ln U}{\lambda}$, where U is a uniformly distributed random variable on $[0, 1]$. Use a computer algebra system to simulate a realization of a Poisson process with $\lambda = 1$.

2. Suppose a population is of fixed size N at $t = 0$ and undergoes a pure birth process with Kolmogorov equations (4.10).

   a) Show, by direct substitution, that

   $$p_n(t) = \binom{n-1}{N-1} e^{-Nbt} (1 - e^{-bt})^{n-N}, \quad n = N, N+1, N+2, ...$$

   satisfies the Kolmogorov equations.

   b) Perform and sketch three simulations of the birth process when $N = b = 1$ over the interval $[0, 5]$. (Use Exercise 1 with $\lambda = bn$.)

   c) Plot the densities $p_n(t)$ for $n = 0, 1, 2, ...$ at times $t = 0, 1, 2, 3$. What does the graph of $p_n(t)$ vs. t look like for a fixed n?

   d) Show that the mean and variance of X_t are Ne^{bt} and $Ne^{bt}(e^{bt} - 1)$, respectively. (Hint: Refer to the negative binomial distribution.)

REFERENCES AND NOTES

Discrete models have come to play an increasingly important role in biological dynamics, and there are many recent books with excellent chapters discussing those models. Only a few are referenced here. These books include, from elementary to advanced, Mooney and Swift (1999), Allman and Rhodes (2004), Edelstein–Keshet (1988), Britton (2003), Kot(2001), and Mangel (2006). The classical article by May (1976) is required reading. There is no word describing these books other than *outstanding*. The text by Allen (2009) focuses on stochastic aspects of discrete models, and Caswell (2001) is the standard reference on matrix models. Of the many outstanding texts on probability and random processes, the books by Grimmett and Welsh (1986) and Ross (2007) are general accounts and are highly recommended. Regarding stochastic differential equations and Brownian motion we refer to Soong (1973), Chapter 8 of Allen (2007), Allen (2009), and the article by Higham (2001).

A direct continuation of the material in this text, particularly on stochastic differential equations, can be found in the last chapter of Logan and Wolesensky (2009).

Finally, with particular attention to elementary mathematical modeling in biology, both discrete and continuous from biologist's viewpoint, Otto and Day (2007) is an excellent resource. There are many other sources on applications to finance and engineering, for example, random vibrations, too numerous to cite.

Allen, E. 2007. *Modeling with Itô Stochastic Differential Equations*, Springer-Verlag, New York.

Allen, L. J. S. 2009. *An Introduction to Stochastic Processes with Applications to Biology*, 2nd ed., Springer-Verlag, New York.

Allman, E. S. & Rhodes, J. A. 2004. *Mathematical Models in Biology*, Cambridge University Press, Cambridge.

Britton, N. 2003. *Essential Mathematical Biology*, Springer-Verlag, New York.

Caswell, H. 2001. *Matrix Population Models*, 2nd ed., Sinauer Associates, Sunderland, MA.

Edelstein-Keshet, L. 1988. *Mathematical Models in Biology*, McGraw-Hill, New York (reprinted by SIAM, Philadelphia, 2003).

Grimmett, G. & Welsh, D. 1986. *Probability: An Introduction*, Oxford Science Publications, Oxford.

Higham, D. J. 2001. An algorithmic introduction to numerical simulation of stochastic differential equations, *SIAM Review* **43**(3), 525–546.

Kot, M. 2001. *Elements of Mathematical Ecology*, Cambridge University Press, Cambridge.

Logan, J. D. & Wolesensky, W. R. 2009. *Methods of Mathematical Biology*, Wiley–Interscience, New York.

Mangel, M. 2006. *The Theoretical Biologist's Toolbox*, Cambridge University Press, Cambridge.

May, R. M. 1976. Simple mathematical models with very complicated dynamics, *Nature* **261**, 459–467.

Mooney, D. & Swift, R. 1999. *A Course in Mathematical Modeling*, Mathematical Association of America, Washington, DC.

Otto, S. P. & Day, T. 2007. *A Biologist's Guide to Mathematical Modeling in Ecology and Evolution*, Princeton University Press, Princeton.

Ross, S. M. 2007. *Introduction to Probability Models*, 9th ed., Academic Press, Boston.

Soong, T. T. 1973. *Random Differential Equations in Science and Engineering*, Academic Press, New York.

Index